Eighth International Symposium on Transportation and Traffic Theory

INTERNATIONAL COMMITTEE

R. E. Allsop, Transport Studies Group, England
N. H. Gartner, University of Lowell, U.S.A.
R. Herman, The University of Texas at Austin, U.S.A.
E. Kometani, Kyoto, Japan
W. Leutzbach, University of Karlsruhe, Germany
G. F. Newell, University of California, U.S.A.
H. G. Retzko, Technical University Darmstadt, Germany
J. C. Tanner, Ministry of Transport, England
R. J. Vaughan, The University of Newcastle, Australia
E. J. Volmuller, Delft University of Technology, The Netherlands
S. Yagar, University of Waterloo, Canada

Proceedings of the Eighth International Symposium on Transportation and Traffic Theory

June 24–26, 1981
Toronto, Canada

Edited by
V.F. Hurdle,
E. Hauer, and
G.N. Steuart

University of Toronto Press
Toronto Buffalo London

Toronto Buffalo London
Printed in Canada

ISBN 0-8020-2461-0

Canadian Cataloguing in Publication Data

International Symposium on Transportation and
Traffic Theory (8th : 1981 : Toronto, Ont.).
The 8th International Symposium on Transportation
and Traffic Theory

ISBN 0-8020-2461-0

1. Traffic engineering - Congresses. 2. Transportation
- Congresses. I. Hauer, E. (Ezra). II. Hurdle, V.F.
III. Steuart, G. IV. Title.

HE332.I66 1981 380.5 C83-098547-6

CONTENTS

ACKNOWLEDGMENTS

The staging of the Eighth International Symposium on Transportation and Traffic Theory and the publication of its proceedings required the combined effort of many individuals and institutions. The editors are grateful to them all. Our greatest debt is to the 122 referees who so generously donated their efforts, and to Professors R. Allsop, M. Florian, and J.-M. Rousseau who helped us administer the refereeing process.

Our thanks also go to Mrs. Anna Hill, who handled the massive amount of correspondence required for the refereeing process, and provided the typing skills needed to edit the final manuscripts. Special thanks go to the symposium secretary, Catherine Riendeau, whose thoroughness and patience guided us through the organization of the meeting, and whose editing skills have made a major contribution to the preparation of this book. The publication of this book and the symposium itself was made possible by financial help from the Ontario Ministry of Transportation and Communications, the Natural Sciences and Engineering Research Council of Canada, the Canadian International Development Agency, and Transport Canada.

Finally, thanks are due to the University of Toronto for hosting the meeting, to the University of Toronto/York University Joint Program in Transportation for providing assistance and a home, and to the Department of Civil Engineering, University of Toronto, for providing secretarial services and for encouraging three of its academic staff to undertake this project.

The Editors

A New Framework for the Microeconomic Analysis and Classification of Transportation Costs

Alan Abouchar*
University of Toronto

Traditional microeconomic theory must be viewed as very oversimplified to the extent of leading to erroneous conclusions, many of which tend to bias investment inefficiently towards the public sector. For example, amortization expenses, properly conceived as fixed costs in the private sector and included in the revenue target, are usually omitted from the public sector revenue goal for many public sector transport activities. Part of the problem arises through the inability of the traditional microtheoretic cost framework (TMCF) to distinguish among the very different types of cost that do arise in reality and its restriction to two-dimensional problems.

In this paper we present a new framework to analyze cost. Its advantages over the TMCF are: 1) its ability to deal simultaneously with cost variation over time and with cost variation per unit time, something which the TMCF does not do, and 2) its generally more faithful representation of reality. The cost configuration proposed is a disconnected 2-dimensional smooth manifold with boundary.

---

* The author expresses his gratitude to two anonymous referees for their valuable comments but reserves to himself responsibility for any errors that may remain.

1. Difficulties in the Traditional Analysis

The traditional microeconomic framework (TMCF) is too restricted in its cost definitions to be directly applied in many real world situations, including most transport activities. Applications require, variously, both adaptions and modifications of the definitions, and simplifications of reality. But often the modifications are different for private sector and public sector activities.

One example is the treatment of fixed costs. Since as economists we like to think of costs as representing the true values of physical inputs, rather than financial accounting expressions, a fixed cost should represent a new physical input required to undertake production in the current year. But in applications we allow fixed costs to include certain financial categories and other types of expenditure which cannot be construed as inputs of new resources for the current year's production, in the private sector at any rate. Here fixed costs may include an annual amortization component of a basic investment, as well as repayment of financial instruments and/or depreciation of capital stock which deteriorates with use or with time. Fixed cost may also include a periodically recurring input of new resources such as certain plant maintenance. Of course, forward planning requires that this cost be covered in each year by having price equal annual average cost. In the usual U-shaped situation, this average cost also equals long- and short-run marginal cost so that a nice set of relationships emerges, except under disequilibrium conditions. Under a downward sloping demand curve for the firm, i.e. monopolistic conditions, the efficiency problem is usually conceived to be to ensure that the monopolist does not restrict output to the point at which marginal revenue equals marginal cost. Beyond that, if perfect discrimination is possible, price-marginal cost is the target and if not, price-average cost equalization with amortization of fixed cost components as stated above.

For the public sector, on the other hand, which in many countries includes a wide range of transport activities, an amortization component is traditionally excluded from the fixed cost category and omitted from the pricing base. This applies to financial contractual payments and to annual depreciation. Moreover, given this precedent, inputs which truly do depreciate with use but for which the functional relationship is hard to specify and which, in consequence, are to be treated as a financial rubric with an annual depreciation being charged, are also omitted from the cost recovery base. Futhermore, most transport activities are much too complicated to fit into a single homogeneous activity framework; for networks growing over time, for example, it is not at all obvious what constitutes fixed cost or variable cost, long-run marginal cost, or long-run average cost. Accordingly, almost any capital-type expenditure, even though it clearly is consumed in proportion to output, is omitted from the price base. We have not even mentioned further problems of defining output-vehicle-miles, tonnage, or vehicle-size-weighted ton-miles which render the TMCF suspect.[1]

This fundamental difference in the treatment of costs and in the resulting price structure will lead to an underpricing of the public sector services and a higher level of activity as compared with private sector activities. Public sector transportation will be underpriced with serious consequences for the regional distribution of industry, the urban-suburban residential and work patterns, and intermodal split.

This approach to the treatment of transport costs and the omission of certain capital expenditures from the price base even though they clearly do deteriorate with output has a long tradition which finds false roots in the work of the 19th century French civil engineer, Jules Dupuit, who is thought to have proved the wisdom of this approach. The fact is usually overlooked, however, that Dupuit was dealing in his main example with a particularly simple kind of project, a footbridge, which by its nature would not be part of an interurban network, which would not deteriorate as a function of ouput per unit of time, and for which only a single design variant would be considered.[2] Thus: i) the

---

1. For an attempt to evaluate a highway user charge structure making use of the AASHO experimental damage-size coefficients see the author's "A New Approach to the Evaluation and Construction of Highway user Charges," Eastern Economic Journal, April/July, 1974.

2. A fuller discussion of this subject is given in my paper "Dupuit's Bridges and the Theory of Marginal Cost Pricing," History of Political Economy 1976, No. 2 (summer).

question of what is meant by marginal costs in an expanding network framework did not arise; ii) the costs which vary with output per unit would be zero; iii) there being but one technological variant whose design and cost would be dictated by river currents and breadth, weather, local building materials and skills, etc., even the definition of long-run marginal cost for a specific local activity would be very simple; iv) long-run marginal costs would be zero; and finally v) the average time incremental costs could in principle be zero or very low.[3]

I think that the main source of the inconsistent theoretical treatment of costs is the ambiguity in the concept of time in economic analysis. On the one hand, "time" and its related notion of long-run supply refer to chronological sequences and, on the other, the dynamics refer to different drawing-board situations. Much of the inconsistency in the treatment of public and private sector capital expenditure stems from this ambiguity: in the public sector, capital expenditures are treated as fixed costs which then become transmuted into sunk costs, and attention is then focused only on the subsequent costs which will vary with the passage of time or with respect to output

3. Dupuit's main contribution involved the consumer surplus triangle as a means to value the benefit. Later on, welfare economists argued that it would be foolish to install a toll-collector and charge a positive price since the toll collector would himself represent an unnecessary social cost and requiring payment would forestall passage by those who valued the crossing positively but at less than average cost (of toll-keeper and bridge amortization). But this is not what Dupuit argued. (Continued)

per unit time. This applies both to those capital expenditures which will in fact deteriorate with use and those which deteriorate with time and also to those elements which do not deteriorate at all. In the private sector, since all three of these items must be recovered if the firm is to continue operation, they are all cast into the form of periodically recurring costs.

There is yet another objection to the traditional cost theory--its assumption that output per unit time is continuous and continuously differentiable. But the output of most production activities is not continuous. For example, production of automobiles, houses, or books is integer-valued. Moreover, even when production is continuous, as in some process industries, to regard the first derivative of the total cost function as continuous would require scale of plant to be continuously variable, and no industry or technology has such a scale parameter--no more than 4 or 5 scales usually constitute the reasonable range of choice. For example, we might consider a gravel road a soil-cement or light macadam surface, a bituminous

---

He said, rather, that the goal should be ". . . to produce the greatest possible utility and at the same time a revenue sufficient to cover the cost of upkeep and interest on capital." But after developing his analysis in terms of this very simple example, he seems to have been deceived into asserting more and he transformed his argument uncritically to a carriage bridge, a project for which all five types of problem stated in the text would indeed arise. (The exact quote appears on p. 40 of Jules Dupuit, "Public Works and the Consumer," in Transport, Denys Munby, editor, Penguin Books, Baltimore, Md. 1968, p. 40)

asphalt road, a wider two- lane rigid or flexible surface with good base and sub-base, or a three-lane higher grade road. Most people recognize this as a minor problem at most and are willing to brush it aside.

But the result of the continuity assumption is to create a bias against the inclusion in the price base of the cost jumps when moving from one scale to another and to reinforce the custom of excluding such capital costs from the price base. The reasoning here is that since output is continuous by assumption, any cost difference in moving from a light-grade road, for example, must represent a fixed cost-sunk cost element which is not part of the marginal cost which by assumption is expressible as a derivative. Also the cost of the lowest grade technology is also to be regarded as a sunk cost and is not amortized into periodically recurring fixed costs. Recognizing that in private sector activities, as well, there are only a limited number of choices would force us to abandon the idea of the sanctity of a smooth long-run average or marginal cost curve. Accordingly, we might begin to treat one-time capital expenditures (e.g., grading of rail right-of-way) differently from output-related wear of the rails (which are part of the initial capital expenditure) or the output-related ballast cleaning operation (yet another form of periodically recurring expenditure).

Thus, the traditional cost theory does not provide sufficient wealth of concepts to analyze many important transportation and other activities. This is to be expected since it is two dimensional and we need at least three dimensions--output, time, and cost--even for a simplified case of homogeneous output. The cost framework proposed below is an attempt to overcome some of these difficulties.

2. A Time-Variable, Output-Variable Cost Framework

The notion of continuously differentiable l.r.t.c. and l.r.a.t.c. curves is very artificial since the scale factor is not continuous. Instead, the design capacity of plants differs by jumps. Thus, the first change to introduce in the TMCF must be to let the l.r.a.c. curve coincide with the dominated segments of a finite family of s.r.a.c. curves rather than be tangent to each of an uncountably infinite number of such curves. However, to simplify the presentation, we continue to assume that short-run cost curves are continuous (although, as noted above, even this is a false, but innocuous, assumption since most production activities deal with discrete units).

Before we can introduce the next and most important modification of the TMCF we must be precise in our definitions of investment expenditure, fixed costs, and variable cost. First, the investment expenditure must be divided into what we will call "isochronic" cost components and variable capital cost components, the new term signifying the true essence of such inputs, viz. that they are the same with respect to output over time and with respect to output per unit time.[4] To motivate the distinction between isochronic costs and variable capital cost, consider a railroad. Here the construction of tunnels or road bed is an activity which will be independent of use but the purchase of rail, locomotives, or rolling stock will not be, i.e., these latter are all output-variable costs (although the expenditure to acquire them recurs periodically). Periodically recurring costs will also include some labour elements (e.g., administration), and some intermediate capital components (e.g., fuel to heat buildings), as well as some items relating to the original implementation decision (e.g., the annual land site rent, reflecting the opportunity cost of the land or the interest on the total investment expenditure including the expenditure on variable capital components).

---

4. This type of input has been traditionally referred to as a "sunk cost". The new term is proposed for two reasons: 1) it is often useful in discussing projects to talk about those costs which will be one-time costs once made but which, until then, we are very much at liberty to avoid; and 2) the term helps to focus on the essential feature of what may be only a small part of total capital expenditures, forcing us to distinguish between them and those inputs which will either vary with ouput per unit time (rolling stock) or with time (weather-related pavement deterioration).

Finally, to keep things simple, we abstract from purely monetary phenomena, we assume that there is no inflation and none is expected: the interest rate is competively set and reflects the productivity of capital, unlimited borrowing may take place at the market rate of interest, and the borrowing and lending rates are equal. Various levels of difficulty would attach to successive adjustments to take account of departures from these assumptions but none of them affects the following analysis in an essential way.

Let the discrete plant scales be indexed by an index set I. The total cost function for plant i is then given by

$$C = C_i\ (Q,\ t,\ k_i),\ i \in I,$$

where C is the total cost of all production up to time t, Q is the total output per unit time $(Q > 0)$, t is time, and k_i is a discrete plant scale parameter. The unit of time will be taken to be one year. Time is permitted to vary continuously but C_i will have a jump at integer values of t. We will generally work in three dimensional space, but as the discussion proceeds we will sometimes look at slices of the whole picture defined by sections parallel to the cost-output plane or to the cost-time plane.

The picture for the first year with three plant sizes is shown in Figure 1. For each plant, i, the surface showing the cost associated with each value of t and each level of output is a smooth manifold with boundary. This can be regarded as a short-run total cost function. We will designate it by $M_{i,\ [1]}$.[5] This manifold intersects the cost-time plane at the level corresponding to the sum of the isochronic costs and the fixed costs of the first year. The latter may include capital and labor elements. For example, administrative personnel, who are hired by the period, would be a fixed labour cost; fuel to heat the buildings would be a circulating capital fixed cost as would be time-related pavement maintenance or snow removal. Some of these accounts would be fixed over sub-intervals within the year and this could easily be accommodated by a more refined partition of the time axis. We stress particularly that the capital expenditures to acquire those inputs which deteriorate with time and/or with respect to output per unit time are not included in the fixed costs (i.e., they are not included as part of the dollar amounts given at the intersection of the closure of $M_{i[1]}$ and the cost-time plane).

---

5. [·] is the greatest integer function. In general $M_{i,\ [t+1]}$ is the sub-manifold for the period of time from the end of the year t up to the end of the year t+1.

In this structure, the long-run total cost manifold is the union of the dominated portions of the individual short-run total cost sub-manifolds.

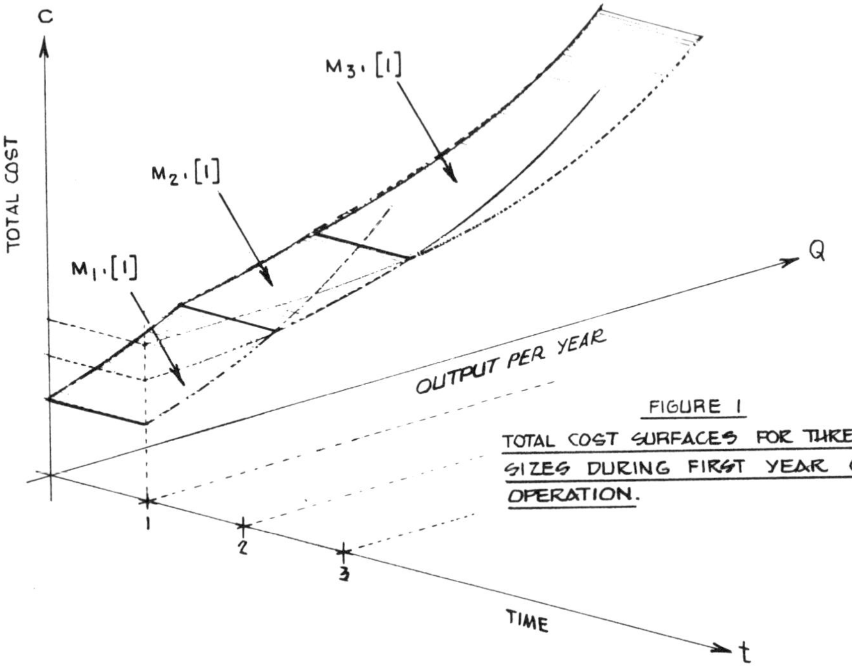

FIGURE I

TOTAL COST SURFACES FOR THREE PLANT SIZES DURING FIRST YEAR OF OPERATION.

During the second year for each plant scale a new fixed cost will be incurred whatever the level of production. For each scale this is equal to the difference between the total cost at the intersection of the closure of its total cost surface and the cost-time plane during the second year and the value of C_i at the corresponding intersection during the first year. The difference between the total output-variable cost of the first year at any output and for the second year at the same output need not be the same for all outputs. The situation for the first three years for plant scale k_1 is shown in Figure 2.

FIGURE 2

(a) TOTAL COST AS OF YEAR END. (PLANT SCALE k_1)

(b) ANNUAL INCREMENTAL FIXED COST AT Q=0

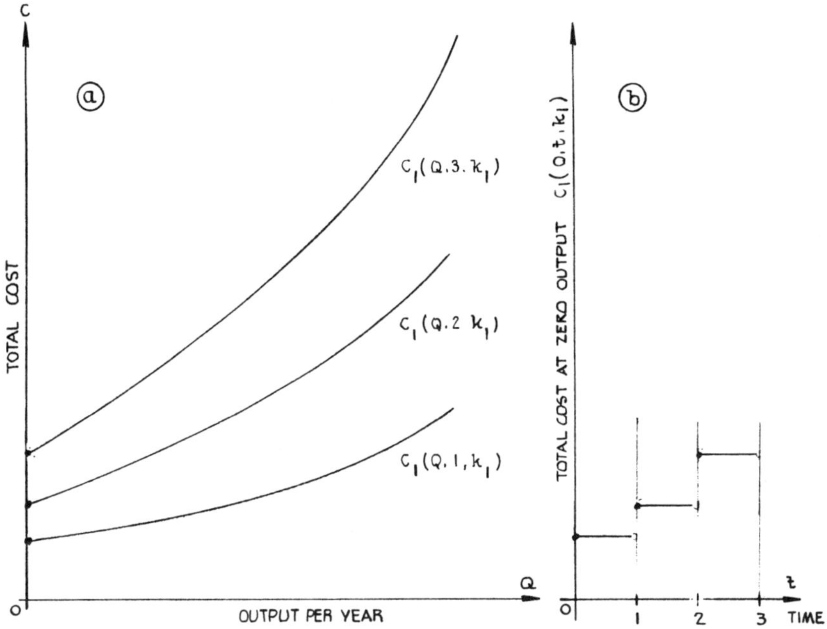

We recall that the initial investment expenditure includes isochronic cost elements as well as physical capital assets which wear with use or with time. In the second (and each successive) year interest must be paid on these as well as on the isochronic items. This will be part of the fixed cost of that year. Again, note that the time-variable costs will also include periodically recurring fixed costs such as administration which are not part of the initial investment expenditure.

It probably will happen--at least for some output ranges of adjacent plant scales--that the output at which the annual total cost manifolds intersect will not be the same over time. In this case it will usually happen that the output level at which, for any pair of plants, the larger plant scale starts to produce at lower cost than the smaller plant shifts toward the cost-time plane. The picture is shown for just two plants in Figure 3.

The import of this shifting phenomenon is that we cannot in general be sure that for any definite annual output a given plant scale will be the lowest cost producer throughout the life of plants: two, three, or even more plants might achieve a lower total cost at different times

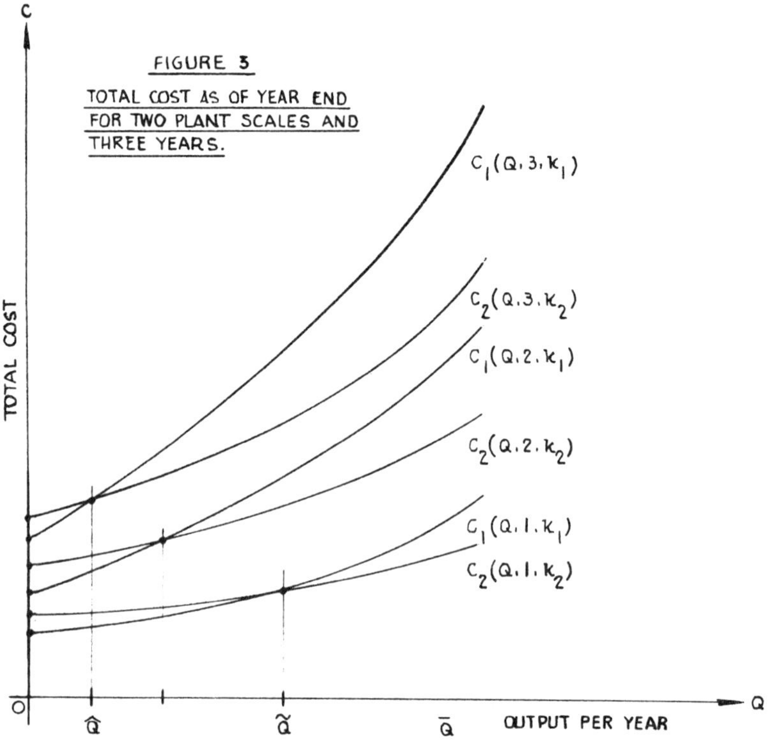

for that output. And just as a given scale will begin to produce the smaller quantities at lower cost than the next smaller plant, so it may be displaced as the lowest cost producer by the plant next larger.

It could happen that a plant might have no range of unambiguous superiority over time; this could occur if the larger plant eventually began to have lower total costs than the next two lower cost producers. Of course, the larger plant could not replace the smaller plant once the latter was in operation and produce at the same cost at which it could have been producing if it had been implanted in the first place. Whether it would be wise to build the larger plant in the first place depends upon product price expectations. We can, however, determine the quantity ranges over which capacity and output can be unambiguously given. These will be the portions of the manifold which remain after we delete all the intervals of transition, since it is only in the remaining intervals that one plant will dominate in respect to total cost minimization over the entire period. (The deleted intervals are closed.) Within the intervals of output rates that remain, plants have lower cumulative production costs over the whole time period horizon. This is shown for two plants and three years in Figure 4 as a two dimensional disconnected smooth manifold with boundary. Consideration of more plant design possibilities is straightforward.

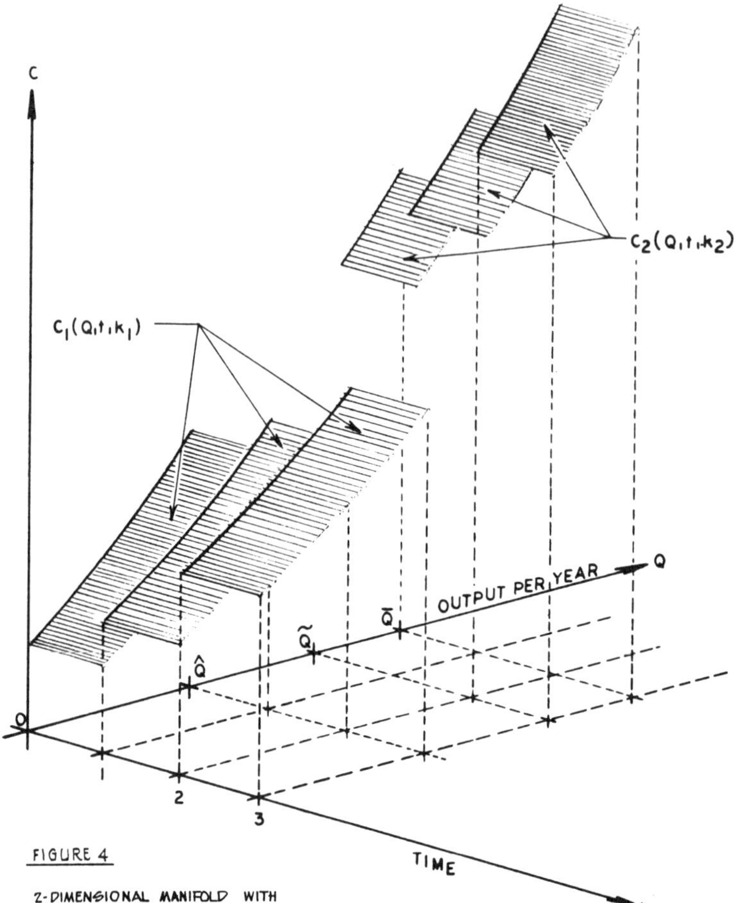

FIGURE 4

2-DIMENSIONAL MANIFOLD WITH BOUNDARY SHOWING TOTAL PRODUCTION COST OVERTIME (OUTPUT LEVELS UP TO $\bar{Q}$ ONLY AND FIRST THREE YEARS OF PRODUCTION.)

3. Conclusion. Advantage of Proposed Framework

The proposed analytical framework has important substantive implications. As we observed earlier, the traditional analysis requires that every input be construed as fixed or variable, and, while this permits neat blackboard solutions, in applied problems the characterization of inputs according to these definitions can be quite arbitrary. The tendency is to regard non-inventory capital items as fixed--indeed, as sunk cost--whether they wear principally with time (buildings) or with use (rolling stock, rail, highway surface), or possibly, do not even wear out at all, or do so over extremely long periods (grading of basic right-of-way). Capital increments associated with certain traffic components--signalling for a new service, air traffic control equipment for a new airport--may also come to be regarded as fixed costs. Since there is a tendency to regard so much as fixed cost, price-marginal cost equilization leads to a situation of very low prices and high subsidies out of general revenues when transportation services are performed in the public sector. Opinions do vary on the precise separation of fixed and variable cost, but virtually all agree, using the traditional theory, that little can be recommended in the way of pricing policy other than to say that if the service is not operating at a financial loss, it must be providing an inefficient level of service, inefficient in the economist's sense that welfare gains could be achieved if demand were expanded since demand price exceeds incremental cost.

The cost classification we are proposing, on the other hand, encourages greater discrimination among cost elements and provides a more useful guide to their inclusion in the price base. For example, while rolling stock may be part of the initial capital expenditure, our framework does not count it as a cost until--and in proportion as--it renders service. This forces us to recognize its true variable cost nature and encourages its inclusion in the price base, in sharp contrast with current practices in a number of countries or transport systems.

This more versatile and realistic cost framework is able to deal with a wide range of cost types whose distinctions are blurred when squeezed into the traditional framework. That the five main cost categories which we have used in the presentation above are all necessary is best seen by the following representative examples:

Cost Type	Exemplified by
1. Isochronic cost	Basic grading for highways or culverts
2. Periodically recurring (discretely time-variable cost)	Central administration; snow removal; heating of buildings
3. Continuously time-variable cost	Heating of sluice gates for the St. Lawrence Seaway for winter operation; fan operation for expulsion of noxious fumes from tunnels
4. Output-variable cost	Deterioration of highway surface (this is partly a time-variable cost); rail cleaning and wear; fuel for operations; stevedoring labour
5. Investment expenditure	Total expenditure undertaken as the project is created (may include cost element categories 1, 3, and 4)

This classification shows clearly the overlap in cost elements. Variable costs can include labor costs as well as capital costs and, moreover, capital costs which are either initially made as part of the investment expenditure or undertaken in closer connection with the production of the output. With this approach the basic incremental cost-price equalization rule leads us to include the annual or monthly recurring costs in the price base, not to speak of encouraging clearer focus on the true nature of those output-variable capital inputs which are purchased as part of the investment expenditure. Thinking in this manner would be a welcome change from past practice in which, starting with the notion that sunk costs need not enter the price base, in a highway cost analysis for example, sunk costs become confused with capital costs, and with the result that all but the "shortest-run" variable costs such as non-administrative and non-maintenance labour, fuel, and materials inputs are to be excluded from the price base.

POINT PROCESS MODELS FOR FREEWAY INCIDENT DETECTION

SAMIR A. AHMED
Oklahoma State University, Stillwater, Oklahoma

ALLEN R. COOK
University of Oklahoma, Norman, Oklahoma

A new approach to the problem of freeway incident detection is proposed, based on point process models of the form ARIMA (0,1,3) that describe the dynamics of traffic occupancy observations. This approach utilizes real-time estimates of the variability in traffic occupancies as detection thresholds, thus eliminating the need for threshold calibration and lessening the problem of false-alarms. The performance of the proposed detection algorithms is evaluated in terms of detection rate, false-alarm rate, and average time-lag to detection, using a total of 1,692 minutes of occupancy observations recorded during 50 representative traffic incidents.

1. INTRODUCTION

It has become widely recognized among the agencies charged with planning and operating urban freeways that the effective management of freeway incidents (accidents, vehicle breakdowns, spilled loads, etc.) permits greater traffic flows to be accommodated more safely on the existing freeway system with considerable savings in travel time to motorists. To effectively manage an incident situation, the information required by the surveillance and control system includes the prompt detection of the incident both in time and space, the magnitude of capacity reduction resulting from the full or partial lane blockage, and the extent of congestion generated by the incident. With this quantitative information, entrance ramp metering rates and variable message signs can be controlled in real-time and the needed services can be efficiently dispatched to the incident scene.

Among the different detection techniques that have been considered, computer-based incident detection seems to be the only method able to detect all incidents and to provide numerical estimates of the magnitude of capacity reduction. Computer-based incident detection, however, has two inherent problems: false alarms and inability to determine the nature of the detected incident. Closed circuit television surveillance can be successfully used in this regard to supplement computer-based incident detection by providing a human operator with some form of visual validation and assessment of incidents.

In this paper we discuss the development of incident detection algorithms based on the time series analysis techniques described by Box and Jenkins [2]. First, an autoregressive integrated moving average (ARIMA) model is presented for the dynamic characterization of freeway traffic occupancy measurements. Second, the ARIMA model is applied to the detection of the sudden changes in traffic stream behavior associated with incidents.

2. POINT PROCESS MODELS FOR FREEWAY TRAFFIC

The basic state variable used in the surveillance and control of freeway traffic is lane occupancy, the percentage of time a detector is occupied by vehicles. It is a surrogate measure for density that provides information about all the important aspects of the traffic stream. Lane occupancies in the ranges 0%-20%, 20%-30%, and 30%-100% indicate free flowing, oscillatory (impending congestion), and congested operations, respectively [9]. Most surveillance and control applications require an average lane occupancy which is computed from the different lane occupancies at each detector station. These averages are continuously updated every 20, 30, or 60 sec depending on the system design.

The dynamic behavior of occupancy observations sampled at uniformly spaced points in time can be regarded as a realization of a discrete time point process $\{X_t\}$. Recent research results (see Ahmed and Cook [1]) have indicated that the process $\{X_t\}$ can be represented by an autoregressive integrated moving average model of the form ARIMA (0,1,3), i.e.,

$$\nabla X_t = (1 - \Theta_1 B - \Theta_2 B^2 - \Theta_3 B^3) a_t \tag{1}$$

where:

∇ = difference operator which has the property $\nabla^d X_t = X_t - X_{t-d}$,

B = backwards operator which has the property $B^m a_t = a_{t-m}$,

$\Theta_1, \Theta_2, \Theta_3$ = parameters of moving average operator of order 3,

and

a_t = white noise variables assumed to be independently distributed as $N(0, \sigma_a^2)$.

The form of the model described by Equation 1 has been found representative for 84 time series data sets from three freeway systems with varying detector configurations and data aggregation time intervals. However, values of the moving average parameters Θ_1, Θ_2, Θ_3 vary from one detector station to another and over time. The differences in parameter estimates are attributed to variations in driving situations, roadway geometry, time of day, and environmental and pavement conditions.

Using the ARIMA (0,1,3) model, the minimum mean-square error predictor of the next observation X_{t+1}, denoted $\hat{X}_t(1)$, is given by:

$$\hat{X}_t(1) = X_t - \Theta_1 \cdot e_{t-1}(1) - \Theta_2 \cdot e_{t-2}(1) - \Theta_3 \cdot e_{t-3}(1) \tag{2}$$

and the approximate 95% confidence limits for X_{t+1} would be

$$\hat{X}_t(1) - 2\,\hat{\sigma}_a \leq X_{t+1} \leq \hat{X}_t(1) + 2\,\hat{\sigma}_a \tag{3}$$

where $e_{t-1}(1)$, $e_{t-2}(1)$, and $e_{t-3}(1)$ are the forecast errors made at times t-1, t-2, and t-3, respectively, and $\hat{\sigma}_a$ is an estimate of the standard deviation of the white noise variables.

The prospects for applying Equation 2 to the real-time forecasting of X_{t+1} may not appear very encouraging since, after all, the moving average parameters vary by time period. One extreme approach to overcome this problem is to use fixed average values of the parameters [7]. The other extreme is to update the parameter estimates in real-time. It is believed, however, that a rapid adjustment of the parameter estimates - each observation interval, for example - may degrade the overall forecasting performance of the ARIMA model. Past experience with adaptive exponential smoothing models has shown that successively changing the smoothing constant values over time yielded larger forecast errors than those resulting from Brown's original exponential-smoothing models [11,3]. One intermediate approach between these two extremes is to update the parameter estimates only occasionally, for example, at the beginning of peak and off-peak periods.

3. APPLICATION TO INCIDENT DETECTION

Several algorithms have been proposed for freeway incident detection. These algorithms can be categorized as pattern recognition algorithms [5,8,10,12,14] and smoothing algorithms [4,6,13]. The first category of algorithms attempts to distinguish between incident and incident-free conditions by means of one or more traffic features which are significantly different in the two situations. A traffic feature can be a single state variable like occupancy or a function of the key state variables (flows, occupancies, and speeds) that is more sensitive to incident patterns. Incident detection is achieved by comparing the observed value of the traffic feature against a preselected threshold level for this feature. The process of determining the values of detection thresholds is known as algorithm calibration.

The second category of algorithms employs short-term

forecasts of the state variables and a set of calibrated thresholds to detect the sudden perturbations in traffic stream behavior generated by incidents. This is similar to the problem of signal detection in noisy communications systems. The most appealing characteristic of this second category of algorithms is that past trends of state variables can be used to predict recurrent congestion which takes some time to build up, whereas nonrecurrent congestion due to incidents is unpredictable.

To evaluate the performance of an incident detection algorithm, three measures of effectiveness are usually used: detection rate, false-alarm rate, and mean time-lag to detection. These measures are highly pertinent to the operational utility of any incident detection algorithm in automated freeway surveillance systems. Detection rate can generally be increased, but only at the expense of an increase in false-alarm rate. Two definitions of false-alarm rate appear in the literature: an on-line definition and an off-line definition [8]. The on-line definition is the percentage of false incident messages out of the total incident messages generated by the algorithm. Alternatively, the off-line definition is the percentage of incident messages generated by the algorithm when applied to incident-free data.

All previously developed algorithms share two major problems: high false-alarm rates and threshold calibration requirements. In fact, the two problems are strongly related, because the threshold levels for detection cannot allow for the multiplicity of factors causing variations in traffic flow (e.g., time of day, geometrics, pavement and environmental conditions). Intuitively, the use of real-time estimates of the variability in state variables in the setting of detection thresholds should lessen the false-alarm problem and should improve the overall performance of the detection algorithm.

Motivated by the preceding remarks, the ARIMA algorithm described here has been structured so that it includes real-time estimates of the variability in traffic occupancy. An incident is detected if the observed occupancy value lies outside the confidence limits constructed two standard deviations away from the corresponding point forecasts. This approach was applied to a total of 50 lane blockage incidents (18 accidents, 28 stalls and breakdowns, 2 instances of debris on the road, and 2 short maintenance operations) which took place on a 2-mile (3.2-km) section of the Lodge Freeway in Detroit. This section contained a closed network of ultrasonic vehicle presence detectors located at four surveillance stations in addition to fourteen television cameras. Most of the incidents took place during the afternoon peak period and had duration times ranging from 1 to 19 minutes. Prevailing occupancies prior to the occurrence of incidents ranged from 9 to 45 percent, while observed flow rates varied from 1,200 to 2,000 vph/lane. Historical records of 1-minute average occupancies were observed at the nearest upstream detector station for each of the 50 incidents. Each record commenced about 10 minutes preceding the television log time of incident occurrence and continued until about 10 minutes after the logged time of removal of the incident or until the dissipation of congestion. This provided a total of 1,692 minutes of observations on occupancy data associated with capacity-reducing incidents.

As pointed out earlier, estimates of the moving average parameters of the ARIMA (0,1,3) model vary from one detector station to another and over time. To explore the effect of this variation on the performance of the ARIMA incident detection algorithm, the 95% confidence limits of Equation 3 were computed once using constant parameter values estimated from representative incident-free data, and a second time using parameters estimated from the observations recorded during the particular incident under consideration. Figure 1 illustrates the performance of the ARIMA occupancy algorithm during one of the 50 incidents used in the analysis. The solid line represents the observed upstream occupancy while the broken lines are the 95% confidence limits of the point forecasts computed from the second method. The algorithm's response to this incident is indicated by the deviation of

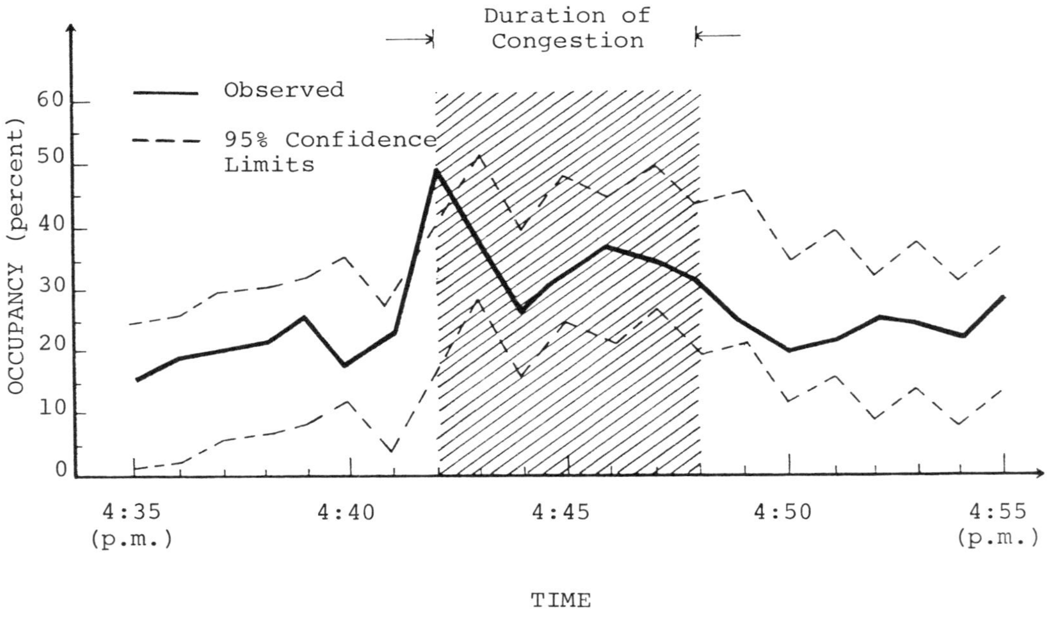

FIGURE 1

PERFORMANCE OF ARIMA OCCUPANCY

ALGORITHM DURING AN INCIDENT

the observed occupancy value from the corresponding confidence limits at the onset of congestion.

The overall performance of the ARIMA occupancy algorithm is shown in Figure 2 by operating characteristic curves relating the detection rate to the false-alarm rate. These curves were obtained by changing the width of the constructed confidence intervals from 3 to 5 standard errors in increments of one standard error. As noted, the algorithm detected all of the 50 incidents with 1.4% and 2.6% false-alarm rates when variable parameter estimates and constant parameter estimates, respectively, were used in computing the confidence intervals. Importantly, these false-alarm rates correspond to the on-line definition discussed earlier, since they were estimated using occupancy observations closely associated with actual incidents. In addition, the average time-lag to detection was found to be 0.58 minute when constant parameter estimates were used and it decreased to 0.39 minute with variable parameter estimates.

Although the performance of the ARIMA occupancy algorithm improves when the estimates of the moving average parameters are updated over time, the question of whether or not to update these parameters must be answered by the operating authority. A tradeoff between the benefits gained from improving the algorithm performance and the computational requirements for parameter updating should be made. It is important in this regard to note that the current trend toward increased use of microprocessors and distributing the computational capabilities in large scale control systems will lead to increased applications of modern control logic. Additional work is required to determine the computational and memory requirements for implementing the ARIMA occupancy algorithm using microprocessors.

4. CONCLUSIONS

Existing incident detection algorithms have two major problems: high false-alarm rates and threshold calibration requirements. These two problems are strongly related. A new approach to the problem of incident detection has been

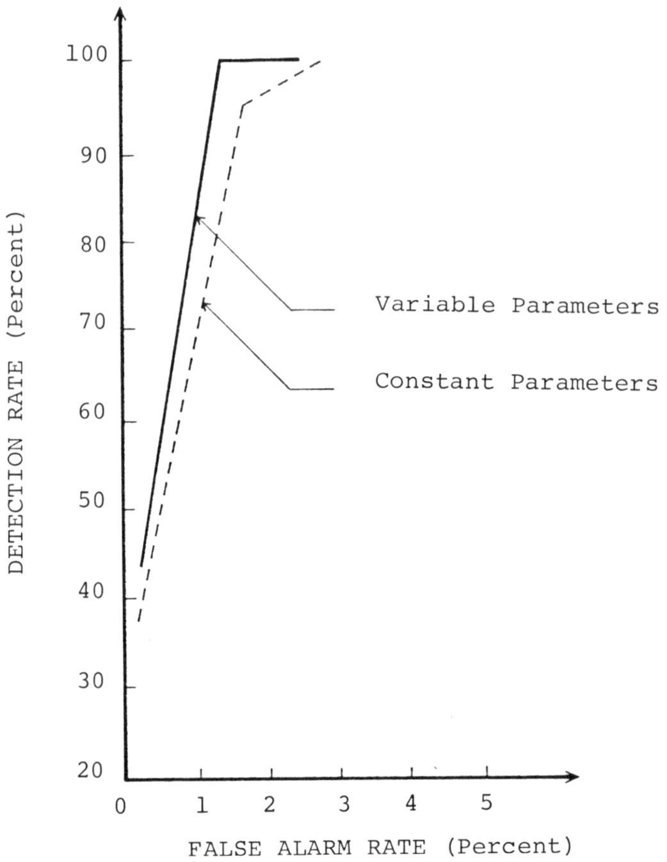

FIGURE 2

OPERATING CHARACTERISTIC CURVES FOR
ARIMA INCIDENT DETECTION ALGORITHM

proposed. This approach has three promising characteristics:

1. Accuracy - The performance results in Section 3 indicate that the proposed algorithms detect the sudden changes in traffic stream behavior associated with traffic incidents accurately and quickly. The reason for this superior performance is that the proposed algorithms are based on sound models that represent the stochastic and dynamic nature of traffic stream behavior. Firm conclusions, however, cannot be drawn without extensive comparison with other proposed algorithms.

2. Freedom from Calibration - As mentioned in the text it is practically impossible to predetermine threshold levels for detection that allow for all the sources of variation in traffic stream behavior. The proposed algorithms employ real-time estimates of this variation as detection thresholds.

3. Computational Efficiency - Although the occasional updating of the moving average parameters may appear to be complicated, the actual on-line estimation method is simple and requires only the execution of a nonlinear least-squares routine.

REFERENCES

1. S. A. Ahmed and A. R. Cook. "Analysis of Freeway Traffic Time Series Data by Using Box-Jenkins Techniques." Transportation Research Record 722, Washington, D.C., 1979, pp. 1-9.

2. G. E. P. Box and G. M. Jenkins. Time Series Analysis: Forecasting and Control. Holden-Day, San Francisco, 1976.

3. R. G. Brown. Smoothing and Prediction of Discrete Time Series. Prentice-Hall, Englewood Cliffs, N.J., 1963.

4. A. R. Cook and D. E. Cleveland. "The Detection of Freeway Capacity-Reducing Incidents by Traffic Stream Measurements." Transportation Research Record 495, Washington, D.C., 1974, pp. 1-11.

5. K. G. Courage and M. Levin. A Freeway Corridor Surveillance, Information and Control System. Research Report 488-8, Texas Transportation Institute, Texas A & M University, College Station, Texas, 1968.

6. C. L. Dudek, C. J. Messer, and N. B. Nuckles. "Incident Detection on Urban Freeways." Transportation Research Record 495, Washington, D.C., 1974, pp. 12-24.

7. M. Eldor. "Demand Predictors for Computerized Freeway Control Systems." Proceedings of the 7th International Symposium on Transportation and Traffic Theory (Edited by Saska, T. and Yamaoka, T.). Institute of Systems Science Research, Kyoto, Japan, 1977, pp. 341-370.

8. M. Levin and G. M. Krause. "Incident Detection - A Bayesian Approach." Transportation Research Record 682, Washington, D.C., 1978, pp. 52-58.

9. J. M. McDermot. "Freeway Surveillance and Control in Chicago Area." Transportation Engineering Journal, ASCE, Vol. 106, No. TE3, 1980, pp. 333-348.

10. H. J. Payne and S. C. Tignor. "Freeway Incident Detection Algorithms Based on Decision Trees with States." Transportation Research Record 682, Washington, D.C., 1978, pp. 30-37.

11. D. E. Smith. "Adaptive Response for Exponential Smoothing: Comparative System Analysis." Operations Research Quarterly, Vol. 25, No. 3, 1974, pp. 421-435.

12. J. T. West. "California Makes its Move." Traffic Engineering, Vol. 41, No. 4, 1971, pp. 12-18.

13. R. H. Whitson, J. H. Buhr, D. R. Drew, and W. R. McCasland. "Real-Time Evaluation of Freeway Quality of Traffic Service." Highway Research Record 289, Washington, D.C., 1969, pp. 38-50.

14. A. S. Willsky, E. Y. Chow, S. B. Gershwin, C. S. Greene, P. K. Houpt, and A. L. Kurkjian. "Dynamic Model-Based Techniques for the Detection of Incidents on Freeways." IEEE Transactions on Automatic Control, Vol. AC-25, No. 3, 1980, pp. 347-360.

ON NETWORK EQUILIBRIA WITH GENERALIZED COST FUNCTIONS
(with application to fuel consumption impacts)

André Babin, Michael Florian and Heinz Spiess
Centre de recherche sur les transports
Université de Montréal

ABSTRACT

We consider in this paper some single mode network equilibrium models with generalized cost functions that incorporate time and distance. The structure of the models and solution algorithms are analyzed. The use of these models for fuel consumption impacts and route choice based on fuel costs is discussed.

INTRODUCTION

The motivation for the formulation and analysis of the models presented in this paper stems from empirical experiences with equilibrium traffic assignment methods. It has been observed (see Florian and Nguyen [4]) that certain routes that are shorter in time but longer in distance attract fewer trips in reality than the number predicted by an equilibrium assignment based on volume-delay curves which only model the travel time. While this empirical observation may have more than one explanation, it is reasonable to postulate that in certain situations travelers may also perceive distance as a component of their total travel cost. In the following, we shall refer to path cost functions which incorporate both time and distance components as generalized path cost functions. Such functions, if successfully incorporated in an equilibrium assignment method, have important application for measuring the impact of energy costs on route choice or the impact of traffic management measures on total energy consumption in a city. Empirical studies (see for instance [2]) show that the fuel consumption of private cars is related to their speed and evidently to the total distance travelled. Thus, it is possible to apply these empirical findings to the time and distances realized by the application of the equilibrium assignment to particular scenarios.

We present, within the framework of an aggregate model of equilibrium traffic assignment, several ways of incorporating measures of distance in the perceived costs of links and paths. The most novel approach postulates a nonlinear perception of the length of the entire path, which permits the modelling of threshold effects: the distance travelled is not perceived as important up to a certain value and beyond this value the total distance is perceived as a component of cost. This model may be formed with any nonlinear monotone function of the total distance of a path and although an equivalent convex cost minimization problem exists, the uniqueness of the arc flows is not ensured and the computation of the equilibrium flows is difficult, except for some special cases, which are analyzed in detail.

The paper is organized in the following way. In the next section we consider generalized cost functions which imbed the travel distance in an additive way. The perception of distance is given by weights which may be common to all the travellers, or vary by origin-destination pairs. Then we formulate and analyze these models, focusing in particular on the model which imbeds a nonlinear perception function of the length of the entire path and develop the details of the solution algorithm; finally we discuss the way in which this model may be calibrated and used for analyzing the impacts of energy costs on route choice and total energy consumption.

1. Traffic Equilibrium Models With Generalized Cost **Functions**

We consider a network used by a single type of traffic. The nodes n, $n \in N$, represent origins, destinations or intersections of arcs. The arcs a, $a \in A$, represent the infrastructure of the networks. The travel time on each link of the network is given by a volume/delay function $t_a(v_a)$ which is a convex and monotone function of the link flows v_a. On the network, the fixed demand g_{pq} for each origin-destination pair (pq) $p,q \in N$, subdivides into path flows $h_{k,pq}$, where k is a path index. The indicator function $\delta_{ak,pq}$ identifies arcs a that belong to path k,pq. Each link is of length d_a.

It is assumed that travellers perceive both time and distance in their route choice, and that these perceived costs are equal on each of the used paths between a given origin and destination, according to Wardrop's [11] "user optimized" principle. The simplest model results when it is postulated that there is a constant θ which converts the distance of each link d_a into a perceived time equivalent, that is

$$\begin{aligned} \sum_a \delta_{ak,pq}\{t_a(v_a)+\theta d_a\} &= u_{pq} \quad \text{if} \quad h_{k,pq} > 0 \\ &\geq u_{pq} \quad \text{if} \quad h_{k,pq} = 0 \end{aligned} \tag{1}$$

subject to flow conservation and nonnegativity of the flows

$$\begin{aligned} &\sum_k h_{k,pq} = g_{pq}, \quad \forall\ (pq)\ ,\ h_{k,pq} \geq 0,\ \forall\ k,pq \\ &\text{and } v_a = \sum_{pqk}\sum\sum \delta_{ak,pq} \cdot h_{k,pq}, \quad \forall\ a \end{aligned} \tag{2}$$

This is the model studied by Dow and Van Vliet [10] and Hamelin [6] which has the equivalent minimization problem

$$\sum_a \int_0^{v_a} \{t_a(x)\,dx + \theta d_a\}dx, \quad \text{subject to (2).} \tag{3}$$

The solution of this model may be obtained by direct application of the adaptation of the Frank and Wolfe [5] algorithm (see [1,7,8]), which will be denoted in the following as the LAP (Linear Approximation) method, since $s_a(v_a)=t_a(v_a)+\theta d_a$ is a generalized cost function which depends only on the link v_a and attribute d_a. This formulation may be somewhat simplistic since the constant θ may be different for various pairs (pq). For instance, it is reasonable to postulate that short distances are perceived differently than longer distances, *but not in a proportional way*, and the spatial separation differs by origin-destination. Thus it may be reasonable to postulate that the perception of distance varies with (pq) and is given by θ_{pq}. The route choice equilibrium conditions are then

$$\begin{aligned} \sum_a \delta_{ak,pq}\{t_a(v_a)+\theta_{pq}d_a\} &= u_{pq} \quad \text{if} \quad h_{k,pq} > 0 \\ &\geq u_{pq} \quad \text{if} \quad h_{k,pq} = 0 \end{aligned} \tag{4}$$

subject to (2).

Proposition 1: The equivalent minimization problem of (4), (2) is given by:

$$\text{Min} \sum_a \int_0^{v_a} t_a(x)\,dx + \sum_a d_a \sum_{pq}\sum \theta_{pq}\; v_{a,pq} \tag{5}$$

where $v_{a,pq}$ is the flow on arc a from p to q which is equivalent to

$$\text{Min} \sum_a \int_0^{v_a} t_a(x)\,dx + \sum_{pq}\sum \theta_{pq} \sum_k d_{k,pq}\, h_{k,pq} \qquad (6)$$

$$\text{where } d_{k,pq} = \sum_a \delta_{ak,pq} \cdot d_a$$

Demonstration: (5) implies (6) since $f_{a,pq} = \sum_k \delta_{ak,pq} h_{k,pq}$ and the second term of (5) is equivalent to

$$\sum_a d_a \sum_{pq}\sum \theta_{pq} \sum_k \delta_{ak,pq}\, h_{k,pq}$$

$$= \sum_{pq}\sum \theta_{pq} \sum_k \left(\sum_a d_a\, \delta_{ak,pq}\right) h_{k,pq}$$

$$= \sum_{pq}\sum \theta_{pq} \sum_k d_{k,pq}\, h_{k,pq} \quad .$$

It suffices to apply the Kuhn-Tucker conditions to (6) in order to obtain (4). □

This model is a special case of a class of network equilibrium models studied by Daganzo [3] and has the following property, which we state formally and demonstrate for the completeness of the presentation.

Proposition 2 (Daganzo) - The link flows $v_a^* = \sum_{pq}\sum v_{a,pq}^*$ of the solution to (4), (2) are unique, hence the costs u_{pq} are unique as well.

Demonstration: Let $\{v_a\}$ be any other feasible flow that satisfies (2). Since $\{v_a^*\}$ solves (5) subject to (2), this implies that the directional derivative of (5) is larger or equal to zero, or that

$$\sum_a t_a(v_a^*)(v_a - v_a^*) + \sum_{pq}\sum \theta_{pq} d_a (v_{a,pq} - v_{a,pq}^*) \geq 0 \qquad (7)$$

Now, suppose that there exist two distinct solutions v_a^1 and v_a^2; then (7) may be rewritten twice, once with $v_a^* = v_a^1$, $v_a = v_a^2$ and vice versa $v_a^* = v_a^2$, $v_a = v_a^1$. By adding the two resulting inequalities one obtains

$$\sum_a (t_a(v_a^1) - t_a(v_a^2))(v_a^1 - v_a^2) \leq 0 \qquad (8)$$

since the linear terms of (7) cancel out. As we have assumed $t_a(v_a)$ to be convex and monotone this implies that

$$\sum_a (t_a(v_a^1) - t_a(v_a^2)) \ (v_a^1 - v_a^2) > 0 \tag{9}$$

and a contradiction results. It follows then that the link costs $(t_a(v_a^*) + \theta_{pq} d_a)$ are unique as well. Since the origin to destination costs may be obtained as the lengths of the shortest paths with these link costs, it follows that they are also unique.□

Another way to model the different perceptions of the distance is to postulate that travellers perceive the total length of a path in a nonlinear way. A monotone nondecreasing function $F_{pq}(d_{k,pq})$ gives the perceived cost and it depends on the total length of the directed path $d_{k,pq}$. The corresponding route choice equilibrium conditions are

$$\begin{aligned} \sum_a \delta_{ak,pq} \cdot t_a(v_a) + F_{pq}(d_{k,pq}) &= u_{pq} \quad \text{if } h_{k,pq} > 0 \\ &\geq u_{pq} \quad \text{if } h_{k,pq} = 0 \end{aligned} \tag{10}$$

subject to (2).

By applying the Kuhn-Tucker conditions it is straightforward to demonstrate the following proposition:

Proposition 3: The equivalent minimization problem of (10), and (2) is

$$\text{Min} \sum_q \int_0^{v_a} t_a(x)\,dx + \sum_{pqk}\sum\sum F_{pq}(d_{k,pq})\,h_{k,pq} \tag{11}$$

subject to (2).

The first term of the objective function (11) is not a strictly convex function of the path flows $\{h_{k,pq}\}$ and since the second term is a linear function of $\{h_{k,pq}\}$, the objective (11) is not a strictly convex function of the path flows and no affirmation may be made regarding the uniqueness of the path or link flows or origin to destination travel costs.

In the following section we present the adaptation of the LAP method for solving the models presented in this section. The solution of the problem (5), (2) is relatively straightforward, but computationally cumbersome and the solution of (11), (2) is, in general, difficult.

2. Solution Algorithms Based On The "LAP" Method

The models that we consider are convex programming problems with linear constraints. One of the simplest methods for solving such problems is the LAP method which is based on the linearization of the objective function and has been used in practice for numerous applications. A version of this method is imbedded in U.M.T.A.'s transportation planning codes [9]. The general algorithm is in the Appendix and we proceed to describe directly its adaptation.

A typical iteration of this algorithm requires the solution of a linear program that determines the descent direction and the solution of a one dimensional minimization problem that determines the step size for a given descent direction. Consider now the application of the algorithm at iteration ℓ, when a current feasible solution v_a^ℓ is known. The descent direction for the objective function (6) is obtained by solving for f_a and $y_{k,pq}$ which

$$\text{Min} \sum_a t_a(v_a^\ell)\cdot f_a + \sum_{pqk}\sum\sum \theta_{pq} d_{k,pq}) y_{k,pq} \tag{12}$$

subject to

$$\begin{aligned} &\sum_k y_{k,pq} = g_{pq}, \quad \forall\ (pq) \qquad y_{k,pq} \geq 0 \quad \text{and} \\ &f_a = \sum_{pqk}\sum\sum \delta_{ak,pq} y_{k,pq} \ . \end{aligned} \tag{13}$$

By substituting the expression for f_a into (12) one obtains

$$\min \sum_{kpqa}\sum\sum\sum \delta_{ak,pq}\{t_a(v_a^\ell) + \theta_{pq}d_a\} \quad y_{k,pq} \tag{14}$$

The solution of the problems (14)(13) is obtained by computing shortest paths for each pair (pq) with link costs $\{t_a(v_a^\ell) = \theta_{pq}d_a\}$. This procedure may be combersome computationally, since one has to consider each pair (pq) in turn and it is not possible to compute a shortest path tree from the origin to all destinations. Let $k_{*}^{}pq$ be the shortest directed path so found for pair (pq) and let

$$\gamma_{a,pq} = \begin{cases} 1 & \text{of } a \in k_{*}pq \\ 0 & \text{otherwise} \end{cases}$$

The solution to (14), (13) is then $y_{k*,pq}=g_{pq}$, $y_{k,pq}=0$, $k \neq k*$ and $f_a=\sum_{pq}\sum \gamma_{a,pq} \cdot g_{pq}$.

Thus, in principle, the solution to the model (6), (2) may be obtained by incurring the cost of computing more shortest paths than for the simple network equilibrium model.

The descent direction for the objective (11) is obtained by solving for f_a and $y_{k,pq}$ in

$$\text{Min} \quad \sum_a t_a(v_a^\ell) f_a + \sum_{pqk}\sum\sum F_{pq}(d_{k,pq}) y_{k,pq}) \tag{15}$$

which, by using a similar development as before, is equivalent to

$$\text{Min} \sum_{pqk}\sum\sum \{\delta_{ak,pq} t_a(v_a^\ell) + F_{pq}(d_{k,pq})\} \, y_{k,pq} \tag{16}$$

subject to (13).

This problem is much more difficult to solve than the previous one, due to the following property.

<u>Proposition 4</u>: The solution to (16), (13) is given by allocating the demand g_{pq} to the least total cost path between

each pair (pq); however these directed paths do not necesarily form a directed tree.

Demonstration: The first part of the proposition is evident. To show the second part consider the following example. The network consists of one origin and two destinations with (time, distance) indicated on its three links as shown below.

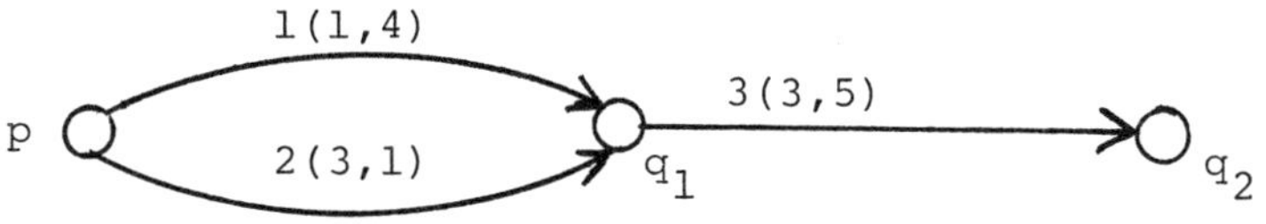

The perception function $F_{pq_1}(\cdot) = F_{pq_2}(\cdot) = F(\cdot)$ and is given by

$$F(d) = \begin{cases} 0, & 0 \le d < 5 \\ d, & 5 \le d \end{cases}$$

The shortest path p to q_1 is link 1, while the shortest path p to q_2 is links 2,3. It is evident that these directed paths do not form a tree. □

The solution to (16), (13) may not, therefore, be obtained by the application of a shortest path algorithm. We conclude that, in practice, it may be preferable to work with network equilibrium models that incorporate time and distance such as in (6). We omit the complete statement of the algorithm for these problems, which, except for the modifications required to find the descent direction and the steplength, is akin to the standard algorithm which considers only time in the link cost functions. Evidently, the arc lengths d_a are required while solving for the descent direction. In the following section we consider the use of the models analyzed above.

3. APPLICATIONS OF GENERALIZED COST FUNCTION MODELS

The evident application of this generalized cost model is to convert the time and distance into fuel consumption units. The work of Chang et al. [2] provides simple ways of relating fuel consumption to the time and distance travelled. In particular they calibrated equations of the form

$$\Phi = k_1 + k_2 \bar{t} \qquad (17)$$

where Φ is the fuel consumption per unit distance, $\bar{t}$ is the average time per unit distance and k_1, k_2 are parameters which depend on the vehicle type. By assuming that the "typical" car for a given city may be properly represented by the constant k_1 and k_2, the fuel consumption on link a is given by

$$\Phi \cdot d_a = \left\{k_1 + k_2 \cdot \frac{t_a(v_a)}{d_a}\right\} \cdot d_a \text{ which is equivalent to}$$

$$\Phi \cdot d_a = k_1\, d_a + k_2\, t_a(v_a) \quad . \qquad (18)$$

The same authors discuss the "home-to-work" trip, which is usually initiated with a "cold start". In [2] a monotone function is calibrated which gives the excess fuel consumed from a cold start as a function of the *total distance* travelled; the colder the outside temperature, the higher this excess fuel consumption for a given trip length. Using the function $F(d_{k,pq})$ would result in total path functions of the form

$$\hat{F}(d_{k,pq}) = F(d_{k,pq}) + k_1 \cdot d_{k,pq} \qquad (19)$$

and link cost functions of the type

$$\hat{t}_a(v_a) = k_2\, t_a(v_a) \qquad (20)$$

Thus, if one wished to determine the impact of traffic management measures on total energy consumption during the peak period, this would be accomplished by finding the solution to the standard network equilibrium model with link costs $t_a(v_a)$ and accounting for the total fuel consumed by using (19) and (20). Another use would be to assume that drivers choose routes according to perceived costs (18) and simulate route choice based on shortest fuel consumption paths.

ACKNOWLEDGMENT

Carlos Daganzo pointed out the relevance of reference [3].

APPENDIX

Consider the problem of minimizing a continuously differentiable convex function subject to linear constraints:

$$\text{Min } F(X), \text{ subject to } A X = b, X \geq 0.$$

The detailed steps of the Frank and Wolfe algorithm for solving this problem are:

Step 1: Given a feasible solution X^1, set $\ell = 1$.

Step 2: Determine Y^ℓ that minimizes

$$\nabla F(X^\ell) Y, \quad \text{subject to } A Y = b, Y \geq 0.$$

Step 3: Set the descent direction $d^\ell = Y^\ell - X^\ell$.

If $|\nabla F(X^\ell) d^\ell| \leq \varepsilon$, terminate, where ε is a suitable convergence parameter. (X^ℓ is the optimal solution.)

Step 4: Find the optimal step length λ^ℓ that minimizes

$$F(X^\ell + \lambda d^\ell) \quad \text{for} \quad 0 \leq \lambda \geq 1.$$

Step 5: Revise the current solution

$$X^{\ell+1} = X^\ell + \lambda^\ell d^\ell ;$$

set $\ell = \ell + 1$ and return to Step 2.

REFERENCES

[1] BRUYNOOGHE, N., A. GIBERT and M. SAKAROVITCH, "Une méthode d'affectation du trafic", in Fourth Symposium on the Theory of Traffic Flow, Karlsruhe (1968).

[2] CHANG, M.F., L. EVANS, R. HERMAN and P. WASILIEWSKI, "Gasoline Consumption in Urban Traffic", Transportation Research Record 599, 25-30 (1976).

[3] DAGANZO, C.F., "Multicost Traffic Equilibrium", Working Paper UCB-ITS-WP-81-4, Institute of Transportation Studies, University of California, Berkeley (1981).

[4] FLORIAN, M. and S. Nguyen, "An Application and Validation of Equilibrium Trip Assignment Methods", Transportation Science 10, 374-389 (1976).

[5] FRANK, M. and P. WOLFE, "An Algorithm of Quadratic Programming". Naval Res. Log. Quart. 3, 95-110 (1956).

[6] HAMELIN, P., "Calibration d'une courbe de volume-délai", M.Sc. project, Département d'informatique et de recherche opérationnelle, Université de Montréal (1980).

[7] LeBLANC, L.J., E.K. MORLOK and W.P. PIERSKALLA, "An Efficient Approach to Solving the Road Network Equilibrium Traffic Assignment Problem", Transportation Research 5, 309-318 (1975).

[8] NGUYEN, S., "A Unified Approach to Equilibrium Methods for Traffic Assignment". in Traffic Equilibrium Methods, Proceedings 1974 (M. Florian, Editor), Vol. 118, Lecture Notes in Economics and Mathematical Systems, Springer Verlag, 148-182 (1976).

[9] UMTA, Transportation Planning System (UTPS) User's Guide, U.S. Dept. of Transportation, Urban Mass Transportation Administration, Office of Transit Planning, Planning Methodology and Technical Sup. Div. (1976).

[10] DOW, P. and D. VAN VLIET, "Capacity Restrained Road Assignment", Traffic Engineering and Control, 261-273 (1979).

[11] WARDROP, J.G. "Some Theoretical Aspects of Road Traffic Research", Proc. Inst. Civil Engineers, Part II, 1, 352-378 (1952).

FUNDAMENTAL ECONOMIC ANALYSIS FOR TRANSIT ROUTES: DEMAND-COST RELATIONSHIPS

James H. Banks

San Diego State University

Past attempts to devise rational design methods for transit services (such as the "total cost minimum") have encountered the difficulty that they require the assumption of a constant demand for service despite the fact that demands are ordinarily assumed to vary with levels of service. If a design rule is employed which specifies at least one service attribute (say, headway) as a strictly monotone function of route patronage, this difficulty can be overcome by expressing the demand function in terms of this attribute. We then have the actual patronage as a function of the service attribute and the service attribute as a function of the anticipated patronage, and may solve for equilibrium points at which actual patronage equals anticipated patronage. For most plausible combinations of headway design rules and demand functions, two cases are likely: there may be either two equilibrium points or none. Examination of the way in which trip disutility should vary with anticipated patronage reveals that trip disutility is not very sensitive to errors in demand estimates for high-patronage routes, but that sensitivity increases as patronage decreases. As it appears that rational design methods would be most beneficial for low-patronage systems, lack of accurate route-specific demand models constitutes an important barrier to their use.

I. INTRODUCTION

Over the past decade, a number of studies have appeared which deal with the development of rational design methods for transit systems[1-7]. Most of these studies have as their object the optimization of some aspect of transit route location or scheduling, such as line spacing, stop spacing, or headway. In addition, most are based on the minimization of a "total cost function" which combines the transit operator's money costs with the time costs experienced by the passengers -- although this is by no means the only conceivable design rule.

There is a logical difficulty associated with these rational design methods as they have been developed hitherto. This is that the optimal designs depend on the demand for service, and this demand is treated as a constant. Actually, the demand should vary with the fare and the level of service, and since the level of service varies with the design, "optimal" designs may not be internally consistent.

The purposes of this paper are (1) to propose a simple conceptual approach to the problem of incorporating the variation in demand into rational designs; (2) to assess the practicality of using such a method, given the current state of the art in demand and cost analysis for transit services; and (3) to ask under what circumstances (if any) it is really important that the demands assumed in rational designs agree with those resulting from them.

The analysis of demand and cost relationships which follows will be limited to individual local bus routes with predetermined locations and stop spacing, so that the headway is the only design issue. Restriction of the analysis to individual routes is justified on the grounds that individual routes are the fundamental units of design and scheduling for conventional transit systems. Exclusion of route spacing and system configuration issues is necessary because current design theories do not permit solutions to problems of this sort under realistic conditions[7]. The issue of optimal stop spacing is also omitted because it would complicate the analysis (models of the effect of stop spacing tend to be quite complex [2,8,9]) but would not add anything significant to the main thrust of the paper.

The remainder of the paper is organized as follows: Section II presents a conceptual framework for dealing with the interaction of costs and demand; Section III discusses headway design rules; Section IV discusses "patronage equilibria" in which the patronage levels assumed in designs agree with those resulting from them; Section V discusses some practical considerations in applying rational design methods; and Section VI discusses the question of the circumstances under which accurate assumptions concerning patronage levels are really important to the success of rational design methods.

II. CONCEPTUAL FRAMEWORK

Consider a transit property which employs a rational rule to design services for a route. For every anticipated level of patronage, this rule specifies a particular design, and this, in turn, implies a vector of service attributes

$$\tau = (\tau_1, \tau_2, \tau_3 \ldots \tau_n)$$

where τ_1 = headway (or some other convenient attribute)

τ_i = value of some other attribute, such as fare, $i = 2, 3, \ldots n$.

Since the rule implies that a given value of each attribute corresponds to each level of anticipated patronage, we have

$$\tau_i = f_i(p_s) \tag{1}$$

where p_s = anticipated patronage in passengers per unit distance and time. Specifically,

$$\tau_1 = f_1(p_s). \tag{1a}$$

If we further assume that f_1 is a strictly monotone function and f_i, $i = 2, \ldots n$, are known functions of any sort, we can determine all other attributes from τ_1, so that there clearly exists some set of functions $\{\tilde{f}_i\}$ such that

$$\tau_i = \tilde{f}_i(\tau_1). \tag{1b}$$

These are known functions if the rational rule is known.

Meanwhile, we assume that each potential passenger has a utility function such that

$$u^j = U^j(\tau, \beta^j, \varepsilon) \tag{2}$$

where u^j = the value of individual j's utility for a particular trip,

τ = the vector of service attributes for that trip,

β^j = a parameter whose value depends on the mode, the value of the trip to the individual, and the individual's socioeconomic characteristics, and

ε = a stochastic term.

We further assume that there is some threshold utility $\hat{u}^j$ for each individual such that a trip is made if, and only if, $u^j > \hat{u}^j$. If no choice of mode is available, $\hat{u}^j$ might be taken to be zero (i.e., net utility for the trip must be positive); if some other mode is available, $\hat{u}^j$ might also depend on the service attributes of the competing mode. The shape of the cumulative distribution function for u^j, and, consequently, the probability that $u^j > \hat{u}^j$, depends on the distribution assumed for ε -- common assumptions might be the Weibull Distribution, which results in the logit demand model, or the Normal Distribution, which results in the probit.

Assuming that we know the value of β^j and the distribution of ε, u^j can be stated entirely in terms of τ, which, in turn, can be stated in terms of τ_1. Thus

$$u^j = \tilde{U}^j(\tau_1). \tag{2a}$$

Also, if we know the distribution of ε, we have the disaggregate demand function

$$p^j = d(u^j) = d^j(\tau_1) \tag{3}$$

where p^j is the probability that individual j makes a trip. The expected number of trips from a group of individuals is the sum of the individual probabilities that a trip is made, so we also have an aggregate demand function

$$p = \sum_j p^j = D(\tau_1) \tag{4}$$

where p is the patronage level resulting from a particular design. Thus (in principle, at least) we can state the aggregate demand function in terms of τ_1.

From expression (1a) we have τ_1 as a function of p_s, and from expression (4) we have p as a function of τ_1. If we can find values of τ_1 for which $p = p_s$, we will have an "equilibrium" in which the patronage resulting from a design coincides with the patronage assumed in the design -- thus fulfilling the assumption underlying the design. Provided we can quantify expressions (1a) and (4) we can determine values of p which equal p_s either graphically or numerically.

We might, of course, go one step further and combine expression (1a) with expression (4). Then

$$p = D(f_1(p_s)) = g(p_s) \tag{5}$$

so that p is stated directly as a function of p_s. Again, the assumptions underlying the design are fulfilled if $p = p_s$.

The main features of the direct relationship between p and p_s are not intuitively obvious; however, if the headway is used as an intervening variable, we can gain considerable insight into the circumstances under which patronage equilibria might occur. On the one hand, most rational design rules proposed to date involve fairly simple monotonic relationships between headways and anticipated patronage (although it is possible to imagine rules which do not); on the other hand, it is easy to make reasonable assumptions about the general features of the relationship between headway and aggregate demand. Consequently, we will use headway as an intervening variable in the analysis which follows, and restrict ourselves to design rules which give the headway as a monotonic function of p_s.

The sections which follow will give examples of the derivation of headway design rules, discuss the general features of patronage equilibria, show how headway design rules can be related to individual utility functions, and discuss some implications for the practical use of rational design rules.

III. HEADWAY DESIGN RULES

This section will recapitulate the derivation of two types of headway design rules, which may serve as examples of the sort of rational rules which have been (or might be) proposed for the design of transit services.

The following symbols will be used in the description of these rules:

M = average passenger load corresponding to capacity operation at the maximum load point.

γ_o = operating cost per vehicle per unit time.

γ_r = value of riding time (in-vehicle time) as determined by the transit operator.

γ_w = value of walking and waiting time (out-of-vehicle time) as determined by the transit operator.

p = patronage in passengers per unit time and distance.

h = headway.

ℓ = passenger's trip length (assumed identical for all passengers).

v = bus speed, exclusive of stops to pick up and/or discharge passengers.

t_s = delay for picking up or discharging one passenger (assumed constant).

a = passenger's average waiting time at a bus stop, as a fraction of the headway.

t_a = passenger's combined access and egress time for the route (assumed constant).

J = "total cost" per passenger, including operator's money costs and passengers' time costs.

In order to derive rules for the determination of headways, it is necessary to make assumptions about the objectives of the transit operator. Two possibilities readily present themselves as extremes: (1) a transit operator who behaves "selfishly" will wish to minimize his average operating cost per passenger; and (2) a more "socially oriented" transit operator will wish to minimize the total average cost per passenger, including both his operating costs and the passengers' time costs.

Objective (1) leads to a very simple headway design rule -- frequency of service should be adjusted so that all vehicles are just filled to capacity at the maximum load point on the route. If M is the average passenger load which corresponds to capacity operation at the maximum load point, and $ph\ell$ represents the average number of passengers on board, the optimum headway h* is:

$$h^* = M/(p\ell). \qquad (6)$$

Objective (2) leads to a more complicated design rule known as the <u>total cost minimum</u> (TCM). Several versions of this rule may be found in the literature. One, which may serve for purposes of illustration, is the one introduced by

Mohring[2] as his "steady state case". The total cost function outlined below is a modification of Mohring's, in which some of Mohring's speed variables have been replaced with time and distance variables to more clearly isolate the factors influencing the total cost.

Consider a transit route of infinite length. Patronage on this route is p passengers per unit time and distance, uniformly distributed. Vehicles traverse this route at a constant speed v (exclusive of stops to pick up and discharge passengers) and cost γ_o to operate. Each passenger has a trip length ℓ and values in-vehicle and out-of-vehicle time of γ_r and γ_w respectively. A constant delay of t_s occurs each time a passenger is picked up or discharged. Also, each passenger spends a constant time t_a in gaining access to and egress from the route, and waits at the stop for the vehicle for some fraction a of the headway h.

The operator's average cost for transporting a passenger on this route is:

$$\gamma_o/(phv) + 2\gamma_o t_s \, .$$

Meanwhile, the amount of time each passenger spends on board the vehicle is the sum of the non-stop travel time for a distance ℓ plus the delays occasioned by picking up and discharging other passengers over a distance ℓ. In the steady-state case, each passenger experiences $ph\ell$ pickups and $ph\ell$ discharges, so that the total cost of a passenger's in-vehicle time is:

$$2\gamma_r t_s ph\ell + \gamma_r(\ell/v).$$

The cost of out-of-vehicle time for each passenger amounts to

$$a\gamma_w h + \gamma_w t_a$$

which results in a total cost per passenger J of

$$J = \gamma_o/(phv) + 2\gamma_o t_s + 2\gamma_r t_s ph\ell + \gamma_r(\ell/v) + a\gamma_w h + \gamma_w t_a . \quad (7)$$

Collecting terms which do not involve h and setting them equal to k, we can rewrite expression (7) as:

$$J = \gamma_o/(phv) + 2\gamma_r t_s ph\ell + a\gamma_w h + k \quad (7a)$$

where $k = 2\gamma_o t_s + \gamma_r(\ell/v) + \gamma_w t_a$. The minimum cost headway can be determined by setting $\partial J/\partial h$ equal to zero, provided this does not result in a load in excess of the capacity of the vehicle; if this would overload the vehicle, expression (6) (the so-called "capacity headway") is the optimum headway. Thus the minimum cost headway for the steady-state case is:

$$\begin{cases} h^* = [\gamma_o/(pv(2\gamma_r t_s p\ell + a\gamma_w))]^{\frac{1}{2}} & \text{if } ph^*\ell \leq M \\ h^* = M/(p\ell) & \text{otherwise} \end{cases} \quad (8)$$

and J*, the total cost per passenger corresponding to a headway of h*, is

$$\begin{cases} J^* = 2[\gamma_o(2\gamma_r t_s p\ell + a\gamma_w)/(pv)]^{\frac{1}{2}} + k & \text{if } ph^*\ell \leq M \\ J^* = \gamma_o\ell/(Mv) + 2\gamma_r t_s M + a\gamma_w M/(p\ell) + k & \text{otherwise} \end{cases} \quad (9)$$

IV. PATRONAGE EQUILIBRIA

The relationships between patronage and headway implied by expression (6) and unconstrained version of expression (8) are plotted in Figure 1 for purposes of comparison. These curves are based on the parameter values given in Table 1, which are believed to be typical of the San Diego

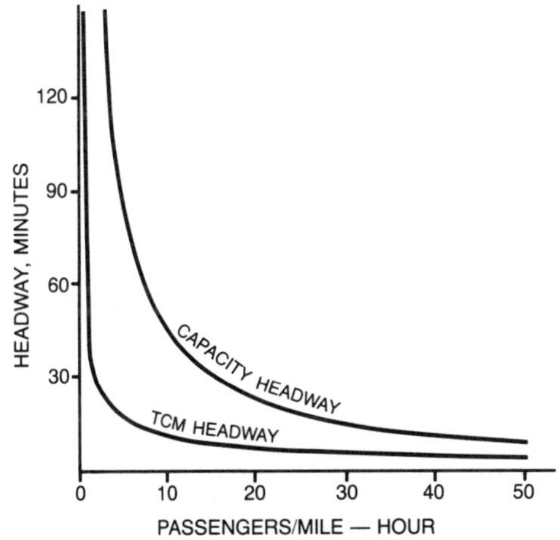

FIGURE 1 HEADWAY DESIGN CURVES, FOR PARAMETER VALUES GIVEN IN TABLE 1

Table 1 Parameter Values Used in Example Calculations

Parameter	Assumed Value
γ_o	\$40/hr.
γ_r	\$ 3/hr.
γ_w	\$10/hr.
ℓ	4 Miles
v	20 MPH
t_a	.092 Hr. (5.5 minutes)
t_s	.0056 Hr. (20 Seconds)
a	.50

region. Note that both curves are negatively sloped and convex to the origin.

Meanwhile, we would expect demand to decrease as headways (and hence, passengers' average waiting times) increase, so that the aggregate demand curve (expressed as a function of $\tau_1 = h$) should also be negatively sloped. If we further assume that there is some headway so great that no trip would be made if it is exceeded, and that there is an upper limit to the number of trips that would be made, no matter how small the headway, the result is the two basic cases shown in Figure 2. If demand is great enough, as in Case I, the demand curve will be above the headway design curve for some range of p. In this case, owing to the shape of the headway design curve and the limits assumed for the demand curve, there will be two intersections and hence two equilibrium points for p. In Case II, the demand curve is below the headway design curve for all

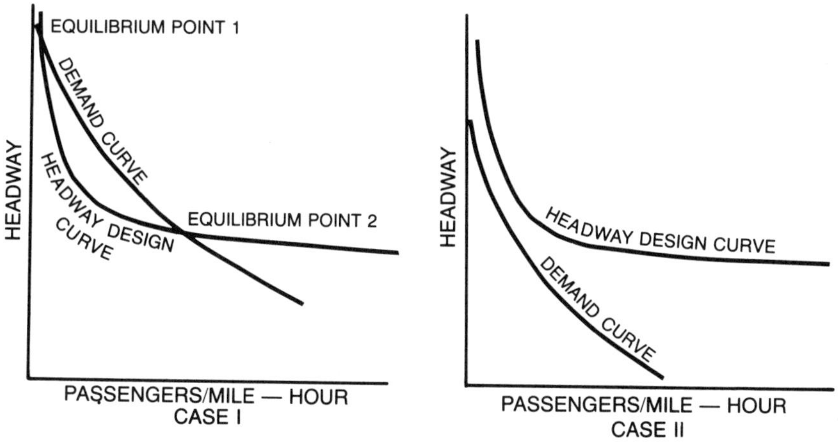

FIGURE 2 RELATIONSHIPS BETWEEN TYPICAL HEADWAY DESIGN AND DEMAND CURVES

values of p, and no solution exists. Obviously, there are other possibilities: for instance, Cases I and II are contiguous, so we might find a case in which the two curves are just tangent at one point, so that one equilibrium point exists; or if the demand curve is sufficiently irregular, we might find even more than two equilibria. However, Cases I and II seem the most likely.

Thus, so long as we know enough about the demand for a given route to derive an aggregate demand curve in terms of the headway, we should be able to determine equilibrium values of p graphically -- provided equilibria exist. Even if we are unable to draw the demand curve, however, repeated applications of the design and price rules should lead to the correct equilibrium value under certain circumstances. We assume that in seeking equilibrium in the face of an unknown demand function, a transit operator would begin by estimating the equilibrium value of p and then setting the fare and headway according to its headway and price rules. The patronage which actually results from this fare and headway combination may be greater or less than the assumed value. The operator will then try a new headway-fare combination based on the actual patronage, note the resulting increase or decrease in patronage, and repeat the process until p converges with the assumed value p_s. Use of a diagram such as Figure 3, which shows the shifts in headway and demand encountered in such a process, indicates that in Case I this procedure will cause the value of p to converge at Equilibrium Point 2, so long as the initial value of p_s exceeds the value of p at Equilibrium Point 1. In Case II, however, this process cannot result in an equilibrium, since none exists. In this case, its repeated application would eventually drive the value of p to zero.

V. PRACTICAL CONSIDERATIONS

The primary purpose of the foregoing discussion has been to show how demands which vary with price and level of

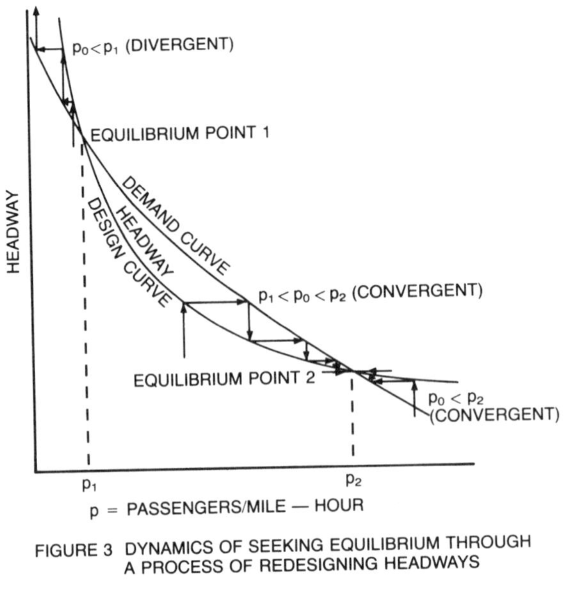

FIGURE 3 DYNAMICS OF SEEKING EQUILIBRIUM THROUGH A PROCESS OF REDESIGNING HEADWAYS

service can be incorporated into rational design methods for transit services. The ability to address this issue in theoretical terms removes one barrier to the practical application of rational design methods, but it by no means insures their feasibility, since several other problems remain.

In the first place, it is beyond the scope of this paper to judge the merits of any particular design or price rules. Obviously, the particular rules employed by a given operator ought to be consistent with that operator's goals; also, they ought to be based on realistic assumptions concerning the utility functions of potential customers and the relationships that exist among patronage levels, design parameters (such as the headway) and service attributes. Since the goals of transit operators are usually complex and subject to change, and since there are many unresolved questions concerning the appropriateness of various utility and cost/travel time functions, it is not possible to say at this time that particular transit operators ought to be employing particular "rational" design rules. At the least, before any such rule is used, the operator should be satisfied that the results are superior to those of its current methods, whatever they may be.

Second, considerable practical difficulty remains in predicting demands for individual transit routes. It is highly unlikely that the aggregate demand function for a transit route can be represented accurately by a simple mathematical expression. Although in principle it should be possible to derive the aggregate function from individual utility functions, this is impractical except under extremely restrictive conditions because it involves the need to integrate the disaggregate function over the distribution of individual characteristics. This does not preclude a relationship between p and the service attributes, nor does it mean that aggregate behavior cannot be predicted approximately using disaggregate models -- only that the aggregation procedure is apt to be cumbersome.

The literature of transportation demand analysis[10,11] suggests several ways of accomplishing the aggregation, and there appears to be no fundamental reason why such methods could not be applied to individual transit routes. It does not appear, however, that anyone has actually devised and validated a route-specific demand model and aggregation procedure; even if such a model were available, it might be difficult to obtain the necessary data at the required level of geographical detail. In the absence of a reliable route-specific demand estimating procedure, graphical methods of solving for patronage equilibria appear to be impractical.

If graphical methods cannot be used owing to a lack of reliable route-specific demand models, it is still possible to locate patronage equilibria, provided they exist, by the experimental methods outlined in Section IV. The amount of experimentation required depends on how closely the equilibrium must be approximated and on how close the initial value of p_s is to the equilibrium point. Given that most transit properties would employ only a limited set of headways and fares, it might be possible to discover the most appropriate headway-fare combination for each route without excessive experimentation. Of course, any experimentation introduces instability into the service offered, and might do more harm than good. Moreover, if no equilibrium exists, an attempt to locate one experimentally could be quite wasteful. Since it would be impossible to tell in advance whether or not an equilibrium exists in the absence of a reliable demand function, experimental methods appear to be quite risky where there is reason to believe no equilibrium exists.

VI. SENSITIVITY OF TRIP DISUTILITY TO PATRONAGE ASSUMPTIONS

At this point, we are faced with a dilemma. We have a method for determining patronage equilibria, but direct application of this method may be impractical for the present because of a lack of reliable demand estimating procedures.

We may also locate patronage equilibria experimentally, but this is risky (especially if no equilibrium exists) and may do more harm than good because of the instability it introduces.

Given these practical difficulties, one possible approach to the use of rational design methods is to try to determine whether accurate patronage estimates are really important; if we can determine circumstances in which they are not, we can still employ the design methods -- at least, under these circumstances.

In order to address this issue, we need to be able to tell how the "costs" perceived by individual passengers vary with the patronage levels assumed for purposes of design. In order to do so, let us isolate the portion of the individual's utility function that varies with the service attributes. In line with much of the recent literature in transportation demand analysis, we will consider three service attributes: fare, in-vehicle time, and out-of-vehicle time. Since these all appear to the passenger as "costs" or disutilities, we will refer to the portion of u^j depending on them as the <u>trip disutility</u>. If we further assume that the trip disutility may be approximated in the region of interest by a linear function of these attributes, we have

$$T^j = F + \gamma_r^j t_r + \gamma_w^j t_w \qquad (10)$$

where T^j = individual j's trip disutility (in some monetary unit)

F = fare

γ_r^j = value of in-vehicle time to individual j

γ_w^j = value of out-of-vehicle time to individual j

t_r = in-vehicle time

t_w = out-of-vehicle time

If we assume a rational design rule, t_r, t_w, and (possibly) F depend on p_s, so we have

$$T^j = S^j(p_s). \qquad (11)$$

$S^j(p_s)$ plays a role analogous to that of a supply function in conventional supply-demand analysis, since it gives the disutility level (price) at which a trip will be produced as a function of the number of trips; consequently, it will be referred to as a parasupply function.

Note that the introduction of γ_r^j and γ_w^j implies that the values of time assumed in the derivation of the TCM design rule in Section III need not apply to all individuals. Rather, as used in deriving the design rule, they might best be regarded as the value the transit operator is willing to place on the passengers' time in various uses -- which may differ from that of any or all of the passengers. To simplify the derivation of parasupply functions based on various design and price rules, we will initially consider an individual whose values of time coincide with those assigned by the transit operator; later, we will consider the effect of assuming other values of time on some aspects of the parasupply functions.

In general, the parasupply function is the sum of the fare and the passengers' time costs. If we adopt the same simplifications and assumptions as in Section III in deriving the time costs,

$$T = F + 2\gamma_r t_s p_s h\ell + \gamma_r(\ell/v) + a\gamma_w h + \gamma_w t_a \,. \tag{12}$$

The headway design rules discussed in Section III give relationships between h and p (such as expressions (6) and (8)) which may be substituted for h in expression (12).

A number of pricing rules are also possible, and their results may be substituted for F in expression (12). Transit operators often charge uniform fares throughout their systems, which means that fares for individual routes appear as fixed charges which do not vary with the cost of providing service. In theory, however, other pricing strategies are possible.

The most obvious of these is average cost pricing. In this case, fares vary from route to route, depending on the average cost per passenger for each route. This fare amounts to $\gamma_o/(p_s hv) + 2\gamma_o t_s$. In this case, the parasupply function and the total cost function are identical, and there is no subsidy. Average cost pricing can also be combined with various subsidy schemes. Among these are fixed subsidies per passenger and fixed subsidies per vehicle-hour of operation.

Another pricing strategy that has some theoretical support is the short run marginal cost (SRMC) fare discussed by Mohring[2]. In this case, the fare is equal to the cost to the system of carrying one additional passenger. This cost is equal to the cost of picking up and discharging the passenger, and consists of money costs to the transit operator and time costs to the other passengers. In the steady-state case, this amounts to $2\gamma_o t_s + 2\gamma_r t_s p_s h\ell$.

Once the relationships between h and p and h and F are known, the actual parasupply function may be determined by substituting for F and h in expression (12). Table 2 presents parasupply functions based on a variety of price and headway rules. Parasupply functions based on the parameter values of Table 1 are also graphed for purposes of comparison in Figure 4.

Although parasupply functions are peculiar to individuals and depend on a large number of parameters, those considered here display one striking feature in common. Note that all the curves in Figure 4, which were derived by applying different design and price rules to a common set of parameter values, display the same general shape: all are negatively sloped, all are convex to the origin, and all display rather abrupt curvature, so that for small values of p_s they represent trip disutilities as being high and relatively sensitive to changes in p_s and for large values of p_s they represent disutilities as being somewhat lower and relatively insensitive to p_s.

One way of quantifying this phenomenon is to determine the elasticity of T with respect to p_s. This elasticity will be designated by $E_p(T)$ and is defined as follows:

$$E_p(T) \equiv -(p_s/T)(\partial T/\partial p_s). \qquad (13)$$

In the case of TCM headways combined with average cost pricing, for instance,

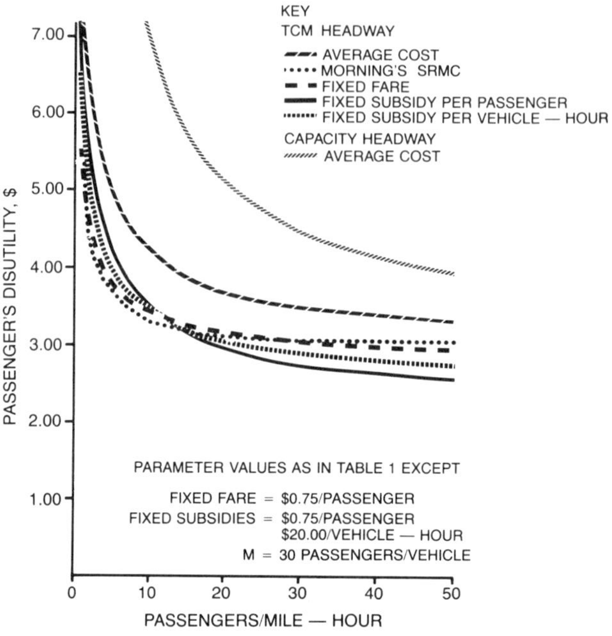

FIGURE 4 PARASUPPLY CURVES

$$E_p(T)=\frac{a\gamma_w\gamma_o}{2\gamma_o(2\gamma_r t_s p_s \ell+a\gamma_w)+k(p_s\gamma_o v)^{\frac{1}{2}}(2\gamma_r t_s p_s \ell+a\gamma_w)^{\frac{1}{2}}} \qquad (14)$$

with k as defined in expression (7a). Figure 5 is a graph of this function. Note that T is inelastic with respect to p for all positive values of p_s, being equal to 0.50 at $p_s = 0$ and approaching zero as p_s approaches infinity. Note further that these features do not depend on the values of the various parameters, although the rate at which the elasticity declines does depend on them. Elasticity curves for other TCM-based parasupply curves are similar and elasticity curves for capacity-headway parasupply functions differ only in that $E_p(T) = 1.0$ when $p_s = 0$.

These characteristics imply that all parasupply functions of the sort considered here have a relatively flat region in which changes in the value of p used for purposes of design will have little effect on the trip disutility. If patronage for a route is high enough to place most potential

Table 2 Parasupply Functions

Headway Design Rule	Price Rule	Parasupply Function
Total Cost Minimum	Average Cost	$T = 2[\gamma_o(2\gamma_r t_s p_s \ell + a\gamma_w)/(p_s v)]^{1/2} + k$
	Mohring's SRMC	$T = [4\gamma_r t_s p_s \ell + a\gamma_w][\gamma_o/(p_s v(2\gamma_r t_s p_s \ell + a\gamma_w))]^{1/2} + k$
	Fixed Fare of F	$T = [\gamma_o(2\gamma_r t_s p_s \ell + a\gamma_w)/(p_s v)]^{1/2} + \gamma_r(\ell/v) + \gamma_w t_a + F$
	Fixed Subsidy per Passenger of S_p	$T = 2[\gamma_o(2\gamma_r t_s p_s \ell + a\gamma_w)/(p_s v)]^{1/2} + k - S_p$
	Fixed Subsidy per Veh-Hr. of S_v	$T = 2[\gamma_o(2\gamma_r t_s p_s \ell + a\gamma_w)/(p_s v)]^{1/2} + k - S_v[(2\gamma_r t_s p_s \ell + a\gamma_w)/(\gamma_o p_s v)]^{1/2} - 2s_v t_s$
Capacity	Average Cost	$T = \gamma_o \ell/(Mv) + 2\gamma_r t_s M + a\gamma_w M/(p_s \ell) + k$

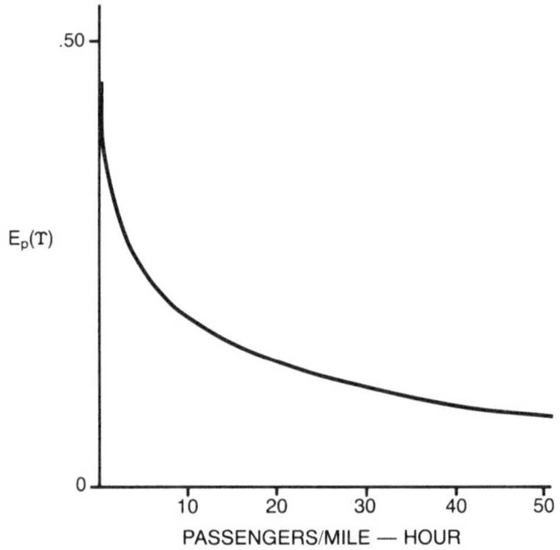

FIGURE 5 ELASTICITY OF PASSENGER DISUTILITY WITH RESPECT TO DEMAND

passengers on this relatively flat part of their parasupply curves, it should make little difference (to the passengers, at least) what value of p_s is used in the design, so long as it is reasonably high and so long as the resulting headway does not overload the vehicles.

It also appears that the parasupply functions of all passengers should become "insensitive" at roughly the same values of p no matter what the individual time values. Let us consider an individual whose values of in-vehicle and out-of-vehicle time are $\gamma_r^j = b\gamma_r$ and $\gamma_w^j = b\gamma_w$, respectively. For such an individual, $E_p(T^j)$ amounts to

$$\frac{(1+b)a\gamma_o\gamma_w}{2[(1+b)\gamma_o(2\gamma_r t_s p_s \ell + a\gamma_w) + (bk_1+k_2)(p_s\gamma_o v)^{\frac{1}{2}}(2\gamma_r t_s p_s \ell + a\gamma_w)^{\frac{1}{2}}]}$$

where $k_1 = \gamma_r(\ell/v) + \gamma_w t_a$ and $k_2 = 2\gamma_o t_s$. It is difficult to make direct comparisons between the value of this function and that of expression (14), but numerical sensitivity analysis indicates that the two elasticities do not differ greatly, even for quite large values of b (say, b = 10). In fact, it is easy to show that the elasticity of the portion of T^j that varies with the headway (i.e., $T^j - (bk_1 + k_2)$) is not at all sensitive to the value of b, since this elasticity amounts to

$$\frac{(1 + b)a\gamma_o\gamma_w}{2\gamma_o(1 + b)(2\gamma_r t_s p_s \ell + a\gamma_w)}$$

and the (1 + b) terms cancel out.

For the parasupply curves analyzed here, the region of insensitivity begins at comparatively modest values of p_s. For the parameter values shown in Table 1 the steep portion of the parasupply curve occurs for values of p_s of about 5 passengers per mile per hour (3 passengers per kilometer per hour) or fewer and the region of relative insensitivity begins around 20 to 25 passengers per mile per hour (12 to 15 passengers per kilometer per hour).

In conclusion, it would appear that many (if not most) routes in comparatively dense urban areas would generate enough patronage to place them in the region of relative insensitivity for their passengers' parasupply curves. Consequently, rational methods could be used to design their headways without too much concern for the accuracy of the patronage estimates involved. At the same time, however, the benefits of using rational methods under such circumstances might be modest, since non-"optimal" designs might result in much the same trip disutility for the passengers. In low density areas, on the other hand, the patronage levels of many routes would cause them to fall into the more sensitive region of their passengers' parasupply curves. In San Diego, for instance, patronage levels as high as 20 passengers per mile per hour (12 passengers per kilometer per hour) are uncommon, even in peak periods. In these low-density systems, the potential rewards for using rational design methods are considerable, but their application is quite tricky. Consequently, it appears that really practical use of rational design methods must await further refinement of methods for demand and cost analysis of transit services.

REFERENCES

1. G. F. Newell, "Dispatching Policies for a Transportation Route," Trans. Sci. 5, 91-105 (1971).
2. H. Mohring, "Optimization and Scale Economies in Urban Bus Transportation," Am. Econ. Rev. 62, 571-604 (1972).
3. V. F. Hurdle, "Minimum Cost Locations for Parallel Public Transit Routes," Trans. Sci. 7, 340-350 (1973).
4. G. C. Clarens and V. F. Hurdle, "An Operating Strategy for a Commuter Bus System," Trans. Sci. 9, 1-20 (1975).
5. S. C. Wirasinghe, "Assignment of Buses in a Coordinated Rail and Bus Transit System," Proceedings of the Seventh International Symposium on Transportation and Traffic Theory, Tsuna Sasaki and Takeo Yamaoka (Eds.), The Institute of Systems Science Research, Kyoto, Japan, August, 1977.
6. J. H. Boyd, N. J. Asher, and E. J. Wetzler, "Non-Technological Innovation in Urban Transit: A Comparison of Some Alternatives," J. of Urban Econ. 5, 1-20 (1978).
7. G. F. Newell, "Some Issues Relating to the Optimal Design of Bus Routes," Trans. Sci. 13, 21-35 (1979).
8. E. Hauer, "Fleet Selection for Public Transit Routes," Trans. Sci. 5, 1-21 (1971).
9. W. O'Brien, "An Example of the Functional Design and Costing of Transit Routes," in J. H. Shortreed (ed), Urban Bus Transit: A Planning Guide. University of Waterloo, Canada, 1974, 153-206.
10. T. A. Domencich and D. McFadden, Urban Travel Demand. A Behavioral Approach. North-Holland/Elsevier, Amsterdam, 1975.
11. M. L. Manheim, Fundamentals of Transportation Systems Analysis, Vol. 1: Basic Concepts. MIT Press, Cambridge, MA, 1979.

THE ACCURACY OF GRAVITY MODEL PARAMETER ESTIMATES

Michael G.H.BELL
Transport Studies Group, University College London, UK

Procedures for the derivation of standard errors for parameters and fitted values are considered with reference to a particular formulation of the gravity model. Methods based on maximum likelihood estimation are found to have certain disadvantages associated with the large number of parameters typically involved and poor goodness-of-fit, probably due to the omission of pertinent factors from the model. Alternative empirical procedures are described, with particular attention given to a method based on the use of Monte Carlo simulation. Two small example data sets are used to compare three procedures, one based on maximum likelihood estimation, a second on an approximation to maximum likelihood estimation under the assumption of a Poisson error structure, and the third on Monte Carlo simulation. The first example data set is designed to contain Poisson variates and the second to contain non-Poisson variates with a variance-to-mean ratio larger than one.

1. INTRODUCTION

The gravity model is frequently used for the analysis of spatial interaction and for forecasting the quantity of movement between each of a set of origin nodes or zones and each of a set of similarly specified destination nodes or zones. Despite numerous applications of the model, in only a very few cases have standard errors for parameters and fitted values been derived. In the paper, the problem of deriving standard errors for the estimates is considered for a formulation of the gravity model that can be regarded as a model of quasi-independence for a three-dimensional contingency table (see Bishop, Fienberg and Holland[1] for a discussion of the concept of quasi-independence). Approaches based on maximum likelihood estimation are found to have disadvantages in situations where large numbers of parameters are fitted and where the model is typically rejected on the basis of conventional significance tests. Consequently, an approximate procedure and two empirical methods are described. Particular attention is given to an empirical procedure based on the use of Monte Carlo simulation to replicate error.

As a prelude to the consideration of the derivation of standard errors, this paper begins with a discussion of point estimation and some of the problems that might be encountered, such as separability and the partial confounding of parameter estimates.

The penultimate section contains a comparison of the standard errors derived by three procedures (maximum likelihood estimation, an approximation to maximum likelihood estimation under the assumption of a Poisson error structure, and Monte Carlo simulation) for two small data sets. The first example data set contains Poisson variates in the cells of the origin-destination matrix, while the second contains variates with a rather larger variance-to-mean ratio. The paper concludes with a discussion on the use of standard errors.

2. THE GRAVITY MODEL AS A MODEL OF QUASI-INDEPENDENCE

The form of gravity model considered can be represented as follows:

$$t_{ijk} = p_i q_j r_k \delta_{ijk} \qquad (1)$$

where t_{ijk} represents the flow from origin i to destination j falling within the kth interval of separation,

p_i is a parameter for origin i, q_j a parameter for destination j, r_k a parameter for the kth interval of separation, and δ_{ijk} is an indicator function equal to one if trips from i to j fall within the kth interval of separation and zero otherwise. It is customary to assume that for each ij pair, δ_{ijk} is non-zero for only one value of k. The model allows for origin and destination scale effects as well as for the effect of separation on movement. In many applications of the model, the effect of separation is given a specific functional form, for example $f(c_{ij}) = \exp(-\beta c_{ij})$ or $f(c_{ij}) = c_{ij}^{-\alpha} \exp(-\beta c_{ij})$ (where c_{ij} is the separation between origin i and destination j).

The model, in the above form, is fitted to a three-dimensional table in which the first two dimensions represent the origin and destination nodes or zones and the third represents separation (categorised). Since it is conventionally assumed that each origin-destination pair is associated with only one separation interval (namely that corresponding to the minimum separation available to travellers), the table usually consists largely of <u>structural zeros</u>. Moreover, if the set of sampled origin-destination pairs is less than the set of feasible origin-destination pairs, there will be additional structural zeros corresponding to the unsampled origin-destination pairs. In this situation, which typically occurs when the sample of flows is collected by roadside interview, the resulting origin-destination matrix is said to be partial[2] or incomplete[1]. The structure assumed by the locations of the structural zeros in the contingency table can result in separability, discussed in Section 3.

Each dimension of the table is referred to as a factor and each category in the classification of the sample (each origin, destination and separation interval) as a factor level. In the model, there is a parameter for each factor level so each factor is represented by a set of parameters. Thus $\{p_i\}$ represents the origin factor, $\{q_j\}$ the destination factor and $\{r_k\}$ the separation factor. Since the model includes no parameters to allow for interactions between factors, it is a model of quasi-independence.

3. GRAVITY MODEL CALIBRATION

In general, the factors $\{p_i\}$, $\{q_j\}$ and $\{r_k\}$ are fitted by the iterative proportional scaling algorithm which ensures agreement between the fitted and observed values of the marginal totals of the three-dimensional table. This has been shown to be consistent with maximum likelihood estimation when the observed values in the table are Poisson or multinomial variates[3][4]. Other procedures have been suggested, for example Murchland[5] considered the use of an optimum decision objective function. In this approach, parameters are derived so as to minimise the expected loss for some particular representation of sampling error, necessitating the assignation of costs to errors. Some authors, for example Cesario[6] and Wansbeek[7], have considered least squares fitting procedures, but it can be shown that these usually constitute particular cases of maximum likelihood estimation.

The essence of maximum likelihood estimation (see, for example, Edwards[8] for a fuller exposition) is that the likelihood of the hypothesis given the outcome of some trial, $L(H|O)$, is defined to be proportional to the probability of the outcome given the hypothesis, $P(O|H)$, where in this case the outcome O represents the data and the hypothesis H a particular model calibration. Thus, the maximisation of $P(O|H)$ with respect to model parameters is regarded as equivalent to the selection of the most likely calibration of the model. If it is assumed that the model form is correct, in the sense that it correctly and completely represents the processes at work, then it is possible to derive both point estimates and standard errors by this procedure.

In general, the observed values in the three-dimensional table are not Poisson variates. Even if data is collected by a roadside interview process that is correctly described as binomial sampling and if the original traffic volumes from which the samples are drawn are themselves Poisson variates, factoring or the grossing-up of sampled values (perhaps to allow for differential sampling fractions) will cause the observed values to be non-Poisson distributed. Hence, the use of the iterative proportional scaling algorithm for model fitting is usually not in practice consistent with maximum likelihood estimation.

A number of problems may be encountered in point estimation. The first arises quite generally in the fitting of log-linear models to <u>separable</u> forms of incomplete contingency tables (the notion of separability is discussed in Bishop, Fienberg and Holland[1]). Two cells in the three-dimensional table are said to be <u>associated</u> if they do not contain structural zeros and if they have a common origin, destination or separation interval. A set of non-structural-zero cells is said to be <u>connected</u> if every pair of cells can be linked by a chain of cells any two consecutive members of which must be associated. Hence the three-dimensional table is connected if its non-structural-zero cells form a connected set. However, if the table as a whole is not connected, it is separable since by suitable permutations of the origins, destinations or separation intervals, blocks or sub-tables may be formed that have no origins, destinations or separation intervals in common.

A consequence of separability is a reduction of the rank of the <u>design matrix</u> (matrix X in Section 4) corresponding to a reduction in the number of independent parameters and an increase in the number of arbitrary scaling constants in the model. In the absence of separability, the parameter estimates are unique after allowance for the arbitrary scaling that may be applied to any two of the three factors. When separability is present, the sets of parameters corresponding to one or more of the factors split into sub-sets with the creation of further arbitrary scaling constants. In the context of the gravity model, non-identifiability (which is one of the consequences of separability) was first reported by Day and Hawkins[9] and subsequently discussed by Murchland[10] and Kirby[2].

In general, if the origin-destination pairs formed by a sub-set of zones or nodes fall exclusively into a (not necessarily contiguous) sub-set of separation intervals, then $\{r_k\}$, $\{p_i\}$ and $\{q_j\}$ will partition creating additional scaling constants. However, if the origin-destination matrix is separable (perhaps due to the method of data collection) then $\{p_i\}$ and $\{q_j\}$, but not necessarily $\{r_k\}$, will partition. As an example, if the data were collected on a single screenline bisecting a study area into two regions, A and B,

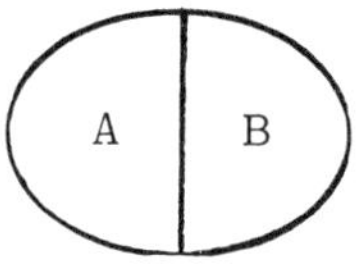

the zones within the study area could be ordered by region so that the sampled origin-destination matrix had the following structure:

$$\left[\begin{array}{c|c} 0 & X \\ \hline Y & 0 \end{array}\right],$$

where the diagonal blocks contain only structural zeros. In this case, the three-dimensional table is separable in two dimensions and $\{p_i\}$ and $\{q_j\}$ each split into two sub-sets. However, if the separation matrix $\{c_{ij}\}$ is symmetric, $\{r_k\}$ will not partition.

In this particular example, there will also be a relationship between the distance of the origin or destination from the screenline and the length of the sampled trip. As a consequence of this relationship, some of the deterrent effect of separation on distribution will be picked up by $\{p_i\}$ and $\{q_j\}$ rather than $\{r_k\}$. When this situation arises (typically when origin-destination data is collected by roadside interview), the origin and destination factors and the separation factor are said to be partially confounded[11].

Another problem, associated with maximum likelihood estimation, is the assumption that the model is correct. It is widely recognised that the gravity model leaves out of account many pertinent factors for the determination of distribution. Hence, the model does not constitute a complete description of the processes at work. The error due to these omitted factors, which in the case of the gravity model tends to be considerable, has been referred to as model mismatch[3]. In practice, attempts are frequently made to reduce model mismatch by sample stratification, thereby indirectly introducing other factors into the model.

The presence of model mismatch accounts for the tendency for the gravity model to be rejected on the basis of conventional tests of significance. Leese[12] has argued that for large sparse origin-destination matrices with low expected values, Pearson's chi-squared statistic $(\chi^2 = \sum(O-E)^2/E)$ is biassed toward rejection even in the absence of substantial model mismatch, possibly contributing to this tendency.

In practice, there is usually a third component of error emanating from incomplete convergence of the tri-proportional procedure. Ill-conditioned problems are sometimes encountered (see, for example, Branston[13]). These are defined by Kirby[2] as situations in which small perturbations of the data result in large perturbations of the parameter estimates and the fitted values. It seems very likely that ill-conditioned problems will be slow to converge.

The problems of separability, partial confounding and the assumption of model correctness affect the derivation of standard errors, as well as point estimation, by maximum likelihood. If the arbitrary scaling constants (including those generated by separability) are not fixed by some form of standardisation, standard errors are undefined. The partial confounding of factors should result in an increase in the size of the standard errors of the parameters affected, as in the equivalent situation of multicollinearity in linear regression.

As far as point estimation is concerned, the iterative proportional scaling procedure has certain attractive features even when there is substantial model mismatch or when the error due to sampling is non-Poisson (perhaps because of the common practice of grossing-up, or factoring, sampled values). Experience has shown that the procedure is almost always convergent. Moreover, for large samples one would expect the sampled and fitted values of the marginal totals in the contingency table to agree irrespective of the form of the error structure since maximum likelihood estimators are consistent in the statistical sense (unbiassed for large samples; see, for example, Edwards[8]). Thus, for large samples:

$$E \sum \tilde{t}_{ijk} = \sum E\tilde{t}_{ijk} = \sum \hat{t}_{ijk} \tag{2}$$

where the summation is over ij , jk or ik , $\tilde{}$ denotes a sampled value and $\hat{}$ a fitted value.

4. STANDARD ERRORS FOR GRAVITY MODEL PARAMETERS

The derivation of standard errors for the parameters of a gravity model is rather more difficult, hence its neglect in most gravity model applications. There has been some interest in the way errors in the inputs affect the fitted values for certain forms of gravity model[14][15].

The estimated covariance matrix for the maximum likelihood estimators of a vector of parameters, $\underline{\theta}$, is equal to

$$\begin{bmatrix} -E\dfrac{\partial^2 L(\underline{\theta}|\underline{\tilde{t}})}{\partial\theta_1^2}, & -E\dfrac{\partial^2 L(\underline{\theta}|\underline{\tilde{t}})}{\partial\theta_1\partial\theta_2}, & \cdots \\ -E\dfrac{\partial^2 L(\underline{\theta}|\underline{\tilde{t}})}{\partial\theta_2\partial\theta_1} & \ddots & \\ \vdots & & \ddots \end{bmatrix}^{-1}$$

where $L(\underline{\theta}|\underline{\tilde{t}})$ is the log-likelihood of $\underline{\theta}$ given observations $\underline{\tilde{t}}$ (see, for example, Edwards[8]). The expected information matrix, which is defined as the Hessian of the log-likelihood function after taking expectations and changing signs, must be non-singular so independence among the parameters must be attained, either by giving several parameters fixed values (and thereby destroying symmetry) or by expressing the parameters as deviations (retaining symmetry but requiring constraints, such as zero sums for sets of parameters).

When a quadratic approximation to the negative of the log-likelihood function at the minimum can be made, there is a weighted least squares problem corresponding to the maximum likelihood problem[16]. In particular, this is so for log-linear models for contingency tables of independent Poisson or multinomial variates. In the absence of separability, the gravity model of Equation (1) may be written as follows:

$$\begin{aligned} \ln t_{ijk} &= \ln p_i + \ln q_j + \ln r_k + \ln \delta_{ijk} \\ &= \alpha + \zeta_i + \mu_j + \gamma_k + \ln \delta_{ijk} \end{aligned} \tag{3}$$

where $$\zeta_1 = \mu_1 = \gamma_1 = 0 \tag{4}$$

and therefore

$$\alpha = \ln p_1 + \ln q_1 + \ln r_1 \tag{5}$$

$$\zeta_i = \ln p_i - \ln p_1 \tag{6}$$

$$\mu_j = \ln q_j - \ln q_1 \tag{7}$$

$$\gamma_k = \ln r_k - \ln r_1 \,. \tag{8}$$

In the matrix notation

$$\underline{\lambda} = X \underline{\xi} \tag{9}$$

where $\underline{\lambda}$ is equal to the elements of $\{\ln t_{ijk}\}$ omitting those elements for which δ_{ijk} is zero, $\underline{\xi}$ is equal to $\{\alpha, \zeta_2, \ldots, \zeta_I, \mu_2, \ldots, \mu_J, \gamma_2, \ldots, \gamma_K\}$ and X is a matrix of ones and zeros, referred to as the design matrix. The covariance matrix for the parameter estimates, $\hat{\underline{\xi}}$, is $(X^T \hat{V}^{-1} X)^{-1}$ while that for the transformed fitted values, $\hat{\underline{\lambda}}$, is $X(X^T \hat{V}^{-1} X)^{-1} X^T$. Matrix $\hat{V}$ is the covariance matrix for the sampled values which, in the case of independent Poisson errors, is taken to be diagonal with elements equal to the fitted values since these are consistent estimates of the expected values.

Both these approaches to the derivation of the covariance matrix for the parameter estimates require matrix inversion, where the order of the matrix to be inverted is equal to the number of parameters to be estimated. The time taken to invert a matrix is approximately proportional to the cube of its order, unless the matrix has certain simplifying characteristics. Since, in general, gravity models involve large numbers of estimated parameters and the matrix to be inverted does not have simplifying characteristics, these two procedures may be costly or even infeasible in the presence of resource constraints.

Another problem associated with these approaches is the assumption that the model is correct in the sense defined earlier. In situations where model mismatch is substantial, leading to the rejection of the model by conventional significance criteria, standard errors derived by maximum likelihood will overstate the accuracy of estimators.

To circumvent some of the limitations to the derivation of standard errors by maximum likelihood, one can either adopt an approximate procedure to reduce the amount of computation required or use an empirical method. Regarding an approximate procedure, it is known that the inverse of a diagonal or block diagonal matrix is also a diagonal or block diagonal matrix consisting of the inverses of the elements or blocks on the principal diagonal (see, for example, Johnston[17]). Thus, if the expected information matrix were diagonal or block diagonal, computation would be greatly reduced. If the covariances for parameters are approximately zero, then the variances are given by:

$$\left[- E \, \frac{\partial^2 L(\underline{\theta} | \underline{\tilde{t}})}{\partial \theta_u^2} \right]^{-1}, \qquad u = 1, \ldots, U,$$

where U is the number of independent parameters.

Concerning empirical methods for the derivation of standard errors, one, referred to as the <u>jackknife method</u>, involves the partitioning of the sample into overlapping sub-samples and the estimation of parameters for these sub-samples. A random partition of the observation for each cell into m parts is required; for sample counts in an unfactored or ungrossed-up form, this is a multinomial distribution with each probability equal to $1/m$. The theory behind the jackknife method is well-developed (see, for example, Miller[18]).

Other empirical methods are based on the use of Monte Carlo simulation to replicate error in the data. Two Monte Carlo procedures may be identified. Both accept the fitted model as if it were true. In the first procedure, a sample is drawn for each cell in the three-dimensional table not containing a structural zero according to an assumption about sampling error. Fitting the model to the resulting samples provides a new estimate for each parameter. Repeating the process yields a series of estimates for each parameter, and the moments of these distributions can be calculated. Standard deviations of the parameter estimates thus calculated represent the sampling component of error but not the mis-specification component.

In the second Monte Carlo procedure, the samples from each cell are drawn from distributions intended to describe residual error (the difference between observed and fitted values). Residual error, which is largely the sum of sampling and specification error, is treated as if it were sampling error and the samples for each cell are drawn accordingly. The standard deviations for the parameter estimates calculated in this way reflect both sampling and specification error. This second Monte Carlo simulation procedure, a version of which is developed in Section 5, is seen as having the following advantages.

(i) When large numbers of origins, destinations and separation intervals are involved, it should involve less computation than alternative maximum likelihood procedures.

(ii) If the model is to be fitted so that the marginal totals of the observed and fitted value contingency tables agree, but if the error structure is not Poisson (either because of the intrinsic behaviour of the sampled variable or because of grossing-up or factoring), then the use of Monte Carlo simulation should give better standard errors than those derived by maximum likelihood estimation based on an incorrectly specified error structure.

(iii) In all situations in which the fitted model is rejected on the grounds of a likelihood ratio test, the corresponding maximum likelihood estimates of the standard errors are likely to be too small. Standard errors derived by the Monte Carlo simulation procedure should be more satisfactory.

5. STANDARD ERROR DERIVATION BY MONTE CARLO SIMULATION

The method explored here is based on the repeated application of the iterative proportional scaling procedure.

Given the observed values of the marginal totals of the three-dimensional table (the origin, destination and separation interval totals), a fitted value origin-destination matrix, $\{\hat{t}_{ij}\}$, is derived by the iterative proportional scaling procedure. Treating these fitted values as unbiassed estimates of the means, a random number generator is used to generate a series of pseudo-sampled matrices in accordance with a specified error structure. For each parameter fitted, a series of estimates is derived by fitting the model (again by the iterative proportional scaling procedure) to the series of pseudo-sampled matrices. Finally, the moments of the distributions of these parameter estimates can be calculated, yielding their covariance matrix.

It was thought that in this case the distribution used for the generation of errors should have the following properties:

(i) non-negativity for the random variate,
(ii) a constant variance-to-mean ratio.

The second desired property arises because there is some evidence[19] that the variance of the residual error tends in practice to be proportional to the size of the fitted value. However, it should be pointed out that sampling error only follows a variance proportional to mean law for small counts (for large counts, a constant coefficient of variation law is followed), and the residuals are a function of the method of fitting.

Let us define the residual index of dispersion, $\hat{\beta}$, as Pearson's chi-squared statistic divided by the number of fitted values minus the number of independent parameters (or the residual degrees of freedom). Assuming that the residuals follow a variance proportional to mean law, $\hat{\beta}$ is an estimate of the constant of proportionality. If the error structure is Poisson and the model correct, $\hat{\beta}$ should be in the vicinity of one. However, the gravity model is typically rejected by a chi-squared significance criterion because $\hat{\beta}$ is significantly greater than one. To allow for this in the derivation of standard errors by Monte Carlo simulation, the simulated variance-to-mean ratio, β , is set equal to $\hat{\beta}$.

The chosen distribution for the simulated error structure was the log-normal. This distribution is positive, continuous and can accommodate a constant variance-to-mean ratio of β . A log-normally distributed variate is one whose logarithm is normally distributed. Thus, if $Y \sim N(\eta, \sigma^2)$ and $Y = \ln X$, then X is a log-normal variate. The distribution of X possesses moments of any order[20]. The mean and variance of X are

$$E(X) = \exp(\eta + \tfrac{1}{2}\sigma^2) \tag{10}$$

$$Var(X) = \exp(2\eta + \sigma^2)(\exp(\sigma^2) - 1). \tag{11}$$

In simulation, $E(X)$ would be equal to the fitted value $\widehat{t_{ij}}$ and $Var(X)$ equal to $\beta\widehat{t_{ij}}$. Thus

$$\sigma^2_{ij} = \ln(\beta/\widehat{t_{ij}} + 1) \tag{12}$$

$$\eta_{ij} = \ln \widehat{t_{ij}} - \sigma^2_{ij}/2 = \ln \widehat{t_{ij}} - \ln(\beta/\widehat{t_{ij}} + 1)/2. \tag{13}$$

The random number generator for the normal distribution can then be used to create the pseudo-observed flows t^s_{ij} ; $t^s_{ij} = \exp(y_{ij})$, where $y_{ij} \sim N(\eta_{ij}, \sigma^2_{ij})$.

6. A COMPARISON

In this section, the standard errors for the parameters derived by three procedures are compared. The first procedure, based on Monte Carlo simulation and the iterative proportional scaling procedure, is the one described in Section 5. The log-normal distribution is used to generate 30 pseudo-observed origin-destination matrices. The second procedure is the approximation to maximum likelihood estimation, described in Section 4, which neglects the covariance between parameters. Finally, the third procedure corresponds to maximum likelihood estimation. The statistical program GLIM[21], which, if the model and error structure hypothesis are correct, derives maximum likelihood estimates and their standard errors exactly (after allowance for rounding error or incomplete convergence) by an iterative weighted least squares algorithm, was used. Under a Poisson error structure hypothesis, the point estimates yielded by GLIM and the iterative proportional scaling algorithm should be equal. However, Tables 2 and 3 reveal that there were differences in practice, it is thought principally due to incomplete convergence in the case of the iterative proportional scaling algorithm.

6.1 Two synthetic data sets

Given a set of 10 origin zone totals, 10 destination zone totals and 8 separation interval totals, a matrix of fitted, or mean, values was derived such that these totals were preserved (the GLIM program was used because of its greater accuracy). To generate the first data set, a Poisson random number generator was used to produce a pseudo-observed origin-destination matrix with a Poisson error structure. For the second data set, a normal random number generator was used with a variance-to-mean ratio set equal to two. The random numbers generated were truncated to give integer values; negative variates were given zero values. As a consequence of these modifications to the random variates, the residual index of dispersion, $\hat{\beta}$, resulting from the fitting of the model to the variates was larger than two.

In Table 1, the number of the separation interval, the mean used for the generation of the random variates, the Poisson variate and the non-Poisson variate are given for each zone pair in a vertical four-element array. For example, in the case of zone pair (1,8), the separation interval index is 6 , the mean used for the generation of the random variates is 3.2, the Poisson variate is 5 and the non-Poisson variate is 10.

Table 1: The separation interval number, mean value and variates of the two synthetic origin-destination matrices

1	2	3	5	7	6	5	6	5	8
2186.0	851.1	59.6	4.7	6.3	5.2	7.8	3.2	3.5	3.5
2191	836	52	4	5	7	13	5	3	6
2093	867	66	15	39	18	11	10	19	0
2	1	3	5	7	5	5	6	5	8
1201.0	12530.0	169.5	13.3	17.8	25.5	22.3	9.2	10.1	10.0
1183	12508	153	8	18	26	21	11	6	14
1154	12336	136	7	12	18	25	3	6	4
3	3	1	3	5	3	3	4	4	5
91.1	183.5	1745.0	36.7	7.0	70.5	61.6	32.9	20.9	2.3
102	194	1731	38	4	81	68	29	27	1
85	163	1696	32	7	86	57	42	25	19
6	5	3	1	3	3	4	5	5	6
36.2	125.7	321.4	4749.0	242.9	344.4	226.0	30.0	19.1	6.5
37	131	309	4683	261	350	236	28	25	6
30	118	342	4734	245	309	219	36	25	27
8	7	5	3	1	3	4	6	6	8
48.0	56.6	20.4	81.2	2920.0	156.1	102.4	7.9	5.0	8.6
47	54	22	85	2919	155	101	5	4	8
39	64	18	63	2895	159	87	11	13	1
6	5	3	2	3	1	1	2	4	6
13.7	47.6	121.7	347.3	92.0	3458.0	3021.0	415.1	38.6	2.4
9	42	106	326	96	3533	3020	427	34	4
0	40	121	396	115	3457	3135	396	42	19
6	5	3	4	4	1	1	2	4	5
11.3	39.3	100.3	42.0	57.0	2852.0	2491.0	342.4	31.9	3.5
12	54	102	39	51	2891	2528	330	31	4
0	42	103	38	52	2836	2526	334	31	8
7	7	5	6	6	4	2	1	3	4
39.7	79.9	28.9	9.3	12.7	165.5	986.7	3634.0	87.0	38.3
51	86	20	4	10	173	1020	3696	79	41
32	61	30	5	11	141	944	3721	82	43
5	6	4	5	6	5	4	3	1	3
25.7	30.0	99.3	10.3	8.1	19.9	92.9	88.0	1481.0	32.7
30	36	92	6	2	20	92	88	1472	38
17	28	97	1	37	13	87	75	1510	31
8	8	6	7	8	6	5	3	3	1
45.2	91.0	11.1	10.6	24.7	11.9	18.0	91.3	57.9	899.1
47	98	12	13	25	7	11	83	62	894
42	88	0	0	34	7	22	91	54	883

6.2 Results

In the case of both data sets, the parameter estimates are defined after allowing for two arbitrary scaling constants. In practice, a Grand Mean α was included in the model (in linear form) and ζ_1, μ_1 and γ_1 were set equal to zero to remove the arbitrary scaling. The goodness-of-fit of the model fitted by GLIM and the iterative proportional scaling procedure to the two data sets is indicated by the following values for Pearson's chi-squared statistic:

	First data set (Poisson)	Second data set (Non-Poisson)	Degrees of freedom
GLIM	58.51	606.1	74
Tri-proportional procedure	64.87	589.6	74

Using a conventional chi-squared significance test, the model would not be rejected in the case of the data with Poisson variates whereas it would be in the case of the data with non-Poisson variates (for example, $\chi^2_{0.99} = 105.21$ for 74 degrees of freedom). The values of $\hat{\beta}$ were thus 0.88 and 7.97 respectively (the value of 0.88 is clearly an underestimate indicating that the model has partially fitted the sampling errors).

The point estimates and their standard errors for the two synthetic data sets are given in Tables 2 and 3. In the case of the approximation to maximum likelihood estimation, where the off-diagonal elements of the information matrix are assigned zero values, the error structure is assumed to be Poisson. Thus, the vector of approximate variances for $\{\hat{\alpha} ; \hat{\zeta}_2 , \ldots , \hat{\zeta}_I ; \hat{\mu}_2 , \ldots , \hat{\mu}_J ; \hat{\gamma}_2 , \ldots , \hat{\gamma}_K\}$ is given by

$$\Big\{ 1 / \sum_{ijk} \tilde{t}_{ijk} ; 1 / \sum_{jk} \tilde{t}_{2jk} , \ldots , 1 / \sum_{jk} \tilde{t}_{Ijk} ;$$

$$1 / \sum_{ik} \tilde{t}_{i2k} , \ldots , 1 / \sum_{ik} \tilde{t}_{iJk} ; 1 / \sum_{ij} \tilde{t}_{ij2} , \ldots , 1 / \sum_{ij} \tilde{t}_{ijK} \Big\} .$$

Table 2: The point and interval estimates for the origin-destination matrix with Poisson variates

Parameter	Iterative proportional scaling procedure				Maximum likelihood estimation by GLIM		
	Point estimate	Simulated S.E. β=0.88	Simulated S.E. β=1.00	Approximate S.E.	H_1 Point estimate	H_1 S.E.	H_2 S.E.
α	7.694	0.023	0.025	0.004	7.691	0.020	0.018
ζ_2	1.045	0.026	0.027	0.009	1.041	0.024	0.022
ζ_3	0.134	0.029	0.031	0.021	0.192	0.041	0.038
ζ_4	1.660	0.035	0.037	0.013	1.755	0.045	0.041
ζ_5	0.822	0.037	0.039	0.017	0.918	0.050	0.046
ζ_6	0.644	0.028	0.030	0.012	0.744	0.043	0.039
ζ_7	0.455	0.027	0.029	0.013	0.555	0.043	0.039
ζ_8	1.185	0.035	0.037	0.014	1.286	0.045	0.041
ζ_9	0.747	0.041	0.044	0.023	0.838	0.055	0.050
ζ_{10}	0.704	0.050	0.053	0.028	0.798	0.065	0.060
μ_2	0.697	0.026	0.028	0.008	0.700	0.023	0.021
μ_3	–0.371	0.036	0.038	0.020	–0.428	0.040	0.036
μ_4	–0.903	0.038	0.041	0.014	–0.995	0.045	0.041
μ_5	–0.542	0.041	0.044	0.017	–0.632	0.049	0.045
μ_6	–0.178	0.032	0.034	0.012	–0.273	0.042	0.038
μ_7	–0.317	0.029	0.031	0.012	–0.413	0.042	0.038
μ_8	–0.667	0.030	0.032	0.015	–0.764	0.045	0.041
μ_9	–1.145	0.048	0.051	0.024	–1.231	0.055	0.050
μ_{10}	–1.599	0.054	0.058	0.031	–1.686	0.067	0.061
γ_2	–1.652	0.018	0.019	0.016	–1.651	0.018	0.016
γ_3	–3.270	0.024	0.026	0.020	–3.276	0.023	0.021
γ_4	–3.593	0.024	0.026	0.033	–3.594	0.035	0.031
γ_5	–5.237	0.048	0.052	0.046	–5.271	0.048	0.044
γ_6	–5.833	0.077	0.083	0.076	–5.875	0.079	0.073
γ_7	–5.124	0.062	0.066	0.066	–5.200	0.073	0.066
γ_8	–4.564	0.068	0.072	0.064	–4.636	0.076	0.069

Table 3: The point and interval estimates for the origin-destination matrix with non-Poisson variates

Parameter	Iterative proportional scaling procedure				Maximum likelihood estimation by GLIM		
	Point estimate	Simulated S.E. β=1.00	Simulated S.E. β=7.97	Approximate S.E.	H_1 Point estimate	H_1 S.E.	H_2 S.E.
α	7.665	0.025	0.070	0.005	7.663	0.020	0.056
ζ_2	0.977	0.027	0.077	0.009	0.975	0.024	0.067
ζ_3	-0.052	0.031	0.091	0.021	-0.020	0.041	0.115
ζ_4	1.471	0.037	0.107	0.013	1.523	0.044	0.124
ζ_5	0.566	0.040	0.114	0.017	0.618	0.049	0.140
ζ_6	0.514	0.030	0.089	0.011	0.570	0.041	0.117
ζ_7	0.300	0.029	0.085	0.013	0.356	0.042	0.119
ζ_8	0.977	0.037	0.109	0.014	1.033	0.044	0.125
ζ_9	0.502	0.044	0.132	0.023	0.551	0.054	0.152
ζ_{10}	0.433	0.054	0.151	0.029	0.482	0.063	0.182
μ_2	0.777	0.028	0.079	0.009	0.779	0.024	0.065
μ_3	-0.160	0.039	0.113	0.020	-0.192	0.040	0.112
μ_4	-0.674	0.042	0.118	0.014	-0.724	0.044	0.125
μ_5	-0.254	0.044	0.125	0.017	-0.303	0.048	0.137
μ_6	-0.027	0.035	0.099	0.012	-0.080	0.041	0.117
μ_7	-0.134	0.032	0.091	0.012	-0.187	0.041	0.117
μ_8	-0.439	0.033	0.096	0.015	-0.493	0.044	0.125
μ_9	-0.843	0.052	0.150	0.024	-0.890	0.054	0.153
μ_{10}	-1.294	0.059	0.162	0.031	-1.339	0.064	0.187
γ_2	-1.647	0.019	0.053	0.016	-1.647	0.018	0.048
γ_3	-3.293	0.026	0.070	0.020	-3.295	0.024	0.065
γ_4	-3.604	0.026	0.077	0.033	-3.604	0.035	0.096
γ_5	-5.158	0.050	0.139	0.045	-5.171	0.046	0.132
γ_6	-5.508	0.074	0.205	0.068	-5.526	0.069	0.226
γ_7	-5.091	0.067	0.194	0.069	-5.127	0.074	0.216
γ_8	-4.559	0.078	0.221	0.069	-4.592	0.078	0.224

The value of β in the two tables corresponds to the value of the variance-to-mean ratio used in the generation of random variates for the simulation of standard errors. Comparisons between Tables 2 and 3 of the standard errors (S.E.) derived suggest, as one might expect, a proportionality between the value obtained and $\sqrt{\beta}$.

In the case of the first example data set, in which the variates are known to be Poisson distributed, the standard errors derived by maximum likelihood estimation under a Poisson error structure hypothesis (referred to as H_1) may be used as a standard of comparison. Table 2 therefore suggests that the approximate procedure often seriously under-estimates the standard errors, especially for $\hat{\alpha}$,$\{\hat{\zeta}_i\}$ and $\{\hat{\mu}_j\}$, presumably because of the omission of covariance. The Monte Carlo simulation procedure also shows a tendency to under-estimate the standard errors for $\{\hat{\zeta}_i\}$ and $\{\hat{\mu}_j\}$, both when β is equal to $\hat{\beta}$ (in this case, $\hat{\beta} = 0.88$) and when β is equal to its 'true' value, namely one. However, there is a better accord between the three procedures for the standard errors of $\{\hat{\gamma}_k\}$.

In Table 3, the standard errors yielded by maximum likelihood estimation under H_1 and the approximate procedure (also based on H_1) remain substantially unchanged in relation to their corresponding values in Table 2, despite the markedly poorer goodness-of-fit in the case of the second data set. This illustrates the fact that maximum likelihood estimates (including those for the standard errors) are based on the assumed rather than the actual error structure. In this case, the standard errors are under-estimated because of the exclusion of the misspecification component of error from the error structure.

If the error structure hypothesis used for the derivation of standard errors by maximum likelihood estimation still assumed a proportionality between the sizes of error variance and the expected values but if the constant of proportionality was not assumed to have a value of one, then the estimates of the standard errors would tend to be larger in the case of the second example data set. To demonstrate this, a second error structure hypothesis (referred to as H_2) was formulated. Under this hypothesis, the dependent variable is normally distributed with constant variance after a square root transformation. It can be shown by means of the Taylor series approximation for the variance of a function that

$$\mathrm{var}(\sqrt{x}) \simeq \frac{\mathrm{var}(x)}{4\eta} \qquad (14)$$

(where η is the expected value for x) and that therefore this use of the square root transformation facilitates a constant but fitted variance-to-mean ratio.

Maximum likelihood point estimation under H_2 does not, however, imply the conservation of the marginal totals of the contingency table, as in the case of H_1 . Maximum likelihood estimation under the assumption of normal errors with constant variance can be shown to be equivalent to unweighted least squares estimation. Thus, if

$$L = \sum_{ijk} (\sqrt{t_{ijk}} - \sqrt{\tilde{t}_{ijk}})^2 \qquad (15)$$

where $t_{ijk} = p_i q_j r_k \delta_{ijk}$, then

$$\frac{\partial L}{\partial p_i} = \frac{2}{\hat{p}_i} \sum_{jk} (\sqrt{\hat{t}_{ijk}} - \sqrt{\tilde{t}_{ijk}})\sqrt{\hat{t}_{ijk}} = 0 \qquad (16)$$

for some i if $\hat{p}_i \neq 0$

which implies that

$$2\sum_{jk}(\sqrt{\hat{t}_{ijk}} - \sqrt{\tilde{t}_{ijk}})\sqrt{\hat{t}_{ijk}} = \sum_{jk} (\sqrt{\hat{t}_{ijk}} - \sqrt{\tilde{t}_{ijk}})(\sqrt{\hat{t}_{ijk}} + \sqrt{\tilde{t}_{ijk}})$$

$$+ \sum_{jk} (\sqrt{\hat{t}_{ijk}} - \sqrt{\tilde{t}_{ijk}})^2 = 0. \qquad (17)$$

This in turn implies that

$$\sum_{jk} (\hat{t}_{ijk} - \tilde{t}_{ijk}) = - \sum_{jk} (\sqrt{\hat{t}_{ijk}} - \sqrt{\tilde{t}_{ijk}})^2 < 0. \qquad (18)$$

Hence, the fitted value marginal total for origin i will be less than the corresponding observed value marginal total. A similar result obtains for the marginal total corresponding to each parameter.

In Table 2, the standard errors derived by maximum likelihood estimation under H_2 are similar to those derived by this procedure under H_1 . However, in Table 3, those derived by large sample maximum likelihood estimation under H_2 are markedly larger than those derived by this procedure under H_1 and are roughly commensurate with those derived by the Monte Carlo simulation procedure when $\beta = \hat{\beta} = 7.97$.

In conclusion, Monte Carlo simulation offers a method of obtaining standard errors when more conventional statistical techniques requiring matrix inversion are either not available or are too costly. The synthetic data set containing Poisson variates suggests that the Monte Carlo simulation procedure developed in Section 5 gives better estimates of standard errors than the approximation in which covariances are assumed to be absent. Moreover, the Monte Carlo simulation procedure only requires a random number generator for the normal distribution and the iterative proportional scaling algorithm which is usually used for fitting a gravity model of the form considered here. Furthermore, since the standard errors obtained make allowance for both sampling and misspecification error, the Monte Carlo simulation procedure may be used when (as is commonly the case) the gravity model is ill-fitting on the basis of a likelihood ratio or equivalent test. Nonetheless, it would be interesting to extend the comparison made here to include the jackknife method and other approximations (another approximation is suggested by Gunn[22]).

7. THE USE OF STANDARD ERRORS FOR PARAMETERS

In general, when the gravity model is used for forecasting, the standard errors for $\hat{\alpha}$, $\{\hat{\zeta}_i\}$ and $\{\hat{\mu}_j\}$ are of little interest since values for these parameters are recalculated so that the origin and destination totals given by the fitted values correspond to exogenous estimates. However, $\{\hat{\gamma}_k\}$, reflecting the revealed deterrence relationship, is assumed to be unchanged or modified in some simple way (for example, to allow for an increased mean trip length).

The standard errors for $\{\hat{\gamma}_k\}$ may be used to define confidence intervals on the assumption that the estimators are normally distributed (maximum likelihood estimators are approximately normally distributed for large samples; see, for example, Edwards[8]). When using the Monte Carlo simulation procedure, it is in principle not necessary to make an assumption of normality in order to derive confidence intervals since they can be calculated directly by enumeration. However, in order to do this, rather more than the 30 pseudo-observed origin-destination matrices, generated in the case of the two example data sets, would be required, implying correspondingly more computation.

Confidence intervals for $\{\hat{\gamma}_k\}$ can be used to help to identify possible smooth continuous forms of separation function. Moreover, the standard errors facilitate an examination of the empirical support given to a variety of separation functions. For example, the functions $f(c_{ij}) = \exp(-a_0 c_{ij})$, $f(c_{ij}) = c_{ij}^{-a_1}$ and $f(c_{ij}) = \exp(-a_2 c_{ij}) c_{ij}^{-a_3}$ linearise under a log transformation. Consequently, linear regression can be used to fit these functions to the points $\hat{\gamma}_1, \ldots, \hat{\gamma}_K$ by weighted least squares, using the inverse of their variances as weights, and the weighted sum of squared errors compared with the percentage points of a chi-squared distribution with appropriate degrees of freedom.

Figures 1 and 2 show the 99 per cent confidence intervals for $\{\hat{\gamma}_k\}$ for the two example data sets described in Section 6. They are based on the standard errors derived by Monte Carlo simulation when the simulated variance-to-mean ratio β is

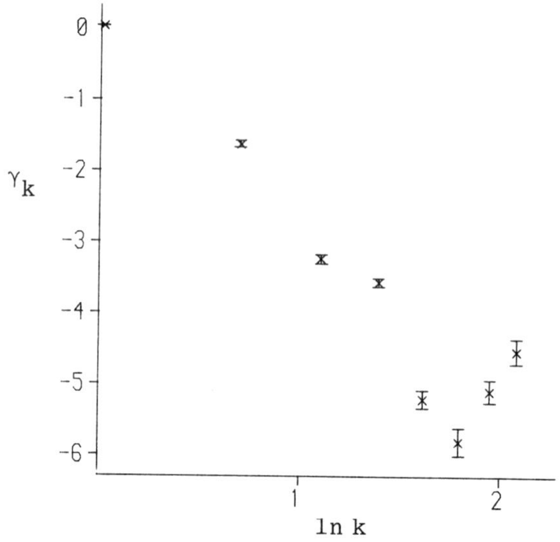

Figure 1: Confidence intervals derived by Monte Carlo simulation for the first example data set

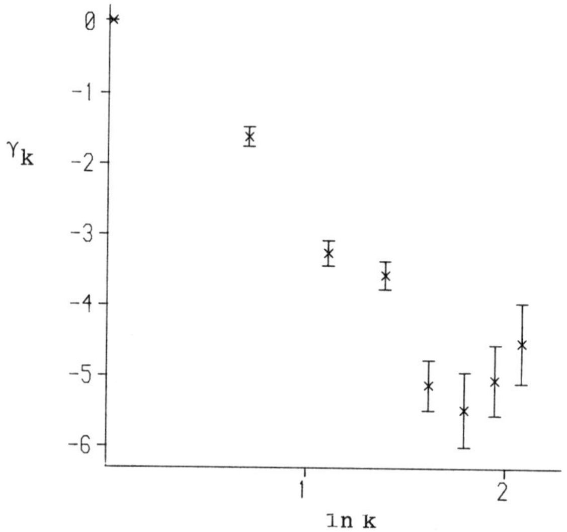

Figure 2: Confidence intervals derived by Monte Carlo simulation for the second example data set

equal to the residual index of dispersion $\hat{\beta}$ and assume normally distributed estimators. In the figures, $\hat{\gamma}_k$ and its confidence interval is plotted against $\ln k$. Hence, if $f(c_{ij}) = c_{ij}^{-a_1}$ is in fact the true relationship and the separation intervals are of equal width, a linear function should be plausible. However, both figures suggest a non-monotonic relationship, although in the case of Figure 2 there is more room for doubt.

ACKNOWLEDGEMENTS

Preliminary work on this paper was carried out at the University of Leeds with the financial support of the Rees Jeffreys Road Fund. The author is grateful for discussions with Mr. H.R.Kirby and Mr. H.F.Gunn at the University of Leeds and with Mr. J.P.Nicholl and Dr. J.D.Murchland formerly at University College London, as well as for the helpful suggestions of Professor R.E.Allsop and an anonymous referee.

REFERENCES

1. Y.M.M.BISHOP, S.E.FIENBERG and P.W.HOLLAND, 'Discrete Multivariate Analysis'. MIT Press, London, 1975.

2. H.R.KIRBY, 'Partial matrix techniques', Traffic Engineering and Control, 20, 422-428, 1979.

3. H.R.KIRBY, 'Theoretical requirements for calibrating gravity models', Transportation Research, 8, 97-104, 1974.

4. H.R.KIRBY and M.N.LEESE, 'Trip distribution calculations and sampling error: some theoretical aspects', Environment and Planning, A10, 837-851, 1978.

5. J.D.MURCHLAND, 'What are the errors we should be seeking to minimise in the estimation process?' Institute for Transport Studies, University of Leeds JDM-318.1, 1979.

6. F.J.CESARIO, 'Least squares estimation of trip distribution parameters', Transportation Research, 9, 13-18, 1975.

7. T.J.WANSBEEK, 'Least squares estimation of trip distribution parameters : a note', Transportation Research, 11, 429-431, 1977.

8. A.W.F.EDWARDS, 'Likelihood', University Press, Cambridge, 1972.

9. M.J.L.DAY and A.F.HAWKINS, 'Partial matrices, empirical deterrence functions and ill-defined results', Traffic Engineering and Control, 20, 429-433, 1979.

10. J.D.MURCHLAND, 'Non-identifiability in tri-proportional problems', Institute for Transport Studies Technical Note 23, University of Leeds, 1979.

11. P.W.M.JOHN, 'Statistical Design and Analysis of Experiments', MacMillan, New York, 1971.

12. M.N.LEESE, 'A statistical test for trip distribution models', 9th Annual Conference Universities Transport Study Group, Universities of Glasgow and Strathclyde, January 1977. (unpublished)

13. D.M.BRANSTON, 'The sensitivity of trip matrices produced by the partial trip matrix technique to errors in the assumed trip end constraints', TPG Note 54, Transportation Planning Group, County Engineer's Department, Greater Manchester Council.

14. J.D.MURCHLAND, 'Perturbation of the bi-proportional problem', Transport Studies Group, University College London JDM-305.1, 1978.

15. T.KNUDSEN, 'Assessing uncertainties in trip distribution forecasting'. Proceedings PTRC Annual Summer Meeting, University of Warwick, July, Session M: Transportation Models, 1979. PTRC Education and Research Services Ltd., London.

16. J.A.NELDER and R.W.M.NEDDERBURN, 'Generalised linear model models', Journal of the Royal Statistical Society A, **135**, 370-384, 1972.

17. J.JOHNSTON, 'Econometric Models', McGraw-Hill Kogakusha, Tokyo, 1972.

18. R.G.MILLER, 'The jackknife - a review', Biometrika, **61**, 1-15, 1974.

19. J.C.HASKEY, 'Empirical methods for judging the efficacy of mathematical models for transportation studies', Proceedings PTRC Seminar on Urban Traffic Models, University of Sussex, June, **1**, 1973. PTRC Education and Research Services Ltd., London.

20. J.AITCHISON and J.A.C.BROWN, 'The Log-Normal Distribution', University Press, Cambridge, 1969.

21. R.J.BAKER and J.A.NELDER, 'The GLIM system; release 3. General Linear Interactive Modelling', Numerical Algorithms Group, Oxford, 1978.

22. H.F.GUNN, 'An approximation to errors in gravity model trip ends', Institute for Transport Studies, RHTM Trip Distribution Investigation Working Note 27, University of Leeds, 1979.

A DECISION-MAKING FRAMEWORK FOR THE EVALUATION OF CLIMBING LANES ON TWO-LANE TWO-WAY RURAL ROADS

Jan L. Botha and Adolf D. May
University of California, Berkeley, California

ABSTRACT

Research was conducted at the University of California at Berkeley to investigate the cost-effectiveness of climbing lanes on two-lane two-way rural roads. Specifically, two issues were of concern:

a) given a specific upgrade with specific traffic conditions and a budget constraint (which allows for the construction of a specific length of climbing lane), the question arises as to where the climbing lane(s) should be located on the upgrade to be the most cost effective; and

b) given a number of candidate upgrades for the construction of climbing lanes and a budget constraint, the question arises as to the upgrades on which climbing lanes should be constructed and what the lengths of the climbing lanes should be.

A microscopic simulation model was developed to evaluate the performance of a climbing lane. Field measurements were obtained at two climbing lane sites using four coordinated time-lapse cameras. One data set was used for model calibration and the second data set was used for model validation. Comparison of individual travel times and lane changes by segment and vehicle-type gave close agreement between field measurements and model predictions.

A two-lane design of experiment was developed which employed the model in the investigation of the two critical issues identified above. The first level was the investigation of the cost-effectiveness of a climbing lane with a specific length at various locations on a specific upgrade. The second level was the investigation of the relative cost-effectiveness of a climbing lane with a specific length and a specific location on upgrades with various geometric and traffic conditions.

These investigations have led to specific conclusions, policy guidelines, and identification of further research needs. The conclusions and guidelines have been distributed by the California Department of Transportation to its various district offices and currently are providing additional guidance in the planning and design of climbing lanes in California. A new research project has been funded to undertake further research with particular attention to additional measures of effectiveness and extending policy guidelines.

1. BACKGROUND

In the past, problems with traffic flow on two-lane two-way rural roads were solved by constructing four-lane roads or freeways. Increased demand for funds for the improvement of urban transportation facilities and for maintenance of existing roads has eroded funds traditionally available for road construction. Environmental considerations and local resistance may also prevent expanding two-lane roads to four-lane roads. This had led to serious consideration of improving the use of existing facilities as an alternative to new construction and also to increased emphasis on the cost-effectiveness of investments in road construction.

One of the possible ways in which better use can be made of a rural road is to construct climbing lanes. A climbing lane is defined by AASHTO [5] as an upgrade lane on the upgrade edge of a two-lane highway meant for the use of vehicles moving slowly uphill so that other vehicles using the normal lane to the right of the centerline are not impeded.

Research was conducted at the University of California at Berkeley to investigate the cost-effectiveness of climbing lanes on two-lane two-way rural roads. Specifically, two issues were of concern:

a) Given a specific upgrade with specific traffic conditions and a budget constraint (which allows for the construction of a specific length of climbing lane), the question arises as to where the climbing lane(s) should be located on the upgrade to be the most cost-effective.

b) Given a number of candidate upgrades for the construction of climbing lanes and a budget constraint, the question arises as to the upgrades on which climbing lanes should be constructed and what the lengths of the climbing lanes should be.

In order to address these issues, a measure of effectiveness was selected to compare the performance of climbing lanes with specific lengths on specific upgrades under specific traffic conditions. Subsequently, a method was developed in order to evaluate the performance of climbing lanes. Finally, an investigation was conducted to determine general guidelines for the design of climbing lanes with respect to length and location.

2. MEASURE OF EFFECTIVENESS

Travel time savings was selected as the measure of effectiveness for the performance of a climbing lane. It was recognized that a complete determination of the cost-effectiveness of a climbing lane should include the benefits of increased safety and savings (or increases) in vehicle operating cost; however, research budget limitations did not allow for the inclusion of these measures of effectiveness. Provision was, however, made for extension of the research to incorporate these measures of effectiveness and this extension is currently under investigation in Berkeley.

3. EVALUATION METHOD

An investigation of the possible methods for the evaluation of the performance of a climbing lane revealed that with available knowledge, microscopic simulation was the most suitable. This method involves keeping track of the movement of individual vehicles on the road. The primary reason for selecting this method of evaluation was that it allows for the explicit modelling of non-steady-state non-uniform conditions which usually prevail in the vicinity of climbing lanes on two-lane roads.

An existing microscopic model, developed and validated by the Midwest Research Institute [6], was modified for use in the research. The existing model had the capability to simulate the behavior of different types of vehicles, i.e., three classes of trucks, seven classes of recreational vehicles and three classes of automobiles, with different types of drivers, under various geometric and traffic conditions on a two-lane two-way rural road. This model was modified to include the simulation of driver behavior in climbing lane sections. Observation of traffic behavior on road sections with climbing lanes experiencing rather low traffic volumes indicated that the major proportion of lane changes occurs at the entrance to a climbing lane and at the exit from a climbing lane. At higher traffic volumes, the amount of lane changing in the middle part of the climbing lane section varies. As a result of these observations, the traffic operation in the climbing lane was modeled by simulating vehicle movements in three zones in the climbing lane section:

Entrance Zone - the zone where traffic controls (lane markings and signs) instruct or guide the driver as to how to act when entering the climbing lane section.

Terminal Zone - the zone where traffic controls (lane markings and signs) instruct or guide the driver as to how to act when leaving the climbing lane section.

Intermediate Zone - the zone between the entrance and terminal zone.

In order to ensure that the model predictions conformed to real-life situations, data were collected to calibrate and validate the model and also to check whether the modelling of the traffic operations in the climbing lane section was realistic. The model calibration and validation will be described in the next section.

4. MODEL CALIBRATION AND VALIDATION

Data were collected at two sites with widely differing characteristics, using time-lapse photography. The locations of the two sites in Northern California are shown in Figure 1 and the major characteristics of the two sites are shown in Table 1. Three time-lapse cameras were used at each site. One camera was located at the entrance to the climbing lane to observe vehicle movements there, and another was located at the exit from the climbing lane. The third camera was positioned near the midpoint of the climbing lane to observe lane changes in the intermediate zone. Although the three cameras did not cover the entire climbing lane only a very small portion of the intermediate zone was out of view. Since the greatest amount of lane changing was expected to occur in the entrance zone and in the terminal zone, it was felt that the amount of information lost would be too small to justify using another camera. When calibrating the model with respect to the number of lane changes which occur in the intermediate zone, it was taken into account that not all lane changes occurring in this zone were recorded on film. Two additional cameras (one located upstream and one located downstream of the climbing lane) were also used, but the results will be

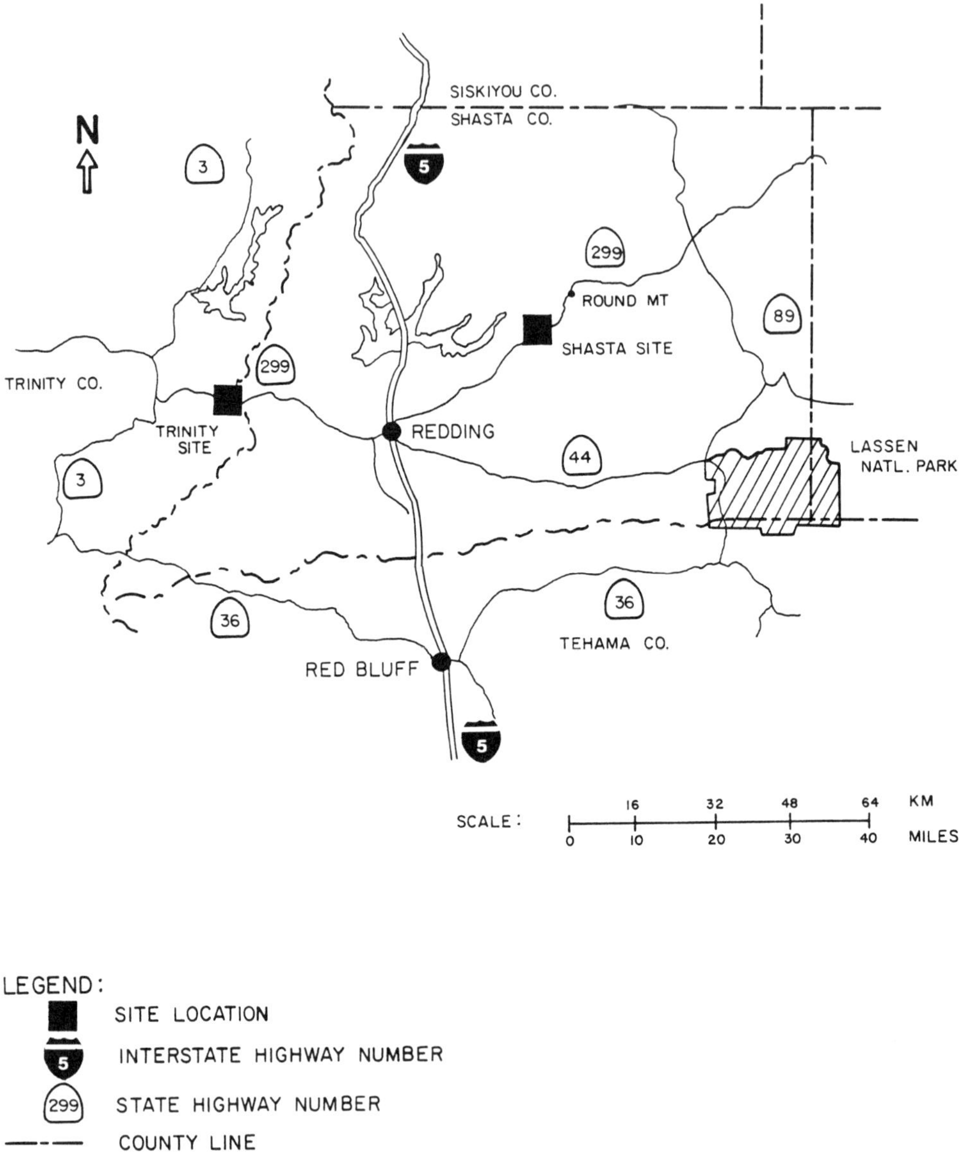

Figure 1 - Location of Data Collection Sites in California.

Table 1 - Characteristics of Data Collection Sites

SITE	COUNTY	ROUTE	POST MILE	LENGTH OF CLIMBING LANE	LENGTH OF GRADE	GRADE (%)	TERMINAL LOCATION	TRAFFIC VOLUME (ADT)
Calibration	Shasta	299	50.93-51.21	0.28 mi (0.45 km)	1.04 mi (1.66 km)	6	Grade	2250
Validation	Trinity	299	70.58-72.17	1.59 mi (2.54 km)	2.27 mi (3,63 km)	5	Crest	2300

analyzed in a future study. Each camera at each site was operated for approximately twelve hours (six hours per day for two days) and time-lapse photos were taken at a rate of one frame per second.

The data obtained at the Shasta site were used to calibrate the model while the data obtained at the Trinity site were used to validate the model. The model was calibrated with respect to the proportion of lane changes in each zone within the climbing lane section for each vehicle type (trucks, recreational vehicles and automobiles) as well as for the travel times by vehicle type and for all vehicles combined, through the climbing lane section. The model was adjusted until the results conformed to the observed data with reasonable accuracy. The comparison of the travel times for all vehicles, as predicted by the model, and the travel times observed in the field at the Shasta site is shown in Figure 2. It can be seen that the simulated travel times are generally lower than the observed travel times in the lower ranges of travel times, while the opposite is true for the higher ranges of travel times. It can also be seen that there is a discrete jump in the simulated travel times at 52 seconds. This is due to the use of discrete vehicle classes in the simulation model and specifically the class of low performance trucks.

The model was validated using the data obtained at the Trinity site. The validation was carried out with respect to the proportion of lane changes for each vehicle type in each zone and also for the travel times by vehicle type and for all vehicles combined, through the climbing lane section. The simulation results and the field data compared satisfactorily. The comparison of the travel times for all vehicles combined is shown in Figure 3. It can be seen that the simulated travel times are generally lower than the observed travel times. If, however, the simulated travel times were increased by approximately 5 percent, the simulation and observed results would be very similar.

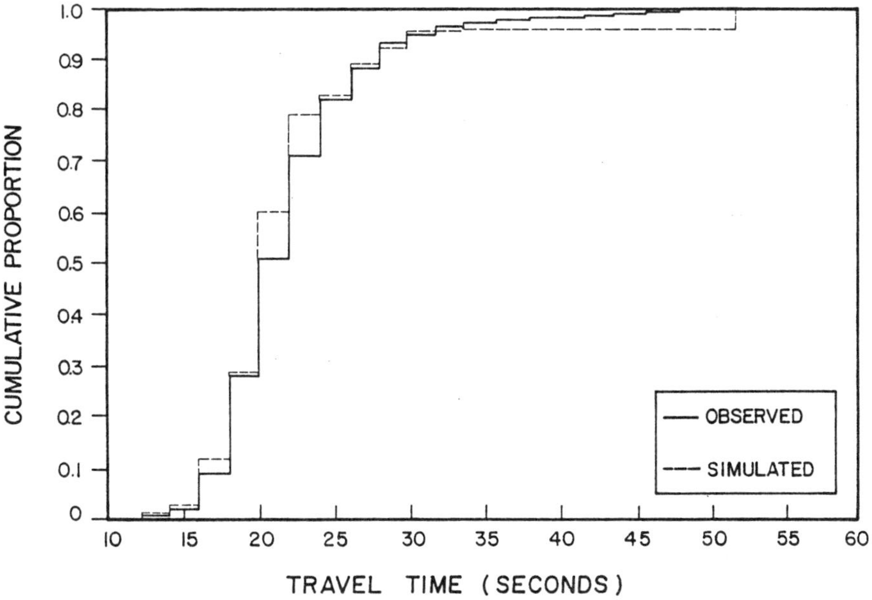

Figure 2 - Comparison of Observed and Simulated Travel Times for All Vehicles - Shasta County Site

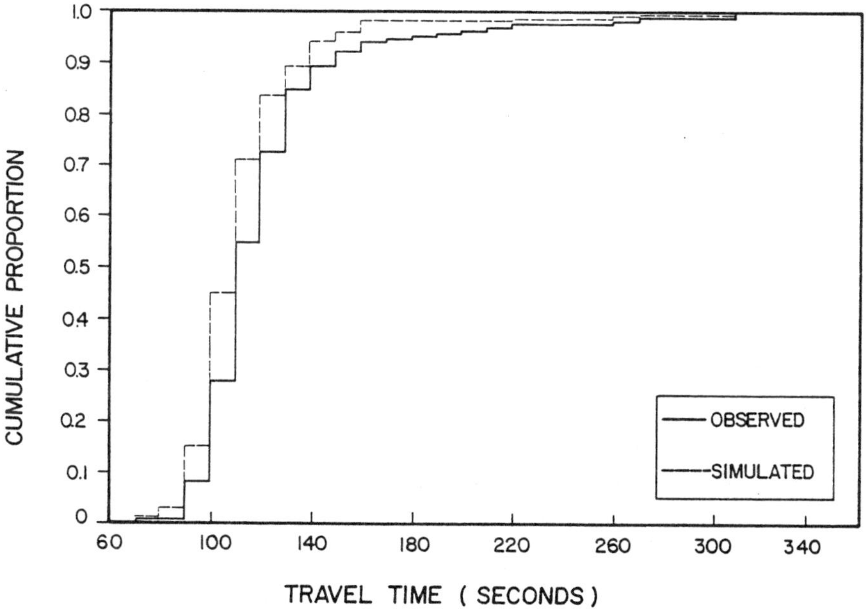

Figure 3 - Comparison of Observed and Simulated Travel Times for All Vehicles - Trinity County Site

5. RESULTS

The investigation to determine general guidelines for the construction of climbing lanes, with respect to length and location, was carried out with the aid of the simulation model. The investigation consisted of two parts:

a) The investigation of the cost-effectiveness of a climbing lane with a specific length at various locations on a specific upgrade.

b) The investigation of the relative cost-effectiveness of a climbing lane with a specific length and a specific location on upgrades with various geometric and traffic conditions.

5.1 The Cost-Effectiveness of a Climbing Lane on a Specific Upgrade

The characteristics of the upgrade used for investigating the cost-effectiveness of a climbing lane on a specific upgrade are presented in Table 2. It should be noted that these characteristics were selected to represent rather severe upgrade conditions in order to bring out any large differences in the performances of climbing lanes at various locations. In order to simplify the experiment and reduce costs, only traffic operations in the uphill direction were examined. Simulating traffic in the downhill direction would have greatly increased computer costs. To compensate for the absence of oncoming traffic, no passing was allowed in the uphill direction outside of the climbing lane section.

Initially, two lengths of climbing lanes were used to determine the most cost-effective location of climbing lanes:

a) A climbing lane with a length of 1500 ft (460 m). This length corresponds to the 15th percentile level of the lengths of a sample of existing climbing lanes in California.

b) A climbing lane with a length of 1 mile (1610 m). This length corresponds to the 85th percentile level of the lengths of a sample of existing climbing lanes in California.

Table 2 - Summary of Characteristics of the Base Situation Used for Comparison of the Performance of Climbing Lanes

CHARACTERISTICS	CONDITIONS INVESTIGATED
Horizontal Alignment	Tangent
Vertical Alignment	
Gradient	6.5%
Length of Grade	4 miles (6.4 km)
Upstream Geometrics	
Gradient	0%
Length of Section	2500 ft (762 m)
Downstream Geometrics	
Gradient	-3%
Length of Section	3000 ft (914 m)
Passing, No-Passing Section	No Passing
Sight Distance	3000 ft (914 m)
Speed Limit	55 mph (90 kph)
Directional Split	Only uphill traffic considered
Traffic Volume (uphill)	400 vph
Percentage Recreational Vehicles	15%
Percentage Trucks	15%

The relative performances of these climbing lanes are shown in Figure 4. (Data points in Figure 4 indicate the location of the beginning of the climbing lane.) Cost-effectiveness, in terms of travel time savings per mile of climbing lane, is measured on the vertical axis. A curve has been drawn for each climbing lane length, 1500 ft (460 m) and 1 mile (1610 m), illustrating how cost-effectiveness varies with climbing lane location. The peak of this curve for the 1500 ft (460 m) length appears to be at a point (read from the horizontal axis) just below the midpoint of the grade. This indicates the most cost-effective location on the grade to begin a 1500 ft (460 m) climbing lane. A 1500 ft (460 m) climbing lane constructed at this point would straddle the midpoint of the grade. The curve drawn for a 1 mile (1610 m) climbing lane can be interpreted in a similar manner. Its peak also identifies a point below the midpoint of the grade as the optimum location to begin a 1 mile (1610 m) climbing lane. A 1 mile (1610 m) climbing lane constructed here would again straddle the midpoint of the grade. These results indicate that a climbing lane should be located symmetrically on the upgrade and should be biased towards the lower side of the grade if field conditions do not permit the symmetrical location. It should be kept in mind, however, that lower gradients would result in less slowing down of vehicles at the start of the grade and in such cases consideration should be given to biasing the location of the climbing lane towards the higher side of the midpoint of the grade.

Having determined the optimum climbing lane location it was then possible to investigate additional climbing lane lengths and configurations. Three other climbing lane configurations were investigated:

c) A climbing lane with a length of one-half mile (805 m) located with its beginning halfway up the grade.

d) A climbing lane with a length of 20,620 ft (6286 m) with its beginning located 1500 ft (460 m) up the

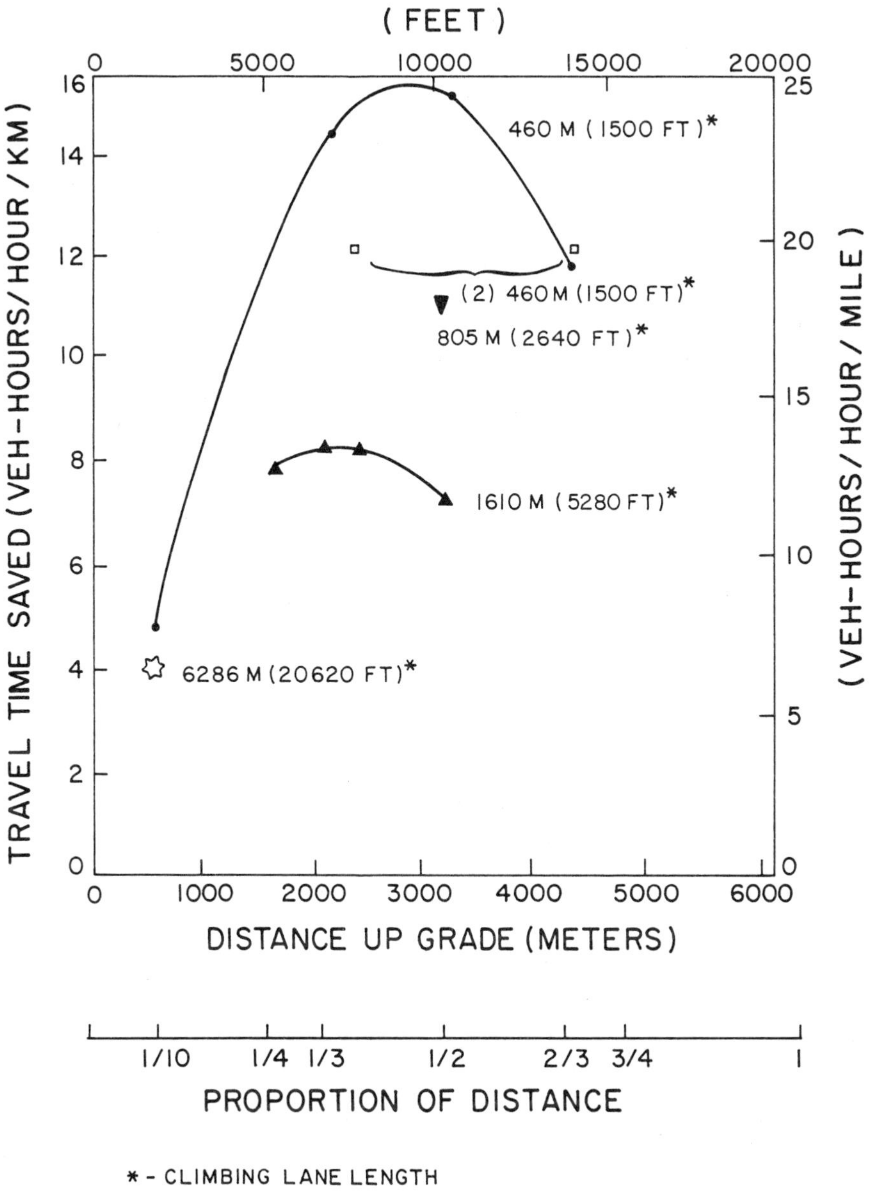

Figure 4 - Relative Performance of Climbing Lanes with Different Lengths and Locations.

grade and extending beyond the crest of the grade. This climbing lane corresponds approximately to the current AASHTO standard [5].

e) Two climbing lanes, each with a length of 1500 ft (460 m), located at the one-third and two-thirds points up the grade.

The relative performance of these climbing lanes is again shown in Figure 4. A comparison of all five sets of data in Figure 4 indicates that a climbing lane with a length of 1500 ft (460 m) located symmetrically on the upgrade is the most cost-effective. Shorter climbing lanes may be more cost-effective, but were not investigated, since safety implications may preclude their use. Investigation of shorter lengths of climbing lanes is left for future research. Another important conclusion is that it is more beneficial to construct one 1500 ft (460 m) climbing lane on two separate upgrades than to concentrate on one upgrade.

It should be noted that the results indicate only the relative performance of the climbing lanes, since the results of the simulation model are random, as a consequence of the random ingredients of the simulation model. In order to obtain the absolute travel time savings, several replications of the experiment should be undertaken with different random number sequences.

5.2 The Relative Cost-Effectiveness of Climbing Lanes on Different Upgrades

In order to investigate the relative cost-effectiveness of climbing lanes on different upgrades, the gradient, length of grade, traffic volume and percentage trucks were varied and the performance of a climbing lane with a length of 1500 ft (460 m) located with its beginning halfway up the grade was evaluated. The other geometric and traffic variables were kept constant at the values shown in Table 2. Budget limitations did not allow for variation of all factors. Only the factors mentioned above, which were considered to be most important with regard to the performance of the climbing lane, were varied.

The investigation of the influence of these factors was conducted in two parts. First, each factor was varied, keeping the values of the other factors constant at the values shown in Table 2. The results of this investigation are shown in Figure 5.

It can be seen that the benefit derived from the construction of a climbing lane is most sensitive to the variation in the gradient. It can also be concluded that very little benefit can be derived from the construction of a climbing lane on a gradient of less than four percent under these other fixed conditions.

The results obtained for the variation of the length of grade indicate an almost linear relationship between benefits and the length of grade. If this linear relationship were to be extrapolated, it may be concluded that very little benefit would be derived at a length of grade shorter than one mile under these other fixed conditions.

The investigation of the effect of traffic volume on the performance of a climbing lane indicated that the benefits are not very sensitive to the volume at high values, but are more sensitive at the lower values. At the higher volumes, headways are short and, therefore, a greater proportion of vehicles are impeded and subsequently a greater number of vehicles benefit from the climbing lane.

The investigation of the effect of the percentage of trucks indicated that there is a great degree of "noise" present in the results. The reason for this is that although truck percentages may be high, there are actually very few trucks on the road. The results are, therefore, very much dependent not only on the total number of trucks, but also on the class of truck and the headway. Specifically, the headways of low performance trucks have the largest influence. In order to predict the influence of trucks more accurately, it would be desirable to increase the number of truck classes to capture better the actual effects of different trucks' performance. It would also be desirable

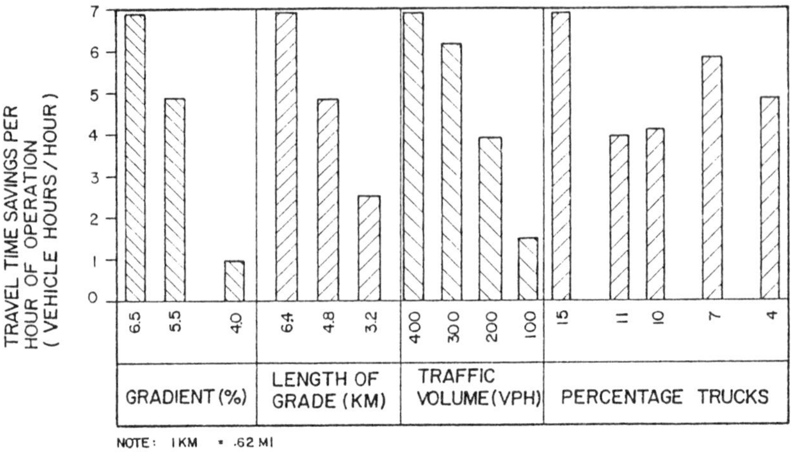

Figure 5 - Effects of the Variation of Factors.

to simulate for very long periods of time and with several replications of the experiment using different random number sequences. On the average, however, the results indicated that the performance of the climbing lane is not very sensitive to the percentage of trucks. It is recommended that the effect of trucks be evaluated by considering whether large proportions of low performance trucks or heavily loaded trucks are actually present and to evaluate this directly with the aid of the computer model. Future expansion of the number of truck classes in the model will enhance this evaluation.

The general conclusion from this analysis is that the performance of a climbing lane on an upgrade is most sensitive to the gradient. Steep upgrades will, therefore, be the most likely candidates for the construction of climbing lanes. In order to obtain more specific guidelines for the selection of upgrades where most benefits can be obtained from the construction of a climbing lane, a limited two-factor analysis was undertaken to determine the effects of the gradient, length of grade and traffic volume. This investigation was conducted by varying two of these factors simultaneously. The other geometric and traffic variables were kept constant at the values shown in Table 2. The relative travel-time savings due to adding the specified climbing lane improvements are presented in Figures 6, 7, and 8. The travel time savings calculated and plotted in Figures 6 and 7 were based on the actual traffic volumes observed during the simulation, i.e., the number of vehicles counted crossing the test section finish line per hour of simulation. Because of the stochastic process by which the model generates vehicle headways, this measured flow rate did not always equal the user specified flow rate. When traffic volume was to be considered a constant in an experiment, such as the two-factor analysis presented in Figure 8, the traffic volume used in calculating total travel time was also held constant for all cases. In Figure 8 the user specified volumes were employed. Contours were interpolated for in-between values on two-dimensional planes for the combinations of gradient-traffic volume, length of grade-traffic volume, and gradient-length of grade. Due to budget constraints the number of cases to be simulated was limited. Only severe upgrade and traffic conditions were examined representing likely candidate sites for the construction of a climbing lane. Consequently the curves in Figures 6, 7, and 8 indicate only approximate relationships for the less severe upgrade and traffic conditions and projections into these ranges should be made with caution.

Additional computer runs were not undertaken for the percentage of trucks because of the "noise" present in the results obtained previously and the large number of replications which would have been required. Time and money were not available to make these runs. Since the benefits obtained from the construction of a climbing lane were not very sensitive to the percentage of trucks, and also because it is recommended that the situations for specific trucks be investigated with the aid of the model, it was considered sufficient to indicate the approximate effect of the variation of the percentage of trucks in combination with

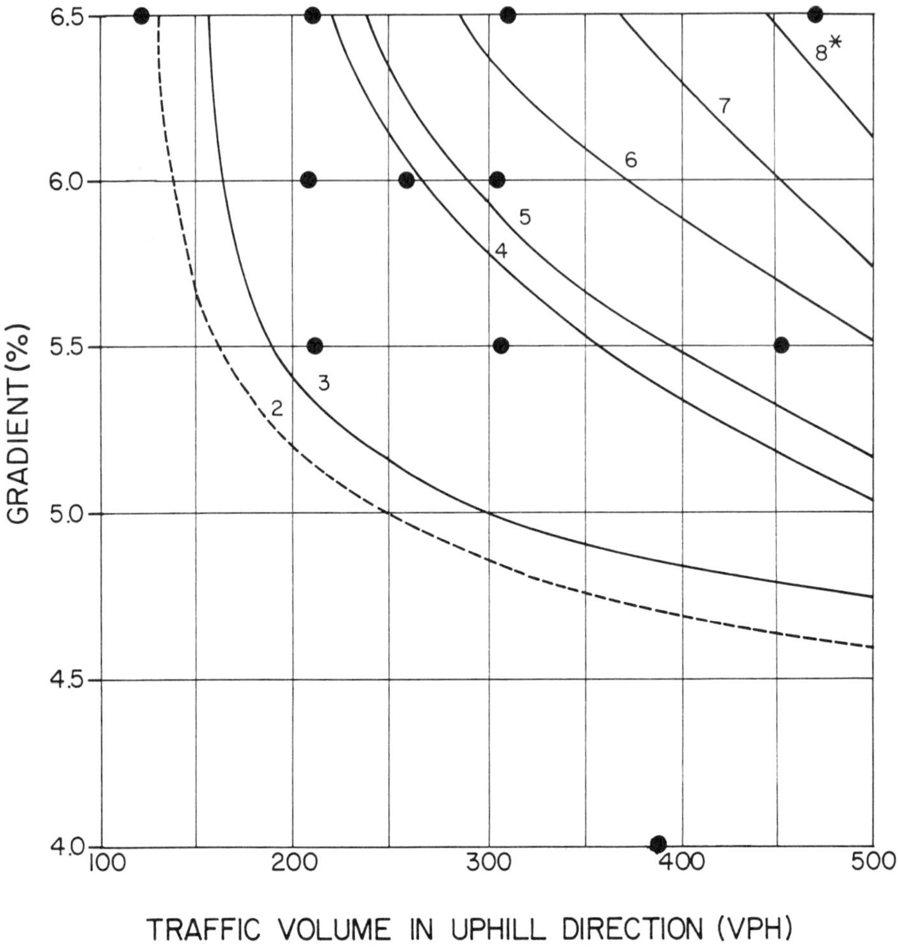

NOTES: 6.4 KM (4 MI) LENGTH OF GRADE, 15% TRUCKS
● INDICATES SITUATION SIMULATED
* INDICATES TOTAL TRAVEL TIME SAVINGS (VEH-HR/HR)
---- INDICATES APPROXIMATE RELATIONSHIP
(CURVES BASED ON ACTUAL TRAFFIC VOLUMES OBSERVED DURING SIMULATION RUNS)

Figure 6 - Relative Travel Time Savings per Hour of Operation for the Combination of Gradient and Traffic Volume.

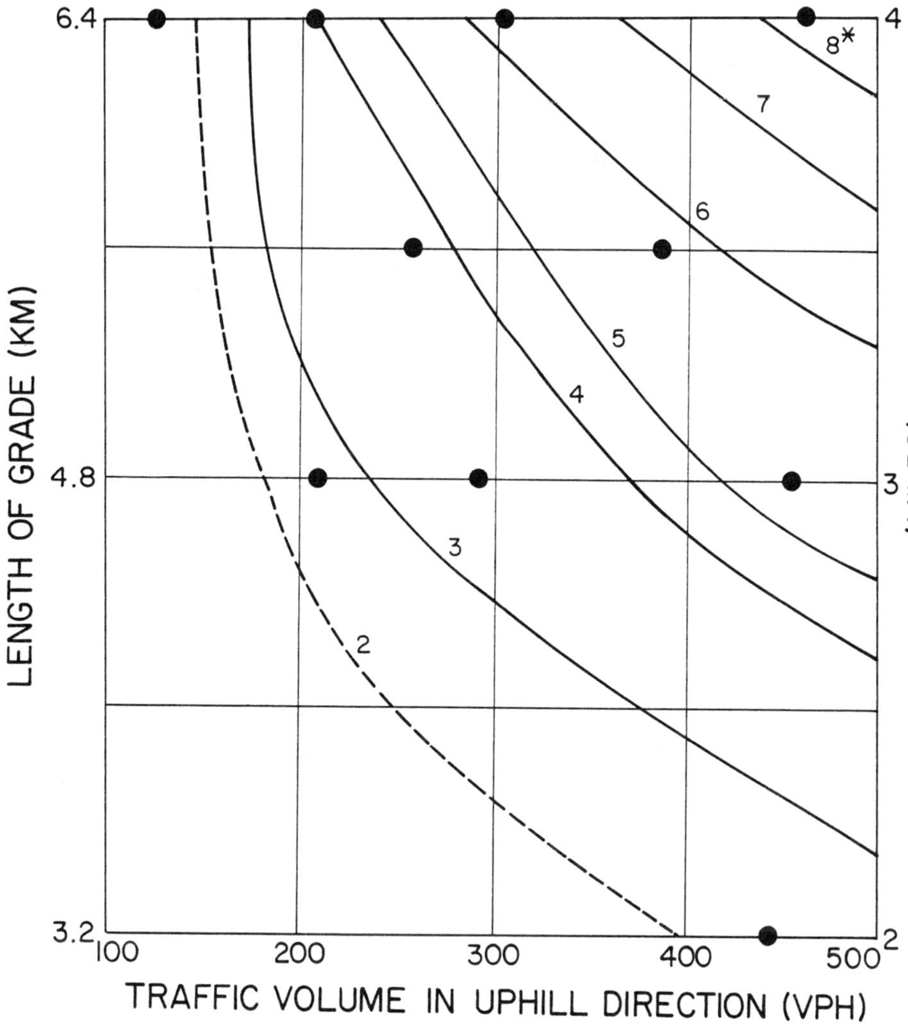

NOTES: 6.5% GRADIENT, 15% TRUCKS
● INDICATES SITUATION SIMULATED
* INDICATES TOTAL TRAVEL TIME SAVINGS (VEH-HR/HR)
---- INDICATES APPROXIMATE RELATIONSHIP
(CURVES BASED ON ACTUAL TRAFFIC VOLUMES OBSERVED DURING SIMULATION RUNS)

Figure 7 - Relative Travel Time Savings per Hour of Operation for the Combination of Length of Grade and Traffic Volume.

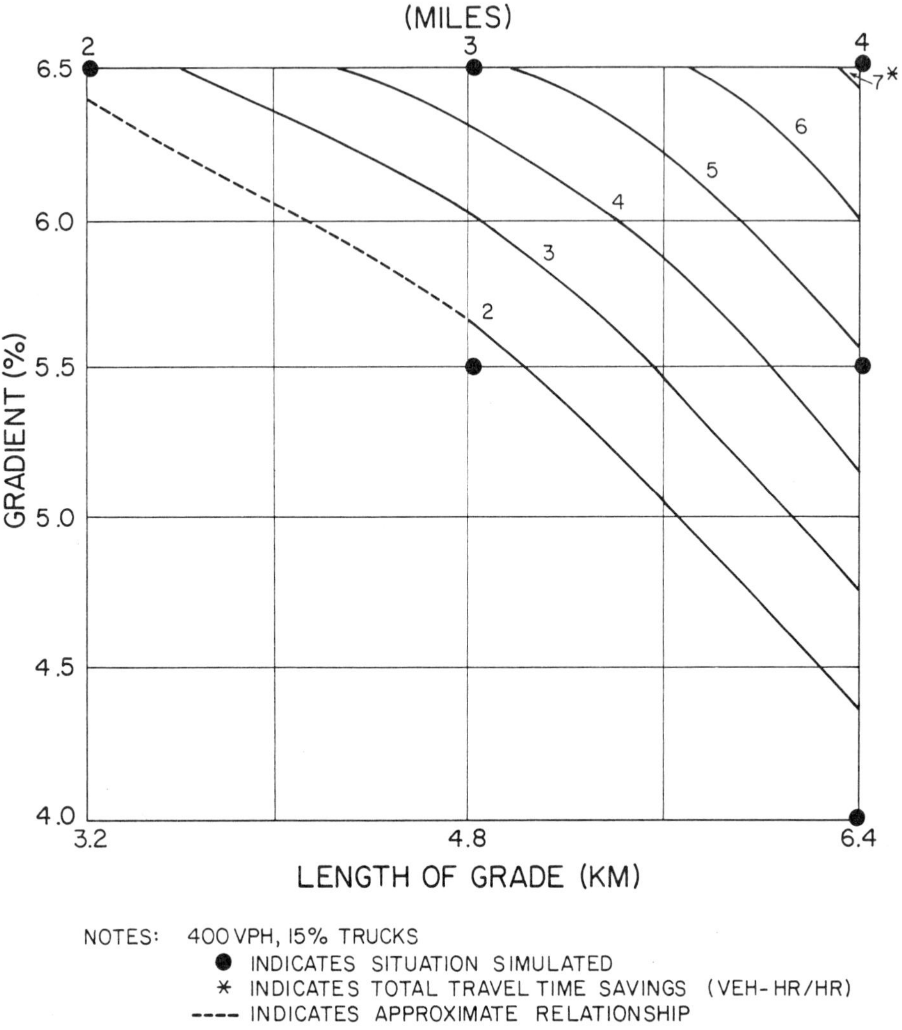

Figure 8 - Relative Travel Time Savings per Hour of Operation for the Combination of Gradient and Length of Grade.

other factors. Planes similar to those presented in Figures 6, 7, and 8 were constructed, using only boundary values and assuming linear relationships between boundaries. These planes are shown in Figures 9, 10, and 11. The boundary values for the effect of the percentage of trucks were obtained by assuming an average straight-line relationship between the percentage of trucks and the travel time savings. Again, due to the small number of cases simulated to produce these curves, the relationships indicated are only approximate and should be used with caution. Future research in this area may help to affirm or modify these relationships.

6. APPLICATIONS

Figures 6 through 11 can be used to select candidate sites for the construction of a climbing lane. For example, if two sites have approximately the same characteristics except for having different gradients and lengths of grade, Figure 8 can be used to determine the relative travel time savings for each site. It should be noted that the travel time savings determined in this manner for the two alternatives are only relative if the base conditions are the same as those used in the experiment, e.g., the traffic volume should be 400 vehicles/hour in the uphill direction with 15% trucks. Using this procedure, the candidate locations for constructing a climbing lane can be narrowed down to a smaller number, which can then be directly investigated with the aid of the model.

It should be noted that the travel time savings determined through use of Figures 6 and 11 should not be considered the absolute travel time savings, since several replications of the experiment would be necessary to determine the absolute travel time savings due to the randomness of the experiment. It should also be noted that these relationships were established assuming that climbing lanes with a length of 1500 ft (460 m) will be constructed, beginning at the midpoint of the grade, and also assuming that the construction cost of the climbing lane is constant

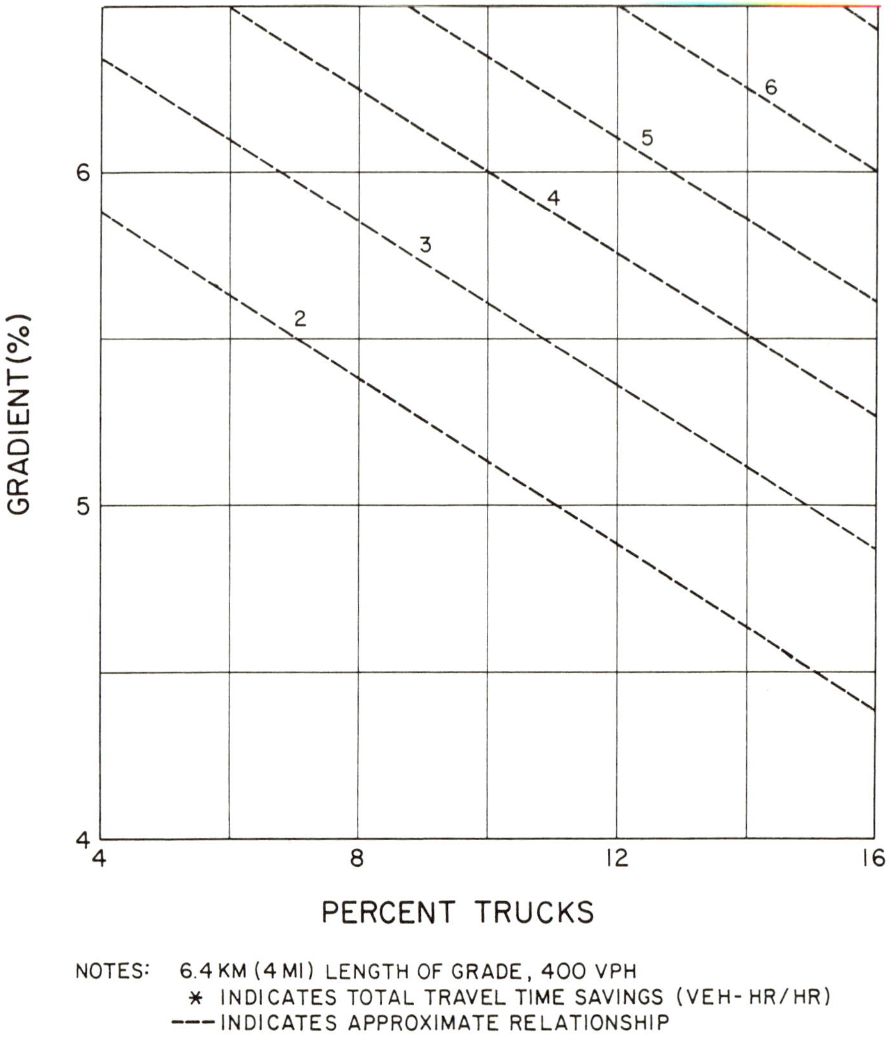

Figure 9 - Relative Travel Time Savings per Hour of Operation for the Combination of Gradient and Percent Trucks.

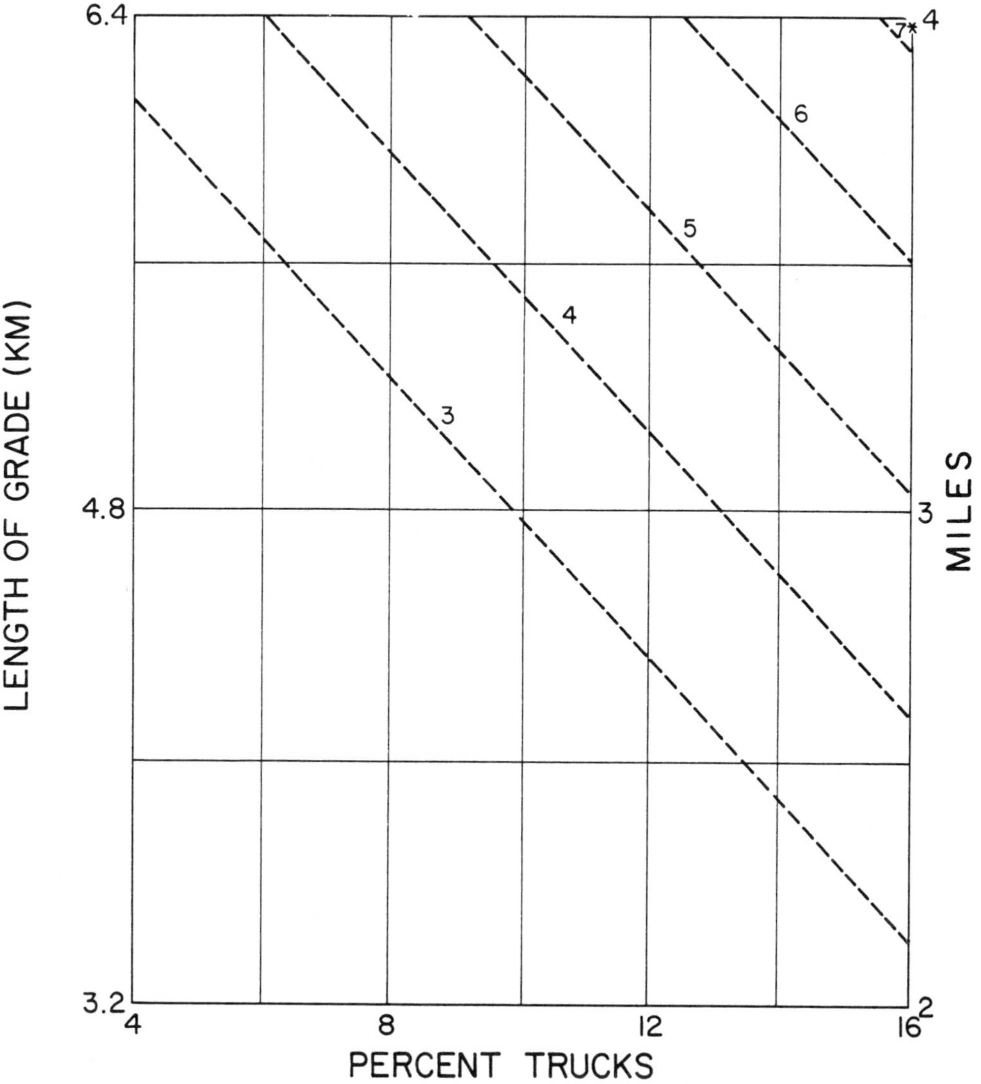

NOTES: 6.5% GRADIENT, 400 VPH
* INDICATES TOTAL TRAVEL TIME SAVINGS (VEH-HR/HR)
---- INDICATES APPROXIMATE RELATIONSHIP

Figure 10 - Relative Travel Time Savings per Hour of Operation for the Combination of Length of Grade and Percent Trucks.

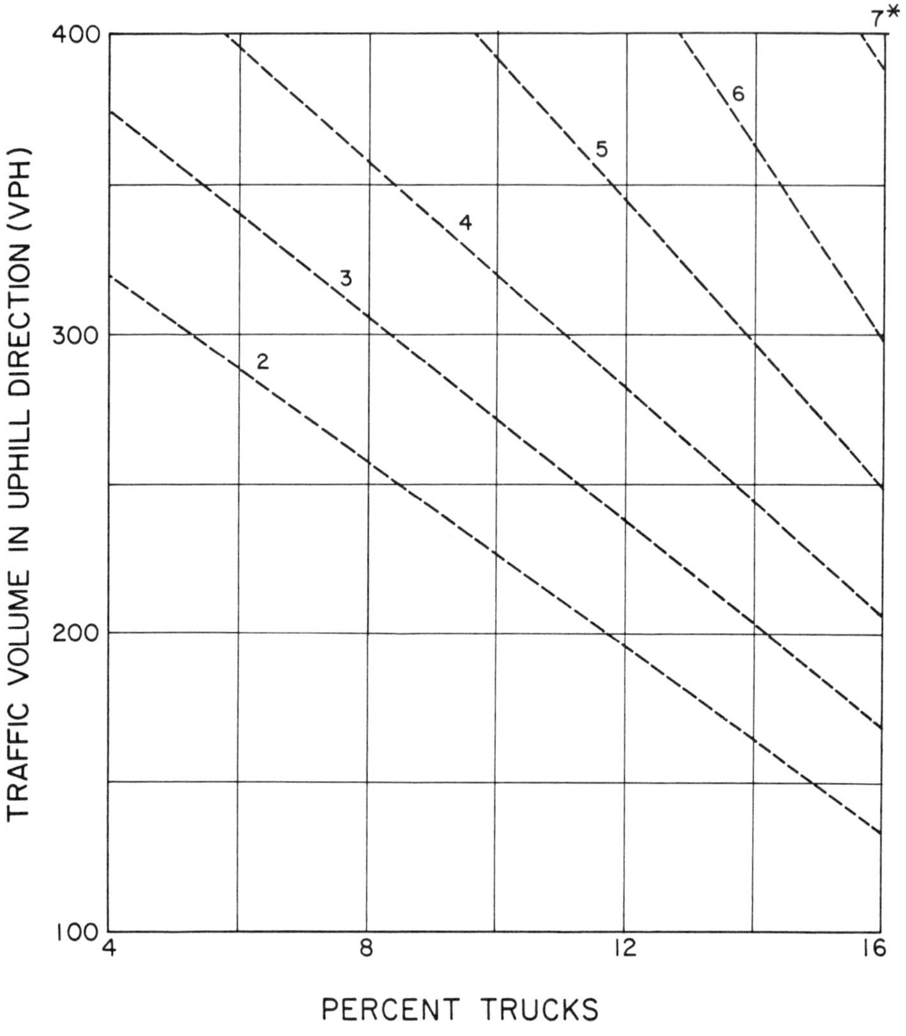

NOTES: 6.4 KM (4 MI) LENGTH OF GRADE, 6.5% GRADIENT
* INDICATES TOTAL TRAVEL TIME SAVINGS (VEH-HR/HR)
--- INDICATES APPROXIMATE RELATIONSHIP

Figure 11 - Relative Travel Time Savings per Hour of Operation for the Combination of Traffic Volume and Percent Trucks.

Table 3 - Characteristics of Alternatives - Example

Alternative Number	Gradient (%)	Length of Grade (Miles)	Traffic Volume in Uphill Direction (VPH)
1	6	3.0 (4.8km)	250
2	5	3.0 (4.8km)	300
3	6	2.5 (4.0km)	300
4	5	2.5 (4.0km)	300

per unit length. An example of the selection of upgrades from among candidate upgrades, with the purpose of making optimum use of funds, follows.

For the purpose of the example, four upgrades are considered, all with limited passing opportunities, approximately the same upstream and downstream geometric alignments, comparable horizontal alignment, and comparable percentages of trucks and recreational vehicles. The other characteristics are as outlined in Table 3. Funds are available for the construction of three climbing lanes, each 1500 ft (460 m) long. These climbing lanes will be constructed at the mid-point of each of three upgrades. In order to select three upgrades out of the four candidate upgrades, the relative performance of a climbing lane on specific upgrades is compared with the performance of a climbing lane on each of the other upgrades. To accomplish this, Figures 6 through 9 are used.

Alternative 1 vs. 2

Use Figure 6, since length of grade is the same.
Relative travel time savings for alternative 1 = 3.8.
Relative travel time savings for alternative 2 = 3.0.
Alternative 1 is the better alternative.

Alternative 1 vs. 3

Use Figure 7, since gradients are equal.

Relative travel time savings for alternative 1 = 3.1.

Relative travel time savings for alternative 3 = 2.6.

Alternative 1 is the better alternative.

Alternative 1 vs. 4

Temporarily assume that traffic volumes are similar for two alternatives.

Use Figure 8.

Relative travel time savings for alternative 1 = 3.1.

Relative travel time savings for alternative 4 = <<2.

Alternative 1 appears to be the better alternative from this comparison. However, the answer was biased in favor of alternative 1, since it was assumed that the traffic volumes were similar. Observed differences in the relative travel time for two alternatives must be large before a valid choice can be made when making such an assumption. If doubt exists and the choice between the two alternatives is critical, the model should be used to evaluate both alternatives.

Temporarily assume that the gradients are similar.

Use Figure 7.

Relative travel time savings for alternative 1 = 3.1.

Relative travel time savings for alternative 4 = 2.6.

Alternative 1 still appears to be the better alternative. In cases where the difference between the travel time savings is this small, it may be accepted that alternative 1 is the better alternative since the answer was biased in favor of alternative 4. It was assumed that the gradients were similar. The travel time savings is very sensitive to the gradient and the bias will, therefore, be considerable.

Temporarily assume that the lengths of grade are similar.

Use Figure 6.

Relative travel time savings for alternative 1 = 3.8.

Relative travel time savings for alternative 4 = 3.0.

Again, alternative 1 appears to be the better alternative. In this case the result was favorably biased for alternative 4; therefore, the result may be accepted.

Alternative 1 appears to be better than all other remaining alternatives.

Alternative 2 vs. 3

Use Figure 8.

Relative travel time savings for alternative 2 = << 2.

Relative travel time savings for alternative 3 = 2.

Alternative 3 is the better alternative.

Alternative 2 vs. 4

Alternative 2 is the better alternative, since only the length of grade is different and it is longer for alternative 2.

Alternative 3 vs. 4

Alternative 3 is the better alternative, as only the gradient is different and it is higher for alternative 1.

From the above analysis, the alternatives can be ranked. Alternative 1 ranks first, alternative 3 second, alternative 2 third, and alternative 4 last. The climbing lanes should thus be constructed on upgrades 1, 2, and 3.

It should be kept in mind that this analysis does not indicate whether the climbing lanes should be constructed at all. Further benefit/cost analysis, possibly incorporating safety and vehicle operating costs, should be undertaken to answer this question.

7. CONCLUSIONS

The results and analyses for establishing the most effective length and location of a climbing lane on an upgrade were presented. Also, guidelines were established for selecting the most cost-effective locations for the construction of a climbing lane, when several different upgrades are considered. The following general conclusions were reached.

a) The most cost-effective location for the construction of a climbing lane is near the midpoint of the upgrade.

b) The most cost-effective length of a climbing lane is 1500 ft (460 m). Shorter lanes were not considered, since safety implications of a 1500 ft (460 m) climbing lane were not investigated. The length of 1500 ft (460 m) was used, based on the presence of climbing lanes of this length in the field. This length does not include the entrance and exit tapers.

c) It is more efficient to construct one 1500 ft (460 m) climbing lane on several upgrades, than more than one 1500 ft (460 m) climbing lane on one upgrade.

d) The benefit (travel time) derived from the construction of a climbing lane is most sensitive to the gradient. Upgrades with severe gradient should thus be considered prime candidates for the construction of climbing lanes.

The general conclusions should hold over time, even if there are changes in the performance and mix of the vehicle fleet. These results are primarily based on differences between situations. Gradual changes in the vehicle fleet should not influence the relative differences significantly.

8. FUTURE RESEARCH

The conclusions reached in this report are based on a limited amount of simulation work and should be viewed in that light. The validity of these results may be affirmed by conducting additional research.

It is recommended that future research be carried out in several areas. First, to validate the results presented in this report, a great deal of additional simulation work should be conducted. Longer simulation times and different random number sequences could be employed. Also, additional factor values and combinations could be examined to enhance the results of the single-factor and two-factor analyses presented in Figures 6 through 11.

A second area in which future study is suggested covers additional measures of performance such as safety, vehicle operating costs and queue dispersion. Research is now underway at Berkeley to examine the effects of climbing lane design on accident potential over a segment of highway.

A final area recommended for future research is the development of a macroscopic two-lane, two-way highway model. The macroscopic model could simulate longer sections of roadway and assess impacts of design changes at less cost than the microscopic model. Macroscopic relationships used by the model could be derived using the microscopic model. Work to develop a macroscopic model of this type has also begun at Berkeley.

REFERENCES

1. Jan L. Botha and Adolf D. May, Working Paper No. 1, Literature Search for the Project, UCB-ITS-WP-78-5, September 1978, 37 pp.

2. Jan L. Botha and Adolf D. May, Working Paper No. 2, Preliminary Evaluation of Candidate Models for Use in the Project, UCB-ITS-WP-78-12, December 1978, 28 pp.

3. Jan L. Botha and Adolf D. May, Working Paper No. 3, Final Model Selection, Simulation Methodology and Proposed Data Collection, UCB-ITS-WP-79-12, October 1979, 93 pp.

4. Jan Louis Botha,"A Decision-making Framework for the Evaluation of Climbing Lanes on Two-Lane Two-Way Rural Roads." University of California Dissertation, May 1980, 454 pp.

5. American Association of State Highway Officials, A Policy on Geometric Design of Rural Highways, 1965, 650 pp.

6. A. D. St. John and D. R. Kobett, Grade Effects on Traffic Flow Stability and Capacity, NCHRP Report 185, Transportation Research Board, 1978, 110 pp.

AFTERWORDS

Question from R. H. Hewitt, University of Glasgow.

I find the results shown in Fig. 4 of this paper of great interest, particularly those relating to the case where two separate climbing lanes are provided at one-third and two-thirds of the distance up the slope. It would appear that the amount of travel time saved is proportional to the square root of the length of climbing lane. If the two lanes were placed together at the centre of the grade, making a lane of twice the length, the total travel time saved would therefore increase by a factor of 1.414, but the travel time saved per mile would reduce to half this value, or 0.707. Moving the first lane to the one-third point would reduce its efficiency to about 0.93 of its value at the mid-point, judging from the curve in the diagram. If we assume that time savings are proportional to the length of grade served by the lane, the second lane would have a corresponding time saving factor of 0.67, and the two together would show a saving of 1.60, or a saving per mile of 0.80. This is exactly the same as the ratio obtained from the values in the figure.

This suggests that it may be quite efficient to provide two climbing lanes spaced at equal intervals along the grade rather than build a single lane of twice the length. I should be interested to have the authors' views on this, and on the potential economic benefits of provision of two, or possibly several, short climbing lanes on a long gradient.

Response by Dr. J. L. Botha.

Your assessment of the efficiency of two climbing lanes spaced at equal intervals along the grade versus a single climbing lane of twice the length appears to be correct, based on the results of the research, although, due to lack of sufficient information, I could not fully evaluate your derivation. It should be noted that this result corresponds roughly to the comparison made between a climbing lane with a length of 2640 ft. and the two climbing lanes each with a length of 1500 ft. located at the one-third and two-thirds points on the grade. Please note however, that care should be taken in the generalization of this result, especially in the use of lower gradients and shorter lengths of grade. The reason for this is that in order for the lower of the two climbing lanes to be effective, queuing has to occur.

DYNAMIC IDENTIFICATION OF FLOWS FROM TRAFFIC COUNTS AT COMPLEX INTERSECTIONS

Michael Cremer and Hartmut Keller
University Hamburg Technical University Munich

The magnitude of traffic flows through a complex intersection is basic information needed for the design and control of transport facilities. When these flows cannot be measured directly, they must be estimated from vehicle counts at the entries and exits of the intersection.

Previous methods of estimating the flows used only the accumulated traffic counts, so can only lead to approximate solutions, but it is shown here that a unique solution to the problem of estimating the unknown flows can be obtained by using the additional information available from analysis of the time sequences and transients of the volume counts.

The algorithm presented has a recursive structure that allows it to adjust the flow estimates as the traffic flows increase or decrease - if the changes do not occur too rapidly - and its computational requirements are small enough to allow on-site application using microprocessors. Empirical studies using both simulated and real data show that the new algorithm produces more reliable estimates of the unknown traffic flows than previously available methods.

INTRODUCTION

The study of the volume flows through networks and intersections provides the basic data for planning traffic control, roadway changes and construction of new facilities. The problem analyzed here is the determination of flows through complex intersections from vehicle counts at the facility's entries and exits. For design purposes the long time average traffic flows can be determined off-line, i.e. the data evaluation algorithm could be applied after the collection of the volume data. However, for the control of intersections the actual changes in traffic flows should be detected on-line to be able to adjust signal timing to changing demand. Existing procedures to determine these flows are not satisfactory, because they solve a static problem by using only the sum, i.e. the accumulated traffic volumes during the period of analysis, and ignore that part of the gathered information which is contained in the time sequence of the vehicle counts [1, 2, 3, 4]. They also require high computational efforts, so they can not be used for on-line control, and lead to approximate solutions, which allow no reliable statements about the quality of the estimates.

In this paper a new dynamic algorithm is proposed which processes the time sequences of changes in the volume counts and makes use of the information contained therein without the need for more sites of data collection. One may expect intuitively that by extracting more information from the gathered data the algorithm will provide more reliable estimates for the unknown traffic flows. In fact, it will be shown in this paper that under certain, not too restrictive, conditions the algorithm provides estimates whose expected values converge uniquely to the average values of the real flows.

As an additional advantage, the algorithm can be applied with moderate computational effort, so that it can be implemented on microprocessors at the site of data collection. Moreover, due to its recursive structure it can adjust the flow estimates to changes in the unknown flows over time, if the flows do not change too rapidly.

PROBLEM STATEMENT

Consider a complex intersection with m approaches and n exits as shown in figure 1.

To describe the time-dependent distribution of traffic flows through the intersection the following variables are introduced:

$q_i(k)$: volume which enters entrance i during time interval $kT \le t < (k+1)T$, where $i = 1, \dots m$

$f_{ij}(k)$: that part of volume $q_i(k)$ which leaves the intersection through exit j during $kT + \tau_j \le t < (k+1)T + \tau_j$

$y_j(k)$: volume which leaves exit j during time-interval $kT + \tau_j \le t < (k+1)T + \tau_j$, where $j = 1, \dots n$.

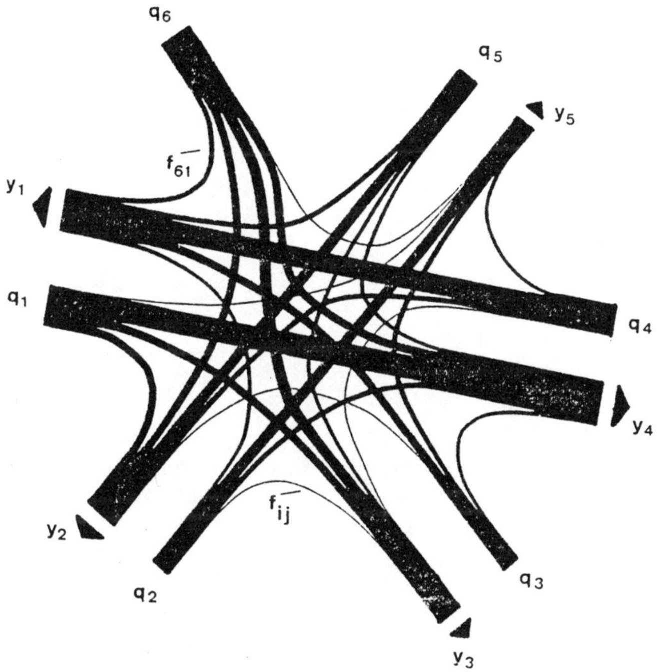

Fig.1: Volume flows through a complex traffic facility.

T is the sampling time interval, which has to be chosen long enough to make the introduced volumes meaningful and short enough so that the introduced variables will be able to follow the time variations and dynamic transitions of the system. In practical cases T may vary from half a minute to half an hour depending on the complexity of the intersection or the network. If the intersection is signal controlled, T should be an integer number of signal cycles. The parameter τ_j denotes the average travel time a vehicle needs to pass from any approach to exit j. In case that the procedure is applied to a network, it might become necessary to introduce individual times τ_{ij} describing the average travel time from entry i to exit j. In this case we have to define volumes $y_{ij}(k)$ which leave exit j during time-interval $kT + \tau_{ij} \leq t < (k+1)T + \tau_{ij}$. Thus for $i = 1, \ldots n$ samples are taken from the leaving vehicle counts which are supposed to contain that part of volume $q_i(k)$ which leaves the network through exit j.

From the law of conservation of vehicles it follows that each exit volume $y_j(k)$ is the sum of the corresponding parts f_{ij} of the entering volumes:

$$y_j(k) = \sum_{i=1}^{m} f_{ij}(k). \qquad (1)$$

Since each volume f_{ij} is a certain portion of the i-th entering volume $q_i(k)$, it may be written

$$f_{ij}(k) = q_i(k) \cdot b_{ij}(k). \qquad (2)$$

Here, $b_{ij}(k)$ is a split parameter which is obviously bounded by

$$0 \leq b_{ij}(k) < 1 \quad \text{for all } i,j,k. \qquad (3)$$

In case the traffic flow through the intersection is stationary we can write

$$b_{ij}(k) = \bar{b}_{ij} + \delta_{ij}(k), \qquad (4)$$

where $\bar{b}_{ij}$ is a constant mean value and δ_{ij} is a random variable with zero mean whose values are small as long as $q_i(k)$ is a representative sample of the whole ensemble of vehicles entering through entry i (the sample interval T should not be chosen too small to justify this assumption).

Inserting eqn.(2) into eqn.(1) yields

$$y_j(k) = \sum_{i=1}^{m} q_i(k) \cdot b_{ij}(k) = \underline{q}'(k) \cdot \underline{b}_j(k) \qquad (5)$$

where $\underline{b}_j(k)$ is the m x 1 column vector of the distribution coefficients $b_{ij}(k)$ and $\underline{q}'(k)$ is the 1 x m row vector of the m entering volumes $q_i(k)$ (the ' denotes transposition of a vector).

Combining the equations (5) for j=1,...n to form a vector equation gives for every k

$$\underline{y}'(k) = \underline{q}'(k) \cdot \underline{B}(k). \qquad (6)$$

Here $\underline{y}'(k)$ is the 1 x n row vector of measured exit volumes $y_j(k)$, and $\underline{B}(k)$ is the m x n matrix of the unknown coefficients $b_{ij}(k)$. It should be pointed out that one obtains the so called origin-destination-matrix of the intersection for the k-th time interval if the elements of the first row of $\underline{B}$ are multiplied by $q_1(k)$, the elements of the second row are multiplied by $q_2(k)$ and so on. (It is because of this relationship that we have chosen a row vector notation in eqn.(6).)

Under the assumption that every vehicle which has entered the intersection through approach i between kT and (k+1)T leaves it through its exit between $kT+\tau_j$ and $(k+1)T+\tau_j$ (this assumption may be justified if T is not too small) the balance of vehicles gives

$$\sum_{j=1}^{n} f_{ij}(k) = q_i(k) \qquad \text{for } i=1,\dots,m \qquad (7)$$

or equivalently

$$\sum_{j=1}^{n} b_{ij}(k) = 1 \qquad \text{for } i=1,\dots,m. \qquad (8)$$

Furthermore it is reasonable to assume that no vehicle leaves the intersection through the approach through which it has entered. Then we have the additional relation

$$b_{ii}(k) = 0 \qquad \text{for } i=1,\dots \min(m,n). \qquad (9)$$

For intersection design as well as for traffic responsive intersection control, it is important to know the partial flows $f_{ij}(k)$. For the design of intersections, their long time means are of interest. In case of traffic responsive control, the actual, mostly time dependent values, or at least their short time means, are of importance.

However, the empirical determination of these partial flows is often possible only with great technical effort or not in time. Then the task is to estimate the unknown flows $f_{ij}(k)$ or - what is the same - the unknown distribution coefficients $b_{ij}(k)$ from the sequence of counted traffic volumes $q_i(k)$ entering and $y_j(k)$ leaving the intersection.

Now eqn.(6) with eqn.(7) gives only n-1 linearly independent relations and eqn.'s (8) and (9) give additionally $m + \min(n,m)$ equations for the $n \cdot m$ unknown parameters $b_{ij}(k)$. Consequently, the number of unknown parameters already exceeds the number of linearly independent equations when an intersection has three approaches and three exits. In existing procedures, the set of equations (6) to (9) is built up only from the accumulated traffic counts q_i and y_j and is treated as a linear and highly underdetermined set of static equations which is solved using pseudo inverses [2] or other techniques [3,4,5]. However, this treatment needs considerable computational effort, while the solutions give only approximations to the real values of the unknown parameters b_{ij} and it is difficult to give reliable statements about the quality of the estimates.

The purpose of the following development is to propose a recursive estimation algorithm which uses for each time interval the additional information contained within the new measurement $q_i(k)$, $y_j(k)$ to generate a successively improved estimate $\hat{b}_{ij}(k)$ for each of the unknown parameters $b_{ij}(k)$. As will be shown in the following section, the statistical expectation of this estimate will approach with increasing time index k the mean value of the real parameter $b_{ij}(k)$, no matter how large the initial estimation error was. Furthermore, the estimate is even able to track movements of the real parameter $b_{ij}(k)$ (or its short time mean) if they are not too rapid , which is an important feature for real time application to signal control. As an additional benefit, the computational effort of the algorithm is rather low.

To proceed, we will first consider a very simple special case in order to make it evident that using the time sequences of collected data can provide additional information. For this purpose it is assumed here that the split coefficients are constant over time $b_{ij}(k) = \bar{b}_{ij}$, that is we make the simplifying, but unrealistic, assumption that the volumes $q_i(k)$ split in the same proportions in each period k. Then we write for the time varying entering and leaving volumes

$$q_i(k) = \bar{q}_i + \Delta q_i(k), \qquad y_j(k) = \bar{y}_j + \Delta y_j(k), \tag{10}$$

where $\bar{q}_i$ and $\bar{y}_j$ are the mean values of the volumes and $\Delta q_i(k)$ and $\Delta y_j(k)$ are their actual deviations from the mean values. Inserting these relations into eqn.(6) we obtain

$$\left[\underline{\bar{y}}' + \Delta \underline{y}'(k)\right] = \left[\underline{\bar{q}}' + \Delta \underline{q}'(k)\right] \cdot \underline{B} \,. \tag{11}$$

At this point we make the additional assumption that the intersection behaves as a linear causal system. Then applying the law of superposition eqn.(6) holds for the mean values as well as for the deviations

$$\underline{\bar{y}}' = \underline{\bar{q}}' \, \underline{B} \quad \text{and} \quad \Delta \underline{y}'(k) = \Delta \underline{q}'(k) \cdot \underline{B} \,. \tag{12}$$

In reality, when the split parameters $\underline{B}(k)$ are time dependent this assumption holds for the expectation values rather than for the individual sample values.

Combining these equations for k=1,2,....,K yields

$$\begin{bmatrix} \Delta \underline{y}'(1) \\ \Delta \underline{y}'(2) \\ \cdot \\ \cdot \\ \cdot \\ \Delta \underline{y}'(K) \end{bmatrix} = \begin{bmatrix} \Delta \underline{q}'(1) \\ \Delta \underline{q}'(2) \\ \cdot \\ \cdot \\ \cdot \\ \Delta \underline{q}'(K) \end{bmatrix} \cdot \underline{B} \tag{13}$$

This matrix equation can be solved uniquely for the unknown matrix $\underline{B}$ of the split coefficients whenever we have m linearly independent rows $\Delta \underline{q}'(k)$ within the brackets on the right side.

This procedure shows that processing the time sequences of the collected traffic counts provides additional information which can be used to determine the unknown split coefficients. In order to make use of this information, it was necessary that at each entry there be independent non-zero deviations $\Delta q_i(k)$, which may not be totally correlated with the deviations at other entries. The procedure has several drawbacks: (i) it works only under the unrealistic assumption that the coefficients $b_{ij}(k)$ are strictly constant, (ii) it still needs considerable computational effort involving the inversion of a matrix, (iii) it is rather intransparent how measurement errors, small changes $\delta_{ij}(k)$ in the coefficients $b_{ij}(k)$ and round off errors will affect the solution.

We will now propose a recursive estimation algorithm for the unknown split coefficients $b_{ij}(k)$ which has not the above mentioned disadvantages. As in the procedure above the algorithm uses the fact that there are generally time variable deviations $\Delta q_i(k)$ from the mean values $\bar{q}_i$ at all entries which generate according to the actual values of the split parameters $b_{ij}(k)$ corresponding deviations $\Delta y_j(k)$ from the mean values $\bar{y}_j$ at the exits of the intersection.

We start with the assumption that we have at the beginning of the k-th interval estimates $\hat{b}_{ij}(k-1)$ for the unknown split parameters which were computed on the basis of the measurements until the (k-1)-th time interval. Then the problem is to improve these estimates with the new measurements $\underline{q}(k)$, $\underline{y}(k)$ available at the end of the k-th time interval. Using the measured vector $\Delta\underline{q}(k)$ of the actual deviations of the entering volumes, we can first compute for each j a prediction $\Delta\hat{y}_j(k)$ for the deviation of the volume at exit j on the basis of the actual parameter estimates $\hat{b}_{ij}(k-1)$:

$$\Delta\hat{y}_j(k) = \Delta\underline{q}'(k)\cdot\hat{\underline{b}}_j(k-1). \qquad (14)$$

Using this estimate, we propose the following discrete time estimation formula by which a new estimate $\hat{b}_{ij}(k)$ for the parameter $b_{ij}(k)$ is computed from each set of new measurements recursively

$$\hat{b}_{ij}(k) = \hat{b}_{ij}(k-1) + \gamma \cdot \Delta q_i(k)\cdot\left[\Delta y_j(k) - \Delta\hat{y}_j(k)\right], \qquad (15)$$

where γ is a gain factor which has to be chosen appropriately (see next section).

The effect of this formula may be understood by considering the following points:

(i) It can easily be seen that $\hat{b}_{ij}(k) = b_{ij}(k)$ for $i=1,\ldots m$ is an equilibrium point of this recursive estimation scheme, since then $\Delta\hat{y}_j(k) = \Delta y_j(k)$ and the bracket term vanishes.

(ii) If $\hat{b}_{ij}(k) \neq b_{ij}(k)$, an entrance deviation $\Delta q_i(k)$ will cause different reactions $\Delta y_j(k)$ and $\Delta\hat{y}_j(k)$. Then multiplication of the exit volume error $\Delta y_j(k) - \Delta\hat{y}_j(k)$ by $\Delta q_i(k)$ effects that the correction of the estimate is performed in the right direction. Moreover, the larger the deviation $\Delta q_i(k)$ is, the clearer is the causal relationship between $\Delta q_i(k)$ and $\Delta y_j(k)$ and, consequently, the larger is the correction of the estimate. If, on the other hand, $\Delta q_i(k)$ is zero, we have no causal dependency between a non zero $\Delta y_j(k)$ and $\Delta\hat{y}_j(k)$, so the correction term in eqn.(15) vanishes.

(iii) The values $\Delta y_j(k)$ and $\Delta\hat{y}_j(k)$ are the superpositions of the system's and the model's reactions due to changes $\Delta q_i(k)$ in entering volumes at all entries according to eqn.(5). Thus it may happen that in some time interval an exit volume error $\Delta y_j(k) - \Delta\hat{y}_j(k)$ which is mainly caused by an entrance volume deviation $\Delta q_1(k)$ at entry 1 will lead, after multiplication with $\Delta q_i(k)$, to a correction of the estimate in the wrong direction. Actually the parameter estimation of formula (15) is a random process because of the randomly varying volume changes $\Delta q_i(k)$ at all entries. To see the long time effect of the estimation scheme, let us consider the sum of the correction term over K time intervals:

$$\gamma \cdot (K-1) \cdot \left[\frac{1}{K-1} \sum_{k=1}^{K} \Delta q_i(k) \cdot (\Delta y_j(k) - \Delta\hat{y}_j(k)) \right].$$

Now the term within the brackets is approximately (since summation is taken over a finite number of intervals) the discrete time cross correlation function of the signal $\Delta q_i(k)$ and the exit volume error $\Delta y_j(k) - \Delta\hat{y}_j(k)$ [6]. From this it becomes apparent that the factor $\Delta q_i(k)$ in the estimation formula has the additional important effect that over a longer period only that part of $\Delta y_j(k) - \Delta\hat{y}_j(k)$ which is correlated with the variations $\Delta q_i(k)$ at the i-th entry leads to a correction of the estimate, i.e. only that part which is related to the partial flow $f_{ij}(k)$.

When the estimation algorithm is to be applied, the mean volumes $\bar{q}_i$ and $\bar{y}_j$ in eqn.(10) are to be computed beforehand from the total data set in the case of an off-line application. For an on-line application these values might be obtained, for example, by taking the average values of the past K volume measurements $q_i(k)$ and $y_j(k)$, respectively.

So far eqn.(15) has been explained heuristically. In the next section it will be shown systematically by a proof of convergence that, in the case of constant split parameters b_{ij}, any initial estimation error will be reduced to zero as time increases. This means that the equilibrium point $\hat{b}_{ij} = b_{ij}$ is globally asymptotically stable and thus the algorithm provides a unique solution to the estimation problem. (It should be mentioned that the flow identification methods so far known could not provide a unique solution even under this idealized, unrealistic assumption.) It will be shown further that, in the more realistic case of randomly varying real split parameters $b_{ij}(k)$, the algorithm provides estimates $\hat{b}_{ij}(k)$ whose expected values approach asymptotically the averages of the real parameters.

To proceed, we combine equations (15) for i=1,...m into a vector difference equation for each column vector $\hat{\underline{b}}_j(k)$:

$$\hat{\underline{b}}_j(k) = \hat{\underline{b}}_j(k-1) + \gamma \cdot \Delta\underline{q}(k) \cdot [\Delta y_j(k) - \Delta\hat{y}_j(k)] \ . \quad (16)$$

Since $\Delta y_j(k)$ and $\Delta\hat{y}_j(k)$ depend only on the parameter vectors $\underline{b}_j(k)$ and $\hat{\underline{b}}_j(k-1)$ according to eqn.(5) and (14), it becomes apparent that the estimation procedure consists of n decoupled vector difference equations of type (16) for j=1,...n.

The parameter estimation scheme as described by eqn.(16) has the structure of a feedback loop as shown in figure 2.

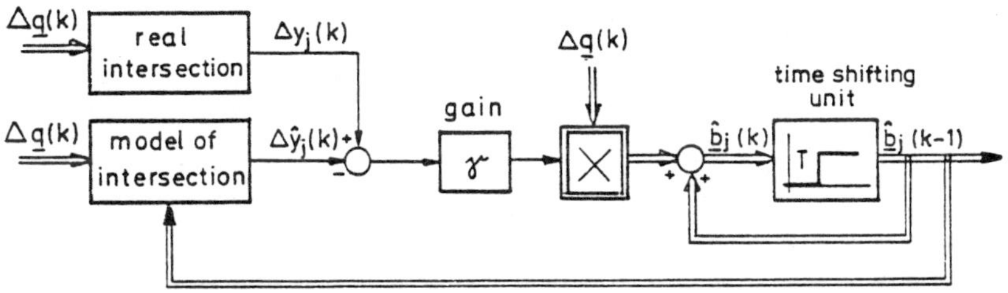

Fig. 2: Feedback structure
of the parameter estimation algorithm

PROOF OF CONVERGENCE

In this section, a proof of convergence based on a stability analysis of eqn.(16) is given for the parameter estimation procedure. Convergence will first be proven under the simplifying assumption that the unknown coefficients $b_{ij}(k)$ of the volume splits have fixed values, i.e.

$$b_{ij}(k) = b_{ij} \text{ for } i=1,\ldots,m;\ j = 1,\ldots,n; \text{ and all } k;$$

It will be shown that under this assumption any initial estimation error $\hat{b}_{ij}(0) - b_{ij}$ is reduced to zero by the algorithm as the time index k increases, whereby the rate of convergence with which this reduction takes place depends on the choice of the gain factor γ . Thus, not only does the algorithm converge, but it is able to track time-varying movements of the coefficients b_{ij} so long as the change is sufficiently slow compared to the algorithm's rate of convergence. The assumption that $b_{ij}(k) = b_{ij}$ for every k will later be dropped and we will show that in the more general case the expectation of the estimate $\hat{b}_{ij}(k)$ will converge to the expected value of the real split parameter $b_{ij}(k)$.

Inserting equations (12) and (14) into the estimation formula (16) gives

$$\underline{\hat{b}}_j(k) = \underline{\hat{b}}_j(k-1)+\gamma\cdot\Delta\underline{q}(k)\cdot\Delta\underline{q}'(k)\cdot\left[\underline{b}_j-\underline{\hat{b}}_j(k-1)\right]. \tag{17}$$

Subtracting both sides from the constant parameter vector $\underline{b}_j$ gives, then, the following vector difference equation:

$$\underline{\beta}_j(k) = \left[\underline{I}_m - \gamma\cdot\Delta\underline{q}(k)\cdot\Delta\underline{q}'(k)\right]\cdot\underline{\beta}_j(k-1)\ , \tag{18}$$

where $\underline{I}_m$ is the m x m identity matrix and

$$\underline{\beta}_j(k) = \underline{b}_j - \underline{\hat{b}}_j(k) \tag{19}$$

is the vector of the estimation errors of the distribution coefficients in the j-th column of the matrix $\underline{B}$. It can be seen that eqn.(18) is a homogeneous difference equation for the estimation error which has the equilibrium point $\hat{\underline{b}}_j = \underline{b}_j$. The adaptation law is then called asymptotically stable if any initial estimation error $\underline{b}_j - \hat{\underline{b}}_j(0) = \underline{\beta}_j(0)$ approaches $\underline{0}$ as time index k grows to infinity.

The homogeneous difference equation (18) has the system matrix

$$\underline{I}_m - \gamma \cdot \Delta\underline{q}(k) \cdot \Delta\underline{q}'(k) \qquad (20)$$

which is a stochastic matrix since $\Delta\underline{q}(k)$ is a stochastic vector. As a consequence, asymptotic stability cannot be investigated by an eigenvalue analysis of eqn.(20) directly as is usual in linear determinstic systems. At this point we take statistical expectations of both sides of eqn.(18). Under the realistic assumption that the actual estimation errors $\beta_{ij}(k-1)$ and the actual variations $\Delta q_i(k)$ of the entry volumes are statistically independent, i.e.

$$E\left\{\underline{\beta}(k-1) \cdot \Delta\underline{q}'(k)\right\} = \underline{0}\ , \qquad (21)$$

we obtain

$$E\left\{\underline{\beta}_j(k)\right\} = \left[\underline{I}_m - \gamma \cdot E\left\{\Delta\underline{q}(k) \cdot \Delta\underline{q}'(k)\right\}\right] \cdot E\left\{\underline{\beta}_j(k-1)\right\}. \qquad (22)$$

This is a homogeneous vector difference equation for the expected values of the estimation errors which now has a constant system matrix $\underline{I}_m - \gamma \cdot E\left\{\Delta\underline{q}(k)\Delta\underline{q}'(k)\right\}$. To keep the subsequent development transparent we make the additional realistic assumption that the variations $\Delta q_i(k)$, $\Delta q_l(k)$ at different entries are statistically independent, that is

$$E\left\{\Delta\underline{q}(k) \cdot \Delta\underline{q}'(k)\right\} = \mathrm{diag}\left\{\rho_1^2, \rho_2^2, \rho_3^2, \ldots, \rho_m^2\right\} \qquad (23)$$

is a diagonal matrix where $\rho_i^2 = E\left\{\Delta q_i^2(k)\right\}$.

(If this assumption is dropped, asymptotic stability can still be established by a treatment similar to this as long as the covariance matrix $E\{\underline{\Delta q}(k)\ \underline{\Delta q}'(k)\}$ is positive definite, that is a long as the variations $\Delta q_i(k)$ at one entry are not totally correlated with the variations at other entries.)

In this case, eqn.(22) decomposes into m decoupled scalar difference equations,

$$E\{\beta_{ij}(k)\} = (1 - \gamma \cdot \rho_i^2) \cdot E\{\beta_{ij}(k-1)\}\ , \tag{24}$$

from which it is seen that the expectation of the estimation error is reduced to zero if γ is chosen according to

$$0 < \gamma < 2/\rho_i^2 \qquad \text{for } i=1,\ldots,m.$$

Moreover, it follows from this equation that the rate of convergence of the algorithm depends on the choice of the gain factor γ within the given boundaries.

However, convergence of the expected value of the estimation error is only a necessary condition for the estimation error to go to zero for $b_{ij}(k) = b_{ij}$. According to the concept of asymptotic stability in the mean square (see for example [7]), we have additionally to guarantee that the covariances of the estimation errors converge to zero with increasing time index:

$$\lim_{k\to\infty} E\{\underline{\beta}(k) \cdot \underline{\beta}'(k)\} \to \underline{0}\ . \tag{25}$$

From eqn.(18) we get for each j:

$$\begin{aligned}\underline{\beta}_j(k)\underline{\beta}_j'(k) = {} & \underline{\beta}_j(k-1)\underline{\beta}_j'(k-1) - \gamma \cdot \underline{\beta}_j(k-1)\underline{\beta}_j'(k-1) \cdot \underline{\Delta q}(k) \cdot \underline{\Delta q}'(k) \\ & - \gamma \cdot \underline{\Delta q}(k)\ \underline{\Delta q}'(k)\ \underline{\beta}_j(k-1)\ \underline{\beta}_j'(k-1) \\ & + \gamma^2 \cdot \underline{\Delta q}'(k)\underline{\Delta q}'(k)\underline{\beta}_j(k-1)\underline{\beta}_j'(k-1)\underline{\Delta q}(k)\underline{\Delta q}'(k)\end{aligned} \tag{26}$$

Taking expectations of both sides of eqn.(26), we obtain under assumptions (21) and (23):

$$
\begin{aligned}
E\{\underline{\beta}_j(k)\cdot\underline{\beta}_j'(k)\} = {} & E\{\underline{\beta}_j(k-1)\cdot\underline{\beta}_j'(k-1)\} \\
& - \gamma\cdot E\{\underline{\beta}_j(k-1)\cdot\underline{\beta}_j'(k-1)\}\cdot\text{diag}\{\rho_i^2\} \\
& - \gamma\cdot\text{diag}\{\rho_i^2\}\cdot E\{\underline{\beta}_j(k-1)\cdot\underline{\beta}_j'(k-1)\} \\
& + \gamma^2\cdot E\{(\Delta\underline{q}'(k)\cdot\underline{\beta}_j(k-1))^2\cdot\Delta\underline{q}(k)\cdot\Delta\underline{q}'(k)\} .
\end{aligned}
\tag{27}
$$

This is a set of m^2 linear coupled difference equations for m^2 expectation values $E\{\beta_{rj}(k)\cdot\ \beta_{sj}(k)\}$ with $r=1,..m$, $s=1,...m$. If γ is chosen sufficiently small,

$$0 < \gamma \ll 1/\rho_i^2 , \tag{28}$$

the last term on the right side of eqn.(27) vanishes and we obtain for each j m^2 scalar decoupled equations

$$E\{\beta_{rj}(k)\cdot\ \beta_{sj}(k)\} = (1-\gamma\rho_r^2-\gamma\rho_s^2)E\{\beta_{rj}(k-1)\cdot\beta_{sj}(k-1)\}$$

$$\text{for } r=1,...m;\ s=1,...m. \tag{29}$$

From this it is evident that any initial estimation error converges to zero in the mean square sense if γ is chosen according to (27), since the eigenvalues $1-\gamma\rho_r^2-\gamma\rho_s^2$ lie inside the unit circle.

A more detailed analysis reveals that eqn.(27) still gives, without the simplifying restriction (28), a set of $m\cdot(m-1)$ decoupled scalar difference equations for $r\neq s$ and m interacting difference equations for the covariances with $r=s$, which might be investigated by an eigenvalue analysis with moderate computational effort to obtain a precise upper bound for γ. This bound lies generally below the value $2/\rho_i^2$ which was found considering only the expected mean error in eqn.(24). This shows that convergence of the covariances imposes a stronger restriction on the admissible range of γ.

As a consequence of the proven global asymptotic stability in the mean square, it can be concluded that, under the assumption that the real parameters b_{ij} are constant and that the gain factor γ is chosen sufficiently small, the estimation formula (16) converges to the unique solution $\hat{b}_{ij}(k) = b_{ij}$ as time increases.

Now we drop the restriction $b_{ij}(k) = b_{ij}$ and assume more realistically random variations of the real split parameters according to eqn.(4)

$$b_{ij}(k) = \bar{b}_{ij} + \delta_{ij}(k) .$$

In addition, we omit the unrealistic assumption of strict linearity which has led to eqn.'s(12). That is, we omit the assumption that in each time-interval the deviations $\Delta q_i(k)$ of the entry volumes split in exactly the same proportions as the mean values $\bar{q}_i$ do. Then we may write:

$$\Delta f_{ij}(k) = f_{ij}(k) - \bar{f}_{ij} = \Delta q_i(k) \cdot b_{ij}^*(k), \qquad (30)$$

with

$$b_{ij}^*(k) = b_{ij}(k) + \eta_{ij}(k) = \bar{b}_{ij} + \delta_{ij}(k) + \eta_{ij}(k), \qquad (31)$$

where $\eta_{ij}(k)$ is another random variable.

Combining $\delta_{ij}(k)$ and $\eta_{ij}(k)$ to a new random variable $\varepsilon_{ij}(k)$, we may write

$$b_{ij}^*(k) = \bar{b}_{ij} + \varepsilon_{ij}(k) \qquad (32)$$

Under the reasonable assumption that the expectation values of both $b_{ij}(k)$ and $b_{ij}^*(k)$ are equal to $\bar{b}_{ij}$, the random variable $\varepsilon_{ij}(k)$ has zero mean, that is

$$E\left\{\varepsilon_{ij}(k)\right\} = 0. \qquad (33)$$

To proceed, we introduce a vector of estimation errors similar to eqn.(19)

$$\underline{\beta}_j(k) = \underline{\bar{b}}_j - \underline{\hat{b}}_j(k). \tag{34}$$

A similar development as above leads to the following vector difference equation for the expected estimation errors, which corresponds to eqn.(22):

$$E\{\underline{\beta}_j(k)\} = \left[\underline{I}_m - \gamma E\{\Delta\underline{q}(k)\, \Delta\underline{q}'(k)\}\right] \cdot E\{\underline{\beta}_j(k-1)\}$$

$$- \gamma\{E\, \Delta\underline{q}(k)\, \Delta\underline{q}'(k)\, \underline{\varepsilon}_j(k)\}\ . \tag{35}$$

If we make now the additional assumption that the random variations $\varepsilon_{ij}(k)$ of the split coefficients are statistically independent of the variations $\Delta q_i(k)$, i.e.

$$E\{\underline{\varepsilon}_j(k)\, \Delta\underline{q}'(k)\} = \underline{0} \quad \text{for } j=1,\ldots n, \tag{36}$$

eqn. (35) reduces to eqn.(22). This shows that the expected values of the parameter estimates still converge to the expected values of the real split parameters.

Investigating the covariances of the estimation errors in a similar way as above leads in this case to a difference equation of the form of eqn.(27) which contains on the right side an additional constant term including the covariance matrix of the random variations $\varepsilon_{ij}(k)$ with i=1,...m. This means that one now obtains a nonhomogeneous difference equation; consequently the covariances of the estimation error will converge to a non-zero steady state value. However, it can be shown again that this remaining covariance is decreased when γ is decreased. On the other hand, it can be seen from equation (24) that the rate of convergence decreases when γ is decreased. In a practical application, the value of γ is to be chosen to yield a good compromise between dynamic speed of adaptation and sufficient smoothing properties - a common situation in designing a control loop.

RESULTS

The parameter identification algorithm presented here was tested in numerous simulation runs with artificial data as well as with real, measured volume flows. It was the objective of these investigations to test the efficiency of the algorithm under various realistic situations. In particular, it was to be determined how quickly initial estimation errors can be reduced and how the parameter γ should be chosen, within the limits mentioned in the previous section, so that a satisfactory speed of adaption is obtained while random variations of the real parameters $b_{ij}(k)$ are sufficiently smoothed.

A representative result of a simulation run is given in figure 3. Here the results with different values of γ are shown for the case where the real split coefficients are slowly drifting with time according to a half sine wave. The intersection had four entries and four exits, which yields, with respect to eqns. (8) and (9), a total of eight unknown parameters $b_{ij}(k)$ to be identified, while the vector equation (6) formulated for the accumulated traffic counts would provide only three linearly independent equations. The adaptation process is shown for parameter $\hat{b}_{13}$ to illustrate a typical time response. It is apparent that the higher the value of γ the better the estimate tracks a time varying real parameter.

For larger values of γ the adaptation process was less smooth and showed greater overshooting. The algorithm finally became unstable at last when γ surmounted the upper bound which follows from eqn.(29). Thus, it was confirmed that the concept of asymptotic stability in the mean square, used in the previous section, is suitable for the evaluation of the stability of the estimation algorithm.

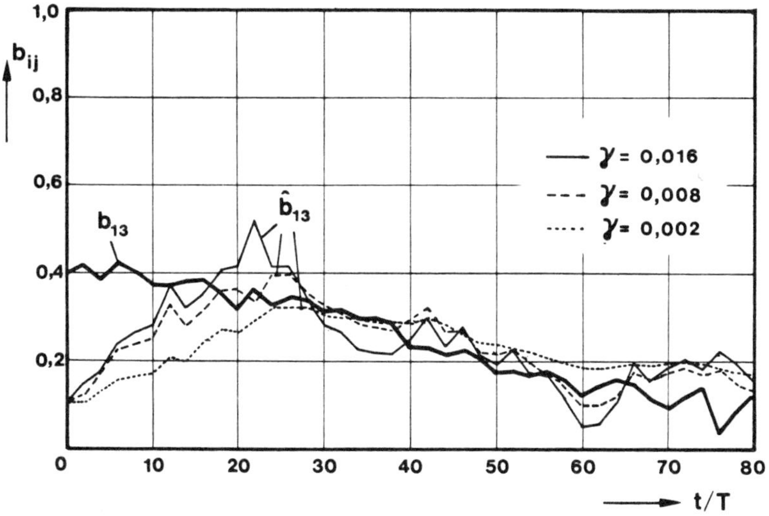

Figure 3: Time response of parameter estimate $\hat{b}_{13}(k)$ with different values of γ for a time varying real parameter $b_{13}(k)$

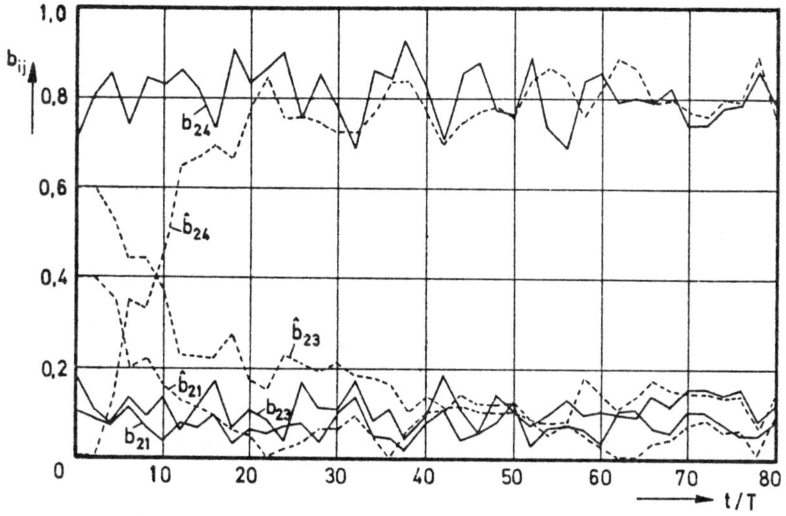

Figure 4: Time response of parameter estimates $\hat{b}_{ij}(k)$ for an observed sequence of parameters $b_{ij}(k)$

The test of the algorithm with real, observed data was supposed to demonstrate particularily the effects of stochastic variations $\varepsilon_{ij}(k)$ in the split parameters $b_{ij}(k)$ as they appear always in reality. For this purpose an inventory of the input volumes $q_i(k)$, the output volumes $y_j(k)$ and the partial flows $f_{ij}(k)$ was made at an intersection in the city of Munich during a time period of 90 minutes. At this intersection, a street with two-directional traffic meets with a one-way street, so there are three entries (i=1,2,3), three exits (j=1,3,4) and seven partial flows $f_{ij}(k)$. For this intersection, eqns.(6) and (7) provide only two linearly independent relations for four unknown split parameters when treated as a static relation. In the test the algorithm was fed only by the sequence of measurements $q_i(k)$, i=1,2,3 and $y_j(k)$, j=1,3,4 while the measured partial flows $f_{ij}(k)$ were used only to compute the real split parameters for comparison with the estimated parameters.

Figure 4 shows the sequence of some estimates for the split coefficients, which adjust themselves to the real parameters in spite of completely false initial values (here for T=140 sec, γ=0.01). While the estimates follow the long time averages for the flow parameters, they cannot follow their short time stochastic variations. This, however, is not necessary for an inventory of the long time distribution of flows for the design of roadway changes or new facilities nor for traffic responsive control of intersections with traffic signals. For the latter purpose it would be adequate to rely on the mean of the $\hat{b}_{ij}(k)$ over the period of the last K estimates.

CONCLUSIONS

A new method is presented with which the distribution of partial flows within a complex intersection can be computed from the traffic counts at the entries and exits of the intersection. The method is distinguished from the possibilities considered so far in that it uses not only the sum of the accumulated traffic volumes over a longer time period, but also evaluates the time sequence of the counted traffic volumes and thus receives more information about the process.

It is based on a recursive parameter identification algorithm which needs less computational effort, so can be implemented on microprocessors at the site of data collection.

By a systematic proof of stability, it was shown that, for a proper choice of the gain factor γ, the estimates asymptotically converge to the expected values of the real parameters for every initial estimation error. Thus, it was made apparent that the problem of determination of partial flows in a complex intersection can be solved uniquely by a dynamic approach, while it is not usually possible to obtain a unique solution by treating the problem as static, as is done in existing procedures.

It will be the problem of future studies to investigate how the method works for more complex facilities, when the travel time through the intersection or network is greater and where the causal dependency of the exit flows on the entry flows is less direct. At the same time, an analysis must be made of the size of variations in the real parameters that can be accepted without loss of efficiency of the proposed algorithm and how the method would work, if flows at different entries were partly correlated. A challenging problem would be the flow identification in a district from cordon counts.

REFERENCES

[1] J. SCHLUMS, "Verkehrserzählungen an Straßenknoten", Zeitschrift für Verkehrswissenschaft, H2, 228-238 (1954).

[2] D. BEIL, Matrixmodelle zur Ermittlung von Teilströmen aus Querschnittsmessungen, Dissertation, Univ. Karlsruhe (1979).

[3] H.R. KIRBY, "Partial Matrix Techniques", Traffic Engineering and Control, 20, 422-428 (1979).

[4] H.J. VAN ZUYLEN AND L.G. WILLUMSEN, "The most likely Trip Matrix estimated from Traffic Counts", Transpn.Res, 14B, 281-293 (1980).

[5] E.HAUER AND B-Y.T. SHIN, "Origin-Destination Matrices from Traffic Counts: Application and Validation on simple Systems", Traffic Engineering and Control, 118-121 (1981).

[6] K.R. GODFREY, "Correlation Methods", Automatica, 16, 527-534 (1980).

[7] L.ARNOLD, Stochastische Differentialgleichungen, OldenbourgVerlag, München (1972).

CONVERGENCE OF A NETWORK DECOMPOSITION ALGORITHM FOR THE TRAFFIC EQUILIBRIUM MODEL*

Stella Dafermos
Lefschetz Center for Dynamical Systems
Brown University

Abstract

In this paper, we establish the convergence of a network decomposition algorithm by origin-destination pairs designed to solve the general asymmetric network equilibrium problem. Using the variational inequality formulation of the equilibrium conditions we establish linear convergence under the assumption that the paths joining different origin-destination pairs are weakly coupled.

1. Introduction

The principal objective of the network equilibrium problem is the computation of user-optimized patterns characterized by the property that, once established, no user has any incentive to alter his travel arrangements. In general, the incentive is measured in terms of a disutility function and a demand function which depend on the traffic volume (congestion effect).

Mathematically, the state of equilibrium is characterized by equilibrium conditions (e.g. [8]) which can be written as a variational inequality ([11], [8]).

In the special "symmetric" case where the cost and demand functions satisfy certain symmetry conditions (see, e.g. [8]) the equilibrium conditions can be interpreted as the Kuhn-Tucker conditions of a convex minimization nonlinear programming problem. Hence this case is amenable

*This work was supported by the Program of Univ. Research, U.S. Dept. of Transpn. (Project number DTRS5680-C-00007).

to the powerful convex nonlinear programming techniques. In the more general and very realistic "asymmetric" case, the state of affairs is less satisfactory. Although a number of algorithms which attempt to reduce the original problem into simpler ones have been discussed ([10], [1], [2], [8]), a lot of theoretical study and computational experience is still necessary in order to test their efficiency.

The complexity of the equilibrium conditions is due to the interaction among travel units with different origin-destination (O/D) pairs which, however, share common links. Therefore, it is reasonable to attempt to compute the overall network equilibrium pattern by decoupling the equilibrium conditions and then successively computing, for each O/D pair, the corresponding equilibrium pattern, keeping fixed the flows induced by all other O/D pairs. Such decomposition schemes were applied in [4], [5], [6] to a number of traffic equilibrium models and convergence was established for "symmetric" problems with fixed O/D demands. The proof relies heavily on the equivalence of the user-equilibrium problem to a convex minimization nonlinear programming problem. Subsequently, decomposition by O/D pairs was applied in [2] to a number of "asymmetric" problems. However, convergence in this case has not yet been established.

It is clear that a decomposition algorithm by O/D pairs will converge (and in fact will terminate in just one cycle) when the O/D pairs are decoupled, i.e., when no paths joining different O/D pairs share common links. It is thus reasonable to expect that the algorithm will generally converge when paths joining different O/D pairs are weakly coupled in that they share a relatively small number of links. Indeed, we establish here linear convergence of the decomposition scheme by O/D pairs, for the general single modal traffic equilibrium model and under the assumption of weak coupling. Of course, this assumption is only sufficient for convergence and probably far from necessary.

The intent of the paper is not so much to provide the practitioner with a precise rule for deciding, in advance, whether the decomposition scheme by O/D pairs converges for a particular problem, but rather to justify the intuitive basis for the decomposition scheme and to reveal the type of networks for which convergence is plausible.

Of course, similar results can be obtained for the multimodal traffic equilibrium problem, either directly or by using the fact that the multimodal model for a network may be visualized as a single modal model for an enlarged network [6].

The convergence of a decomposition algorithm by modes, for the multimodal model, has been established in [9] by similar techniques.

2. The Algorithm and Its Convergence

In this section we describe a decomposition algorithm by origin-destination (O/D) pairs of nodes and derive conditions for convergence together with a priori estimates of the rate of convergence. The analysis is carried out in the framework of the general single modal asymmetric traffic equilibrium model with elastic demands. Of course, analogous results can be readily derived for the multimodal model as well.

We briefly outline the traffic equilibrium problem for the general single modal model with elastic demands.

We consider a network G with N links, M paths and J O/D pairs of nodes. Links are denoted by a,b etc., paths by p,q etc., and O/D pairs of nodes by w,ω etc. Let L be the set of all links, P the set of all paths and W the set of all O/D pairs. Let P_w be the set of paths connecting the O/D pair w.

The nonnegative flow on path p is denoted by F_p. We group together the path flows into a vector F in R^M. The flow vector F induces a demand vector d with components defined on every O/D pair w by

$$d_w = \sum_{p \varepsilon P_w} F_p, \; w \varepsilon W \tag{2.1}$$

or, in matrix form,

$$d = Bf \tag{2.2}$$

where B is the $J \times M$ matrix whose (w,p) entry is 1 if p connects w and 0 otherwise.

The flow vector F includes a load vector f with components f_a defined on every link a by

$$f_a = \sum_{p \varepsilon p} \delta_{ap} F_p, \; a \varepsilon L \tag{2.3}$$

or, in matrix form,

$$f = \Delta F, \tag{2.4}$$

where Δ is the $N \times M$ link-path incidence matrix whose (a,p) entry is 1 if link a is contained in path p and 0 otherwise.

In what follows, it will be important to distinguish this portion of a flow which is generated by traffic associated with an O/D pair w. We thus associate with every $w \varepsilon W$ the vectors

$$F_w = \{F_p : p \varepsilon P_w\}, \; w \varepsilon W \tag{2.5}$$

$$f_w = \Delta_w F_w \; , \; w \varepsilon W \tag{2.6}$$

where Δ_w is the link-path incidence matrix of the subnetwork consisting of all the paths connecting the O/D pair w. Δ_w is not generally invertible so there may be many flows F_w which induce a fixed load f_w. Since it will be expedient below to have one-to-one correspondence between f_w and F_w, we will normalize F_w by requiring that it lie in the range of a fixed left inverse* E_w of the

*In practice this condition is implemented by requiring that a number of components of F_w be zero.

matrix Δ_w so that

$$F_w = E_w f_w \ , \quad w \varepsilon W. \tag{2.7}$$

A pair (F,d) will be called a *normalized feasible flow pattern* if F lies in the positive orthant of R^M and for $w \varepsilon W$, F_w lies in the range of E_w and d is related to F through (2.2). The set of all normalized feasible flow patterns is a closed convex subset of R^{M+J}.

Now we discuss the cost structure of the network. The travel disutility associated with the O/D pair w is denoted by v_w. We group together the travel disutilities v_w into a vector v in R^J. The travel cost on the link a is denoted by c_a. We group together the c_a into a vector c in R^N. We also assume that a user traveling on path p incurs travel cost

$$C_p = \sum_{a \varepsilon L} \delta_{ap} c_a, \ p \varepsilon P \tag{2.8}$$

and we group together the C_p into a vector C in R^M.

The travel disutility v is related to travel demand d through a smooth function

$$v = \hat{v}(d) . \tag{2.9}$$

The travel cost c is related to the load f through a smooth function

$$c = \hat{c}(f) \tag{2.10}$$

and, in view of (2.4) and (2.8), the travel cost C is related to the flow F through a smooth function

$$C = \hat{C}(F) . \tag{2.11}$$

A normalized feasible flow pattern $(\bar{F},\bar{d})$ (or the corresponding load pattern $(\bar{f},\bar{d})$) is in equilibrium if, once established, no user has any incentive to alter his travel arrangements. This state is characterized by equilibrium conditions which, as shown in [11] and [8], can be expressed as a variational inequality. There are different equivalent formulations of this variational inequality, namely, in terms of link flows,

$$\hat{c}(\bar{f})^T(f-\bar{f}) - \hat{v}(\bar{d})^T(d-\bar{d}) \geq 0 , \qquad (2.12)$$
$$\text{for all admissible } (f,d)$$

or, in terms of path flows [3],

$$\hat{C}(\bar{F})^T(F-\bar{F}) - \hat{v}(\bar{d})^T(d-\bar{d}) \geq 0 , \qquad (2.13)$$
$$\text{for all } (F,d)\varepsilon K .$$

In turn, it is easily seen that (2.13) is equivalent to the following system of J variational inequalities:

$$\sum_{p\varepsilon P_w} \hat{C}_p(\bar{F})(F_p-\bar{F}_p) - \hat{v}_w(\bar{d})^T(d_w-\bar{d}_w) \geq 0 , \qquad (2.14)$$
$$\text{for all admissible } (F_w,d_w),\ w = 1,..,J .$$

We will describe a decomposition scheme which reduces the solution of the system of J coupled variational inequalities (2.14) to the iterative solution of a system of J decoupled variational inequalities.

First we construct the functions $\tilde{v}(d,d')$ defined for feasible pairs of demands (d,d') by

$$\tilde{v}_w(d,d') = \hat{v}_w(d_1',..,d_w,..,d_j') , \quad w\varepsilon W . \qquad (2.15)$$

Next, we define the function $\tilde{C}(F,F')$ by

$$\tilde{C}_p(F,F') = \sum_{a\varepsilon L} \delta_{ap}\hat{c}_a(\tilde{f}_w(F, F')),\ p\varepsilon P_w , \qquad (2.16)$$

where w is the O/D pair joined by p and

$$\tilde{f}_w(F,F') = f_w + \sum_{\omega\neq w} f_\omega' \qquad (2.17)$$

with f_w, f'_ω the loads induced by F_w, F'_ω, respectively.

We observe that

$$\tilde{C}(F,F) = \hat{C}(F),\ \tilde{v}(d,d) = \hat{v}(d)\ , \qquad (2.18)$$
for all $(F,d) \varepsilon K$.

We now fix $(F',d') \varepsilon K$ and consider the system of J variational inequalities:

$$\sum_{p\varepsilon P_w} \tilde{C}_p(\bar{F},F')(F_p-\bar{F}_p) - \tilde{v}_w(\bar{d},d')(d_w-\bar{d}_w) \geq 0\ , \qquad (2.19)$$
for all normalized admissible (F_w,d_w), $w = 1,\ldots,J$.

We observe that in (2.16) $\tilde{C}_p(F,F')$ for $p\varepsilon P_w$ solely depends on F_w and, in (2.15), $\tilde{v}_w(d,d')$ solely depends on d_w; therefore, the system (2.19) is decoupled and for each w we can independently solve the corresponding variational inequality. As a matter of fact, any solution to (2.19), for a particular w, is a user equilibrium flow pattern for the O/D pair w, under the assumption that the flow on every path $q \not\varepsilon P_w$ is fixed and equal to F'_q and the demand associated with every O/D pair $\omega \neq w$ is fixed and equal to d'_ω.

Below we will make assumptions on the cost and demand functions that will guarantee that the variational inequalities (2.19) possess a unique solution (F_w,d_w)* for every $w\varepsilon W$. Under these conditions we define the map

$$T : K \to K \qquad (2.20)$$

which carries $(F',d') \varepsilon K$ into the solution $(\bar{F},\bar{d})$ of the variational inequalities (2.19).

We are now ready to describe the algorithm.

---

*In the standard literature one usually infers uniqueness of the load pattern (f_w,d_w) but generally nonuniqueness of the flow pattern (F_w,d_w). We recall, however, that here we require that the flow satisfy the normalization condition (2.7) so that uniqueness of (f_w,d_w) implies also uniqueness of (F_w,d_w).

Algorithm

Starting with an arbitrary $(F^{\circ},d^{\circ})\varepsilon K$ we construct the sequence
$\{(F^{\nu},d^{\nu})\}\varepsilon K$ by $(F^{\nu},d^{\nu}) = T(F^{\nu-1},d^{\nu-1}) = \ldots = T^{\nu}(F^{o},d^{o})$.

We now proceed to show that the algorithm converges, provided that the cost and demand functions as well as the network satisfy appropriate conditions, namely,

(a) For every $w\varepsilon W$,

$$[\hat{c}(f) - \hat{c}(\overline{f})]^{T}(f-\overline{f}) - [\hat{v}(d) - v(\overline{d})](d-\overline{d}) \geq \alpha_w |f-\overline{f}|^2 + \beta_w |d-\overline{d}|^2 \qquad (2.21)$$

for all admissible (f,d), $(\overline{f},\overline{d})$ with the property $F_\omega = \overline{F}_\omega$, $\omega \neq w$, $d_\omega = \overline{d}_\omega$, $\omega \neq w$; here α_w, β_w are positive constants.

(b) For every $w\varepsilon W$,

$$\sum_{\omega \neq w} \frac{e_\omega^2 n_\omega^2}{\alpha_\omega} \leq k^2 \frac{\alpha_w}{e_w^2} \qquad (2.22)$$

$$\sum_{\omega \neq w} \frac{\ell^2}{\beta_\omega} \leq k^2 \beta_w \qquad (2.23)$$

where $0 < k < 1$, e_w is the norm of the matrix E_w, entering in (2.7) (i.e., e_w^2 is the maximum eigenvalue of the matrix $E_w^T E_w$) or an upper bound thereof, ℓ_w^2 is an upper bound, over the feasible set, of $\sum_{\omega\neq w} (\frac{\partial \hat{v}_w}{\partial d})^2$ and n_w is the norm, or an upper bound thereof, of the matrix A^w defined by

$$A_{pq}^{w} = \frac{\partial \tilde{c}_p}{\partial F_q'} = \sum_{a,b} \delta_{ap}\, \delta_{bp} \frac{\partial \hat{c}_a}{\partial f_b}, \quad p\varepsilon P_w,\ q\notin P_w, \qquad (2.24)$$

i.e., n_w^2 is an upper bound of the maximum eigenvalue of the matrix $(A^w)^T A^w$ over the feasible set.

The interpretation of the above assumption will be discussed later. For the time being, we just mention that the monotonicity assumption (2.21) guarantees the existence and uniqueness of solutions to the system of variational inequalities (2.19) and therefore the map T is well defined.

Theorem

Under the above assumptions there is a unique normalized feasible pattern $(\bar{F},\bar{d}) \in K$ which solves the original variational inequality (2.14) (or the equivalent (2.13)). Furthermore, for arbitrary starting point $(F^\circ,d^\circ) \in K$, the sequence $\{(F^\nu,d^\nu)\}$ converges linearly to $(\bar{F},\bar{d})$ and we have the estimate

$$||(F^\nu,d^\nu) - (\bar{F},\bar{d})|| < \frac{k^\nu}{1-k} ||(F^\circ,d^\circ) - (F',d')||. \quad (2.25)$$

Proof

From (2.19) and (2.18) it is clear that $(\bar{F},\bar{d})$ is a solution to (2.14) (or (2.13)) if it is a fixed point of the map T. The assertion of the theorem will then follow from the Banach fixed point theorem if it is shown that T is a contraction with constant k on K equipped with an appropriate norm $||.||$. To this end, we fix (F',d'), $(F'',d'') \in K$, and we set $(\bar{F},\bar{d}) = T(F',d')$, $(\bar{\bar{F}},\bar{\bar{d}}) = T(F'',d'')$. We fix an O/D pair w and write (2.19),

$$\sum_{p \in P_w} \tilde{C}_p(\bar{F},F')(F_p-\bar{F}_p) - \tilde{v}_w(\bar{d},d')(d_w-\bar{d}_w) \geq 0 \text{ ,}$$
$$\text{for all normalized admissible } (F_w,d_w) \quad (2.26)$$

$$\sum_{p \in P_w} \tilde{C}_p(\bar{\bar{F}},F'')(F_p-\bar{\bar{F}}_p) - \tilde{v}_w(\bar{\bar{d}},d'')(d_w-\bar{\bar{d}}_w) \geq 0 \text{ ,}$$
$$\text{for all normalized admissible } (F_w,d_w). \quad (2.27)$$

We now apply (2.26) with $(F,d) = (\bar{\bar{F}},\bar{\bar{d}})$ and (2.27) with $(F,d) = (\bar{F},\bar{d})$ and add the resulting inequalities, thus obtaining

$$\sum_{p\varepsilon P_w} [\tilde{C}_p(\bar{F},F') - \tilde{C}_p(\bar{\bar{F}},F'')](\bar{F}_p-\bar{\bar{F}}_p)$$
$$- [\tilde{v}_w(\bar{d},d') - \tilde{v}_w(\bar{\bar{d}},d'')](\bar{d}_w-\bar{\bar{d}}_w) \leq 0 ,$$
for all normalized admissible (F_w,d_w) . (2.28)

We rewrite (2.28) in the form

$$\sum_{p\varepsilon P_w} [\tilde{C}_p(\bar{F},F') - \tilde{C}_p(\bar{\bar{F}},F')](\bar{F}_p-\bar{\bar{F}}_p)$$
$$= [\tilde{v}_w(\bar{d},d') - \tilde{v}_w(\bar{\bar{d}},d'')](\bar{d}_w-\bar{\bar{d}}_w) \leq$$
$$- \sum_{p\varepsilon P_w} [\tilde{C}_p(\bar{\bar{F}},F') - \tilde{C}_p(\bar{\bar{F}},F'')](\bar{F}_p-\bar{\bar{F}}_p)$$
$$- [\tilde{v}_w(\bar{\bar{d}},d') - \tilde{v}_w(\bar{\bar{d}},d'')](\bar{d}_w-\bar{\bar{d}}_w)$$
for all normalized admissible (F_w,d_w). (2.29)

Using (2.16), (2.6) and (2.17)

$$\sum_{p\varepsilon P_w} [\tilde{C}_p(\bar{F},F') - \tilde{C}_p(\bar{\bar{F}},F')](\bar{F}_p-\bar{\bar{F}}_p)$$
$$= \sum_{p\varepsilon P_w} \sum_{a\varepsilon L} \delta_{ap}[\hat{c}_a(\tilde{f}_w(\bar{F},F')) - \hat{c}_a(\tilde{f}_w(\bar{\bar{F}},F''))](\bar{F}_p-\bar{\bar{F}}_p)$$
$$= \sum_{a\varepsilon L} [\hat{c}_a(\tilde{f}_w(\bar{F},F')) - \hat{c}_a(\hat{f}_w(\bar{\bar{F}},F''))](\bar{f}_{w,a}-\bar{\bar{f}}_{w,a})$$
$$= \sum_{a\varepsilon L} [\hat{c}_a(\tilde{f}_w(\bar{F},F')) - \hat{c}_a(\tilde{f}_w(\bar{\bar{F}},F'))][\tilde{f}_{w,a}(\bar{F},F')$$
$$- \tilde{f}_{w,a}(\bar{\bar{F}},F')] . \qquad (2.30)$$

Therefore, in view of (2.21), the ℓ.h.s. of (2.29) is minorized by

$$\alpha_w|\tilde{f}_w(\bar{F},F') - \tilde{f}_w(\bar{\bar{F}},F')|^2 + \beta_w|\bar{d}_w-\bar{\bar{d}}_w|^2 . \qquad (2.31)$$

We finally note that on account of (2.7),

$$|\tilde{f}_w(\bar{F},F') - \tilde{f}_w(\bar{\bar{F}},F')|^2 = |\bar{f}_w - \bar{\bar{f}}_w|^2 \geq$$

$$\frac{1}{e_w^2}|\bar{F}_w - \bar{\bar{F}}_w|^2 \qquad (2.32)$$

where e_w is the norm (or an upper bound thereof) of the matrix E_w, entering in (2.7).

We now work on the r.h.s. of (2.29). Applying the mean value theorem and the Cauchy-Schwarz inequality

$$- \sum_{p\in P_w} [\tilde{C}_p(\bar{\bar{F}},F') - \tilde{C}_p(\bar{\bar{F}},F'')](\bar{F}_p - \bar{\bar{F}}_p)$$

$$= - \sum_{p\in P_w} \sum_{q\notin P_w} \frac{\partial \tilde{C}_p}{\partial F'_q} (F'_q - F''_q)(\bar{F}_p - \bar{\bar{F}}_p)$$

$$\leq \frac{e_w^2 n_w^2}{2\alpha_w} \sum_{\omega\neq w} |F'_\omega - F''_\omega|^2 + \frac{\alpha_w}{2e_w^2} |\bar{F}_w - \bar{\bar{F}}_w|^2 \qquad (2.33)$$

and

$$- [\tilde{v}_w(\bar{\bar{d}},d') - \tilde{v}_w(\bar{\bar{d}},d'')](\bar{d}_w - \bar{\bar{d}}_w)$$

$$\leq \frac{1}{2\beta_w} \sum_{\omega\neq w} \left(\frac{\partial \hat{v}_w}{\partial d_\omega}\right)^2 \sum_{\omega\neq w} (d'_\omega - d''_\omega)^2 + \frac{\beta_w}{2} (\bar{d}_w - \bar{\bar{d}}_w)^2$$

$$< \frac{\ell_w^2}{2\beta_w} \sum_{\omega\neq w} (d'_\omega - d''_\omega)^2 + \frac{\beta_w}{2} (\bar{d}_w - \bar{\bar{d}}_w)^2 . \qquad (2.34)$$

Combining (2.29), (2.31), (2.32), (2.33) and (2.34), we deduce

$$\frac{\alpha_w}{e_w^2}|\bar{F}_w - \bar{\bar{F}}_w|^2 + \beta_w|\bar{d}_w - \bar{\bar{d}}_w|^2$$

$$\leq \frac{e_w^2 n_w^2}{\alpha_w} \sum_{\omega\neq w} |F'_\omega - F''_\omega|^2 + \frac{\ell_w^2}{\beta_w} \sum_{\omega\neq w} (d'_\omega - d''_\omega)^2 . \qquad (2.35)$$

Summing up (2.35) with respect to w and rearranging terms on the r.h.s. we obtain

$$\sum_{\omega \in W} \left\{ \frac{\alpha_w}{e_w^2} |\bar{F}_w - \bar{\bar{F}}_w|^2 + \beta_w \; |\bar{d}_w - \bar{\bar{d}}_w|^2 \right\}$$

$$\leq \sum_{\omega \in W} \left\{ \left\{ \sum_{w \neq \omega} \frac{e_w^2 n_w^2}{\alpha_w} \right\} |F'_\omega - F''_\omega|^2 + \left\{ \sum_{w \neq \omega} \frac{\ell_w^2}{\beta_w} \right\} |d'_\omega - d''_\omega|^2 \right\} . \qquad (2.36)$$

We now norm the feasible set K through the norm

$$||(F,d)||^2 \overset{\text{def}}{=} \sum_{w \in W} \left\{ \frac{\alpha_w}{e_w^2} |F_w|^2 + \beta_w |d_w|^2 \right\} . \qquad (2.37)$$

Then using (2.22) and (2.23), (2.36) yields

$$||(\bar{\bar{F}},\bar{\bar{d}}) - (\bar{F},\bar{d})|| \leq k||(F'',d'') - (F',d')|| \qquad (2.38)$$

which shows that T is indeed a contraction with constant k on K equipped with the norm (2.37) and thus the proof of the theorem is complete.

We now discuss our assumptions (2.21), (2.22) and (2.23).

A sufficient condition for (2.21) is that the function $\hat{v}_w(d)$ be uniformly monotone decreasing in d_w , i.e., for all feasible d ,

$$-\frac{\partial \hat{v}_w(d)}{\partial d_w} \geq \beta_w > 0 , \qquad (2.39)$$

and at the same time the cost functions $\hat{c}$ are monotone in f , in the sense

$$[\hat{c}(f) - \hat{c}(\bar{f})]^T (f-\bar{f}) \geq \alpha |f-\bar{f}|^2 , \qquad (2.40)$$

for all feasible pairs $(f,\bar{f})$. Indeed, (2.39) and (2.40) imply condition (2.21) with $\alpha_w = \alpha$ for all O/D pairs w . Assumption (2.39) is of course reasonable and is expected to hold in all cases. For further information on condition (2.40) we refer to [7] where it is shown, in particular, that (2.40) is satisfied if the matrix $[\frac{\partial \hat{c}(f)}{\partial f}]$ is positive definite and that α can then be chosen as the minimum eigenvalue (or a lower bound thereof), over the feasible set, of the symmetric part of this matrix.

Condition (2.23) means that the functions $\hat{v}(d)$ are diagonally dominant; i.e., although $\hat{v}(d)$ may depend on the entire set of demands d, it mainly depends on the component d_w. This assumption is, of course, reasonable.

The interpretation of condition (2.22) is more complicated. Roughly speaking, this condition requires that the O/D pairs be weakly coupled in that any two paths connecting different O/D pairs only share a small number of links. In order to get a quantitative feeling for the meaning of this condition, let us consider the hypothetical situation where i) the network contains J O/D pairs, ii) each O/D pair is connected by Q paths each one consisting of K links, iii) paths connecting the same O/D pairs share D links $(D < K)$, iv) the cost function on every link a is $c_a(f) = gf_a + h_a$. For this case, (2.21) is satisfied for $\alpha = g$. Moreover, for every O/D pair w we get from (2.24) that A^w_{pq} is a $Q\times[(J-1)Q]$ matrix all entries of which are equal to gD. Therefore,

$$n_w^2 = g^2D^2(J-1)Q^2 . \qquad (2.41)$$

Finally, an easy calculation yields

$$e_w^2 = 1/K . \qquad (2.42)$$

Hence, (2.22) reads

$$\frac{(J-1)QD}{K} \leq k < 1 . \qquad (2.43)$$

We observe that $(J-1)QD$ is precisely the number of links that each path of the network shares with all other paths (if some link is shared with, say, S paths, then it is counted with multiplicity S). Therefore, condition (2.43) states that the number of links that each path of the network shares with all other paths is less than the total number of links of the path. This shows that, as expected, the algorithm is likely to converge when paths joining different O/D pairs share relatively few links.

References

[1] Abdulaal, M. and L. J. Leblanc, "Methods for Combining Modal Split and Equilibrium Assignment Models", Transportation Science, 13, pp. 292-314 (1979).

[2] Aashtiani, H. Z. and T. L. Magnanti, "Equilibria on a Congested Transportation Network", to appear in SIAM Journal on Algebraic and Discrete Methods.

[3] Bertsekas, D. P. and E. M. Gafni, "Projection Methods for Variational Inequalities with Application to the Traffic Assignment Problem", Laboratory for Information and Decision Systems, M.I.T., Cambridge, MA, 1980.

[4] Dafermos, S. and F. T. Sparrow, "The Traffic Assignment Problem for a General Network", J. Res. Nat. Bureau of Standards, 73B, No. 2, pp. 91-118 (1969).

[5] Dafermos, S., "An Extended Traffic Assignment Model with Applications to Two-Way Traffic", Transportation Science, 5, pp. 366-389 (1971).

[6] Dafermos, S., "The Traffic Assignment Problem for Multi-class-User Transportation Networks", Transportation Science, 6, pp. 73-87 (1972).

[7] Dafermos, S., "Traffic Equilibrium and Variational Inequalities", Transportation Science, 14, pp. 42-54 (1979).

[8] Dafermos, S., "The General Multimodal Network Equilibrium Problem with Elastic Demands", to appear in Networks.

[9] Dafermos, S., "Decomposition Algorithms for the General Asymmetric Traffic Equilibrium Problem", to appear in Transportation Science, 16, No. 2.

[10] Florian, M., "A Traffic Equilibrium Model of Travel by Car and Public Transit Modes", Transportation Science, 8, pp. 166-179 (1977).

[11] Smith, M. J., Existence, "Uniqueness and Stability of Traffic Equilibria", Transportation Research, 13B, pp. 295-304 (1979).

PARKING STUDIES, GAP ACCEPTANCE AND THE INTERVENING OPPORTUNITIES MODEL: A UNIFIED QUICK CALIBRATION/ESTIMATION METHOD

Carlos F. Daganzo†

University of California, Berkeley, California

Discrete choice models are statistical tools, akin to regression, in which the dependent variable is categorical. They have been used mainly in connection with modal split and marketing studies, but some other transportation traffic theory studies can also benefit from their application; for example, gap acceptance studies, certain studies of parking durations, and safety studies are areas of traffic engineering that have been or can be studied with discrete choice models. Although discrete choice models usually require a computer and sophisticated software, these needs may be obviated in some special cases. This paper describes a family of such cases and illustrates the results by showing how they can improve substantially the execution of periodic check parking studies and the calibration of the intervening opportunities model.

A discrete choice model is an analytical tool that can be used to study how groups of individuals split themselves into classes. The objective is to see whether the fractions of these groups corresponding to the classes are affected substantially by observable explanatory variables of the classes and the individuals in the group.

For example, in gap acceptance studies one is interested in predicting how motorists decide to cross or not to cross a high-priority road. Explanatory variables that could be considered for study of this choice process include the length of the oncoming gap, vehicle type, oncoming vehicle type, type of maneuver to be performed, age and sex

---

†Research supported by NSF research grant ENG-78-12004 to the University of California, Berkeley.

of the driver, vehicle occupancy, duration of delay at the time of decision, and visibility/pavement conditions. Obviously, not all of these need to be used for a given application, but the application (be it delay or safety studies) will dictate those that deserve some consideration.

The basis of a discrete choice model is the choice-function which relates the explanatory variables, A, to the fractions of the population that choose each alternative:

$$p_i = P_i(A), \qquad i = 1, \ldots, I, \tag{1}$$

where p_i is the fraction choosing class i, and the classes or alternatives have been numbered from 1 to I.

The choice functions of a discrete choice model must satisfy:

$$\sum_{i=1}^{I} P_i(A) = 1 \quad \text{and} \quad 0 \leq P_i(A) \leq 1, \; i = 1, \ldots, I. \tag{2}$$

Although many different forms of these functions have been proposed and used in different contexts, the logit and probit functions are the ones that have received the most attention[1,2].

Because the choice functions are not known a priori, a discrete choice model has to be specified and calibrated. The specification step consists in defining a family of choice functions whose individual members are characterized by the value of a parameter vector, θ:

$$p_i = P_i(\theta, A), \qquad i = 1, \ldots, I, \tag{3}$$

and such that the true, unknown, function can be reasonably expected to be a member of the family with an unknown parameter vector, θ_o. The family defined by Eq. (3) must satisfy Eq. (2) for all values of θ.

The calibration or estimation step consists in finding an approximation of θ_o, $\hat{\theta}$. This is efficiently done from data with the maximum likelihood method, because for a randomly selected individual Eq. (3) has the following probabilistic interpretation:

$$p_i = \Pr\{\text{choose } i \mid \theta, A\}, \quad i = 1, \ldots, I. \tag{4}$$

If the data set consists of N randomly selected observations $(n = 1, \ldots, N)$ consisting of an observed choice $i(n)$ and a vector of attributes (as sometimes the explanatory variables are called), $A^{(n)}$, the likelihood of the sample is:

$$L(\hat{\theta}) = \prod_{n=1}^{N} \Pr\{A = A^{(n)} \text{ and choose } i(n) \mid \theta\}$$

$$= \prod_{n=1}^{N} \Pr\{\text{choose } i(n) \mid \theta, A^{(n)}\} \Pr\{A = A^{(n)} \mid \theta\}, \tag{5}$$

where A represents the random variable from which the $A^{(n)}$ are drawn.

If the sampling distribution of A is independent of the unknown value, θ_o, as happens if the sampling mechanism is random and independent of the choices that have been made, the last term of Eq. (5) is independent of θ and the log-likelihood function is given by:

$$\log L(\hat{\theta}) = \sum_{n=1}^{N} \log P_{i(n)}\left(\theta, A^{(n)}\right), \tag{6}$$

where the additive constant $\sum_{n=1}^{N} \log \Pr\{A = A^{(n)}\}$ has been omitted.

The maximum likelihood estimate is given by the maximum of Eq. (6), $\hat{\theta}$, which for large samples and under the regularity conditions

- the choice functions are 3 times continuously differentiable positive functions of θ and A,
- the random vector, A, is bounded,
- as $N \to \infty$ the matrix of second derivatives of $\log L(\theta)$ (the Hessian, $\nabla^2 \log L(\theta)$) is non-singular at θ_o,

can be visualized as the outcome of a random variable, $\hat{\Theta}$, which is multivariate normally distributed with mean θ_o and covariance matrix:

$$\Sigma_\Theta = -\underset{N\to\infty}{\text{plim}} \, [\nabla^2 \log L(\hat{\Theta})]^{-1} . \tag{7}$$

Equation (7) is useful because it is an easily obtained indicator of the accuracy of $\hat{\theta}$. Furthermore, under the regularity conditions, Σ_Θ cannot be made smaller by using another estimation method.

Because Eq. (6) and its derivatives are hard to calculate by hand when N is large, $\hat{\theta}$ must be found by computer in most instances. A number of computer programs are available for this purpose[3,4,5]. An exception is models with only one parameter and few attributes because for them it may be possible to group the N terms of Eq. (6) into so few terms that the search process can be carried out by hand, as is done with the Berkson-Theil binary probit/logit estimation method[6]. As a matter of fact, in some cases $\hat{\theta}$ and Σ_Θ have a closed form which can be exploited to derive optimal sampling strategies that minimize Σ_θ (see Sec. 2).

The next section introduces a family of one-parameter models that apply to some traffic/transportation problems and shows how they can be treated manually. Section 2 develops closed form formulae for parking studies and Section 3 presents a case study using the intervening opportunities model.

1. ONE-PARAMETER EXPONENTIAL MODELS

A one-parameter exponential model is defined by the following choice function:

$$p_i = \frac{\exp(-\theta A_{i-1}) - \exp(-\theta A_i)}{1 - \exp(-\theta A_I)} , \quad i = 1, \ldots I , \tag{8}$$

where the attributes A_i satisfy $A_i \leq A_{i+1}$, $i = 1, \ldots, I$, and we define $A_0 = 0$. Clearly,

$$0 \leq p_i \leq 1 \quad (i = 1, \ldots, I) , \qquad \sum_{i=1}^{I} p_i = 1 ,$$

and Eq. (8) defines a choice model.

This choice model arises in the following instances, among others:

- The intervening opportunities model[7].
- The periodic check parking study[8].
- Gap acceptance studies with no within-driver variability.

In each of the instances the explanatory variables, A_i, have a different meaning.

The intervening opportunities model of trip distribution as modified by Schneider[9] assumes that a randomly selected shopper (if the model is for shopping trips) from a small zone chooses his shopping destination as if he ranked all the shops by a proximity measure, such as distance or travel time, and then searched for the widget of his choice starting with the closest shop until the first suitable widget was found. See Reference 10 for a textbook level description.

Because shops of different sizes will have different probabilities of containing a suitable widget we define a variable, O_i (e.g., number of parking spaces, floor space), that describes the sales volume of the store and is readily available as an indication of the number of "opportunities" available in it.

It is clear that if for any given origin zone one ranks all the destinations by proximity from $i = 1$ to $i = I$, and one defines π as the probability that any one opportunity satisfies the shopper,

$$p_i = \frac{\pi^{A_{i-1}} - \pi^{A_i}}{1 - \pi^{A_I}} , \quad i = 1, \ldots, I, \tag{9}$$

with $A_o = 0$ and $A_i = \sum_{j=1}^{i} O_j, \forall i$. Note that if one defines a parameter, $\theta = -\log \pi$, Eq. (1) applies. Although the A_i's will change from person to person depending on the location of his residence, which determines the ordering of the alternatives, the requirement $A_{i+1} \geq A_i$, $\forall i$, is always met.

Thus, for the intervening opportunity model the A_i's can be interpreted as the cumulative number of opportunities up to and including the ith closest destination, and θ as the negative logarithm of the probability that an opportunity contains the widget.

The periodic check parking study is another instance where Eq. (8) applies. The reader can find background material on this subject in Refs. 8 and 11 . The object of this study is to determine the average parking duration in an open area where all the parking arrivals and departures cannot be observed simultaneously. The study is carried out by performing checks on the license plates of cars parked at every stall in the study area at regular intervals, and recording the number of times that each and every vehicle was observed.

The variables in this study are its duration, T , the interval between visits to any one given stall (or cycle), C , and the number of stalls surveyed, S . Although in this paper T is assumed to be a multiple of C , this assumption could be relaxed easily. One should select these variables to obtain sufficiently accurate results in the most cost-effective way.

Under the assumption that parking durations are negative exponentially distributed with mean m (this seems reasonable in the light of data[12]) it is possible to write the probability that a vehicle is seen exactly i times (i = 1, 2, . . .), given that it was seen once. Because the negative exponential distribution is memory-less, the probability, ρ , that a vehicle is seen on the next visit does not depend on how long a vehicle has been parked and is given by:

$$\rho = \Pr\{\text{duration} \geq C\}$$
$$= \exp(-C/m) \ .$$

The probability that a vehicle is seen exactly i times given that it was seen once is therefore:

$$p_i = (1-\rho)\rho^{i-1} = [1-\exp(-C/m)]\exp(-C(i-1)/m) \ , \quad i = 1, 2, \ldots \ .$$

This expression assumes an infinitely long study. Consequently it must be modified because there cannot be an infinite number of visits. The maximum number of times a vehicle can be seen, $\tilde{I}$, is equal to the remaining number of visits to its stall from a time immediately before it was first seen. Letting I be the total number of times any stall is observed (note that $I = T/C$ and that $\tilde{I} \leq I$) , the probability that a vehicle is observed exactly i times, given that it can be observed a maximum of $\tilde{I}$ times, is:

$$p_i = \begin{cases} \exp\left(-\dfrac{C(i-1)}{m}\right) - \exp\left(-\dfrac{Ci}{m}\right) , & i = 1, 2, \ldots, \tilde{I} - 1 \\ \exp\left(-\dfrac{C(i-1)}{m}\right) & i = \tilde{I} \\ 0 & \tilde{I} < i \leq I , \end{cases}$$

which by defining

$$\theta = \frac{1}{m} ,$$
$$A_j = j \times C \quad \text{if} \quad j < \tilde{I}$$
$$= \infty \qquad \text{if} \quad j \geq \tilde{I} ,$$

reduces to Eq. (8). Note that the A_j's are not the same for all the vehicles because $\tilde{I}$ will change from vehicle to vehicle.

The last example concerns a hypothetical problem that will not be analyzed in detail but is presented to illustrate yet another context of possible application. Assume that drivers are seen at a stop sign accepting and rejecting gaps and that each driver has a fixed threshold, G , such that if an oncoming gap is longer than G crossing takes place but not otherwise. This threshold is assumed to remain fixed for each driver, so that the accepted gap will always be larger than all rejected gaps, but to vary from driver to driver according to a negative exponential distribution with unknown mean, $\bar{G}$. Studies of gap acceptance can relax these assumptions[13,14] but queueing theory formulas become quite complicated even if G is fixed across drivers.

For each driver, rank the i gaps she considered from 1 to i , starting with the shortest, g_1 , and ending with the longest, g_i . Because G is fixed, g_i is the accepted gap; of course significant departures from this in the data invalidate the procedure to be outlined, but small ones could probably be corrected without harm.

The probability that, given the gap sequence, a driver accepts the ith gap is:

$$p_i = \Pr\{G \geq g_1 , G \geq g_2 , \ldots , G \geq g_{i-1} , G < g_i\}$$

$$= \Pr\{g_{i-1} \leq G < g_i\}$$

$$= \exp(-g_{i-1}/\bar{G}) - \exp(-g_i/\bar{G}) ,$$

which corresponds to Eq. (8) with $\theta = 1/\bar{G}$, $I > i$, $A_I = \infty$, $A_o = 0$, and $A_j = g_j$, $j = 1, \ldots i$. This model is a special case of the maximum likelihood model proposed by Miller (method 9 in Ref. 13).

Currently, heuristic estimation methods are used to calibrate the intervening opportunities model[15] and to perform the periodic check parking study[16]. To perform gap acceptance studies one must typically use computer codes that are not widely available. Clearly, then, hand estimation of the parameter θ in the one-parameter exponential model with maximum likelihood has the potential to improve the accuracy and/or facilitate the calculations of these studies considerably.

The estimation process

The log-likelihood function is:

$$\log L(\theta) = \sum_{n=1}^{N} \log \left[\frac{\exp(-\theta A_{i(n)-1}^{(n)}) - \exp(-\theta A_{i(n)}^{(n)})}{1 - \exp(-\theta A_I^{(n)})} \right], \quad (11)$$

where

$i(n)$ is the choice of the nth observation and

$A_j^{(n)}$ is the jth component of the attribute vector of the nth observation .

It is more convenient to work with the parameter $\pi = \exp(-\theta)$:

$$\log L(\pi) = \sum_{n=1}^{N} \log\left[\frac{\pi^{A_{i(n)-1}^{(n)}} - \pi^{A_{i(n)}^{(n)}}}{1 - \pi^{A_I^{(n)}}}\right]$$

$$= \sum_{n=1}^{N} \log\left[\frac{\pi^{A_{i(n)-1}^{(n)}}\left[1 - \pi^{A_{i(n)}^{(n)} - A_{i(n)-1}^{(n)}}\right]}{1 - \pi^{A_I^{(n)}}}\right]$$

$$= \sum_{n=1}^{N} \left[A_{i(n)-1}^{(n)} \log \pi + \log\left(1 - \pi^{a_{i(n)}}\right) - \log\left(1 - \pi^{A_I^{(n)}}\right)\right],$$

where $a_{i(n)}$ represents $\left(A_{i(n)}^{(n)} - A_{i(n)-1}^{(n)}\right) \geq 0$.

The maximum likelihood estimate, $\hat{\pi}$, is obtained by finding the solution to:

$$\sum_{n=1}^{N} \left(\frac{A_{i(n)-1}^{(n)}}{\pi}\right) - \sum_{n=1}^{N} \left(\frac{a_{i(n)} \pi^{a_{i(n)}-1}}{1 - \pi^{a_{i(n)}}}\right) + \sum_{n=1}^{N} \frac{A_I^{(n)} \pi^{A_I^{(n)}-1}}{1 - \pi^{A_I^{(n)}}} = 0 .$$

This is not easy to do by hand because the left side has many terms and it is cumbersome to calculate it repeatedly. However, we can approximate it very easily.

Select a few positive values, x_ℓ , $\ell = 1, \ldots, L$ (L, small) such that each value $a_{i(n)}$ can be approximated by an x_ℓ. Do the same with the $A_I^{(n)}$ values and another set of positive values y_r , $r = 1, \ldots, R$ (R, small). Then

$\hat{\pi}$ is the solution to

$$\frac{\alpha}{\pi} + \sum_{r=1}^{R} \beta_r \frac{y_r \pi^{y_r - 1}}{1 - \pi^{y_r}} - \sum_{\ell=1}^{L} \gamma_\ell \frac{x_\ell \pi^{x_\ell - 1}}{1 - \pi^{x_\ell}} = 0 ,$$

with $\alpha = \sum_{n=1}^{N} A_{i(n)-1}^{(n)}$,

β_r = number of observations with $A_I^{(n)} \simeq y_r$

and γ_ℓ = number of observations with $a_{i(n)} \simeq x_\ell$.

If both sides are multiplied by $\pi (\pi \neq 0)$, $\hat{\pi}$ is also the solution to:

$$\alpha + \sum_{r=1}^{R} \beta_r y_r (\pi^{-y_r} - 1)^{-1} - \sum_{\ell=1}^{L} \gamma_\ell x_\ell (\pi^{-x_\ell} - 1)^{-1} = 0 . \quad (12)$$

For the intervening opportunities model and the periodic check parking study the values of x_ℓ and y_r are naturally defined.

For the intervening opportunites model $a_{i(n)}$ represents the total number of opportunities in the chosen destination; it can only take I values and, therefore, if the number of destinations is not very large, it is reasonable to set $L = I$ and $x_\ell = O_\ell$. Furthermore, since $A_{I(n)}$ is the total number of opportunities (which is usually fixed), $R = 1$, and

$$y_1 = A_I = \sum_{\ell=1}^{L} O_\ell .$$

Equation (12) has $I + 1$ terms and can be evaluated exactly.

For the periodic check parking study something similar happens. $a_{i(n)}$ is either equal to C or, if $i(n) = \tilde{I}$, infinite, and $A_I^{(n)}$ is always infinite. Therefore, we can naturally select $L = 2$, $x_1 = C$, $x_2 = \infty$, $R = 1$, and $y_r = \infty$.

For the gap acceptance problem one can choose x_ℓ at 1

or 2 second intervals up to a reasonably large value and then use larger intervals. Of course, A_I does not change from observation to observation and we may use $R = 1$, $y_1 = \infty$.

In all three cases, once $\hat{\pi}$ is found it is also possible to determine a large sample approximation to the variance of the MVN approximation to the estimator, $\hat{\Pi}$, because the regularity conditions hold and Eq. (7) yields:

$$\mathrm{Var}\,(\hat{\Pi}) \simeq \hat{\pi}^2 \Big/ \left[\sum_{\ell=1}^{L} \gamma_\ell x_\ell^2 \frac{\hat{\pi}^{x_\ell}}{\left(1 - \hat{\pi}^{x_\ell}\right)^2} - \sum_{r=1}^{R} \beta_r y_r^2 \frac{\hat{\pi}^{y_r}}{\left(1 - \hat{\pi}^{y_r}\right)^2} \right]. \tag{13}$$

Thus, all the calculations can be done by hand and, as shown in the next section, one may even derive optimal sampling schemes.

2. THE PERIODIC-CHECK PARKING STUDY

The periodic check parking study yields closed form formulas. Let $\delta = \pi^{-C} = \rho^{-1}$, and since $y_r \to \infty$, Eq. (12) becomes:

$$\frac{\alpha}{C} = \frac{\gamma_1}{\delta - 1},$$

where, from the definition, α/C is the total number of stall observations yielding a vehicle that was also there at the previous visit, and γ_1 is the total number of different license plates that were observed but had departed before the last check. Note that α/C is given by the total number of occupied-stall observations minus the total number of different license plates seen. Then

$$\hat{\delta} = (\gamma_1 C/\alpha) + 1,$$

and because of the invariance property of maximum likelihood estimators:

$$\hat{\pi} = \left((\gamma_1 C/\alpha) + 1\right)^{-1/C}$$

and

$$\hat{m} = C/\log(1 + \gamma_1 C/\alpha) \ . \tag{14}$$

Note that if $T \to \infty$, so that only a small fraction of all observed cars remain at the end of the study, $\alpha/(\gamma_1 C)$ is the average number of times cars were observed minus one, and Eq.(14) coincides with the results of Ref.16. Eq.(14) is, however, more general because it can be applied in practical cases where T is not large; it corresponds also to a maximum likelihood estimator, $\hat{M}$, so that the asymptotic variance of its normal approximation can be calculated from Eq.(13):

$$\text{var}(\hat{M}) \simeq \text{var}(\hat{\Pi}) \left(\left.\frac{\partial m}{\partial \pi}\right|_{\hat{\pi}}\right)^2 = \text{var}(\hat{\Pi})/[\hat{\pi}(\log \hat{\pi})^2]^2 \ , \tag{15}$$

which after a few algebraic manipulations reduces to:

$$\text{var}(\hat{M}) \simeq \hat{m}^2 \frac{[1 - e^{-C/\hat{m}}]^2}{\left(\frac{C}{\hat{m}}\right)^2 e^{-C/\hat{m}} \gamma_1} \ . \tag{16}$$

Because γ_1 represents the total number of vehicles observed to have departed, it can be expressed as follows:

$$\begin{aligned}\gamma_1 &= (\text{number of vehicles seen per cycle}) \times \\ &\quad (\text{total number of periods minus 1}) \times \\ &\quad (\text{probability that a vehicle departs in any given cycle}) \\ &= (P_o C) \times \left(\frac{T}{C} - 1\right) \times (1 - e^{-C/m}) \ ,\end{aligned}$$

where P_o is the productivity of the data gatherer in license plates per unit time. Replacing γ_1 in Eq. (16) by this expression gives

$$\mathrm{var}(\hat{M}) = \frac{\hat{m}^2}{P_o T}\left[\frac{(e^b - 1)a}{b^2(a - b)}\right] \; ; \; a = \frac{T}{\bar{m}} \; ; \; b = \frac{C}{\bar{m}} \; . \qquad (17)$$

For any value of a it is possible to determine an optimal cycle time by finding the minimum of the parenthetical expression in Eq. (17). This is done in Fig. 1, which plots the optimum b, b_{op}, versus a. Note that if the study lasts muchlonger than one typical parking duration $(a \to \infty)$ the optimal cycle should be comparable to and slightly larger than one mean duration $(b_{op} \to 1.6)$. However, if the study is shorter than one parking duration, the optimal cycle is one-half the duration of the study. Intermediate values can be derived from the figure.

The figure also displays the inefficiency factor in parenthesis in Eq. (17), z, which drops asymptotically to 1.55 as the length of the study relative to the average duration increases.

3. CASE STUDY

This section presents an example of calibration of the intervening opportunities model. The data were gathered during the spring of 1979 in the city of Pinole, California, for a class project that involved the analysis of shopping trips[17]. The procedure involved questioning individuals at their place of residence by phone, eliciting information about their address and their last shopping trip. The data gathering was done by a team of ten students and was completed in one evening session.

The city of Pinole has five shopping centers in it, or its immediate vicinity, and for the purposes of this study was carved into 24 study zones, so that all the points from each zone had all the shopping centers in the same order of proximity. Although this goal could have been achieved with a smaller number of zones, this detailed zoning system was adopted because it had to be used for other purposes for which more zones were needed.

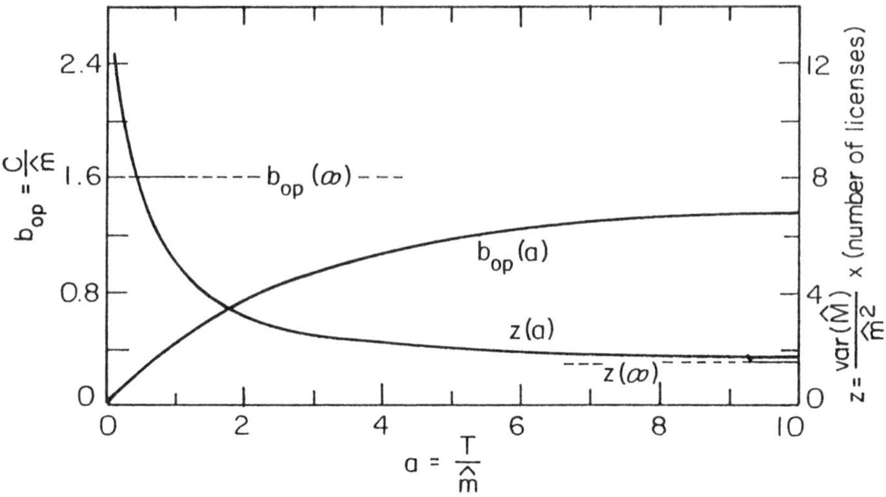

Figure 1 - Optimal cycle time vs. duration of study for the periodic-check parking study

Thus, each shopping trip was characterized unambiguously by the origin zone and the shopping mall that was used. The number of opportunities in each shopping area was determined from a weighting scheme that gave 1 opportunity to a small store, 5 to a supermarket and 10 to a department store. Table I gives the characteristics of the shopping centers and their opportunity sizes, O_i . Table II summarizes the data corresponding to the 65 trips that were recorded. It should be noted that although this is a small data set, it provides accurate enough information when used in connection with the maximum likelihood method. Furthermore, the complexity of the technique about to be illustrated is not influenced by the size of the data set -- only by the number of destinations.

From the information in Table II it is possible to obtain the parameters of Eq. (12):

$L = 5,\ x_1 = 35,\ x_2 = 10,\ x_3 = 30,\ x_4 = 25,\ x_5 = 130$

$R = 1,\ y_1 = 230$

$\alpha = 2{,}495$

TABLE I

Zone Number	Shopping Center	Description	Number of Opportunities
6	Old Town	Several small stores spread out over an area of several blocks	35
10	Tara Hills	Very small	10
17	Pinole Valley	1 Supermarket 5 Gasoline Stations Many small shops	30
21	Appian 80	2 Supermarkets Several restaurants and small stores	25
24	Hilltop	Very large including 3 department stores	130

$\beta_1 = 65$

$\gamma_1 = 9,\ \gamma_2 = 4,\ \gamma_3 = 21,\ \gamma_4 = 21,\ \gamma_5 = 10$.

Eq. (12) then becomes:

$$2495 + \frac{14950}{\pi^{-230}-1} = \frac{315}{\pi^{-35}-1} + \frac{40}{\pi^{-10}-1} + \frac{630}{\pi^{-30}-1} + \frac{525}{\pi^{-25}-1} + \frac{1300}{\pi^{-130}-1} .$$

It yields $\hat{\pi} = 0.9839$ and $\hat{\theta} = 0.0163$. From Eq. (13), and after a change of variable similar to that made in Eq.(15), we obtain the standard error of $\hat{\theta}$, 0.0026, which is not large despite the small data set that was used.

4. SUMMARY

The object of this paper was to show that discrete choice models can be applied to several transportation problems and to demonstrate that in some cases it is possible to calibrate them by hand, perhaps with the help of a programmable pocket calculator.

A family of negative exponential discrete choice models was selected for more detailed analysis because three transportation/traffic engineering problems are special cases of

TABLE II

Origin Zone	No. of Observations	Number of Trips (Number of Opportunities) per Shopping Center				
		Proximity of Center				
		Nearest	Second	Third	Fourth	Farthest
1	-					
2	-					
3	1	0(35)	1(10)	0(25)	0(30)	0(130)
4	1	0(35)	1(25)	0(10)	0(30)	0(130)
5	1	0(35)	0(30)	1(25)	0(10)	0(130)
6	-					
7	8	2(35)	5(30)	1(25)	0(10)	0(130)
8	-					
9	8	5(25)	0(10)	1(35)	0(30)	2(130)
10	-					
11	-					
12	-					
13	-					
14	6	2(25)	3(30)	1(35)	0(10)	0(130)
15	-					
16	-					
17	1	1(30)	0(35)	0(25)	0(10)	0(130)
18	6	3(30)	1(35)	1(25)	0(10)	1(130)
19	5	2(30)	1(35)	1(25)	0(10)	1(130)
20	19	7(30)	3(35)	3(25)	1(10)	5(130)
21	-					
22	7	0(35)	4(25)	0(30)	2(10)	1(130)
23	2	0(35)	2(25)	0(30)	0(10)	0(130)
24	-					

it: the periodic-check parking study, some gap acceptance studies and the intervening opportunities model.

The periodic-check parking study yielded closed form formulas for the mean parking duration and its standard error. From these it was possible to derive a chart that gives the optimal periodic-check frequency and the associated standard error.

A small data set for the intervening opportunities model was used to illustrate the technique when closed form formulas cannot be developed. The calculations were carried out by hand in less than one hour, and despite the smallness of the data set a rather accurate result was obtained. A realistic data set, say ten times larger, would not increase the calculation effort but would have decreased the size of the standard error by a factor of $\sqrt{.1}$.

It is concluded that systematic analysis of data in many transportation studies can result in improved predictions, and that computers are not always needed.

REFERENCES

1. T. A. DOMENCICH and D. McFADDEN, Urban Travel Demand: A Behavioral Analysis. American Elsevier, New York, 1975.
2. C. F. DAGANZO, Multinomial Probit: The Theory and Its Application to Demand Forecasting, Academic Press, New York, 1979.
3. J. BERKMAN, D. BROWNSTONE, G. M. DUNCAN and D. McFADDEN, "QUAIL 3.0 User's Manual," Working paper, Dept. of Economics, University of California, Berkeley, CA (1977).
4. C. F. DAGANZO and L. SCHOENFELD, "CHOMP's User's Manual," Research Report UCB-ITS-RR-78-7, University of California, Berkeley, CA (1978).
5. S. J. HABERMANN, Analysis of Qualitative Data: Vol. 2. New Developments, Academic Press, New York, 1979.
6. J. BERKSON, "A Statistically Precise and Relatively

Simple Method of Estimating the Bio-Assay with Quantal Response, Based on the Logistic Function," J. Am. Stat. Assoc. 39, 357-365 (1953).
7. S. A. STOUFFER, "Intervening Opportunities: A Theory Relating Mobility and Distance," Am. Soc. Rev. 5, 845-867 (1940).
8. D. E. CLEVELAND, "Traffic Studies," in Transportation and Traffic Engineering Handbook, J. E. Baerwald, M. J. Huber and L. E. Keefer (eds.), Prentice-Hall, Englewood Cliffs, N. J., 1976.
9. M. SCHNEIDER, "Panel Discussion on Inter-Area Travel Formulas," Highway Res. Board Bulletin 253, 128-138 (1960).
10. B. G. HUTCHINSON, Principles of Urban Transport Systems Planning, McGraw-Hill, New York, 1974.
11. D. E. CLEVELAND, "Parking Studies," in Manual of Traffic Engineering Studies (3rd ed.), D. E. Cleveland (ed.), Institute of Traffic Engineers, Washington, D. C., 1964.
12. E. MOHR, "Euclid Avenue Parking Study," Student Research Report, Inst. of Trans. and Traffic Eng., University of California, Berkeley, CA, 1954.
13. A. J. MILLER, "Nine Estimators of Gap-Acceptance Parameters," in Traffic Flow and Transportation, G. F. Newell, (ed.), American Elsevier, New York, 1972.
14. C. F. DAGANZO, "Estimation of Gap Acceptance Parameters Within and Across the Population from Direct Roadside Observation," Trans. Res. 15B, 1-16 (1981).
15. E. R. RUITER, "Toward a Better Understanding of the Intervening Opportunities Model," Trans. Res. 1, 47-56 (1967).
16. D. E. CLEVELAND, "Accuracy of the Periodic-Check Parking Study," Traffic Eng. 33, 12, 14-17 (September 1963).
17. C. K. BEE and J. FIEGERMAN, "Pinole City Transportation Study Report," unpublished notes, University of California, Berkeley, CA (1979).

GAP ACCEPTANCE: MYTH AND REALITY

J. Darzentas

Mathematics Department
Royal Holloway College
(University of London)
Egham Hill, Surrey

Three general models of driver gap acceptance behaviour are discussed. They are: (a) a fixed critical gap, (b) consistent individual drivers, and (c) inconsistent individual drivers. The limitations and implications of the models and the restrictions imposed by the nature of the data are discussed.

The model in which each driver has a constant critical gap, distributed over the population, is closely examined, and it is concluded that the distribution of critical gaps cannot be estimated from data available from field observations. It is suggested that use of the distribution of observed accepted lags (or gaps) together with conditional probabilities of acceptance should be used in comparison studies. Truncated Normal and Weibull distributions have been successfully fitted to empirical data on accepted lags and gaps.

1. Introduction

The situation where a driver has to decide whether or not to enter the path of an oncoming vehicle in a priority stream has been the concern of a great number of traffic studies (Adams[1], Ashton[2], Ashworth[3,4], Ashworth and Bottom[5], Blumenfeld and Weiss[6], Blunden et al.[7], Bottom and Ashworth[8], Cooper and McDowell[9], Cooper et al.[10], Darzentas et al.[11], Darzentas et al.[12], Drew[13], Gibbs[14], Herman and Weiss[15], McNeill and Morgan[16], McNeill and Smith[17], Raff[18], Ramsey and Routledge[19], Tanner[20], Troutbeck[21], Weiss and Maradudin[22]). This decision naturally depends on numerous factors such as the type of the intersection, the speed of the oncoming vehicles, the types of vehicles involved, the surrounding environment, and the time of the day.

Yet, whatever the degree of influence of these factors, the decision to act devolves entirely upon the driver,who has to give priority and his cognition of the situation. This decision-making process is referred to in the literature as "gap acceptance behaviour".

The research on gap acceptance behaviour has concentrated on the problem of understanding and describing the driver's behaviour with respect to judgment and acceptability of the relevant risk. A number of gap acceptance models have been developed in the last few decades for the purpose of studying the delay and capacity at the intersections, and also accident risk.

The majority of these models are based on real life observations, where each driver in the non-priority streams contributes to the data one accepted gap and perhaps a number of rejected gaps. A few authors[12,14] conducted test-track experiments using the same subject several times; however, the behaviour of experimental subjects in controlled conditions is always suspect, as a basis for general conclusions about the real life situation. For both real life and test-track experiments, data on gap acceptance are collected, and the accepted and rejected gaps are classified into time or distance bands, from which frequencies are obtained. The time or distance gap, between the arrival of the front of the vehicle in the non-priority stream and that of the first oncoming vehicle (front) in the priority stream at the junction, is called "lag". The time or distance gaps between vehicles in the priority streams are called "gaps". Each turning driver considers only one lag and perhaps a number of gaps presented in a major road stream of traffic. The analysis and discussions throughout this paper will assume that the data available consist of frequencies of two types of lags or gaps: accepted and rejected. These together constitute the "presented" lags or gaps.

The nature of the field observations, as far as the

limited information on any individual driver is concerned, led to the following three general behavioural models of gap acceptance decision-making corresponding to the individual driver:

I. There is a minimum acceptable (critical) lag or gap t which is identical for all drivers in a population. If a presented lag or gap in the major road is greater than or equal to t, then the driver accepts it: otherwise he rejects it.

II. Each individual driver has a constant critical lag or gap which is distributed over the population. This pattern of behaviour implies absolute consistency for each individual.

III. A probability of acceptance is assigned to every presented lag or gap in the major road independently according to a distribution function. The same distribution describes the inconsistent behaviour of all members of the driver population.

Using as base these three general behavioural models, several theoretical and empirical models have been developed by various researchers to examine different operational objectives of intersections, as well as driver population gap acceptance characteristics. It is apparent that the true behavioural pattern is somewhere in between the last two models. The variability within an individual can be explained by the degree of concentration at the moment of judgment, as well as the ability of the drivers to detect small differences between presented gaps or lags. Furthermore, the presence of variability between individuals is well known in every aspect of human behaviour; in this case it is mainly due to the coexistence of "cautious" and "aggressive" drivers.

A number of investigators[6,17] compared these different models, in order to examine their implications on the calculation of delay and capacity, and they found that the resulting differences are significant and cannot be easily ignored.

The ideas presented in this paper are concerned with the limitations of these three general models of gap acceptance behaviour (most noticeable in comparison studies), their true meaning, and the restrictions upon the modelling of gap acceptance imposed by the nature of the data. Another angle of approach to gap acceptance behaviour which might be more useful in studies concerned with safety, rather than delay and capacity, is also suggested.

2. Model I - A Constant Critical Gap for the Population

Variability within and between individuals in certain behavioural processes is synonymous with human nature; thus no one can expect the researchers who adopted this model to believe that the constant critical lag or gap is a true representation of the behaviour of a driver population. However, in model building, any simplification of an intermediate stage of the process is welcome, provided that it does not cause significant damage to the model. A complete and final list of the implications of this assumption on the calculation of delay and capacity of priority intersections is not available yet, but there are strong indications about its effects and limitations.

This modelling concept was introduced by Adams[1], who examined the delay to pedestrians crossing a road. Since then it has been used by a great number of researchers who studied mainly delay and capacity at priority intersections.

In empirical studies, the critical lag or gap is usually estimated from field observations, which are in the form of proportions of acceptance classified into time or distance bands. Raff[18] developed a method of calculating the lag size for which the number of drivers who accepted it equals the number of drivers who rejected it; this he called the "critical lag" of the population of drivers in the non-priority stream. Raff's critical lag is actually the point of intersection of the cumulative curve of the accepted lags with the cumulative curve of the rejected lags, the latter being calculated by starting from the large lags.

Many authors have used Raff's critical lag to compare junctions with different layout or control systems, or even to provide rules for the imposition of certain control systems. However, it should be noted that the critical lag as defined by Raff depends on the major road flow (the distribution of presented lags) and this must be taken into account in comparison studies. The concept of a constant critical lag or gap for a population of drivers has also been used in some theoretical[20] studies for developing delay and capacity models.

3. Model II - Consistent Behaviour

The assumption of absolute consistency of individual drivers has dominated gap acceptance modelling in the last two decades. Nevertheless the "within individual drivers" variance has been pointed out several times by a number of authors who conducted test-track experiments using the same subject several times. Recently, Ashworth and Bottom[5,8] conducted a series of experiments which combined the multiple information on individuals together with the real life conditions. Their results further support the findings of the test-track experiments.

However, while this kind of variability comes as no surprise to common sense, the adoption of the consistency assumption could perhaps be justified by the fact that the "between drivers" variance is much higher than the "within drivers" variance; moreover if the variances of the individuals can be considered identical, then an attempt to estimate the distribution of the mean minimum (critical) acceptable lag or gap would also be justifiable.

Unfortunately, this cannot be the case. Ashworth and Bottom[5,8] showed that the higher the mean critical gap the higher the variance; this indicates that the "cautious" drivers are less consistent than the "aggressive" ones. This may be because the distances or the time gaps they judge as acceptable are very large and the oncoming vehicle is too far away from the junction. However, this assumption can still be useful, if shown to have no significant effect on the calculation of delay, capacity and accident risk. Blumenfeld and Weiss[6] in a comparison study showed that for data produced by an individual gap acceptance function model but analyzed under the assumption of a fixed critical gap for each driver which is distributed over the population, the average delay will be estimated accurately, the probability of no delay and the capacity of the minor road will be slightly over-estimated, while the variance of delay will be seriously overestimated.

Since this distribution of critical lags or gaps for a population of drivers in a non-priority stream of traffic has been so prominent in the study of gap acceptance, and it still is useful in estimating certain measures of performance, it is worth examining more closely. An attempt should be made to understand and determine, if possible, how much the empirically derived distributions correspond to the theoretical distributions of critical gaps.

3.1 The bias

Raff[18] was the first to point out that if all accepted and rejected lags and gaps observed at an intersection were included in the analysis, then the proportions of acceptance will be biased; that is, the proportions of accepted lags and gaps in each band do not represent the proportion of drivers who are prepared to accept that size lag or gap when presented with it. This is a result of the fact that each driver may reject a number of presented gaps before accepting one. If, for example, a driver rejects three gaps, say $t-2$, $t-3$, and t seconds, and finally accepts the next presented gap of say $t+8$ seconds, then his contribution to the proportions of acceptances will be a decrease in the ratios of the bands with centre points $t-3$, $t-2$, and t seconds and an increase in the ratio of the band with $t+8$ seconds centre, while his actual critical gap may very well be in the band with centre $t+2$, say.

The above mentioned example is a typical average case which occurs at a fairly busy, but uncongested, intersection.

This bias has been the concern of several researchers[3] who attempted to remove it from the distribution of critical gaps or avoid it by estimating the proportion of acceptances in different ways[13,16,19,21].

3.2 <u>The proportions of accepted lags</u>

A driver on the minor road, arriving at the junction, is presented with a lag of size t in the major road. He will either accept it, with a probability p, or reject it, with a probability $1-p$; this situation can be considered as a Bernoulli trial. If a sample from the population of drivers in the non-priority stream is offered the same size of lag t, then the proportion of acceptances is actually the maximum likelihood estimator of the parameter p of the binomial distribution underlying the number of acceptances for that presented lag t.

Empirical observations give estimates of the proportions of accepted lags p_i for several different sizes of offered lags t_i usually grouped into appropriate intervals.

An observed proportion p_i, corresponding to an interval with centre t_i, represents the probability p_i with which t_i will be accepted. In terms of critical lags, it says that the probability of a driver's critical lag being less than or equal to t_i is p_i.

Naturally, p_i increases as the size of the presented lag increases. Several methods have been used to fit a cumulative curve to the increasing proportion of acceptances, the most popular one being the probit transformation, which is used to give a linear representation of a cumulative Normal distribution F(t) on a logarithmic axis. It should be emphasized here that F(t) is claimed to be the distribution of the critical lags for the population of the minor road drivers. If that is the case then $F(t) = P(C \leq t)$ where C is the critical lag: consequently, $F(t-dt) = P(C \leq t-dt)$, so that $[F(t) - F(t-dt)]dt$ is the probability that C belongs to the interval $(t-dt, t)$.

However the probit analysis method estimates more precisely the median (50%) critical lag or gap and less precisely the larger or smaller percentages of acceptance, becoming inefficient for the very large or very small percentages.

At this point, it should be noted that probit analysis has been used in biology to determine the distributions of tolerances of certain species as a function of dose level of a drug, and in other similar cases. It has been mentioned in the relevant literature of Bioassay that some objections have been expressed about the concept of tolerance distribution, but that is beyond the scope of this analysis. A few authors have mentioned that, in order to use the method of probit analysis, one should consider only the accepted and rejected lags and not the gaps, because each driver must be considered only once. In the majority of cases, the authors used lags and gaps combined, for, after treating them separately, they found that the estimated parameters and probabilities of acceptance were similar for both lags and gaps.

Nevertheless, in this paper, the discussion refers to lags only, in order to satisfy the assumption that each driver is considered once only, in spite of the possibility that lags and gaps may be considered together in practice.

3.3 The elusive distribution of critical lags

Assuming that each driver has a fixed critical lag, then every time an accepted or rejected lag t_i is observed, the information provided by t_i is $C \leq t_i$ in the case of an accepted lag and $C > t_i$ in the case of a rejected one; t_i is the presented lag in the major road and C the critical lag of the driver in the non-priority stream.

The proportion of accepted lags size t_i out of the total number of offered lags of all sizes represents the probability that a driver has a critical lag size C less than or equal to t_i and the presented lag $\underline{o}$ is t_i, i.e. $p(C \leqq t_i \cap \underline{o} = t_i)$. Let $G(o)$ and $F(C)$ be the distributions of the presented $\underline{o}$ and critical $\underline{C}$ lags respectively. $\underline{o}$ and $\underline{C}$ are independent random variables. The empirical observations are classified into bands and therefore the case where both distributions are discontinuous is considered at this stage; this can be generalized to include the continuous case. For each proportion p_i, two probabilities $p_1(C \leqq t_i \cap o = t_i)$ and $p_2(C > t_i \cap o = t_i)$ are obtained and

$$p_1 + p_2 = g(t_i), \quad \text{where} \quad \sum_{n=1}^{i} g(t_n) = G(t_i).$$

$g(t_i)$ is the marginal proportion of the presented lags distribution for the band with centre t_i.

One way to estimate the distribution of critical lags is by estimating their marginal distribution, a facility which is not provided for by the data, because no information of the form $p(C = t_i \cap o = t_i) \forall_{i,j}$ is available. This is also illustrated in Figure 1. If for a given offered lag t_i, the probability $p(C \leqq t_i \cap o = t_i)$ is divided by $g(t_i) = p(o = t_i)$ then this ratio is the conditional probability:

$$p(C \leqq t_i | o = t_i) = \frac{P(C \leqq t_i \cap o = t_i)}{p(o = t_i)}.$$

i.e. the probability that a driver's critical lag C is less than t_i, given that the lag presented to him is t_i. This illustrates that the proportions of acceptance also represent the conditional distribution of the critical lag being less than a certain size presented lag, given that this lag is presented.

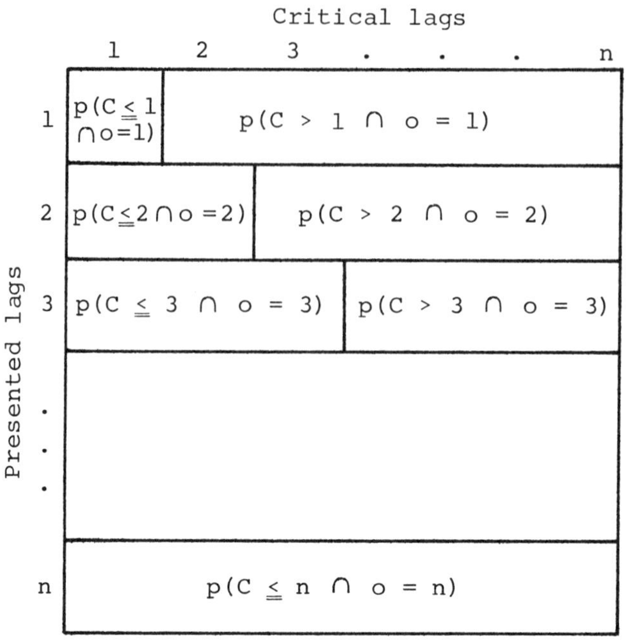

Figure 1. Tabulation of the available information on critical lags derived from observations (see text). These are classified into intervals 1 ... n (time or distance units).

4. Model III - Inconsistent Behaviour

For the third general model about the gap acceptance behaviour of the individual driver, where probability p is assigned independently to a presented lag or gap, the same methods of fitting distributions to the proportions of acceptances have been used. In this case, different drivers are not taken into consideration, and the distribution represents the behaviour of an "average inconsistent driver". Several authors have pointed out that the probability distribution functions of gap acceptance created under the assumptions of consistent and inconsistent behaviour are identical. That is, if a probability of a lag or gap t being accepted is p, p is also the probability that a driver has a critical lag or gap less than or equal to t. Again, the empirically derived proportions of acceptance classified into bands, which are used as the distribution of acceptances, impose restrictions upon this distribution, as far as its interpretation is concerned, similar to the case analysed in section 3.

The actual difference between consistent and inconsistent behaviour is in the generation of the gap acceptance decision-making process. In simulation models, for example, a consistent driver has to wait until an offered gap is greater than or equal to his pre-assigned critical gap, while an inconsistent one might reject a lag of 10 seconds and immediately after accept a gap of 3 seconds. The probability that a driver will accept a lag, size t, under the two models is the same, but, if this lag is rejected, the probabilities of acceptance of the following presented gaps are different.

Adopting, however, the conditional probability of acceptance discussed in section 3, then, if a driver rejects a presented lag of size t, the model is committed to process this particular driver. A reasonable assumption, in this case, might be that since the driver rejected a lag t then his vehicle will be stationary and he will reject all gaps less than t. Hence, in the model, presented gaps less than t should no longer be considered for this driver; he has to decide, therefore, independently about each presented gap greater than or equal to t, and not necessarily accept the first gap larger than a pre-assigned critical gap.

5. The Reality and the Problems of Working with It

5.1 The reality

Consider a population of drivers in a non-priority stream. The gap acceptance behaviour of each individual driver, in terms of his minimum acceptable (critical) lag or gap is a random variable X_i; that is, his minimum acceptable lag or gap varies because of his inability to detect small differences, and perhaps because of many other reasons, such as the degree of concentration at the moment of judgment. Let F_i be the corresponding distribution function with mean μ_i and standard deviation σ_i. The existence of the X_i, F_i, μ_i, σ_i, which would be different for different i's , has implicitly been shown by the work of others, as referred to in section 3.

Thus, if empirical data were actually collected on the minimum acceptable lags or gaps, and that is possible only in test-track experiments, then, if certain conditions are satisfied, and as the sample size $n \to \infty$, the $F_i \to F$, where F can be any distribution which describes the empirical data adequately.

However, the nature of field data, where the accepted lags or gaps do not necessarily correspond to minimum acceptable lags or gaps, dismisses the possibility of convergence of the frequencies of the accepted lags or gaps to a distribution F of the minimum acceptable (critical) lags or gaps.

5.2 All available data versus selected data

If the objective of a traffic researcher is to study an operational characteristic of an intersection which depends on the gap acceptance behaviour, then obviously data have to be collected on that particular behaviour. It has been mentioned before in this paper that, in real life experiments, only two types of lags or gaps can be observed, accepted and rejected, which together are the presented lags or gaps. Whatever models are to be developed, or conclusions to be drawn, they have to originate from the information available. Naturally, researchers want to use all information at their disposal in building a model, because the amount of data they have is always less than the amount they need. And that is exactly the case here, where practically nothing is known reliably about any individual driver.

However, sometimes, in special cases the use of only a part of data available provides more freedom for the researcher, in terms of mutual communication between himself and his model. For example, in comparison studies of the gap acceptance behaviour of driver populations it may be necessary to support the conclusions derived from the existing gap acceptance model with an independent and close examination of one or two of the three types of observable lags or gaps (accepted, rejected, presented).

5.3 Accepted lags or gaps distribution and conditional probabilities

An observed acceptance at a particular intersection, under certain environmental and other conditions, is a result of a combination of two independent random variables: the presented lags or gaps in the major road, and the minimum acceptable (critical) lags or gaps of the turning driver.

However, observed accepted lags or gaps at an intersection constitute a random variable in their own right.

Figure 2 shows a typical frequency distribution of accepted lags and gaps combined, classified into one-second intervals, for a busy but uncongested intersection. As can be seen, the distribution is essentially unimodal, which is to be expected because fewer drivers accept small lags or gaps, even though these are frequently presented, and also because very few drivers wait for very large lags or gaps, which in any case are not very frequent.

When comparison studies are being made, the distributions of accepted lags or gaps have to be compared, as do the conditional probabilities for each time or distance band, between the different cases, with the exception of the case where the distributions of presented lags or gaps are the same. In this last case either can be used. This is illustrated by the following example cases.

Suppose that in comparing two different intersections,

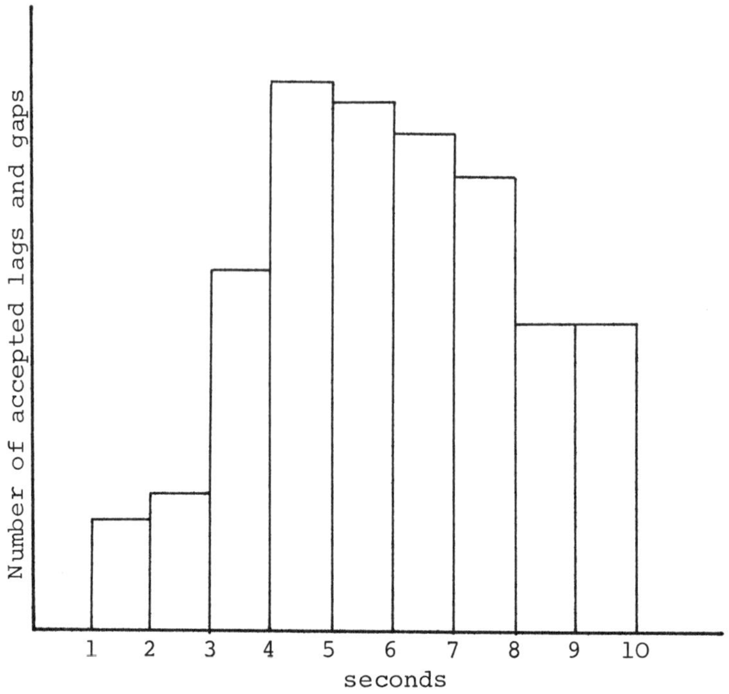

Figure 2. Frequency distribution of accepted lags and gaps combined, classified into 1 second intervals for $1 \leq t \leq 10$ seconds.

(a) and (b), the frequency distributions of the presented lags or gaps are found to be significantly different, i.e.

$$\frac{A_{ia} + R_{ia}}{N_a} \neq \frac{A_{ib} + R_{ib}}{N_b} \quad \text{for some or all } i\text{'s where}$$

A_{ia}, A_{ib} are the numbers of accepted lags or gaps in the interval i for (a) and (b) respectively, R_{ia}, R_{ib} the corresponding numbers of rejected lags or gaps and N_a, N_b the total number of presented lags or gaps for (a) and (b) respectively. Then if the conditional probabilities are the same for (a) and (b), that is if

$$\frac{A_{ia}}{A_{ia} + R_{ia}} \equiv \frac{A_{ib}}{A_{ib} + R_{ib}}, \text{ then } \frac{\frac{A_{ia}}{N_a}}{\frac{A_{ia} + R_{ia}}{N_a}} \equiv \frac{\frac{A_{ib}}{N_b}}{\frac{A_{ib} + R_{ib}}{N_b}}, \text{ which}$$

implies $\frac{A_{ia}}{N_a} \neq \frac{A_{ib}}{N_b}$, because the denominators (i.e. the proportions of presented lags or gaps for i) are different. This means that if, for example, the flow is higher in (a) than (b) and consequently more small lags or gaps will be presented in the case of (a) and also the conditional probabilities of acceptance are the same for both (a) and (b) for some i's, then any difference in the overall performance of the intersection in terms of lags or gaps accepted will be detected by the distribution of accepted lags or gaps.

Similarly, if the distributions of presented lags or gaps are different between (a) and (b) and the distributions of accepted lags or gaps are the same, then the conditional probabilities have to be different. In the cases where the presented lags or gaps distributions are different, as well as either the conditional probabilities or the accepted lags or gaps distributions are different, then a much more careful analysis is required to find the relationship between the distributions of accepted lags or gaps or the conditional probabilities respectively.

However, it is the cases where the distributions of presented lags or gaps are the same which are most appropriate for the study of other possible factors influencing gap acceptance behaviour, such as age and sex of drivers, type of vehicles and daylight-darkness.

In these cases any difference in gap acceptance behaviour detected by the conditional probabilities will be detected also by the accepted lag or gap distributions and vice-versa, e.g. if $\frac{A_{ia} + R_{ia}}{N_a} \equiv \frac{A_{ib} + R_{ib}}{N_b}$ and

$\frac{A_{ia}}{A_{ia} + R_{ia}} \circledast \frac{A_{ib}}{A_{ib} + R_{ib}}$ where $\circledast$ can be any of $\equiv$ or $<$ or $>$, then from $\frac{A_{ia}}{A_{ia} + R_{ia}} \circledast \frac{A_{ib}}{A_{ib} + R_{ib}}$ it follows that

$$\frac{\frac{A_{ia}}{N_a}}{\frac{A_{ia} + R_{ia}}{N_a}} \circledast \frac{\frac{A_{ib}}{N_b}}{\frac{A_{ib} + R_{ib}}{N_b}}$$ and hence that $\frac{A_{ia}}{N_a} \circledast \frac{A_{ib}}{N_b}$.

Furthermore, theoretical distributions can be fitted to the frequency distributions of accepted lags, and a truncated Normal distribution has been successfully fitted to a large number of sets of data. This offers the facility of easy testing for various percentiles of the fitted distribution, thereby overcoming the inflexibility of the probit transformation, which in order to be used in comparison studies requires the slopes of the linearly represented cumulative Normal distributions to be the same.

Obviously, a number of theoretical distributions may be adequate in describing the observed frequencies of accepted lags or gaps, and, in addition to the Normal, a truncated Weibull distribution has been found to provide a very satisfactory fit.

The mathematical form of the fit is not very important in itself, since the fitted distributions do not support any assumptions about how the model represents a real life situation. The task of the distributions is to describe the data as adequately as possible, for example for simulation purposes. What is important is whether or not the accepted lags or gaps are themselves meaningful. The answer to that is that although they are not the solution to every problem concerning gap acceptance behaviour, their study, together with the study of the conditional probabilities of acceptance (for which confidence intervals are readily available through the binomial distribution), offers rewarding insights in the search for factors which positively or negatively influence the level of risk-taking by turning drivers.

References

1. W. F. ADAMS, "Road traffic considered as a random series," J. Inst. Civil Engineers, 121-130 (1936).
2. W. D. ASHTON, "Gap acceptance problems at a traffic intersection," Appl. Stat. 20, 130-138 (1971).
3. R. ASHWORTH, "A note on the selection of gap acceptance criteria for traffic simulation studies," Trans. Res. 2, 171-175 (1968).
4. R. ASHWORTH, "The analysis and interpretation of gap acceptance data," Trans. Sci. 4, 270-280 (1970).

5. R. ASHWORTH AND C. G. BOTTOM, "Some observations of driver gap acceptance behaviour at a priority intersection," Traff. Engng. Control, 18, 569-571 (1977).

6. D. E. BLUMENFELD AND G. H. WEISS, "The effects of gap acceptance criteria on merging delay and capacity at an uncontrolled junction," Traff. Engng. Control, 20, 16-20 (1979).

7. W. R. BLUNDEN, C. M. CLISSOLD AND R. B. FISHER, "Distribution of acceptance gaps for crossing and turning manoeuvres," Proc. Australian Road Research Board 1, Part 1, 188-205 (1962).

8. C. G. BOTTOM AND R. ASHWORTH, "Factors affecting the variability of drivers gap acceptance behaviour," Ergonomics 21, 721-734 (1978).

9. D. F. COOPER AND M. R. C. McDOWELL, "Police effects on accident risk at T-junctions," Traff. Engng. Control, 18, 486-487 (1977).

10. D. F. COOPER, P. A. STORR AND J. WENNELL, "Traffic studies at T-junctions: 4. The effect of speed on gap acceptance and conflict rate," Traff. Engng. Control, 18, 110-112 (1977).

11. J. DARZENTAS, V. HOLMS AND M. R. C. McDOWELL, "Driver behaviour at a T-junction in daylight and darkness," Traff. Engng. Control, 21, 186-189 (1980a).

12. J. DARZENTAS, M. R. C. McDOWELL AND D. F. COOPER, "Minimum acceptable gaps and conflict involvement in a simple crossing manoeuvre," Traff. Engng. Control, 21, 58-61 (1980b).

13. D. R. DREW, "Gap acceptance characteristics for ramp-freeway surveillance and control," Highway Research Record 157, 108-135 Washington, D.C., (1967).

14. W. L. GIBBS, "Driver gap acceptance at intersections," J. Appl. Psych. 52, 200-204 (1968).

15. R. HERMAN AND G. H. WEISS, "Comments on the highway crossing problem," Opns. Res. 9, 828-840 (1961).

16. D. R. McNEIL AND J. H. T. MORGAN, "Estimating minimum gap acceptances for merging motorists," Trans. Sci. 2, 265-277 (1968).

17. D. R. McNEIL AND J. T. SMITH, "A comparison of motorist delays for different merging strategies," Trans. Sci. 3, 239-254 (1969).

18. M. S. RAFF AND J. W. HART, A volume warrant for urban stop signs, Eno Foundation for Highway Traffic Control, Saugatuck, Connecticut, 1950.

19. J. B. H. RAMSEY AND I. W. ROUTLEDGE, "A new approach to the analysis of gap acceptance times," Traff. Engng. Control, 15, 353-357 (1973).

20. J. C. TANNER, "A theoretical analysis of delays at an uncontrolled intersection," Biometrika, 49, 163-170 (1962).

21. R. J. TROUTBECK, "A review of the Ramsey-Routledge method for gap acceptance times," Traff. Engng. Control, 16, 373-375 (1975).

22. G. H. WEISS AND A. A. MARADUDIN, "Some problems in traffic delay," Opns. Res. 10, 74-104 (1962).

Acknowledgements

The author is most grateful to Professor M. R. C. McDowell for his support and guidance and would like to acknowledge the very helpful advice of Professor W. R. Buckland and Mr. K. C. Bowen, and to thank Mr. S. J. Older for his constructive comments.

INFLUENCE OF VEHICLE SIZE AND PERFORMANCE ON INTERSECTION SATURATION FLOW

Leonard Evans
and
Richard W. Rothery
Transportation and Traffic Science Department
General Motors Research Laboratories
Warren, Michigan 48090

ABSTRACT

This research was performed to investigate how vehicle size and performance affect the saturation flow of overloaded signalized intersections. Data were collected on 15 138 vehicles discharging from a signalized intersection when there were always more vehicles waiting to pass through than the green phase would permit. Saturation flow was examined as a function of vehicle length and headway. Vehicle size was also characterized by number of axles. The performance of the lead vehicle for each cycle was characterized by the time it took, after starting from a stop, to clear the intersection.

On the basis of the data analyzed, it is concluded that:

1. Smaller cars give higher intersection saturation flow.
2. When a queue leader is a low performance vehicle, saturation flow is reduced. One vehicle per cycle is lost for each additional two seconds the lead vehicle takes to clear the intersection.
3. Vehicles with low performance reduce saturation flow.
4. Large trucks are over-represented in queue leader positions.
5. Automatic traffic control schemes to minimize total system fuel use should strive to reduce likelihood of large trucks being queue leaders.

6. The fraction of large trucks in the sample observed to "run the red" was larger than the corresponding fraction for passenger cars.

INTRODUCTION

The efficient flow of traffic at signalized intersections has long been recognized as an important factor in yielding overall efficient traffic flow in urban areas. Many theoretical and experimental studies have been conducted on various aspects of this problem (see, for example, references in Pignataro [1]). In the present experimental investigation, detailed information on 15 138 vehicles discharging from a traffic signal was obtained using data collecting traps consisting of two tape switches placed on a roadway. The main aim was to study how vehicle size and performance influence intersection saturation flow.

A number of studies have addressed effects due to the size of passenger automobiles. Test track experiments [2] suggested that small cars yield higher intersection saturation flow than standard-sized cars. Observation of traffic at 12 signalized intersections in the metropolitan area of Toronto found higher saturation flow to be associated with small cars [3]. The magnitudes of the car-size effects in the above studies [2,3] are larger than would result solely from physical differences in car size, suggesting that the small cars are driven differently. Driver behavior studies support this contention. Reduced saturation flow resulted when the driver's view of the roadway was partially masked [4]. The amount of roadway visible to a driver following another vehicle influences his judgment of intervehicular spacing and may cause him to adopt different spacings when driving different cars even if he desires to maintain the same spacing [5]. This effect generally is in the direction of drivers of smaller cars following more closely, which would tend to increase intersection saturation flow for smaller cars. In the present study we seek additional information on the effect of small cars; we also investigate lead vehicle acceleration and effects due to heavy vehicles. Some characteristics of vehicular platoons discharging when traffic signals turn green are also discussed.

To obtain traffic discharge as free as possible from extraneous variability, a site with the following characteristics was sought.

1. The intersection signal should remain over-saturated for a sufficient time to allow efficient data collection.

2. The roadway downstream from the signal should be sufficiently clear to allow relatively unimpeded discharges.

3. All turns should be prohibited. Even if exclusive turning lanes are provided, vehicles entering them when the queue is discharging can leave gaps in the through lanes which influence capacity.

4. As little interference as possible from cross-traffic should occur.

Many intersections were examined with respect to these criteria, but none found which satisfied all of them. As a compromise, the site described in the next section was selected.

DATA COLLECTION AND REDUCTION

A schematic of the site, which was in a suburban location 25 km north of the center of Detroit, is shown in Figure 1. The traffic signal, which had a 41 s green, 4 s yellow and 15 s red phase, controlled egress from a factory. Traffic in the two northbound lanes* of the four lane roadway was studied. The speed limit was 45 miles per hour (72 km/h). The stop line in Fig. 1 indicates where vehicles

*This stretch of roadway now has 6 lanes -- our data were collected in mid 1972.

typically stopped -- no stop line was marked on the roadway. The vehicles were measured 15 m beyond the stop line, a distance representative of that required to clear a typical intersection. This choice yields information on, for example, the effects of vehicular acceleration that is not available at a position a meter or so beyond the stop line, a position chosen for headway measurements in a number of earlier studies [3, 6, 7].

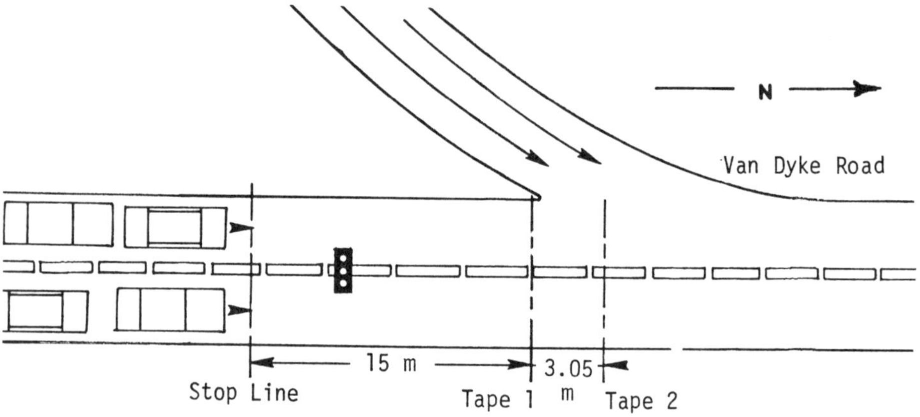

Figure 1. The site.

Data were collected in each lane using a "trap" consisting of a pair of tape switches. These are conducting strips separated by a compressible insulator. A vehicle crossing the tape closes the switch. Each vehicle (or, more generally, each pair of axles) yields four time events when it crosses the two tape switches in its lane. These were synchronously recorded using a multiple channel tape recorder. From these four recorded times the speed, acceleration and wheelbase of the vehicle can be estimated (for a similar approach, see ref. 8).

Calculation of Wheelbase, Speed and Acceleration

Consider a vehicle, wheelbase W, crossing a "trap" of length D, as shown in Fig. 2. Let the front wheels strike the first and second tapes at times t_1 and t_2 respectively; let the rear wheel strike the first and second tapes at t_3 and t_4 respectively.

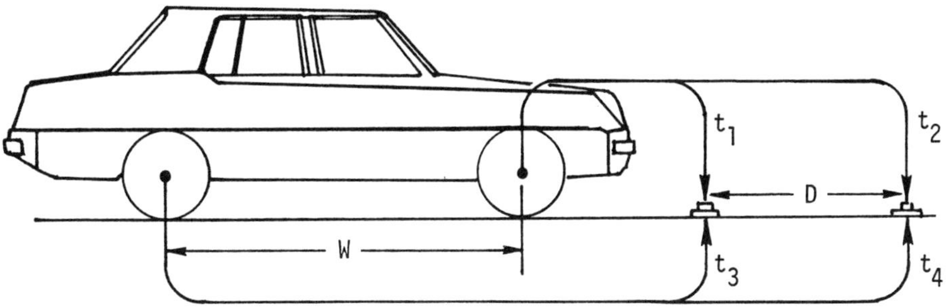

Figure 2. Definition of the four time events.

Let the position of the front axle of the vehicle be given by

$$x = f(t) \tag{1}$$

Then, the rear axle position is given by

$$x_R = f(t) - W \tag{2}$$

A space-time diagram with arbitrarily positioned tape switches is shown in Fig. 3. If the coordinates are transformed so that the position origin is at the first tape and the origin of time is when the first axle crosses the first tape switch, then, with upper case letters denoting the transformed coordinates, we obtain

$$T_1 = t_1 - t_1 = 0, \tag{3}$$

$$T_2 = t_2 - t_1, \text{ etc.} \tag{4}$$

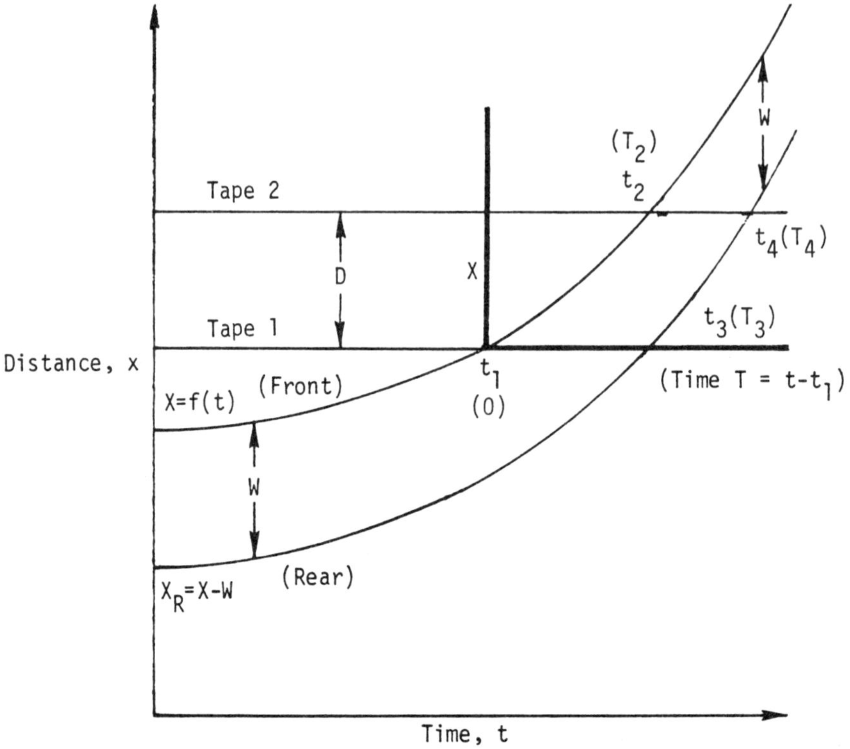

Figure 3. Speed-time history of vehicle traversing two-tape trap.

Let us represent the trajectory of the vehicle by a polynomial in T, and ignore third and higher order terms in T. That is, the highest order effect within the trap is assumed to be a uniform acceleration. Then, because X = 0 at T = 0, we can write

$$X = X(T) = bT + cT^2 \tag{5}$$

Substituting the time and distance values for the other three events into Equation 5 yields

$$D = bT_2 + cT_2^{\,2} \tag{6}$$

$$W = bT_3 + cT_3^2 \quad \text{and} \tag{7}$$

$$D + W = bT_4 + cT_4^2 \tag{8}$$

Solving these equations for b, c, and W gives:

- The speed of the vehicle, v, on entering the trap (that is, b from equation (5)) is

$$v = D \frac{T_2 + T_3 - T_4}{T_2(T_4 - T_3)(T_3 + T_4 - T_2)} \tag{9}$$

- The average acceleration, a (2c from equation (5)), is

$$a = 2D \frac{T_2 + T_3 - T_4}{T_2(T_4 - T_3)(T_3 + T_4 - T_2)} \tag{10}$$

- The wheelbase, W, is

$$W = \frac{DT_3}{T_2} \frac{(T_4 - T_2)(T_2 + T_4 - T_3)}{(T_4 - T_3)(T_3 + T_4 - T_2)} \tag{11}$$

Note that if the front wheels strike the second tape at the same time as the rear wheels strike the first tape (i.e., $T_2 = T_3$), one always correctly infers from equation (11) that W = D even if f(T), contrary to our simplifying assumption, contains third or higher order terms. This would argue in favor of choosing a value of D close to typical values of W.

The value D = 3.05 m was used for the trap length throughout the experiment.

Procedure

In order to check equipment and procedures, traps were laid on two lanes of a non-public road and a subcompact car was driven 18 times through the right lane trap and 17 times through the left lane trap at various speeds and accelerations. The standard deviations of the wheelbase measurements were 6 mm for the right lane and 7 mm for the left lane. This precision, better than 0.3%, indicates high potential of this technique for vehicle identification.

The traps were laid at the observation site just prior to the p.m. peak period. A major problem with tape switches is reliability. After many passages of very heavy vehicles, tape switches are liable to permanently close, and many such failures were encountered. If the trap in one lane became inoperative, data collection continued in the other. Usable data were obtained for 6 afternoon (about 3:45 p.m. to 5:15 p.m.) peak periods.

The authors observed the discharging traffic from a car parked on the soft shoulder near the stopping line. This car also contained the instrumentation. It was necessary to note all vehicles with more than two axles (we will refer to such vehicles as many-axle vehicles) because the instrumentation cannot discriminate between, say, certain vehicles with four axles and two closely following vehicles with two axles each. Untypical vehicles (motorcycle, car hauling boat, etc.) and easily recognizable vehicle makes and models were also noted.

Data Set

The analysis was performed using a data set containing the following information for each of 15 138 vehicles (see Table 1):

- right or left lane used
- number of axles
- particular model or type of vehicle if available from notes

- queue position
- arrival time
- time front axle reaches first tape switch after the onset of the green phase
- clearing time
- time last axle reaches first tape switch after the onset of green phase
- speed on entering the trap
- acceleration
- total wheelbase

TABLE 1. DATA SUMMARY

	RIGHT LANE	LEFT LANE
NUMBER OF TRAFFIC SIGNAL CYCLES (=NUMBER OF QUEUE LEADERS)	408	381
NUMBER OF VEHICLES NOT QUEUE LEADERS	7230	7119
TOTAL NUMBER OF VEHICLES	7638	7500
NUMBER OF VEHICLES WITH THREE OR MORE AXLES	229	12

In addition, the following two quantities were calculated for each vehicle that was not a queue leader:

Headway - defined here as the time between the rear axle of the previous vehicle crossing the first tape and the rear axle of the vehicle in question crossing the first tape. This headway corresponds approximately to how much green time the vehicle consumes.

Gap - defined here as the time between the rear axle of the previous vehicle crossing the first tape and the front axle of the vehicle in question crossing the first tape. Apart from the question of how far vehicles extend beyond their front and rear axles, and the fact that the vehicles were not measured simultaneously, this gap is determined by the chosen following distance and speed of each driver.

RESULTS

Effect of Vehicle Size on Gaps and Headways

In order to study effects associated with vehicle size, the vehicles were divided into seven wheelbase categories as shown in Table 2. The boundaries of these wheelbase categories were chosen to delineate, as far as possible, different body classifications. These wheelbase intervals generally contained a local maximum from the wheelbase distribution. For both lanes, the mode of the wheelbase distribution was at 3.07 m (121 inches). The representative vehicle given in Table 2 for each category is a typical

TABLE 2. GAPS AND HEADWAYS (SECONDS) FOR VEHICLES WITH DIFFERENT SIZE WHEELBASES

SIZE CATEGORY	TYPICAL VEHICLE TYPE	WHEELBASE RANGE, METERS	RIGHT LANE				LEFT LANE			
			N	GAP S.D.	HEADWAY S.D.	pce	N	GAP S.D.	HEADWAY S.D.	pce
1	MOTORCYCLE, ETC.	<2.35	123	1.75 (1.02)	1.92 (1.04)	0.96	95	1.61 (0.90)	1.76 (0.91)	0.88
2	SUBCOMPACT CAR	2.35 to 2.58	635	1.68 (0.73)	1.90 (0.72)	0.95	729	1.75 (0.90)	1.94 (0.89)	0.97
3	COMPACT CAR	2.58 to 2.86	1632	1.73 (0.68)	1.98 (0.68)	0.99	1540	1.79 (1.04)	2.00 (1.04)	1.00
4	INTERMEDIATE CAR	2.86 to 3.01	1424	1.77 (0.77)	2.03 (0.77)	1.01	1388	1.86 (1.34)	2.09 (1.33)	1.04
5	LARGE CAR	3.01 to 3.21	2517	1.77 (0.81)	2.05 (0.80)	1.02	2700	1.78 (0.93)	2.01 (0.92)	1.00
6	LIGHT TRUCK, VAN	3.21 to 3.81	609	1.93 (0.88)	2.22 (0.88)	1.10	633	1.85 (0.98)	2.10 (0.98)	1.04
7	HEAVY TRUCK, ETC.	>3.81	290	3.65 (1.92)	4.69 (2.18)	3.33	34	3.17 (2.50)	3.68 (2.63)	1.83
	ENTIRE POPULATION		7230	1.84 (0.93)	2.13 (1.02)	1.06	7119	1.81 (1.06)	2.03 (1.06)	1.01

member of that category. However, each category contains a variety of vehicle types. For example, category 6 contained some very large cars.

Vehicles in categories 2 through 5 are predominantly passenger cars, including station wagons. For these, clear cut systematic average gap size differences dependent on vehicle size are not apparent from the data in Table 2. Nonetheless, the average gap for category 2 vehicles is less than that for category 4 and category 5 vehicles in both right and left lanes. Indeed, a two-tailed t-test applied to individual pairs of differences without regard to the others yields the following statistically significant differences:

Right lane: average category 2 gap < average category 4 gap ($p < 0.02$);
average category 2 gap < average category 5 gap ($p < 0.02$).

Left lane: average category 2 gap < average category 4 gap ($p < 0.05$).

As the average speed of vehicles entering the trap was 45 km/h, a gap difference of 0.1 s corresponds typically to 1.2 m, suggesting that category 2 drivers chose spacings between their front axle and the vehicle they were following about a meter less than the corresponding spacing for category 4 and 5 drivers. Some portion of this difference is attributable to the trend of increasing front-axle to front-bumper dimension for cars of larger size. Therefore, the present data offer at most a suggestion that drivers of smaller size vehicles adopt smaller intervehicular spacings.

The green time consumed by a vehicle is, approximately, what we have defined as headway; namely, gap plus the time taken for the vehicle to travel the distance of its own wheelbase. The data in Table 2 show that headway increases with vehicle size so that flow, which is the reciprocal of headway, decreases with vehicle size. For example, the right lane flow during the discharge for category 2 vehicles (1895 veh/h) is 7.9% greater than that for category 5 vehicles (1756 veh/h); the corresponding left lane difference is 3.6%. Because of a general increase in rear-axle to rear-bumper distance with increasing car size, these differences are less than would be calculated by using the more appropriate time the vehicle takes to travel a distance equal to its length rather than its wheelbase.

Vehicles in categories 6 and 7 had larger gaps than those in the other categories. This probably arises because, with lower power to weight ratios, they were on balance unable to match the acceleration of vehicles they were following.

The average headways of these larger vehicles are now compared to the average headway of all the vehicles in categories 2 through 5 (i.e. 2.01 s) which we take as the average passenger car headway. For both lanes combined, category 6 and 7 vehicles had average headways 2.16 s and 4.58 s, respectively. Thus, for non-queue leader positions, a category 6 vehicle corresponds to 2.16/2.01 = 1.07 passenger car equivalents (pce) (for a more detailed discussion of pce, see, e.g., p. 353 of Reference 1). The corresponding value for a category 7 vehicle is 4.58/2.01 = 2.28 pce.

All the many-axle vehicles in our sample were category 7. Indeed, almost all category 7 vehicles were many-axle. It has been indicated elsewhere that the number of axles is a good parameter to characterize the weight of heavy trucks, both when loaded and unloaded [9]. In the present paper we examine a number of effects of vehicles discharging from a traffic signal in terms of number of axles.

Average headway and pce inferred from headway, together with other quantities to be discussed later, are given in Table 4 as a function of the number of axles of many-axle vehicles (also see Figure 4). It is apparent that headway generally increases with the number of axles; the average

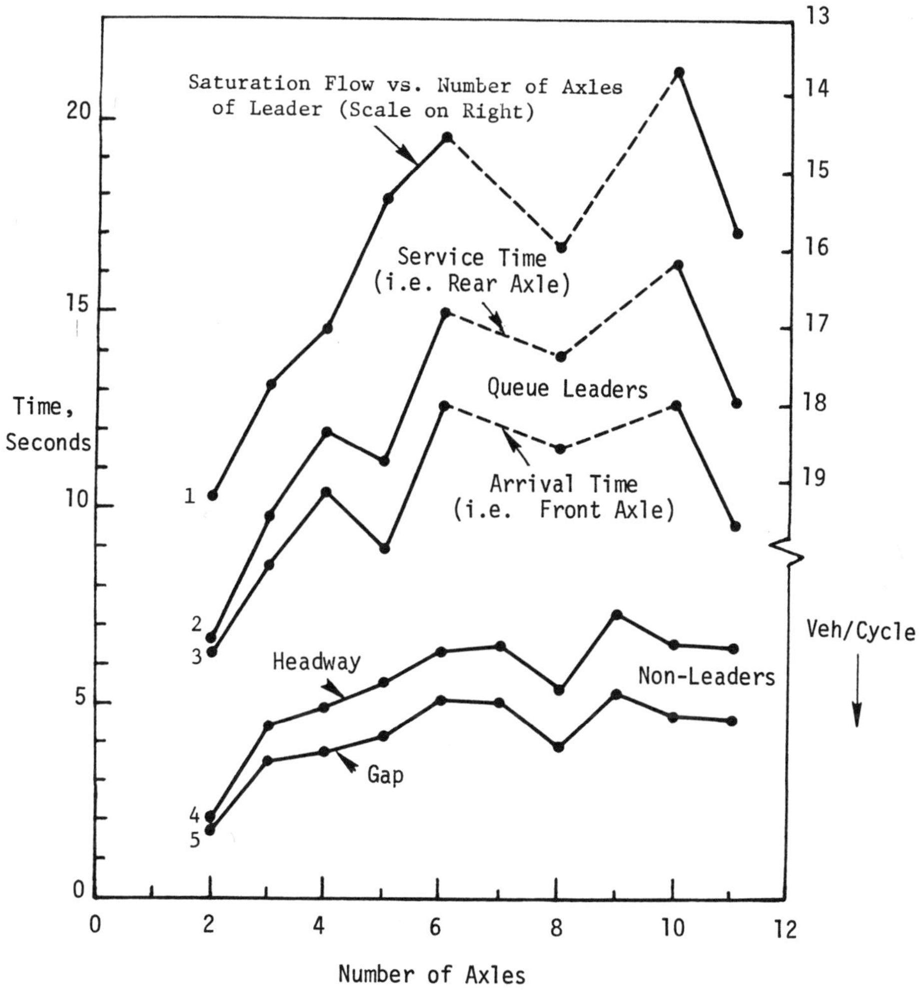

Figure 4. From top to bottom

1. Saturation flow versus number of axles of queue leader.

2 (and 3). Queue leader service time (arrival time) (i.e., time required after onset of green for last (first) axle to reach tape 1) versus number of axles.

4 (and 5). Headway (gap) averaged over all non-leader queue positions for vehicles with different numbers of axles.

headway of many-axle vehicles in a non-queue leader position correspond to 2.63 pce.

Effect of Queue Leaders on Throughput

The average saturation flow (i.e., vehicles per cycle with 45 s green plus yellow) that occurred when vehicles in the different size categories were queue leaders is shown in Table 3. No systematic effects due to size category of leaders are apparent when the leader is a member of categories 2 through 5 (i.e., a passenger car). The average throughput associated with passenger car leaders is 19.05 veh/cycle for the right lane and 19.71 veh/cycle for the left lane. When a category 6 vehicle rather than a passenger car is a leader, throughput is reduced by 0.63 veh/cycle in the right lane and 0.34 veh/cycle in the left lane. To express these throughput reductions in terms of pce we note that (see Table 2) an average non-queue leader vehicle in the right lane is equivalent to 2.13/2.01 = 1.06 pce. The corresponding number for the left lane is 2.03/2.01 = 1.01 pce. Using these values, we infer that, as

TABLE 3. SATURATION FLOW FOR DIFFERENT SIZE QUEUE LEADERS

SIZE CATEGORY	TYPICAL VEHICLE TYPE	WHEELBASE RANGE, METERS	RIGHT LANE: NO. OF CYCLES	RIGHT LANE: AVERAGE ARRIVAL TIME, s	RIGHT LANE: AVG. CAPACITY VEH/CYCLE	LEFT LANE: NO. OF CYCLES	LEFT LANE: AVERAGE ARRIVAL TIME, s	LEFT LANE: AVG. CAPACITY VEH/CYCLE
1	MOTORCYCLES, ETC.	<2.35	7	6.96	18.57	4	6.94	20.50
2	SUBCOMPACT CAR	2.35 to 2.58	33	6.31	19.30	32	5.94	19.25
3	COMPACT CAR	2.58 to 2.86	79	6.35	18.70	85	6.01	19.60
4	INTERMEDIATE CAR	2.86 to 3.01	80	6.29	19.09	73	5.98	20.01
5	LARGE CAR	3.01 to 3.21	146	6.28	19.15	152	5.91	19.72
6	LIGHT TRUCK, VAN	3.21 to 3.81	36	6.60	18.42	35	6.07	19.37
7	HEAVY TRUCK, ETC.	>3.81	27	9.61	16.67	0	--	--
	ENTIRE POPULATION		408	6.56	18.72	381	5.97	19.68

far as the directly measured effect on capacity is concerned, a category 6 vehicle queue leader is equivalent to 1.67 pce in the right lane and 1.34 pce in the left lane. The difference between the lanes might suggest that drivers of category 6 vehicles with comparatively higher acceleration capabilities were choosing the left lane. The values for both lanes are larger than the previously mentioned value of 1.07 pce of a category 6 vehicle that was not a queue leader.

There were no category 7 queue leaders in the left lane data. The capacity when a category 7 vehicle was queue leader in the right lane is 2.38 veh/cycle less (i.e., 19.05 - 16.67 from data in Table 3) than when a passenger car was leader. A category 7 queue leader is hence equivalent to 1 + 2.38 x 1.06 = 3.52 pce when it is a queue leader as compared to previously noted value of 2.28 pce when it is not a queue leader. Thus the capacity penalty for a category 7 vehicle is more than 1 pce greater when it is a queue leader than when it is a non-queue leader.

More detailed information on this effect is given in Table 4, which shows pce equivalents of vehicles with given numbers of axles in lead and non-lead positions.

Intersection capacity is now analyzed in terms of the arrival time (that is, the time after onset of green to travel 15 m) of the lead vehicle. This time is related to the average acceleration initially used to clear the intersection -- longer times being associated with lower acceleration. Average capacity versus the arrival time of the lead vehicle (in 0.4 s cells) is shown in Figure 5 for both lanes combined. Only points that are averages of three or more observations are plotted; a few individual cycles showed erratic departures from the trend in Figure 5. The line is an unweighted least squares fit to the points plotted. The reciprocal of the slope of this line indicates that the capacity is reduced by one veh/cycle for each additional 1.92 s delay that the lead vehicle takes to reach the clearing line. This is consistent with the usual interpretation (see, e.g., p. 351 of Reference 1) that a lead car delay consumes effective green time that would otherwise be available to serve vehicles at the previously determined rate of one passenger car per 2.01 s. Note that the arrival time that corresponds to zero capacity for the regression equation given in Figure 5 is 43.4 s, a value close to the actual effective green time of 45 s for the particular signal being studied.

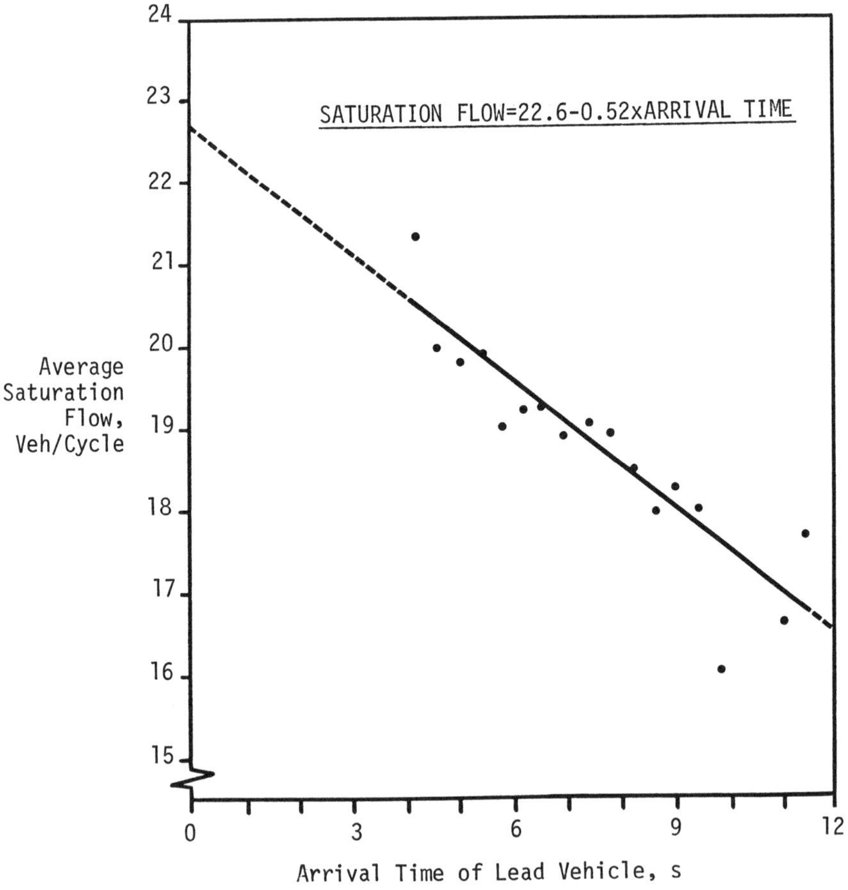

Figure 5. Saturation flow versus arrival time of lead vehicle. The data are grouped in 0.4 second cells. Only cells with three or more data are plotted. The line is an unweighted least squares fit to the points plotted.

Arrival times for different size queue leaders are shown in Table 3. For passenger car queue leaders (categories 2 through 5), there are no effects dependent on car size. However, arrival times are systematically longer for the right lane than the left lane for all size categories, suggesting that faster accelerating drivers choose the left lane.

Equations expressing average instantaneous acceleration of different classes of lead vehicles in terms of time since starting to move are given by Dockerty [10]. Our arrival times (i.e. 6.30 s for passenger cars and 9.61 s for category 7 vehicles), after subtracting 1.5 s to represent a delay between onset of green and the vehicle's starting to move, were substituted into Dockerty's (10) Equation 4 (for cars) and Equation 5 (for "heavies"), to yield the following acceleration estimates: 1.01 m/s^2 for cars and 0.29 m/s^2 for heavies. These may be compared to our directly measured average values of acceleration at the trap for cars and category 7 vehicles, which were 1.38 m/s^2 and 0.39 m/s^2, respectively. The essentially instantaneous value of acceleration measured at the trap does not provide a reliable indication of average acceleration since starting for individual vehicles, especially large vehicles. For example, a gear change by an "accelerating" vehicle while traversing the trap can even lead to a negative acceleration measurement within the trap. However, this effect does not invalidate the comparison of averages given above. A better indicator of individual vehicle performance is arrival time, which is the variable used in Figure 5.

Another approach to determining saturation flow reduction due to lead vehicles is to examine clearing times of different queue leaders. We confine this discussion to the 22 vehicles in the right lane that had more than two axles (all are category 7). The data are shown in Table 4 and plotted in Figure 4. By subtracting the average clearing time for a passenger car queue leader in the right lane (namely, 6.60 s) from these and dividing by 2.01 s the losses in saturation flow shown in Table 4 were determined. The pce of a leader derived from the clearing time is compared to that derived from the directly measured reduction in saturation flow. The two values are in relatively good agreement, and the average is taken as the best estimate of the saturation flow reduction associated with a lead vehicle with a given number of axles. This average value is compared in Table 4 with the pce of vehicles with the same number of axles that were not queue leaders. It can be seen that an additional capacity reduction of 1.42 pce is associated with a many-axle vehicle being a queue leader as compared to its not being a queue leader. This penalty is compounded by the fact that the probability of a random many-axle vehicle in the traffic stream becoming a queue leader is greater than the fraction of many-axle vehicles in the traffic stream, as shown below.

TABLE 4. SATURATION FLOW REDUCTIONS FOR MANY-AXLE VEHICLES IN QUEUE LEADER AND NON-QUEUE LEADER POSITIONS

	QUEUE LEADERS						NON-QUEUE LEADERS			
NO. OF AXLES	N	DIRECTLY OBSERVED VEH/CY	pce FROM VEH/CY	CLEARING TIME, s	pce FROM CLEARING TIME	AVERAGE pce FOR QUEUE-LEADER	N	AVERAGE HEADWAY s	pce FROM HEADWAY	ADDNL. PENALTY WHEN A QUEUE LEADER (pce)
3	6	17.67*	2.46	9.79*	2.58	2.52	65	4.47*	2.22	0.30
4	3	17.00	3.17	12.07	3.72	3.44	57	4.95	2.46	0.98
5	3	15.33	4.94	11.18	3.28	4.11	28	5.56	2.76	1.35
6	2	14.50	5.82	14.95	5.15	5.49	10	6.42	3.19	2.30
7	0	---	---	---	---	---	7	6.58	3.27	---
8	1	16.00	4.23	13.79	4.58	4.40	5	5.37	2.67	1.73
9	0	---	---	---	---	---	3	7.30	3.63	---
10	3	13.67	6.70	16.31	5.83	6.27	5	6.59	3.28	2.99
11	4	15.75	4.49	12.71	4.04	4.26	27	6.47	3.22	1.04
ALL	22	16.0	4.23	12.36	3.87	4.05	207	5.29	2.63	1.42

*THE CORRESPONDING VALUES FOR PASSENGER CARS ARE 19.05 VEH/CYCLE, 6.60 s AND 2.01 s, RESPECTIVELY.

Does Probability of Being Queue Leader or Last Vehicle Depend on Vehicle Size?

Here we address the question: do all types of vehicles have an equal likelihood of becoming queue leaders or last vehicles? If so, then the number of queue leaders of a certain type of vehicle would be proportional to the number of such vehicles in the traffic stream. To address this question passenger cars (i.e. categories 2 - 5) are compared to many-axle vehicles (see Table 5). Because category 1 and 6 vehicles and category 7 vehicles with 2 axles are not included in Table 5, the two columns do not sum to the total sample values in Table 1.

TABLE 5. PROBABILITY OF RANDOM PASSENGER CARS AND RANDOM MANY AXLE VEHICLES IN TRAFFIC STREAM BEING IN FIRST AND LAST POSITION IN RIGHT LANE.

	PASSENGER CARS (CATEGORIES 2 THROUGH 5)	VEHICLES WITH THREE OR MORE AXLES
TOTAL NUMBER OF VEHICLES	6546	229
NUMBER OF QUEUE LEADERS	338	22
PERCENT OF VEHICLES THAT WERE QUEUE LEADER	5.16%	9.61%
NUMBER OF LAST VEHICLES	339	23
PERCENT OF VEHICLES THAT WERE LAST VEHICLE	5.18%	10.04%
AVERAGE HEADWAY (OF NON-QUEUE LEADERS)	2.01 s	5.29 s
AVERAGE GAP (OF NON-QUEUE LEADERS)	1.75 s	4.02 s

Of the total of 6546 passenger cars in the right lane, 338 were queue leaders and 339 were last vehicles, corresponding to 5.16% and 5.18% respectively. That is, a passenger car in the traffic stream had a 5.2% possibility of becoming a queue leader or a last vehicle. Of 229 many-axle vehicles, 22 were queue leaders and 23 were last vehicles. That is, the probability of a many-axle vehicle in the traffic stream becoming a queue leader was 9.16%, and of becoming a last vehicle was 10.04%. Applying a chi-squared test to 2 x 2 contingency tables shows that both these proportions differ from the passenger car proportion at $p < 0.01$.

These probabilities imply that a random many-axle vehicle in the traffic stream is 9.16/5.16 = 1.86 times as likely to become a queue leader and 10.04/5.18 = 1.94 times as likely to become a last vehicle as is a random passenger car.

The probability that a vehicle is stopped by the onset of red to become queue leader of the next cycle should be, other factors being equal, proportional to its headway*. The headway of a many-axle vehicle is 2.63 times (see Table 4) that of a passenger car, considerably more than the 1.86 ratio in which many-axle vehicles were over-represented as queue leaders.

A simple model of the discharge process suggests that the probability of a particular vehicle becoming a last vehicle should not depend on its headway. Hence some additional factor is needed to explain the over-representation of many-axle vehicles in the last position and the fact that their over-representation in the lead position is less than proportional to that indicated by their headways.

The average arrival and clearing times of last vehicles in various categories, including number of axles for many-axle vehicles, is shown in Table 6. Fourteen last vehicles, all two-axle, judged at time of data collection to be not followed by another vehicle, are excluded from the data in Table 6. The average arrival time for many-axle vehicles is 0.80 s later than that for passenger cars. A t-test shows this difference to be significant at $p < 0.05$. Hence, many-axle vehicles showed a greater tendency to enter the intersection after the onset of the red phase.

The clearing times which include the effect due to vehicle length increase with number of axles (see Table 6 and Figure 6). Average clearing time for many-axle vehicles

*Assumes goal is that entire vehicle clears intersection by onset of red. The discussion is similar if gap rather than headway is used, the resulting ratio, 2.30, being somewhat closer to 1.86.

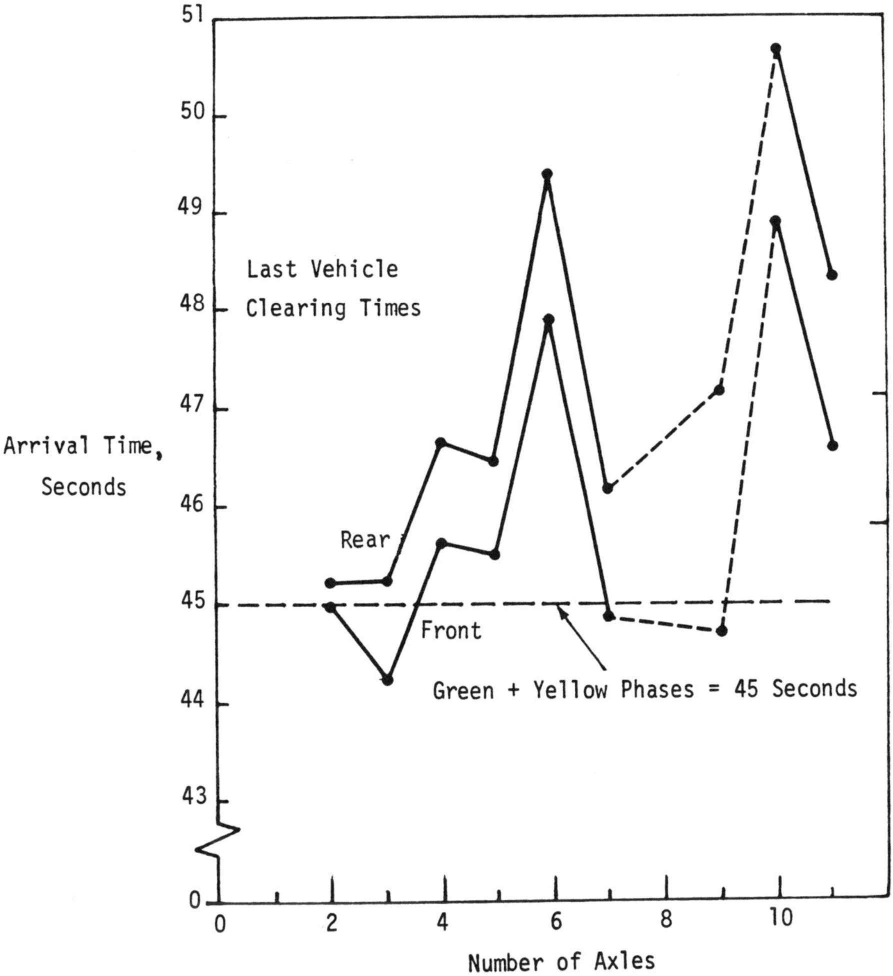

Figure 6. Average clearing (arrival) times (i.e., time required after onset of green for last (first) axle to reach tape 1) for last vehicle in cycle versus number of axles of last vehicle. Points above the dotted line indicate presence in intersection after the onset of the red phase.

was 47.11 s, which is 1.88 s more than the average for passenger cars. Note that if this is subtracted from the average headway for many-axle vehicles given in Table 5, the ratio of the headways of many-axle vehicles to passenger cars becomes 3.41/2.01 = 1.70, a value in much closer agreement with the ratio, namely 1.86, in which many-axle vehicles were over-represented as queue leaders.

Any arrival of a rear axle at a time greater than 45 s may reduce cross-traffic capacity. It is possible that the untypical nature of the signal studied, involving no normal cross-traffic, was conducive to a greater tendency to be in the intersection at the onset of red than would occur at a more typical intersection. A factor that could contribute to the many-axle vehicles' greater probability in this regard is the increase in the size of the so-called "dilemma zone" [11] as vehicle length increases. The average total wheel-base of the many-axle vehicles is 11.6 m, about 8.7 m greater than for a typical car.

There are not statistically significant differences between the arrival times or clearing times of last vehicles that depend on car size for vehicles in categories 2 through 5. This absence of any car-size effects on the probability of being a violator is in agreement with an earlier finding of Evans and Rothery [12].

TABLE 6. ARRIVAL TIMES AND CLEARING TIMES OF LAST VEHICLES (SECONDS SINCE BEGINNING OF GREEN PHASE) FOR RIGHT LANE

VEHICLE SAMPLE	SAMPLE SIZE	AVERAGE ARRIVAL TIME (FIRST AXLE AT TAPE 1)	AVERAGE CLEARING TIME (LAST AXLE AT TAPE 1)
CATEGORY 1*	10	44.21	44.34
2	36	44.87	45.06
3	93	45.04	45.25
4	68	45.09	45.32
5	129	44.98	45.22
6	31	45.32	45.58
7	27	45.80	46.60
PASSENGER CARS (**)	326	45.01	45.23
3 AXLE	3	44.26	45.24
4	8	45.65	46.69
5	2	45.56	46.46
6	1	47.95	49.40
7	2	44.88	46.18
8	0	---	---
9	1	44.74	47.20
10	1	48.92	50.68
11	5	46.59	48.34
ALL MANY AXLE VEHICLES	23	45.81	47.11

*-AS DEFINED IN TABLE I
(**)-AVERAGE OF CATEGORIES 2 THROUGH 5

Vehicle Position Effects

The headways of all vehicles as a function of vehicle position are shown in Table 7. The values for position 1 are lead vehicle clearing times, defined as the time between onset of green and the vehicle's last axle reaching the first tape. As all cycles did not have the same number of vehicles passing through the intersection, a decrease in sample size occurs with increasing vehicle position. For the right lane, the headways in positions two and three are larger than for any other positions, in keeping with published results that the first three vehicles use more green time than others (see, e.g., p. 352 of Ref. 1). In the left lane the second and third position headways are not different from the others. The headways for positions 20 and above are shorter than average because such data arose only for cycles of above average throughput, which were therefore associated with below average headways.

TABLE 7. AVERAGE HEADWAY (SECONDS) VERSUS QUEUE POSITION

VEHICLE POSITION	RIGHT LANE			LEFT LANE		
	SAMPLE SIZE	AVERAGE HEADWAY	STANDARD DEVIATION	SAMPLE SIZE	AVERAGE HEADWAY	STANDARD DEVIATION
(1)	(408)	(6.78)*	(1.48)	(381)	(6.17)*	(1.03)
2	408	2.29	0.91	381	1.95	0.70
3	408	2.26	1.14	381	1.98	0.73
4	408	2.19	1.10	381	1.96	0.62
5	407	2.16	0.92	381	1.94	0.67
6	404	2.15	1.02	380	1.91	0.60
7	403	2.11	0.99	380	2.07	1.52
8	403	2.01	0.81	378	2.07	1.58
9	403	2.04	0.88	377	1.97	0.97
10	402	2.19	1.11	375	2.07	1.27
11	401	2.11	1.05	374	2.09	1.23
12	398	2.21	1.19	373	2.19	1.36
13	393	2.18	1.24	366	2.14	1.29
14	387	2.13	1.06	362	2.17	1.28
15	379	2.16	1.01	354	2.06	0.91
16	362	2.11	0.92	350	2.13	1.12
17	338	2.18	1.12	338	2.07	1.06
18	305	2.07	0.86	323	2.09	0.98
19	246	2.10	1.02	292	2.02	0.81
20	185	1.85	0.81	233	1.87	0.65
21	111	1.92	0.85	169	1.95	0.84
22	57	1.84	0.73	99	1.71	0.58
23	18	1.69	0.59	47	1.73	0.55
24	4	1.43	0.47	18	1.66	0.64
25	0	--	--	6	1.38	0.15
26	0	--	--	1	1.80	--
ALL EXCEPT FIRST	7230	2.13	1.02	7119	2.03	1.06

*THE ENTRY FOR THE FIRST VEHICLE IS CLEARING TIME

Ancker, Gafarian and Grey [7] found successive headways to be independent random variables. We have applied a similar, but less extensive, analysis to our data. An examination of the correlation coefficients $r_{i,i+1}$ between the ith and $i+1^{th}$ headway for all positions, i, led to the conclusion that headways in the present study were essentially independent random variables, in accord with the previously mentioned [7] finding.

The vehicle headways in Table 7 may be converted to passenger car values by dividing the right lane headways by 2.13/2.01 and the left lane headways by 2.03/2.01.

The average speed and acceleration at the trap for vehicles in given positions (including position 1 - the queue leader) are shown versus vehicle position in Figure 7. Note that for high platoon positions, the acceleration is actually negative. This is likely a manifestation of a speed correction effect that has been observed in starting platoons of buses [13] and cars [14] in test track experiments.

DISCUSSION

The present data set offers evidence that drivers of smaller sized cars adopt smaller inter-vehicular spacings than the drivers of standard-sized cars. A dependence of spacing on amount of visible roadway [4,5] might not necessarily lead to car-size effects in the present study. Although, in general, forward obscuration increases with vehicle size for typical passenger cars, there are many counter-examples in other vehicles. Sport cars with long hoods and therefore long forward obscuration are in category 2; vans with very short forward obscuration are in the larger size categories. Smaller vehicles typically have lower power to weight ratios, so that larger gaps could arise from an inability to match lead-car accelerations. Such effects could also bear on an earlier finding that different size passenger car queue leaders did not affect throughput. These effects could also help explain why the

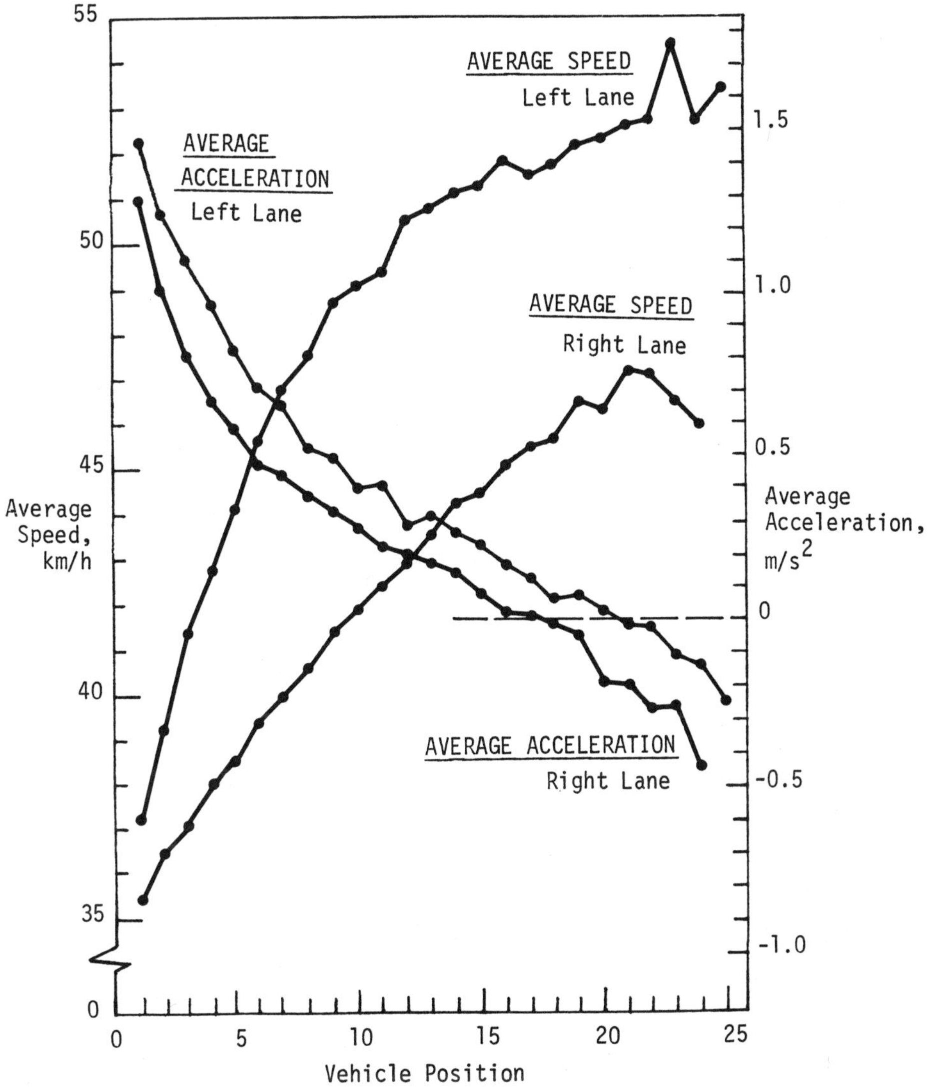

Figure 7. Average speed and acceleration versus position of vehicle in queue.

clear-cut car-size effects observed at a clearing line [3] were not observed in the present study.

Intersection saturation flow was found to be a linear function of the time required for the lead vehicle to travel a typical distance (15 m) to clear an intersection. One passenger car equivalent (pce) of saturation flow was lost for each 1.92 s of additional time the queue leader required to travel this distance. This result may be used to infer effects of low performance vehicles on urban traffic.

Recent experiments [15] have shown that in order to accelerate vehicles from rest to cruising speeds using the least amount of fuel drivers should use lower levels of acceleration than the average values currently encountered in traffic [15]. However, it is an extremely complicated problem to relate how individuals attempting to conserve their personal consumption of fuel by using lower levels of starting acceleration would affect the overall urban traffic system [16,17]. Any reduction in intersection throughput resulting from drivers choosing very low accelerations in an urban traffic system working at near capacity could have highly non-linear deleterious effects on delay, and therefore, on fuel consumption [18, 19]. Indeed, if all drivers in an urban traffic system try to minimize their individual fuel consumption, a possible result might be to increase their collective fuel consumption [15, 16].

Substantial reductions in saturation flow were attributed to large vehicles, particularly those with more than three axles, which in this discussion will, for convenience, be called "heavy trucks". A heavy truck in a non-leader position was equivalent to 2.63 passenger car equivalents (pce). On the other hand, a heavy truck in a queue leader position was equivalent to 4.05 pce. Compounding this adverse effect on saturation flow of the leader was the finding that heavy trucks were over-represented in the more deleterious leader position. Our data indicated that a random heavy truck in the traffic stream was almost twice as likely to become a queue leader as was a random passenger car. Indeed, were it not that the trucks were also more likely to enter the intersection during the red phase the number of trucks becoming queue leaders would have been even greater.

These findings have important implications for automatic control schemes aimed at minimizing fuel consumption. Already computer-controlled schemes in Tokyo [20], Glasgow [21] and Toronto [22] have been described in the literature. In the future, it would seem feasible to identify vehicles in greater detail, perhaps using a trap similar in principle to the one in the present observations. One algorithm would be to change the signal light from green to red when a large gap occurred. However, these results show that such a policy would further increase (beyond its already high value) the probability that heavy trucks would be queue leaders. Stopping a heavy truck as a queue leader has a doubly detrimental effect on system fuel consumption. First, the heavy vehicle will use substantially more fuel [12] idling and regaining its original speed than a passenger car. Secondly, the heavy truck in a queue leader position has a more adverse affect on cycle saturation flow. Hence, control algorithms to minimize system fuel consumption should operate so as to reduce the likelihood of heavy trucks becoming queue leaders.

ACKNOWLEDGMENTS

George Gorday and Stanislaw Koziel are acknowledged here for their valuable technical support. James Stephens is thanked for his skillful work in performing some of the data analysis. Helpful referee comments are incorporated in this version. Finally, grateful acknowledgement is extended for the support and encouragement of Robert Herman, with whom several fruitful discussions have occurred.

REFERENCES

1. L. J. Pignataro, Traffic Engineering - Theory and Practice. Englewood Cliffs, New Jersey: Prentice-Hall, Inc., 1973 .

2. R. Herman, T. Lam and R. Rothery, "An Experiment on Car Size Effects in Traffic," Traffic Engineering and Control, 15, 90-93, 99, June 1973.

3. Steuart, G. N. and Shin, B., "Effect of Small Cars on the Capacity of Signalized Urban Intersections," Transportation Science, 12, 250-263, 1978.

4. J. Almond, "Effect of Forward Visibility, Power/Weight Ratio and Type of Transmission of Cars on the Capacity of Traffic Signals," RRL Report LR 60, Road Research Laboratory, Crowthorne, England, 1967.

5. L. Evans and Richard Rothery, "The Influence of Forward Vision and Target Size on Apparent Inter-Vehicular Spacing," Transportation Science, 10(1), 85-101, February 1976.

6. Briggs, T. "Time Headways on Crossing the Stop-Line after Queuing At Traffic Lights," Traffic Engineering and Control, May 1977, 264-265.

7. C. J. Ancker, Jr., A. V. Gafarian and R. K. Gray, "The Oversaturated Signalized Intersection -- Some Statistics," Transportation Science, 2(4), 340-361.

8. L. C. Edie, R. S. Foote, R. Herman and R. Rothery, "Analysis of Single Lane Traffic Flow," Traffic Engineering, January 1963.

9. "Heavy Duty Vehicle Driving Pattern and Use Survey: Part II - Los Angeles Basin Final Report," U.S. Environmental Protection Agency, Ann Arbor, Michigan, EPT-460/3-75-005, February 1974.

10. A. Dockerty, "Accelerations of Queue Leaders from Stop Lines," Traffic Engineering and Control, 8(3), 150-155, July 1966.

11. D. C. Gazis, R. Herman and A. A. Maradudin, "The Problem of the Amber Light in Traffic Flow," Operations Research, 8, 112-232, 1960.

12. L. Evans and R. Rothery, "Comments on Effects of Vehicle Type and Age on Driver Behavior at Signalized Intersections," Ergonomics, 19(5), 559-570, 1976.

13. R. Herman, T. N. Lam and R. Rothery, "Further Studies on Single Lane Bus Flow: Transient Characteristics," Transportation Science, 4(2), 187-216, 1970.

14. R. Herman, T. N. Lam and R. Rothery, "The Starting Characteristics of Automobile Platoons," Proceedings of the Fifth International Symposium on the Theory of Traffic Flow and Transportation, June 1971.

15. L. Evans and G. M. Takasaki, "Fuel Used to Accelerate Vehicles from Rest to Cruising Speeds," Society of Automotive Engineers, SAE paper No. 810781, June 1981.

16. M.-F. Chang, L. Evans, R. Herman and P. Wasielewski, "Gasoline Consumption in Urban Traffic," Transportation Research Record 599, Transportation Research Board, Washington, D.C., 25-30, 1976.

17. L. Evans, "Driver Behavior Effects on Fuel Consumption in Urban Driving," Human Factors, 21(4), 389-398, 1979.

18. L. Evans and R. Herman, "A Simplified Approach to Calculations of Fuel Consumption in Urban Traffic Systems," Traffic Engineering and Control, 17(8,9), 352-354, August/September 1976.

19. L. Evans, "How Does Traffic Speed Affect Urban Fuel Consumption?", ITE Journal, Institute of Transportation Engineers, June 1979 and September 1979.

20. H. Inose, "Road-Traffic Control with the Particular Reference to Tokyo Traffic Control and Surveillance System," Proceedings of the I.E.E.E., 64(7), 1028-1039, July 1976.

21. J. R. Pierce and K. Wood, "Bus TRANSYT - A User's Guide" TRRL Supplementary Report 266, Transport and Road Research Laboratory, Crowthorne, England, 1977 (also see references cited in this report).

22. L. Rach, J. K. Lam, D. C. Kaufman and D. B. Richardson, "Evaluation of off-line Traffic-Signal Optimization Techniques," Transportation Research Record 538, 48-58, 1975.

23. P. J. Claffey, "Running Costs of Motor Vehicles as Affected by Road Design and Traffic," NCHRP, Report III, 1971.

EFFECTS OF THE CHOICE OF DEPARTURE TIME ON ROAD TRAFFIC CONGESTION

THEORETICAL APPROACH

Paul-Henri FARGIER

Institut de Recherche des Transports, Arcueil, France

Traffic studies do not normally take account of the way road users decide on their time of departure on a journey as a function of the traffic conditions. In this paper, however, an attempt is made to give an initial indication of the possible results of allowing for such a response. This is achieved by making some simple assumptions concerning the behaviour of the road users. In addition to the conventional approach, where the departure time is assumed to be independent of the traffic conditions (case 1), we consider the situation where the road user advances or delays his departure with a view to reducing the journey time. Two cases are considered according to whether the road user takes a decision with respect to a desired departure time (case 2, typical of work to home trips) or a desired arrival time (case 3, typical of home to work trips).

This approach is employed in considering the way in which a queue of vehicles can build up behind what is assumed to be a unique bottleneck in congested traffic conditions. A simple mathematical model for the build-up of the queue of vehicles is found to apply to both case 2 when there is pronounced peak in demand and to case 3 and the results given by the theoretical analysis are compared with those arising in real traffic situations. Finally the implications of the analysis are considered with respect to some particular factors such as the benefits resulting from an increase in road capacity, the staggering of working hours and the control of the access to a motorway.

Most road traffic studies are concerned with static situations where there is no variation in demand with time. Some investigators have, however, studied problems associated with peak traffic conditions by taking account of the dynamic characteristics involved. Thus, as early as 1967, Smeed (1) studied the effects of modifying the concentration of journey starting times. Following this study, other investigators assumed that the demand varied with time but that either the starting or arrival time was predetermined. In this connection, Hurdle (2) studied the allocation of traffic between an express road of limited capacity and an unsaturated road designed to carry slower traffic. Pearce (3) studied the traffic conditions in the case of an urban road system where the times of arrival were predetermined while Fargier, Morin and Ducornet (4) and Morin (5) considered the possibility of controlling the access to a motorway where the overall capacity of the latter was inadequate and where the times of departure were predetermined. Finally, Hurdle (6) dealt with the case of a congested urban motorway assuming that there was an elasticity in the demand at each instant but with no displacement of the demand in time.

The studies mentioned above did not take full account of the way in which users change their times of departure as a function of traffic conditions. Such a response needs to be taken into account for the following reason (particularly in the case of congested traffic conditions) :

- In general, in order to predict the effects of any traffic control measures or of any capital investment, since any change in the traffic conditions can give rise to displacements with respect to time in the demands for travel.

- In particular, to study the effects of any staggering of work starting times since this leads to changes in the desired arrival times and the associated departure times. (It will be seen that the assumption according to which each individual makes arrangements to arrive at the desired time is not applicable in the case of congested conditions.)

Gaver (7) studied the mechanism involved in the choice of the time of undertaking a journey as a function of the required time of arrival, and taking into account a poor knowledge of the duration of the journey and of the random variations that can be involved. Alfa and Minh (8) made similar assumptions in the case of a bottleneck along a particular route and considered the state of equilibrium that was attained for the whole traffic of the route.

In this paper we assume that the road users behave in a similar way to that assumed in the case of papers (7) and (8). We assume, however, that this behaviour arises in a deterministic context so that we are able to make a more detailed study of the consequences of the behaviour. We consider both the case where the action is determined by the time of departure (work to home trips) and that where it is governed by the time of arrival (home to work trips). As was the case for paper (8), we limit our study to the consideration of a simple situation, namely a bottleneck along a particular route.

Following completion of this paper, an article by Hendrickson and Kocur (9) appeared which referred to an approach similar to the one that we describe, except that it is limited to the case where the time of arrival is the determining factor.

It will be possible to compare the results of our theoretical analysis with the way in which queues of vehicles develop in real situations during periods of traffic congestion. We also propose to refer very briefly to the implications of our analysis with respect to some particular factors such as the benefits arising from any increase in road capacity, the staggering of working hours, and the control of the access to a motorway.

1. CONVENTIONAL APPROACH : DEPARTURE TIME INDEPENDENT OF TRAFFIC CONDITIONS

In what we refer to here as the conventional approach it is assumed that the results of taking any action to improve the traffic conditions will not lead to any changes in the times of departure of the road users. This assumption is in fact fully justified in a certain number of cases :

- Where the reasons for deciding on a particular time of departure are sufficiently compelling that it is unlikely that the response will be affected as a result of any changes in the traffic conditions.
- Where it is a matter of responding to random events which cannot be foreseen at the moment of departure. This applies in the case of any incident.

If at a time t_1 the demand exceeds the capacity at a bottleneck, a queue of vehicles will build up and the rate of flow of vehicles will then correspond to this capacity. When the demand subsequently decreases and falls below the capacity of the bottleneck, the length of the queue of vehicles will decrease and the queue will finally disappear at a time t_2 . We refer to the interval of time bounded by t_1 and t_2 as the period of congestion; it has a duration of $T = t_2 - t_1$. Variations in demand or in the capacity of the bottleneck lead to changes in the value of T, to the way in which the queue of vehicles develops and to the total waiting time experienced by the road users. These changes can be directly calculated.

An interesting point to be considered is the "marginal cost" arising due to the arrival of traffic at a time t lying between times t_1 and t_2 . The time t is defined as the instant when the vehicle would have arrived at the bottleneck in the absence of any traffic congestion. This time t is regarded as the beginning of the waiting period to be considered.

Now assume that an additional quantity of traffic δ arrives at time t. This will give rise to a corresponding increase δW in the total delay experienced by the collection of vehicles. The marginal cost at time t is defined as the limiting value of $\delta W/\delta q$ as δq tends towards zero (assuming that the traffic can vary in a continuous manner, an approximating assumption that is often made). The additional delay first of all includes the delay experienced by the additional traffic δq up to time s when it leaves the front of the queue. In addition to this delay, if we assume that the saturation capacity is C vehicles per unit time, then each of the following vehicles will be delayed by a time $\delta q/C$ (as long as the congestion continues), the number of vehicles involved here amounting to $(t_2 - s)C$. Thus δW can be expressed as a function of δq as follows :

$$\delta W = (s - t)\ \delta q + (t_2 - s)\ C \frac{\delta q}{C} = (t_2 - t)\ \delta q.$$

Thus the "marginal cost" is $(t_2 - t)$, i.e. the length of time for which the congestion continues beyond the time t . If t coincides with the beginning of the congestion, this marginal cost is equal to T.

The "marginal cost" can be appreciably greater than the delay experienced by each individual in the queue and can be of considerable duration if the congestion continues for a long period of time. This shows, on the basis of the assumption made in this section of the paper, that there is an advantage in reducing the demand at the beginning of the period of congestion, by providing an alternative route for example, even when this involves a longer journey time. This is an example of the difference between user optimising solutions and system optimising solutions.

2. PROPOSED APPROACH, NOTATION EMPLOYED AND DEFINITION OF THE PROBLEM

As opposed to the approach in the previous section, we now consider the cases where the time of departure is decided as a function of the traffic conditions. The traffic situation to be considered remains the same, a bottleneck where a queue of vehicles begins to build up as soon as the demand exceeds the saturating capacity C and where the rate at which vehicles leave the front of the queue amounts to C for as long as the congestion continues. We also assume that the travel time required for the remainder of the journey does not vary with time for road users who pass through the bottleneck.

2.1 Basis of the assumptions concerning behaviour

We need to make assumptions that, in the simplest possible way, will enable us to take account of the tendency of road users to avoid traffic jams as far as possible. The assumptions are based on the following considerations :

- We suppose that the road users know exactly how the queue of vehicles will develop during the period of time with which they are concerned. This assumption can be justified when we consider that most of the road users concerned regularly travel over the same route, so they gradually acquire a good understanding of the way in which the queue of vehicles develops on average; furthermore, most of the analysis that we are going to perform remains valid if we replace the term "delay" by "mathematical expectation of a delay".

- Two kinds of behavioural response are to be considered. In the first instance it is assumed that road users make a decision with respect to a desired time of departure (usually the time when work finishes) and in the second case with respect to a desired time of arrival at their destinations (usually the time when work starts, with a possible safety allowance to take account of day-to-day variations).

- We define the situation considered in section 1 of the paper, where the time of departure is predetermined with no possibility of any variation, as "case 1". The situation where the road user can advance or delay his departure with respect to the desired departure time in order to reduce the duration of any waiting period is defined as "case 2". The displacement in time for a given reduction in the duration of the waiting period will generally differ according to whether an advance or a delay in the time of departure is involved.

- When the time of arrival is the determining factor there is a temptation to assume a behavioural response where a road user who is aware of the waiting times that can be involved at any moment will change his departure time in order to arrive at his desired time. It is easy to see, however, that it is impossible for each road user to arrive at his desired time when there is high demand, since this would mean a rate of flow of vehicles into the bottleneck that exceeded its capacity. It is for this reason that we define a "case 3" similar to "case 2", i.e., where the subject can advance or delay his departure in order to reduce his waiting time.

- Given that we assume that the duration of the journeys outside the bottleneck are the same at all times, we can neglect the uncongested travel time and consider the time of departure to be identical to the time of arrival at the bottleneck. Similarly the time when a vehicle leaves the front of the queue corresponds to a particular time of arrival at the destination, and the two times may be treated as though they were identical.
- We assume, in order to simplify the analysis, that all road users have the same behavioural response during a given period of time.

2.2 Definition of the behavioural responses

For each road user we write :

e is the time when he arrives at the bottleneck, i.e., the time when he would arrive at the bottleneck location in the absence of any queue of vehicles, given his time of departure,

w is the duration of the waiting period at the bottleneck,

$s = e + w$ is the time when he leaves the bottleneck,

and, assuming that the departure time is the determining factor:

$\hat{e}$ is the desired time of entry into the bottleneck.

Then

$\delta e = e - \hat{e}$,

and

$g(\delta e)$ is the nuisance factor due to the difference δe

or where the arrival time is the determining factor:

$\hat{s}$ is the desired time of leaving the bottleneck,

$\delta s = s - \hat{s}$,

and

$g(\delta s)$ is the nuisance factor due to the difference δs.

For a variable x we write :

$x^+ = x$ and $x^- = 0$ if $x \geqslant 0$,
$x^+ = 0$ and $x^- = -0$ if $x < 0$.

We assume that during a given peak period all users behave in the same way. The different peak periods can be classified as follows :

Case 1 : time of departure predetermined as considered in section 1.
Each road user has a value $\hat{e}$ and chooses e such that $\delta e = 0$.

Case 2 : time of departure the determining factor.
Each road user has a value $\hat{e}$ and chooses e so as to minimise $w + g(\delta e)$ where $g(\delta e) = a\,\delta e^+ + b\,\delta e^-$, the values of the coefficients a and b being independent of the user for the period of time considered and such that $0 \leqslant a \leqslant 1$ and $b \geqslant 0$.

Case 3 : time of arrival the determining factor.
Each road user has a value $\hat{s}$ and chooses e (and hence $s = e + w$) so as to minimise $w + g(\delta s)$

where $g(\delta s) = a\,\delta s^{+} + b\,\delta s^{-}$, the coefficients a and b being independent of the user for the period of time considered and such that $a \geqslant 0$ and $0 \leqslant b \leqslant 1$.

The significance of the equations giving the values of g is as follows : in case 2 (case 3) the road user is willing to depart (arrive at his destination) $\frac{1}{a}$ minutes later or $\frac{1}{b}$ minutes earlier in order to reduce the duration of the waiting period in the bottleneck by 1 minute. Furthermore, the accepted amount of advance or delay is proportional to the reduction of waiting time.

In case 2, $a \leqslant 1$ signifies that a road user who delays his departure cannot expect to arrive at his destination any earlier. In case 3, $b \leqslant 1$ signifies that a road user who arrives at his destination earlier cannot expect to have delayed his departure.

It would appear that two of the assumptions described above are open to criticism, namely, the linear relationship assumed in the equations giving values of g (which can lead to large and therefore somewhat unrealistic values for the advances or delays in departure times in some cases) and the fact that for a given period of time the values of the coefficients a and b do not vary from one road user to another. These assumptions have been made because they lead to a considerable simplification of the analysis. The validity of the results obtained on making these assumptions will be discussed in section 6 of the paper.

2.3 Notation employed (period of congestion)

For a given period of time we consider the following values (functions of time) :

E the flow of traffic at the entry to the bottleneck

S the flow of traffic at the exit of the bottleneck

$\hat{E}$ the flow of traffic at the entry to the bottleneck that would occur if each road user made an entry at his desired time $\hat{e}$ (case 2)

$\hat{S}$ the flow of traffic at the exit of the bottleneck that would occur if each road user made an exit at his desired time $\hat{s}$ (case 3)

Q the number of vehicles in the queue

$\hat{Q}$ the number of vehicles that would be in the queue if the flow at the entry amounted to $\hat{E}$ (case 2)

p the slope of the Q/C versus t curve.

We are interested here in peak periods with congestion, i.e., during which $\hat{E}$ or $\hat{S}$ exceeds the capacity of the bottleneck for some time. We assume that the demand is unimodal, i.e., we assume that the curves showing the way in which $\hat{E}$ and $\hat{S}$ vary with time rise to a maximum and then decrease (with no levelling off during the increase or decrease in value). The problem that we need to consider here is that of determining the number of vehicles Q in the queue and their rate of entry E into the queue.

Given these conditions we also consider :

t_1 the time when the congestion commences and Q becomes positive

t_2 the time when the congestion ceases and Q falls to zero after having been positive

$\tilde{Q}(t)$ the value at time t of $\int_{t_1}^{t} (\hat{E} - C)dt$, for $t_1 \leqslant t \leqslant t_2$

$T = t_2 - t_1$, the duration of the congestion

W the total of the delays experienced by all users

$\hat{t}_1$ the time when $\hat{E}$ exceeds C (case 2)

$\hat{t}_2$ the time when $\hat{Q}$ falls to zero after having been positive (case 2).

2.4 Problem to be solved

This is a matter of finding solutions corresponding to "equilibrium conditions" in terms of game theory, i.e., so that a user cannot benefit from any unilateral change in his selected time of entry into or exit from the bottleneck.

Given that the users all behave in the same way, we can assume, without departing from the general case, that their actual order of entry into or exit from the bottleneck will be the same as the desired order. To be more precise we say that for every equilibrium solution that does not satisfy this condition we can find another having the same characteristics with respect to E, Q and W, but where the order of entry or exit is conserved. This rather intuitive result is justified in appendix I. In what follows, this allows us to confine our attention to solutions that conserve the order of entry into and exit from the bottleneck.

The equilibrium solutions correspond to optimum conditions so far as each individual user is concerned but not for the users as a whole. In fact, in case 2 the attempt of each individual to find his own optimum solution can lead to a situation which is worse than the simple solution $\delta e = 0$ for all concerned (see explanation given in section 4). Before considering the equilibrium solutions (sections 4 and 5) we make some comments on the collective optimum (section 3).

3. COMMENTS ON A COLLECTIVE OPTIMUM SOLUTION

Suppose that the time of departure of each individual could be decided by some central authority. Now consider the situation where a queue of vehicles normally forms each day at a particular bottleneck. We can easily postulate a better situation by supposing that the central authority instructs each road user to depart w later, where w is the time that a particular user ordinarily spends waiting in the queue. Then no queue would form and each individual would enter the bottleneck at the precise time when he would otherwise have reached the front of the queue of vehicles. Thus there would be no waiting and the times at which individuals would come out of the bottleneck (and hence the times at which they would arrive at their destination) would remain the same. The new situation is to be preferred to the preceding one since no individual would suffer and all those who would otherwise have been obliged to wait in a queue of vehicles would gain.

Thus with an optimum solution there is no waiting, we have e = s, and the mathematical models for cases 2 and 3 are the same : we need to minimise $g(\delta e) = a\,\delta e^{+} + b\,\delta e^{-}$. We accordingly arrive at a situation similar to some considered in economic studies : if we divide the period of time into small intervals we can say that each road user tries to maximise an advantage equal to $-\,g(\delta e)$ on consuming an item of goods, namely the right to pass through the bottleneck during one of these small intervals. This implies that we can in theory arrive at an optimum situation by imposing a toll varying with time in such a way that the bottleneck is just saturated when this toll is different from zero. A curious observation arises here : in case 3 we can regard w as this toll and it can be shown that the distribution of the exits from the bottleneck given by our mathematical model minimises the total of the nuisances $g(\delta s)$ experienced by the different road users.

4. QUEUES OF VEHICLES IN CASE 2 (TIME OF DEPARTURE THE DETERMINING FACTOR)

It should be understood that, in terms of the notation given in section 2 of the paper, each road user tries to minimise $w + g(\delta e) = w + a\,\delta e^{+} + b\,\delta e^{-}$ where $0 \leqslant a \leqslant 1$ and $b \geqslant 0$. The desired rate of entry of vehicles into the queue, $\hat{E}$, first increases and then falls, and it exceeds the capacity of the bottleneck, C, during a certain period of time commencing at time $\hat{t}_1$ (see figure 1). We can determine, for each instant of time, the value of $\hat{Q}$, the number of vehicles that would be held up in the queue if each road user entered the bottleneck at his desired time. Assuming that the value of $\hat{t}_2$ is such that

$$\int_{\hat{t}_1}^{\hat{t}_2} (\hat{E} - C)\,dt = 0 ,$$

we have

$$\hat{Q} = 0 \quad \text{when} \quad t \leqslant \hat{t}_1 \quad \text{or} \quad t \geqslant \hat{t}_2$$

and

$$\hat{Q} = \int_{\hat{t}_1}^{t} (\hat{E} - C)\,dt \quad \text{when} \quad \hat{t}_1 \leqslant t \leqslant \hat{t}_2 .$$

4.1 Characteristics of the equilibrium solutions

We need to determine the value of Q, given that the actual rate of entry of vehicles E can be different than $\hat{E}$. The length of time for which an individual will have to wait on joining a queue of Q vehicles is given by $w = Q/C$. Let p be the slope of the curve showing how Q/C varies with time. Given the behavioural response of the road users, this curve can never include slopes of $p > b$ or $p < -a$. In fact, if a road user were to arrive at the queue when the Q/C curve did include such slopes, then it is clear that he should have advanced or delayed his arrival in order to minimise w + g

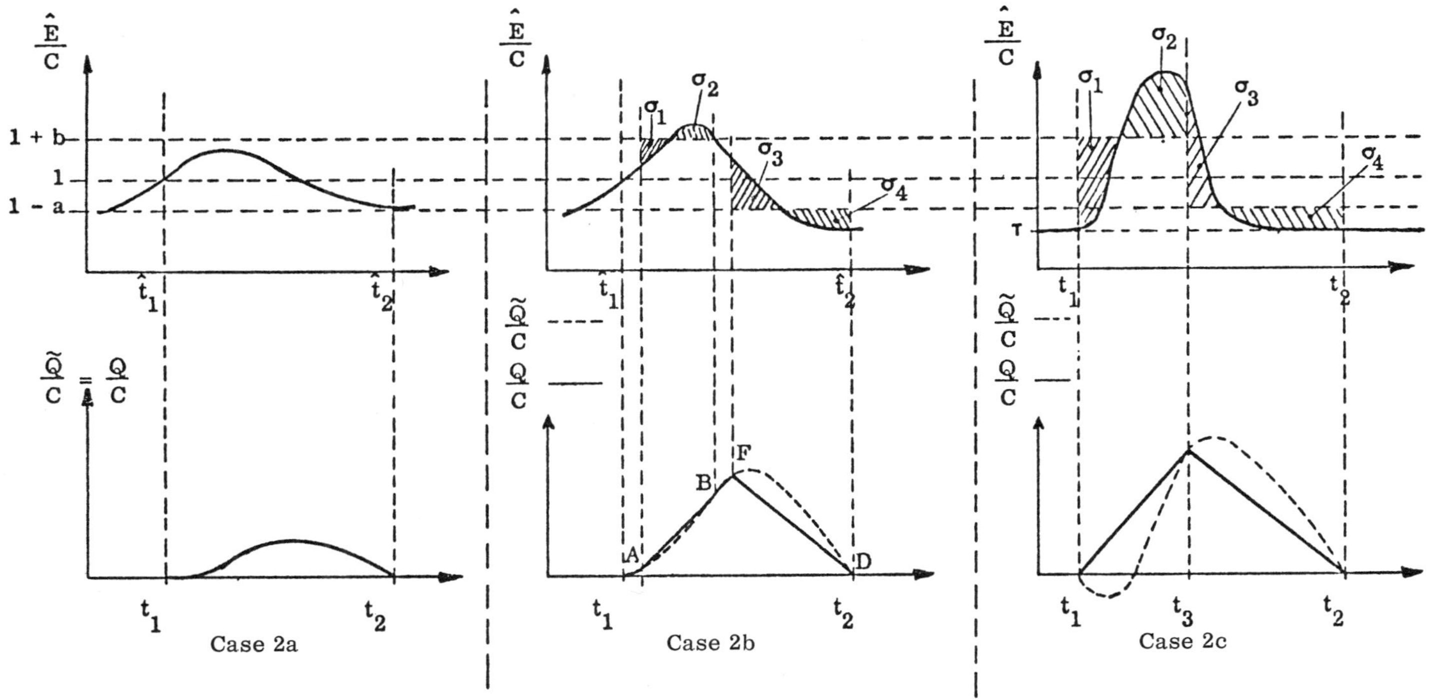

Fig. 1 – Departure time the determining factor

and we will not then have an equilibrium solution. In fact, the curve of Q/C as a function of t corresponds to an equilibrium solution for values of $Q > 0$ if, and only if, it is made up of parts belonging to one of the following three categories :

$$(1) \quad -a \leqslant p \leqslant b \quad \text{with} \quad \delta e = 0$$
$$(2) \quad p = -a \quad \text{with} \quad \delta e \geqslant 0$$
$$(3) \quad p = b \quad \text{with} \quad \delta e \leqslant 0 .$$

Alternatively we can consider the same situation as follows : with our assumptions, $Q = 0$ implies that $\delta e = 0$. As a consequence of this, if t_1 is the time when Q becomes positive for the first time, then Q will again be zero at a time t_2 such that

$$\int_{t_1}^{t_2} (\hat{E} - C)dt = \int_{t_1}^{t_2} (E - C)dt = 0.$$

Given the shape of the curve of $\hat{E}$ as a function of t and knowing the value of t_1, we can now determine the value of t_2 and confirm that the value of Q remains zero beyond time t_2.

Let

$$\tilde{Q}(t) = \int_{t_1}^{t} (\hat{E} - C)dt \quad \text{for} \quad t_1 \leqslant t \leqslant t_2 .$$

We also have

$$\tilde{Q}(t_1) = \tilde{Q}(t_2) = 0.$$

For a given value of t, δe has the same sign and falls to zero at the same instant as

$$\int_{t_1}^{t} (\hat{E} - E)dt.$$

This expression represents the difference between the number of users whose desired time of entry into the queue lies between t_1 and t and the number that actually entered between times t_1 and t (it should be recalled that the order of desired entries is conserved in actual entries).

Given that $$Q(t) = \int_{t_1}^{t} (E - C)dt,$$

the necessary and sufficient condition for the curve of Q/C versus t to correspond to an equilibrium solution can be defined as follows : times t_1 and t_2 being such that

$$\tilde{Q}(t_2) = \int_{t_1}^{t_2} (\hat{E} - C)dt = 0,$$

the function Q(t)/C must be continuous, equal to zero outside the time interval bounded by t_1 and t_2, and positive and made up of parts belonging to any one of the following three categories within this same time interval :

$$(1) \quad Q(t) = \tilde{Q}(t) \quad \text{and} \quad -a \leqslant p \leqslant b$$

$$(2) \quad Q(t) \leqslant \tilde{Q}(t) \quad \text{and} \quad p = -a$$

$$(3) \quad Q(t) \geqslant \tilde{Q}(t) \quad \text{and} \quad p = b.$$

In other words the Q/C curve is made up of parts corresponding to $\tilde{Q}/C$, of straight lines of slope -a located below the $\tilde{Q}/C$ curve and of straight lines of slope b located above this same curve.

Given the curve of $\tilde{Q}/C$ (and hence the values of t_1 and t_2) we can construct the Q/C curve, by proceeding backwards from the point where $t = t_2$ as follows. We retain the $\tilde{Q}/C$ curve whenever this will produce a curve such that $-\grave{a} < p < b$. If, however, the slope, $\tilde{p}$, of $\tilde{Q}/C$ is anywhere less than -a, we instead draw a straight line with slope -a until it intersects the $\tilde{Q}/C$ curve. Similarly, if $\tilde{p} > b$, we draw a straight line with slope b. This procedure is then repeated as necessary until we reach the point where $t = t_1$ as shown in figure 1.

In appendix II we show how we prove the following rather intuitive conclusion : there is always one and only one Q/C curve corresponding to an equilibrium solution. Thus it is sufficient to find just one solution satisfying the condition in order to be sure that it is the only solution.

If the slope of the $\hat{Q}$/C versus time curve lies between -a and b at all points, then we have that $Q = \hat{Q}$, with $e = 0$, and the whole of the curve belongs to slope category (1). We classify this situation as case 2a and it has the same characteristics as case 1. An example of case 2a is shown on the left side of figure 1.

If at certain points the slope of the $\hat{Q}$/C curve is greater than b or less than -a, then at least some parts of the curve belong to slope category (2) or (3), as shown in the other two examples in figure 1.

We make a distinction here between case 2b, for which the Q/C curve includes one or more sections belonging to slope category (1), although the curve as a whole does not fall into this category, and case 2c for which the Q/C curve no longer contains any sections belonging to slope category (1).

4.2 Example of case 2b

Several possibilities exist, but we limit ourselves here to the one illustrated by figure 1. Assuming that $t_1 = \hat{t}_1$ and hence that $t_2 = \hat{t}_2$, then the slope of the $\tilde{Q}$/C curve at point D will be less than -a. We draw the line DF of slope -a. At point B, the slope of the $\tilde{Q}$/C curve is b and the tangent to the curve at this point intersects the $\tilde{Q}$/C curve again at point A, later than time $\hat{t}_1$. The shaded areas on the figure are such that $\sigma_1 = \sigma_2$ and $\sigma_3 = \sigma_4$. The curve obtained in this way satisfies all the conditions. The following two points can be made :

- If there is a straight line of slope $-a$, it must pass through the point where $t = t_2$ since this is the point within the time interval bounded by t_1 and t_2 where the $\tilde{Q}/C$ curve has the lowest slope (this slope is equal to $\hat{E}/C - 1$)
- The users corresponding to the straight line AB leave before their desired departure time. Since this line lies above the $\tilde{Q}/C$ curve it is easy to see that these users are at a disadvantage with respect to the situation that would have applied if they had each been able to start their journey at the desired time. This latter situation would not correspond to an equilibrium solution, however, since a number of users would then benefit individually from an earlier departure.

4.3 Analysis of case 2c

As the demand becomes more concentrated, then the sections of the curve belonging to slope category (1) become less extensive and beyond a certain degree of concentration the Q/C curve is reduced to two straight lines having slopes of $-a$ and b respectively. This case applies when we can identify times t_1, t_2 and t_3 (see figure 1) such that, on letting $T_1 = t_3 - t_1$, $T_2 = t_2 - t_3$ and $T = t_2 - t_1 = T_1 + T_2$, we have :

$$\int_{t_1}^{t_2} (\hat{E} - C)dt = 0$$

and

$$\frac{1}{C} \int_{t_1}^{t_3} (\hat{E} - C)dt = bT_1 = aT_2 \quad \text{(the value of}$$

Q/C at time t_3).

Again, as shown by the shaded areas on figure 1 we have $\sigma_1 = \sigma_2$ and $\sigma_3 = \sigma_4$.

Calculations show that the total waiting time W expressed by $\int_{t_1}^{t_2} Q\,dt$ is given by :

$$W = \frac{1}{2}\,\frac{ab}{a+b}\,T^2 C$$

It is interesting to consider the following particular case that we identify here as case 2c*. We suppose, as shown by figure 1 for case 2c, that the desired rate of entry of vehicles $\hat{E}$ has a constant value of τC (where $\tau < 1$) before and after the peak period and that the times t_1 and t_2 lie within these steady state periods. In these conditions the additional overall demand corresponding to the peak of the curve is given by

$$D = \int_{-\infty}^{+\infty} (\hat{E} - \tau C)\,dt .$$

The duration T is then independent of t_1 and has a value of $T = \frac{D}{C(1-\tau)}$. It can be shown that $T_1 = T\,\frac{a}{a+b}$. The value of t_1 can be determined by locating the point ahead of the peak of the curve which satisfies the following relationship :

$$\int_{t_1}^{t_1+T_1} (\hat{E} - C)\,dt = bT_1 .$$

The other relationships are then also satisified. By substituting T as a function of D we can express the value of the total delay :

$$W = \frac{1}{2}\,\frac{ab}{a+b}\,\frac{D^2}{C(1-\tau)^2} .$$

We have no data concerning the frequency with which the different cases can arise. Case 2c in particular will arise when there is a concentration of desired departures at a particular time (factory closing time) but figure 1 shows that this case can also be applicable when there is a continuous variation in the demand. Case $2c^{*}$ is somewhat artificial but it is nevertheless of interest as an approximation that can be employed to simplify the calculations and the results.

5. QUEUES OF VEHICLES IN CASE 3 (TIME OF ARRIVAL THE DETERMINING FACTOR)

As defined in section 2 of the paper, in case 3 each road user seeks to minimise $w + g = w + a\,\delta s^{+} + b\,\delta s^{-}$ where $a \geqslant 0$ and $0 \leqslant b \leqslant 1$. The demand is given by the desired rate of flow of vehicles out of the bottleneck. This demand first increases and then falls ; it exceeds the capacity C of the bottleneck during a certain period of time (see figure 2).

It should be recalled that our solutions can be limited to situations where the actual order of exit from the queue is identical to the desired order of exit. As a consequence of this, δs is a continuous function of s, the actual time of exit. In these conditions, during the period where $\hat{S}$ exceeds the capacity :

- We will have $\delta s = 0$, at least at one instant s_3, since otherwise users arriving during this period will all be too early or too late, whereas they cannot in fact be too late at the beginning or too early at the end of the period.
- It is not possible for δs to be zero at two distinct instants of time s_3 and s_4 since if $s_4 > s_3$, for example, the desired rate of flow out of the queue of vehicles between

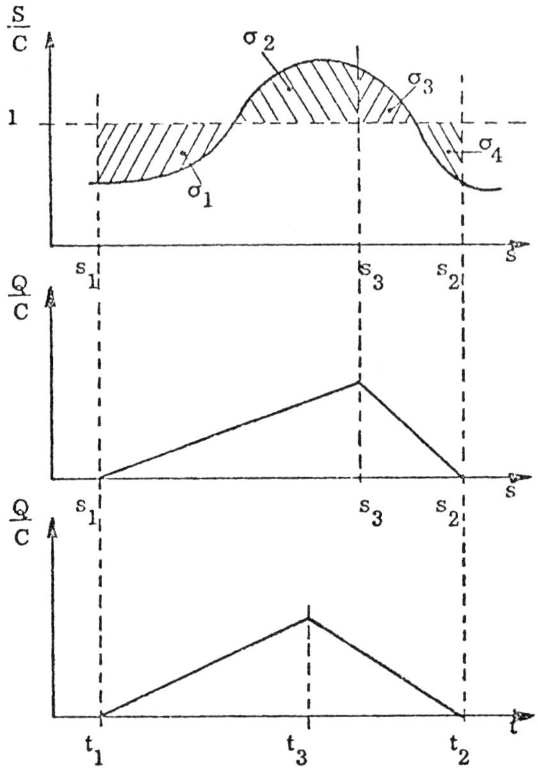

Fig. 2 - Time of arrival the determining factor

times s_3 and s_4, which is equal to

$$\int_{s_3}^{s_4} \hat{S}\, ds,$$ is greater than the maximum possible output of $(s_4 - s_3)C$.

As the rate at which vehicles leave the bottleneck cannot exceed the maximum value C, then all the instants of time before s_3 have a $\delta s < 0$ as far back as an instant s_1, before which there has been no congestion. From time s_3 until time s_2 we have $\delta s > 0$. The period of time from s_1 to s_2 is the period of congestion, where $s_1 = t_1$, $s_2 = t_2$, $T = t_2 - t_1 = s_2 - s_1$.

Consider the curve of $w = Q/C$ as a function of s, i.e. the curve giving the waiting time experienced by a road user coming out of the bottleneck at time s (Q is the number of vehicles that were in the queue when the road user who is now leaving first entered the bottleneck). Let p be the slope of this curve. Allowing for the behavioural response, and following a similar line of reasoning to that considered in section 4, we have :

$$p = -a \quad \text{when} \quad \delta s > 0$$

$$p = b \quad \text{when} \quad \delta s < 0.$$

The curve of $w = Q/C$ as a function of s (see figure 2), where Q/C has a zero value at times s_1 and s_2, is accordingly made up of two straight lines, one having a slope of b between times s_1 and s_3 and the other a slope of -a between times s_3 and s_2. The times s_1, s_2 and s_3 are uniquely defined by the relationship :

$$\int_{s_1}^{s_3} (\hat{S} - C)\,dt = \int_{s_3}^{s_2} (\hat{S} - C)\,dt = 0$$

and

$$b(s_3 - s_1) = a(s_2 - s_3).$$

The shaded areas on figure 2 are such that $\sigma_1 = \sigma_2$ and $\sigma_3 = \sigma_4$.

We can convert the curve giving $Q/C = w$ versus s to a curve of Q/C as a function of time t or, which amounts to the same thing, to a curve giving w as a function of the instant e of entry into the bottleneck by making use of the equation $e = s - w$. This new curve (see figure 2) is accordingly made up of another two straight lines having slopes of b' and -a' where :

$$b' = \frac{b}{1 - b} \quad \text{and} \quad a' = \frac{a}{1 + a}$$

As in section 4 we can derive an expression for the total waiting time W :

$$W = \frac{1}{2} \frac{a'b'}{a' + b'} T^2 C$$

or once again :

$$\boxed{W = \frac{1}{2} \frac{ab}{a + b} T^2 C}$$

It should be noted, however, that a and b do not have the same significance here as they did for case 2.

In the same way as we introduced a case $2c^*$, we also consider a <u>case 3^*</u>, supposing that before and after the peak period the desired rate of flow $\hat{S}$ has a constant value equal to τC (where $\tau < 1$) and that $s_1 = t_1$ and $s_2 = t_2$ lie within the steady state periods. As in section 4 we let D be the additional demand due to the peak :

$$D = \int_{-\infty}^{+\infty} (\hat{S} - \tau C)\,dt.$$

The value of T is then independent of t_1 and equal to $\frac{D}{C(1 - \tau)}$. We can now express the total delay as

$$\boxed{W = \frac{1}{2} \frac{ab}{a + b} \frac{D^2}{C(1 - \tau)^2}}$$

6. COMPARISON OF THE RESULTS OF THE ANALYSIS WITH ACTUAL TRAFFIC SITUATIONS (CASE 3)

A detailed verification of the results of the analysis would call for rather extensive surveys in order to acquire the required information on the desired times of departure or arrival. One of the results can however be more easily compared with real traffic situations and that is the shape of the curve $w = Q/C$ as a function of time for case 3 where the time of arrival is the determining factor.

This curve is in fact made up of two straight lines (figure 2) and we therefore can ascertain if the curves for real traffic situations are of this form and, if so, determine the slopes a and b.

Therefore we looked for traffic situations where the conditions were as close as possible to those postulated in our analysis, namely :

- time of arrival the determining factor (morning home to work peak).
- one and only one traffic bottleneck along the route.

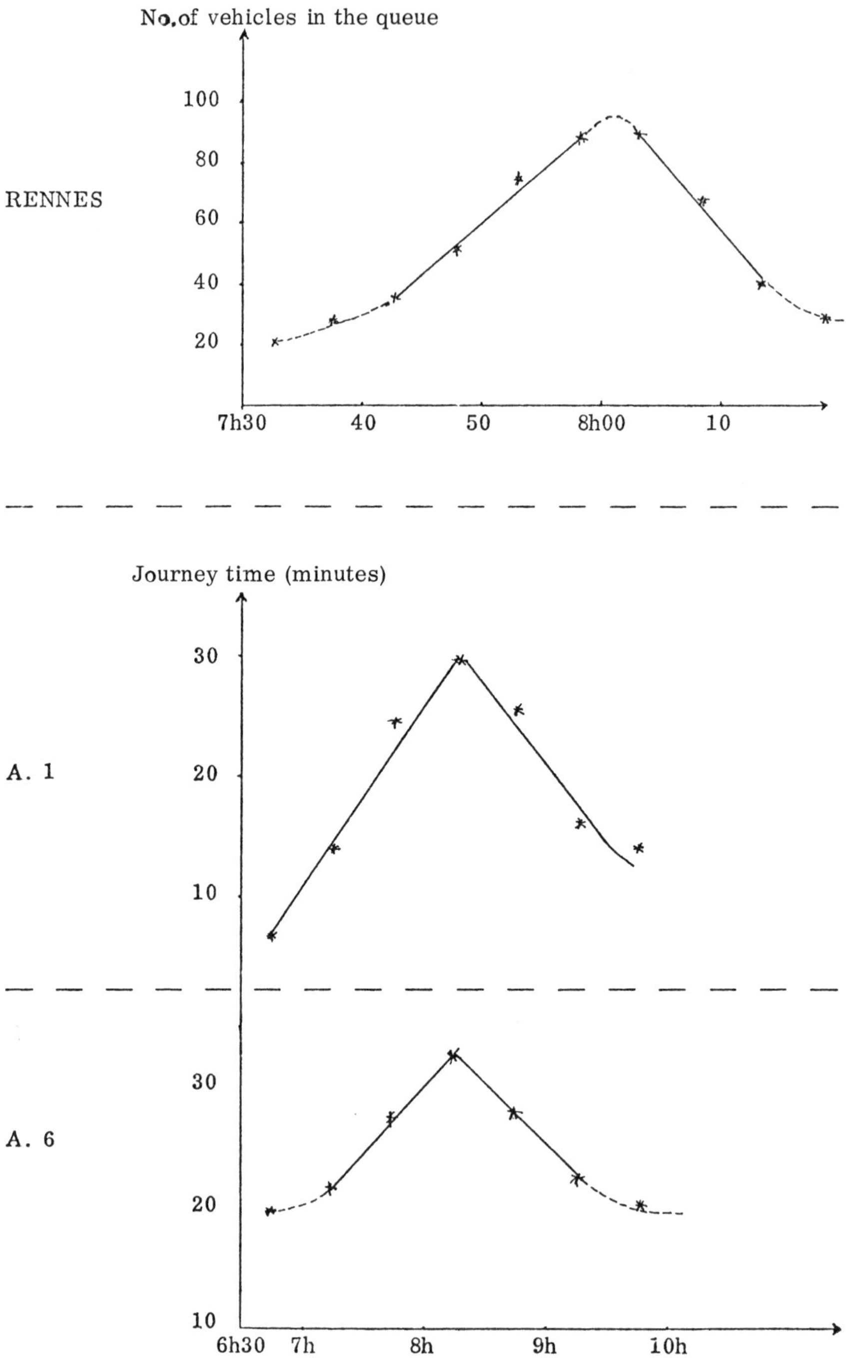

Fig. 3 - Results of measurements

On this basis we considered the results from three locations :

- queue of vehicles at the moment when the traffic lights changed to green at an entry into Rennes at the Nantes Bridge for traffic coming from the south around 0800 hours (average of the results for three days during November 1979).
- journey time during the morning on an 8 km long section of the A1 motorway for traffic going towards Paris (average of the times observed for five days during 1976 for traffic going towards Paris, the destinations being the Boulevard Périphérique Ouest and the Porte de la Chapelle).
- journey times during the morning on the A6 motorway for traffic going from Corbeil to Arcueil, i.e. towards Paris (average of the times observed for five days during 1978).

In the case of the motorways, the measurements were made before the introduction of any control over the access, i.e. in the absence of any waiting at the motorway entries.

The results of the measurements are given in tables 1 and 2 and by the graphs of figure 3.

The waiting time at the Rennes location can be obtained by dividing the number of vehicles in the queue at any instant by the saturation rate of flow of traffic of 33 vehicles per minute. The slopes of the straight lines giving w as a function of time give the values of b' and $-a'$ from which we can in turn determine the values of a and b, making use of the relationship $a = \frac{a'}{1 - a'}$ and $b = \frac{b'}{1 + b'}$.

Table 1

Queues of vehicles at Rennes

Five minute interval commencing at	Number of vehicles in the queue
7 h 30	22
7 h 35	30
7 h 40	36
7 h 45	52
7 h 50	77
7 h 55	89
8 h 00	90
8 h 05	69
8 h 10	42
8 h 15	30

Table 2

Journey times on the A1 and A6 motorways

Thirty minute interval commencing at	Journey times in minutes	
	On the A1	On the A6
6 h 30	6.46	19.25
7 h 00	14.10	21.15
7 h 30	24.77	26.78
8 h 00	29.52	32.10
8 h 30	25.34	26.90
9 h 00	15.77	22.20
9 h 30	14.09	19.67

These values are, of course, approximations. The values for the three different locations are as follows :

Rennes	a = 0.17	b = 0.10
A1 motorway	a = 0.30	b = 0.20
A6 motorway	a = 0.20	b = 0.15

The values of a and b differ from one location to another but it can be seen that the order of magnitude of the values remains the same although the conditions and the duration of the peaks are very different. It is surprising to see the relatively good agreement between the theoretical and the actual results, given the simple assumptions that were made, particularly the two listed below, in conducting the analysis :

- For a given period of time the same values of a and b apply to all road users.
- The equation for the nuisance factor is linear in δs, which can lead to very large, and hence not very realistic, differences between the desired and finally selected times.

In fact it seems possible that the observed slopes for w correspond to "average values" of a and b that arise as a result of the type of responses described below : we only consider the case of b, i.e., the part of the curve preceding the peak value of w :

- Road users for whom the slope is greater than b depart at their desired time but this type of response in itself does not determine the slope of the curve.
- Those for whom the slope approximates to b behave as predicted by the theoretical analysis, providing that the time difference that they are expected to accept is not too great.
- Those for whom the slope b is less than that given by the curve should in theory all depart at the beginning of the congestion but in fact they do not do so since this would involve an unacceptably large time difference (introduction of a δs^2 term for g). The subjects can nevertheless accept fairly large time differences.
- The time differences accepted by the last two categories of subjects compensate for the refusal of those in the first category to accept time differences.

It should be noted that the abrupt change in slope of the curve for w predicted by the theory cannot occur in practice since such a change would involve a significant discontinuity in the rate of flow of vehicles into the bottleneck.

7. SUMMARY OF THE RESULTS OF THE ANALYSIS

We list below the different cases that have been considered together with a reference to the corresponding conditions :

Case 1 : Departure time the determining factor (conventional approach)

Case 2a : Departure time the determining factor, peak not very concentrated

Case 2b : Departure time the determining factor, peak fairly concentrated

Case 2c : Departure time the determining factor, peak very concentrated

Case 3 : Arrival time the determining factor

Cases $2c^*$ and 3^* : Particular versions of cases 2c and 3 respectively. Desired rate of flow of vehicles outside the peak the same before and after the peak. Sufficient concentration of the peak.

The main results obtained for the different cases were as follows :

Cases 1 and 2a : These cases are the same. It should be noted that the "marginal cost" due to the arrival of an additional vehicle during the congestion at time t amounts to $t_2 - t$ (duration of the congestion beyond time t).

Cases 2c and 3 : The curves of w as a function of t have the same shape and are made up of two straight lines for both of these cases (see figures 2 and 3). The total waiting time is given by $W = \frac{1}{2}\frac{ab}{a+b}T^2C$

Case 2b : Lies between the two previously mentioned cases.

Cases 2c* and 3* : If we let C be the desired rate of flow of vehicles outside the peak and D the overall additional demand accounting for the peak we have :

$$T = \frac{D}{C(1 - \tau)}$$

and

$$W = \frac{1}{2}\frac{ab}{a+b}\frac{D^2}{C(1 - \tau)^2}$$

In addition it was found that the order of magnitude of the values of the parameters a and b for case 3 amounted to :

$$a \simeq 0.20$$

$$b \simeq 0.15$$

whence :

$$\frac{ab}{a+b} \simeq 0.10$$

8. IMPLICATIONS IN THE CASE OF SOME PARTICULAR FACTORS

Our purpose in conducting the analysis was not to be exhaustive, but rather to identify the difference that could arise in different circumstances. We accordingly limit our considerations here to two of the situations mentioned in section 7:

- Situation A covering cases 1 and 2a (identical to the case of predetermined departure)
- Situation B corresponding to cases 2c* and 3* where we assume in addition that $\frac{ab}{a+b} = 0.10$ so that we have (D being the additional overall demand due to the peak) :

$$W = \frac{0.10}{2} \quad \frac{D^2}{C(1 - \tau)^2}$$

8.1 Marginal costs of congestion : effects of variation in capacity

The "marginal cost" considered here is the additional total delay due to the arrival of an additional vehicle at time t during the congestion. A knowledge of this cost is of interest to the economists (taxation problems) and it also enables us to determine the consequences of taking limited actions to improve the situation.

In the case of situation A we have seen that this cost amounts to $c = t_2 - t$, the time for which the congestion continues beyond time t. This cost varies from T at the beginning to zero at the end of the congestion.

In the case of situation B the cost is determined by deriving a value for W as a function of D. We then have $c = 0.10 \frac{D}{C(1 - \tau)^2} = 0.10 \frac{T}{1 - \tau}$. Thus the cost is independent of the value of t.

It can be seen that we have very different results for these two situations and in situation B the value of c depends to a large extent on the rate of flow of vehicles on either side of the peak (the value tending towards infinity when τ tends towards 1). This shows that benefit calculations (e.g. for any traffic deviation measures or for increases in capacity) carried out assuming that we are dealing with a situation A when in fact we really have a situation B can lead to completely false results.

In particular, the effect of a small variation in capacity dC amounts to

$$dC \int_{t_1}^{t_2} c\,dt = \frac{T^2}{2}\,dC \qquad \text{for situation A}$$

and to

$$0.10\,\frac{T^2}{1-\tau}\,dC \qquad \text{for situation B.}$$

8.2 Staggering of working hours

If there is little staggering of working hours, then the home to work and the work to home trips can be considered to belong, as a first approximation, to situation B. For a demand D there is a corresponding total waiting time $W = kD^2$ where k is a proportionality coefficient whose value is independent of D. If D is made up of a number of separate demands D_i that are sufficiently spaced in time ($D = \sum_i D_i$) we will have $W = k \sum_i D_i^2 < kD^2$. If, for example, D is made up of two equal parts D_1 and D_2, then the total waiting time will be reduced to one half (see figure 4). If, however, D_1 and D_2 are not sufficiently spaced in time, we can have the situation illustrated by the graph of figure 4b ; the total waiting time

$$W = \int_{-\infty}^{+\infty} w\,ds,$$

which can be evaluated, given the value of T (unchanged) and the slopes a and b of the curve, will then be found to be higher.

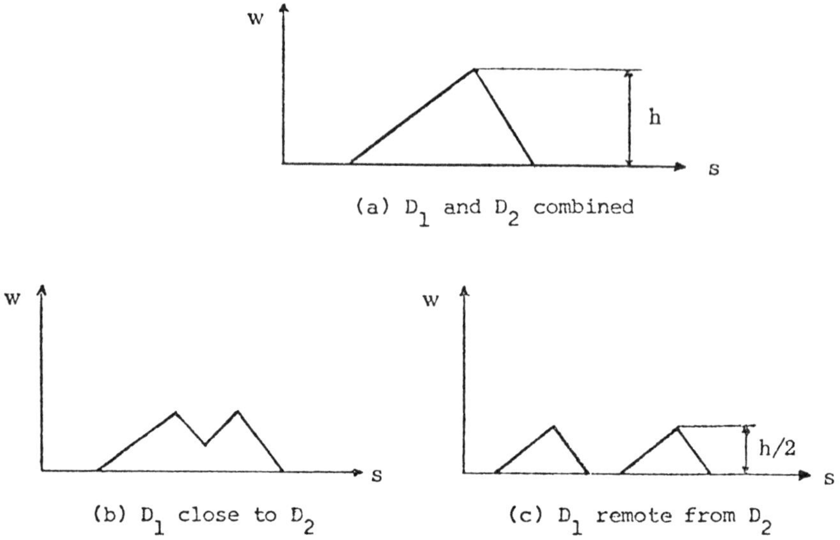

Fig. 4 - Staggering of working hours

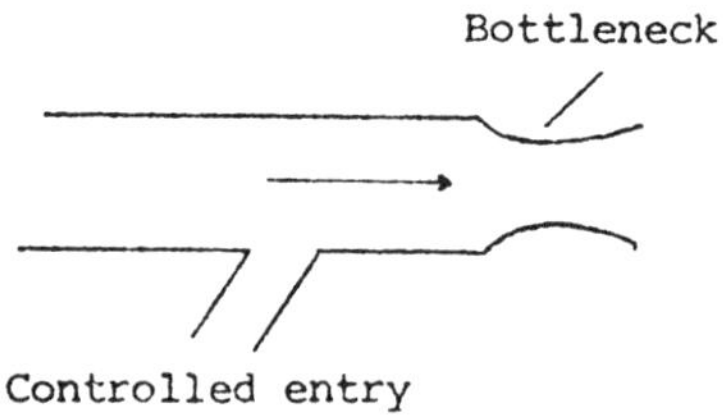

Fig. 5 - Controlled access to a motorway

8.3 Controlling the access to a motorway

Consider a section of a motorway where there is a bottleneck for which the duration of the congestion amounts to T and where there is an important entry upstream of that bottleneck (figure 5). We wish to study the effects of controlling the access to the motorway at this entry in such a way that the flow of traffic through the bottleneck is maintained exactly at its saturating capacity, assuming that there is no deviation of traffic to the surface roads.

In the case of a situation A, the total waiting time W will remain the same. This waiting time will simply be transferred from the bottleneck to the motorway entry and concentrated on a lower rate of flow of traffic. This result can be derived by considering the "motorway section plus the entry" system (including the queue of vehicles). The flow of traffic into and out of this system is not affected by the control of the access to the motorway.

In the case of a situation B, the duration of the congestion T and the corresponding curve of w versus time are transferred, unchanged, from the bottleneck to the entry to the motorway. However, since the individual waiting times w are now suffered by a smaller flow of vehicles, the total waiting time W will be less. It should also be noted that the main flow of traffic, for which there is now no congestion, will be more concentrated ; the resulting shifts in the times of departure of the people making use of the motorway entry can be rather large.

CONCLUSIONS

The results of this theoretical analysis show that, under certain assumptions, it may be necessary to take account of the way in which road users advance or delay their departure times in order to avoid being involved in congestion. In order to have more detailed information on this subject it would be necessary to collect some data on the way in which waiting times develop in various locations where there is congestion and to conduct inquiries concerning the behaviour of the road users. However, we should not overlook other possible effects of traffic congestion on demand, namely the displacement of traffic in space, changes in the mode of travel and the suppression of certain journeys.

REFERENCES

1. R.J. SMEED, "Some Circumstances in which Vehicles will Reach their Destinations Earlier by Starting Later," Trans.Sci.1, 308-317 (1967)
2. V.F. HURDLE, "The Effect of Queueing on Traffic Assignment in a Simple Road Network," Proceedings of the Sixth International Symposium on Transportation and Traffic Theory, Sydney, Australia (1974)
3. C.E.M. PEARCE, "Time Dependence in Commuter Traffic Models," Trans.Sci.9, 289-307 (1975)
4. P.H. FARGIER, J.M. MORIN and D. DUCORNET, "Reducing Travel Time by Freeway Ramp Metering, Especially when Peak Traffic Demand Exceeds Corridor Capacity," Proceedings of the International Symposium on Traffic Control Systems, Berkeley, California (August 1979)

5. J.M. MORIN, "Controlling the Access to Saturated Urban Motorways Corridors (Le contrôle d'accès dans les corridors autoroutiers urbains saturés), T.E.C. (Transport Environnement Circulation) n° 39,44-50 (March-April 1980)
6. V.F. HURDLE, "Equilibrium Flows on Urban Freeways" To appear in Trans.Sci. 15, 255-293 (1981)
7. Donald P. GAVER Junior, "Headstart Strategies for Combating Congestion," Trans.Sci.2, 172-181 (1968)
8. A.S. ALFA and D.L. MINH, "A Stochastic Model for the Temporal Distribution of Traffic Demand - The peak Hour Problem," Trans. Sci.13, 315-324 (1979)
8. C. HENDRICKSON and G. KOCUR, "Schedule Delay and Departure Time Decision in a Deterministic Model," Trans.Sci.15, 62-77 (1981)

APPENDIX I

ORDER OF ENTRY INTO OR EXIT FROM THE QUEUE IN THE CASE OF EQUILIBRIUM SOLUTIONS

We want to prove that, without departing from the general case, it is possible to confine ourselves to equilibrium solutions for which the assumed entries into (or exits from) the queue take place in the same order as would the desired entries (or exits). We deal with case 2, on considering the entries, but the proof readily applies to case 3 if one considers the exits from the queue.

Consider, in the case of an equilibrium solution, two users for which the actual and desired times of entry and the waiting times are e_1, $\hat{e}_1$, w_1 and e_2, $\hat{e}_2$, w_2 respectively and where $e_1 < e_2$ and $\hat{e}_1 > \hat{e}_2$. The behaviour of user 1 being optimum we have :

$$g(e_1 - \hat{e}_1) + w_1 \leqslant g(e_2 - \hat{e}_1) + w_2. \qquad (1)$$

Similarly, for user 2 we have :

$$g(e_2 - \hat{e}_2) + w_2 \leqslant g(e_1 - \hat{e}_2) + w_1. \qquad (2)$$

On adding the individual terms of (1) and (2) we obtain:

$$g(e_1 - \hat{e}_1) + g(e_2 - \hat{e}_2) \leqslant g(e_2 - \hat{e}_1) + g(e_1 - \hat{e}_2). \qquad (3)$$

We also have :

$$e_1 - \hat{e}_1 < e_2 - \hat{e}_1 < e_2 - \hat{e}_2 ,$$
$$e_1 - \hat{e}_1 < e_1 - \hat{e}_2 < e_2 - \hat{e}_2 .$$

Given that the function g is convex, we can derive the inequality inverse of (3). This confirms that the equality within relationship (3) is valid. The same applies in the case of the relationships (1) and (2) and this indicates that if we invert the times of entry into the queue of the two users we will still have an equilibrium solution.

Thus for any equilibrium solution we can, on making the required number of inversions, allow the users to enter the queue in the desired order without affecting the characteristics of the solution (way in which Q and w change, value of W, etc.).

APPENDIX II

EXISTENCE AND UNIQUENESS OF THE EQUILIBRIUM SOLUTION IN CASE 2

We make use of the same notation here as in the main text of the paper.

Let $\hat{t}_3$ be the end of the period when $\hat{E} > C$ (see fig. 6). When $\hat{E} > C$, then it must also be the case that $Q > 0$. Thus $t_1 \leqslant \hat{t}_1$, which implies that $t_2 \leqslant \hat{t}_2$ and $t_2 \geqslant \hat{t}_3$. Thus we have that $\hat{t}_3 \leqslant t_2 < \hat{t}_2$.

Let $$\tilde{Q}^*(t) = \int_{t_1}^{t} (\hat{E} - C)dt.$$

Between times t_1 and t_2 we have $\tilde{Q}^*(t) = \tilde{Q}(t) + \tilde{Q}^*(t_1)$.

We now assume that t_2 is known and construct a Q^*/C relationship based on the $\tilde{Q}^*/C$ curve in the same way as we constructed a Q/C relationship based on the $\tilde{Q}/C$ curve, i.e., by proceeding backwards from the point which has an abscissa value of t_2 (see figure 6) :

- We draw a straight line DF if the slope of $\tilde{Q}^*/C$ at time t_2 is less than -a.
- Starting from point F, if it exists, or from point B, if we have a point where the slope is greater than b, we draw a straight line BA having a slope b (the slope of $\tilde{Q}^*/C$ at B is b).

The curve of $\tilde{Q}^*/C$ includes these straight lines, where they exist, while elsewhere it is identical to the $\tilde{Q}^*/C$ curve.

Providing that t_2 has been properly selected, t_1 must satisfy the relationship $Q^*(t) = Q^*(t_2)$ and correspond to a local minimum of Q^*.

Conversely, if these two requirements are met, then the function Q, which is equal to $Q^*(t) - Q^*(t_1)$ between the times t_1 and t_2 and to zero outside these times, corresponds to an equilibrium solution (a local minimum at time t_1 means that $(E - C)$ has a zero value or that it is passed through such a value).

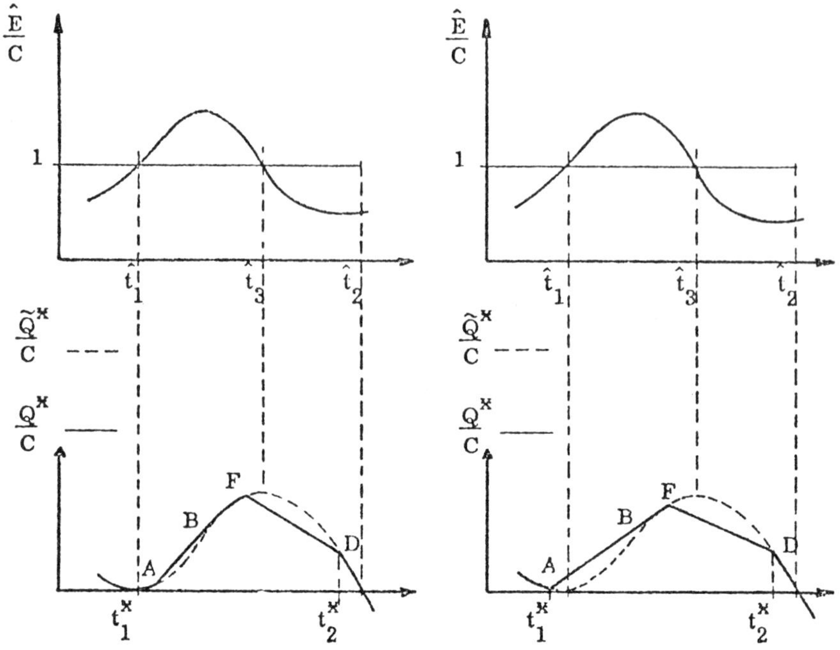

Fig. 6

$\frac{Q}{C}^*$ and $\frac{Q}{C}^*$ curves

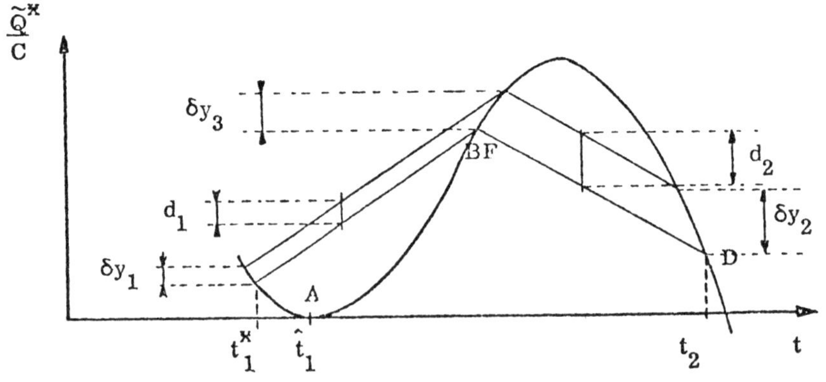

Fig. 7 - Variations in the magnitudes of y_1 and y_2

For a given t_2 we consider :

- t^*_1 the unique value corresponding to a local minimum of Q^*.
- $y_1 = \frac{Q^*}{C}(t)$ and $y_2 = \frac{Q^*}{C}(t_2)$

We now need to prove that there is always one and only one value of t_2 for which $y_1 = y_2$.

When $t_2 = \hat{t}_2$ we have : $y_2 = 0$ $y_1 \geqslant 0$ $(y_2 - y_1) \leqslant 0$,

and when $t_2 = \hat{t}_3$ we have : $(y_2 - y_1) > 0$.

Now we simply need to show that $y_2 - y_1$ increases in a strictly monotonic manner as t_2 decreases from $\hat{t}_2$ to $\hat{t}_3$.

If at least one of the straight lines DF or AB does not exist or if points F and B are distinct, then y_1 does not vary with t_2.

If points B and F coincide, then it can be seen from figure 7 that

$$\delta y_1 < d_1 < \delta y_3 < d_2 < \delta y_2 \qquad \text{i.e.,} \quad \delta y_1 < \delta y_2 .$$

In all cases, $(y_2 - y_1)$ increases in a strictly monotonic manner.

This completes the proof of the existence and uniqueness of the equilibrium solution.

OPTIMAL STATION LOCATION
FOR A TWO HIERARCHY TRANSIT SYSTEM

Masaharu Fukuyama
Instituto Militar de Engenharia
Rio de Janeiro

Abstract

A two hierarchy transit system is operated along a public transportation corridor. The second mode (transit) in the hierarchy offers fast linehaul travel for longer trips. The first mode (access mode), on the other hand, serves the access-trips to transit or local trips. The location of the transit station in such a system is determined.

The objective is to maximize the total benefit, which is defined as the sum of the savings in travel time of all travellers gained by using transit instead of completing their trips only by the access mode. The effect of the transfer cost, however, is not included.

The main emphasis is placed on the development of methods to determine the optimal station locations which reflect the general multiple origin and destination trip distribution pattern.

A Dynamic Programming formulation and an analytical model are developed. A formula that determines the optimal location of stations if the majority of people benefit from transit travel is given. In this computation, the trip characteristics are represented only by the trip origin density and destination density (the marginal distribution of the OD distribution). The Dynamic Program can be applied to more general conditions than that for the formula, but its computation requires the OD distribution table.

1. Introduction

It is common to see two transportation modes coexisting along one public transportation corridor and forming a hierarchy so that the first mode serves as a feeder or the access mode for the second. The second mode, for example, may be an express train, train or bus if the first mode is a local train, bus or walking, respectively.

This study will develop a method to determine the station locations for the second mode provided that the configuration for the first mode, such as the location of stops or travel speed, is given and fixed.

In the following we will refer to the first and second mode as "access mode" and "transit" respectively. It will be assumed that the trip origins and destinations are distributed only along the corridor and that the transfer cost between the two modes is small enough so as not to affect the station spacings (or the station locations). Further, for simplicity, it is assumed that the transit vehicle stops at all the points which are assigned as the stations.

The total benefit associated with the two hierarchy transit system is defined as the total saving in travel time to all travellers gained by using transit instead of completing their entire trip only by the access mode. The objective is to maximize this benefit.

We are especially concerned with the total benefit for the general multi-origin and destination trip distribution and for the general two hierarchy transit system in determining the optimal station location. This type of benefit has not been considered by any other previous studies such as Bendtsen[1], Fukuyama[2], Ghoneim[3], Hurdle and Wirasinghe[5], Landau[6], Tomii and Kawachi[8], Vuchic[9], Vuchic and Newell[10], Wirasinghe, Hurdle and Newell[12], Wirasinghe[11,13,14,15] and others.

2. Benefit

In deriving the benefit, it will be assumed that all travellers will choose a route of minimum travel time. Transit and the access mode operate between the points O and E shown in Fig. 1. The length of this interval will be denoted by L and the (N-1) stations are to be allocated between O and E (O and E are the Oth and Nth station respectively).

The origin and destination points of the travellers and the location of any point will be measured from the point O and be denoted by x, y and z respectively. If the points

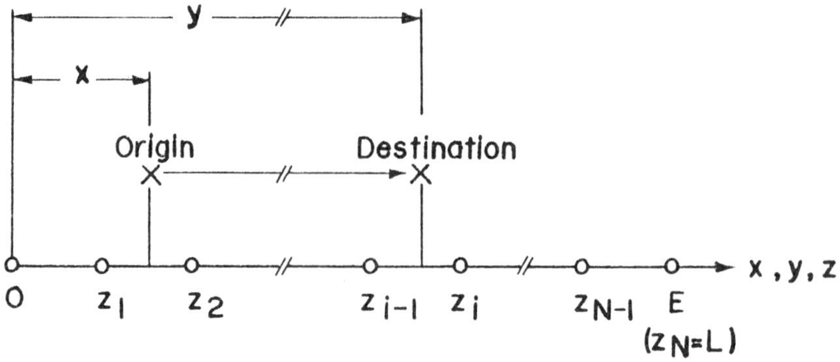

Fig. 1 Co-ordinate system

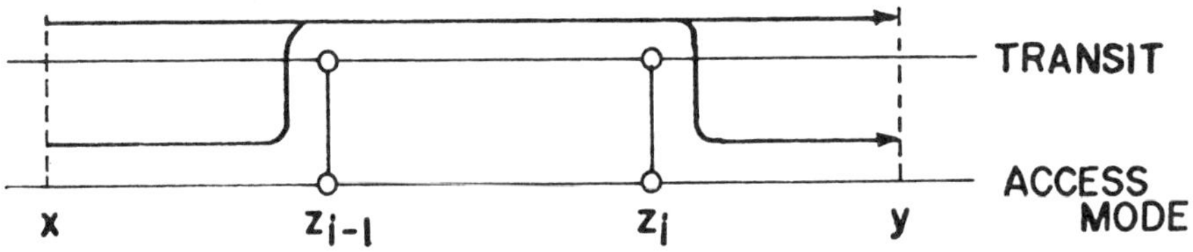

Fig.2 a ss trip ($x \leqq z_{i-1}$, $z_i \leqq y$)

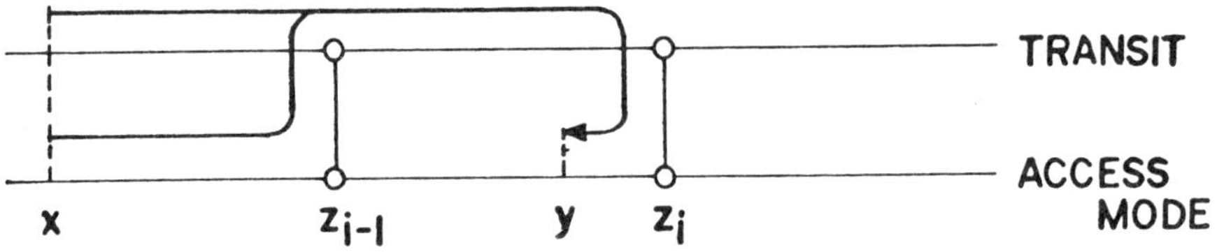

Fig.2 b sb trip ($x \leqq z_{i-1}$, $z_{i-1} < y < z_i$)

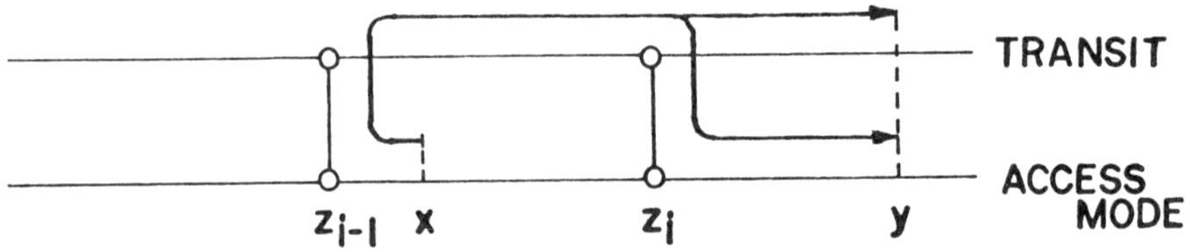

Fig.2 c bs trip ($z_{i-1} < x < z_i$, $z_i \leqq y$)

Fig. 2 Types of trip

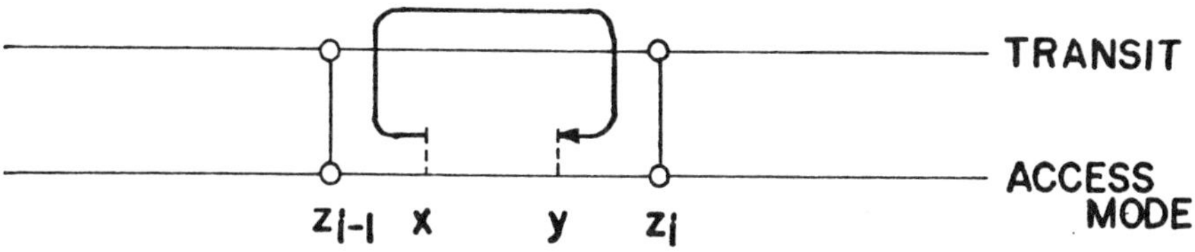

Fig.2 d bb trip ($z_{i-1} < x, y < z_i$)

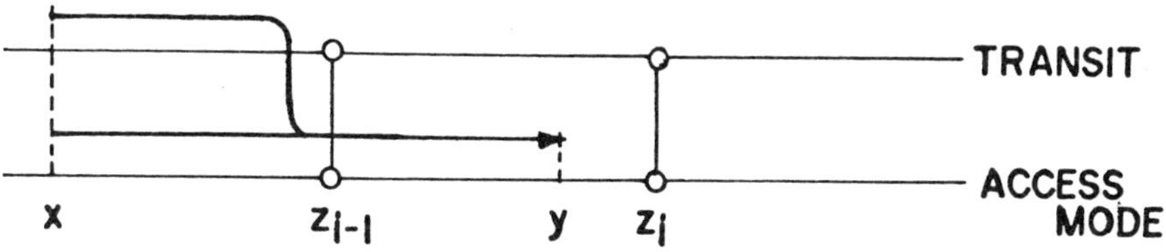

Fig.2 e sn trip

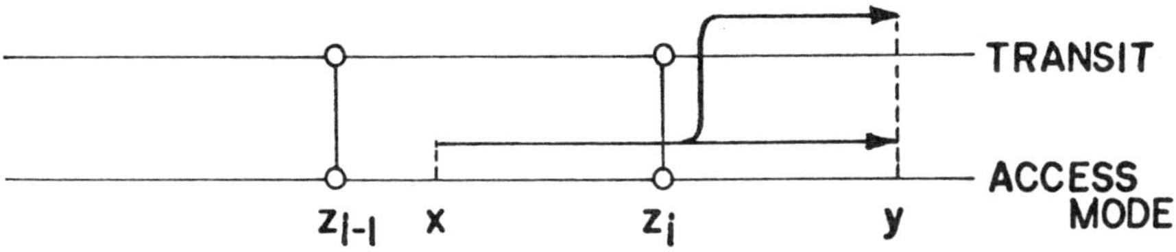

Fig.2 f ns trip

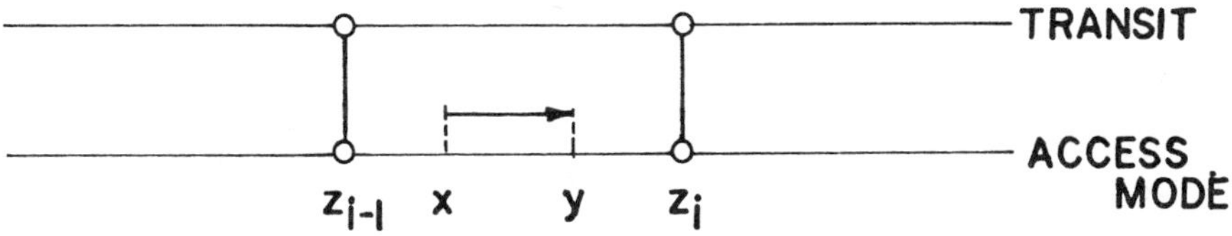

Fig.2 g r trip

Fig.2 Continued

are specified by some subscripts, i, j, k etc., the distance from the point O to these points will be expressed as z_i, z_j, z_k, etc.

The number of trips which originate per unit length at x and terminate per unit length at y per unit period of time will be denoted by $f(x, y)$, $0 \leqq x \leqq L$ and $0 \leqq y \leqq L$.

The average velocity of transit in a station interval will be denoted by V_{jk}, where the first and second subscripts designate the starting and arriving stations of transit in the interval. This velocity may depend on the cruise speed, acceleration and deceleration of transit and also the station spacing. For simplicity, it will be assumed that the velocity of the access mode, v, is constant along the line and for both directions.

To determine the benefit for $x \leqq y$, consider a section between successive stations at z_{i-1} and z_i $(i=1,2,..,N)$. The types of trips through this section by transit will be classified as

- ss trip : trips which travel through the section,
- sb trip : trips which travel to the station at z_i and backtrack to their destination,
- bs trip : trips which backtrack to the station at z_{i-1} and travel beyond the station at z_i,
- bb trip : trips which backtrack at both stations, z_{i-1} and z_i.

These trip types are shown in Fig. 2 a,b,c and d respectively. The regions of x and y for each trip type are specified in the figure. There are three other types of trips associated with this section which do not involve transit. These trip types are shown in Fig. 2 e,f and g and defined respectively as

- sn trip : trips through the point z_{i-1} which terminate within the interval,
- ns trip : trips which originate within the interval and continue beyond the point z_i,
- r trip : trips which originate and terminate within the interval.

The benefit for an ss trip, Bss, between the (i-1)th and ith stations can be written as

$$Bss = \left(\frac{1}{v} - \frac{1}{V_{i-1,i}}\right)(z_i - z_{i-1}) - S_{i-1} \quad (=a_i) \tag{1}$$

where S_{i-1} is the dwell time of transit (which travels to the right in Fig. 1) at the (i-1)th station and is assumed to be given.† The a_i should be non-negative, otherwise the traveller loses his benefit by travelling this section by transit.

The benefit for an sb trip, Bsb, is the difference of the travel time between sn and sb trips,

$$Bsb = \frac{y - z_{i-1}}{v} - \left(\frac{z_i - z_{i-1}}{V_{i-1,i}} + \frac{z_i - y}{v} + S_{i-1}\right)$$

$$= a_i - \frac{2}{v}(z_i - y). \tag{2}$$

The second term in eq. (2) can be interpreted as the penalty caused by backtracking. In the same way, the benefits for a bs trip, Bbs, and bb trip, Bbb, are

$$Bbs = a_i - \frac{2}{v}(x - z_{i-1}), \tag{3}$$

$$Bbb = a_i - \frac{2}{v}(x - z_{i-1}) - \frac{2}{v}(z_i - y). \tag{4}$$

These benefits are defined on the regions of x and y specified in Fig. 2 a,b,c and e. Further, eqs.(2),(3) and (4) are defined only for the region where each equation takes non-negative values. Therefore, we obtain the benefit surface for this station interval as shown in Fig.3 for i=2, 3,..,N-1.

For i=1 and i=N, the regions for ss and sb trips and for ss and bs trips respectively are not defined; consequently these benefits will be cut off from the figure. Especially if there is no intermediate station, only bb trips will be defined.

In general, if the positions of (N-1) intermediate sta-

---

†Ghoneim[3] and Wirasinghe[13,15] have discussed a dependence of stopping time upon the number of people who board and alight at stations.

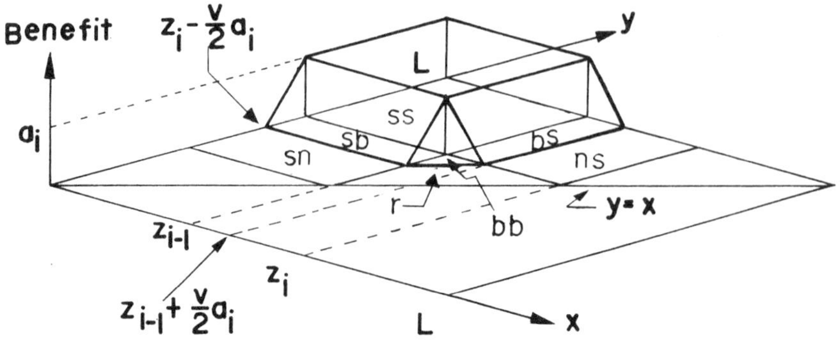

Fig. 3 Benefit surface for station interval between z_{i-1} and z_i

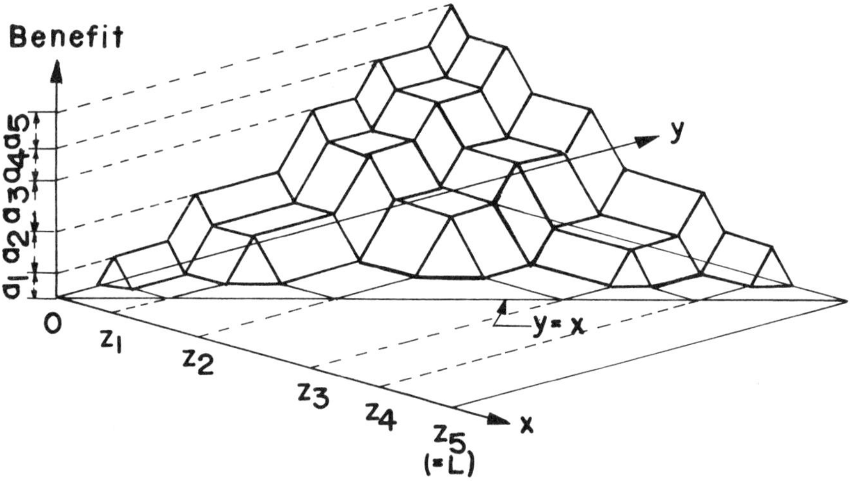

Fig.4 Benefit surface for individual travellers

tions are specified, we have N such figures. The benefit of the travellers who originate at x and terminate at y then can be obtained by considering an analogy. Suppose that one could make the shapes of the benefits for all the intervals from some material which has no vertical shearing resistance, then put them one by one on their corresponding region on the x-y space. By doing so, the benefit surface could be obtained as shown in Fig.4. Note that the benefit surface intersects the x-y plane at the shed boundary between transit travel and non-transit travel.

The total benefit can be obtained either by integrating the product of the benefit at (x, y) and f(x, y) over the region or first by computing the total benefit corresponding to each station interval, then summing it over all the intervals. We will employ the latter because it will yield the simpler form in the expression of the benefit.

Referring to Fig.3, the total benefit associated with the interval between z_{i-1} and z_i, $A_{i-1,i}$, can be written as

$$A_{i-1,i} = \int_{z_i}^{L} \int_{0}^{z_{i-1}} a_i \; f(x, y) \; dx \; dy$$

$$+ \int_{z_i-\frac{v}{2}a_i}^{z_i} \int_{0}^{z_{i-1}} \{a_i - \frac{2}{v}(z_i - y)\} f(x, y) dx \; dy$$

$$+ \int_{z_i}^{L} \int_{z_{i-1}}^{z_{i-1}+\frac{v}{2}a_i} \{a_i - \frac{2}{v}(x - z_{i-1})\} f(x, y) dx \; dy$$

$$+ \int_{z_i-\frac{v}{2}a_i}^{z_i} \int_{z_{i-1}}^{y-(z_i-z_{i-1})+\frac{v}{2}a_i} \{a_i - \frac{2}{v}(z_i - y)$$

$$- \frac{2}{v}(x - z_{i-1})\} f(x, y) \; dx \; dy, \qquad (5)$$

where the first through fourth terms correspond to the total benefit for ss, sb, bs and bb trips respectively. Note that the above expression automatically satisfies the case with i=1 and i=N and also the case with no intermediate station between O and E.

Finally the total benefit of all travellers who travel to the right in Fig. 1 and over all station intervals, T, becomes

$$T = \sum_{i=1}^{N} A_{i-1,i}. \qquad (6)$$

The benefit for $x > y$ is obtained simply by exchanging $V_{i-1,i}$ and S_{i-1} for the rightward travelling transit with $V_{i,i-1}$ and S_i for the leftward travelling transit respectively in eq. (1), and also x,y, dx and dy with y,x, dy and dx respectively, except in f(x, y) in eq.(5). Let $A_{i,i-1}$ be this benefit. The total benefit, then, can be written as

$$T' = \sum_{i=1}^{N} (A_{i-1,i} + A_{i,i-1}). \qquad (6a)$$

Here the objective is to determine z_i, i=1,2,...,N-1 and N, which maximize eq.(6a), given that $z_0 = 0$ and $z_N = L$.

If, however, the trip distribution is symmetrical with respect to y=x and the performance of transit is the same for both directions, eq.(6) can be maximized instead of eq.(6a). Further, if there is no restriction that the stations for both directions must be located at the same position (location of bus stop for example), the two terms in eq.(6a) can be maximized independently.

There may be several ways to attack this problem. First, the necessary condition which maximizes eq.(6a) will be obtained by taking the partial derivatives of the equation with respect to z_i (i=1,2,...,N-1) and by letting them be zero. By doing so, we will have simultaneous equations, each of which contains three successive variables: z_{i-1}, z_i and z_{i+1}. This fact coincides with the results which Landau[6], Vuchic[9], and Vuchic and Newell[10] have derived for models with multiple origins and a single destination. Actually, it can be shown that the condition derived by them is a special case of this model. Vuchic solved his model by using a finite difference method. This model might also be solved in a similar manner.

As can be seen in eq. (5), the benefit gained in one station spacing, i.e., $A_{i-1,i} + A_{i,i-1}$, depends only on the elements which are related to this interval, which, conversely, indicates that this benefit does not depend on how one allocates the stations between O and z_{i-1} or between z_i and E. Therefore one can use Dynamic Programming to maximize eq.(6a). This approach will be discussed in the next section. Tomii and Kawachi[8] employed Dynamic Programming to determine the stops of an express train for a model with multiple origins and a single destination, also considering the effect of transfers. They, however, assumed the same technology for local and express trains; consequently the saving in the travel time was represented by the sum of the stopping penalties. Their model, therefore, cannot be applied to the general two hierarchy transit system.

One may also apply continuum approximations and the calculus to this problem. The application of this method to the determination of some parameters related to the operation of public transportation systems has been indicated by Newell[7]. Hurdle[4], Wirasinghe, Hurdle and Newell[12], Ghoneim[3] and Wirasinghe[11,13,14,15] have discussed some issues related to station location with this method. The benefit derived in this section will be simplified by introducing the continuum approximation and some assumptions in section 4. Explicit formulas for the optimal station spacing will be derived in section 5.

3. Dynamic Programming formulation

For the Dynamic Programming formulation we need to specify a set of possible locations for the stations along the transit line. For example, these points may be the stations of the local transit if the problem is to determine the stops for express transit. If there is no preferred set of points, one may choose the locations and the number of these points arbitrarily. The accuracy of the solution may be increased as one increases the number of points, but the computation time will also be increased. Let M be the number of these points, including the point E.

Suppose that transit stations are allocated at the jth and kth points ($j < k$) with no intermediate station. The benefit for an ss trip for this station interval, a_{jk}, will be obtained by substituting subscripts j and k for i-1 and i in eq.(1). Then the benefit gained from this station interval, A_{jk}, will be obtained by substituting a_{jk} and subscripts j and k for a_i and subscripts i-1 and i respectively in eq. (5). The benefit for the opposite direction, A_{kj}, is also obtained in the same way as discussed previously.

The total benefit to all travellers can be obtained by adding $(A_{jk}+ A_{kj})$ from j=0 to k=M for each of the 2^{M-1} possible choices of intermediate stations. Our objective is to find the set of intermediate stations which gives the maximum total benefit.

We define the optimal value function as

TB(k) = the maximum benefit gained from the section between 0 and k by allocating the stations optimally in this section.

The recurrence formula, then, will be written as

$$TB(k) = \max_{j=0,1,2,\ldots,k-1} \{TB(j)+ A_{jk}+ A_{kj}\}, \quad (7)$$

for k=0,1,2,...,M.

Eq.(7) can be computed, starting with the obvious boundary condition, TB(0) = 0, successively for k=1,2,..,M. The optimal policy function, P(k), contains j which gives the maximum value of the right-hand side of eq.(7) for each k; i.e., P(k) contains the optimal location of the station immediately before k. The maximum benefit gained from this transit line is obtained at TB(M) and the optimal location of the station will be found by tracing back P(k) from k=M to 0.

This Dynamic Program is equivalent to finding the maximum benefit route of the graph shown in Fig. 5. In the figure each arc has the benefit $(A_{jk}+ A_{kj})$ and the total benefits will be those associated with the routes between 0 and M.

If some stations must be located between O and E, j* for example, for reasons other than the maximization of the total benefit as defined here (e.g., j* may be a major transfer point between transit lines), then all the routes pass through the point j*. In such a case, obviously the optimal route from the points O and j* and that from j* and E can be found individually. If there are several points along the transit line which must be stations for some reason, the opti-

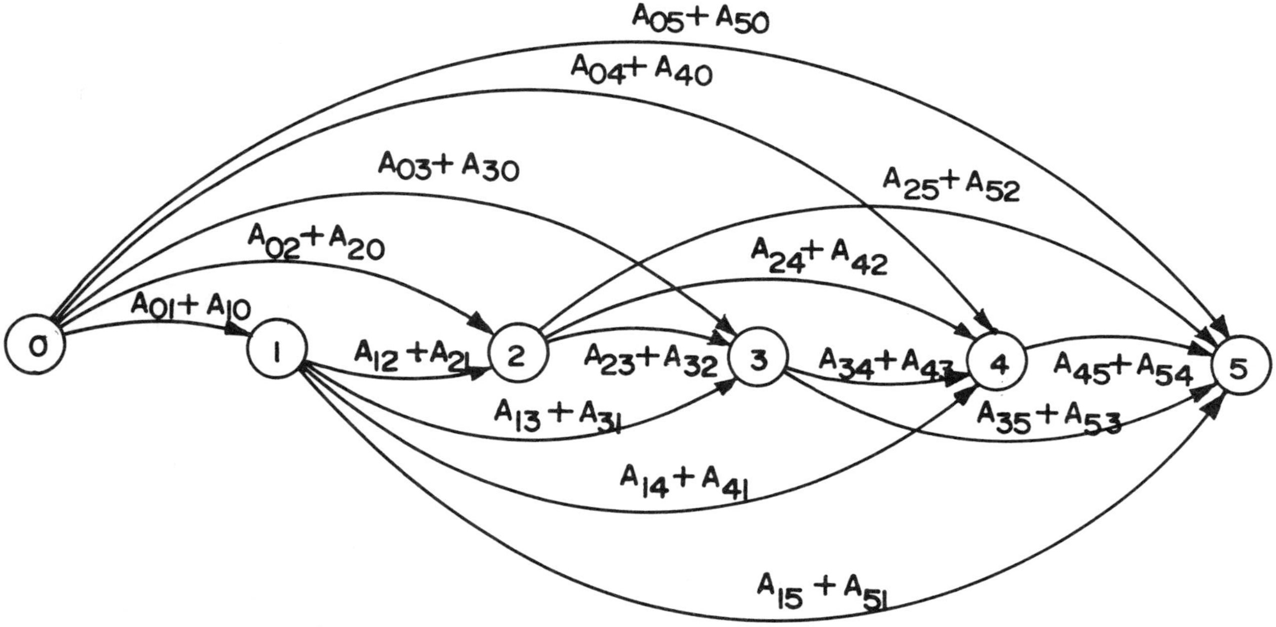

Fig. 5 A graph corresponding to the Dynamic Programming formulation (for M = 5)

mization can be performed for each of the intervals between these points individually. The trips which have their origins and destinations outside of the section may be treated as if they originate and/or terminate at the boundaries of this section.

When an arbitrarily large number of proposed points for the stations are chosen, some of the distances between these points become so short that the saving in the travel time given by the difference in the velocity between the two modes could not compensate for the penalty of stopping, thus giving a negative a_{jk} (or a_{kj}). Since the benefit A_{jk} (or A_{kj}) could not be defined for negative a_{jk} (or a_{kj}), one may remove the arcs that correspond to such a section or give some large negative values to these arcs so that they will not be considered during the optimization.

4. Approximation of the Benefits

First, for simplicity, we assume that:

i) transit and the access mode operate on an infinitely long line,

ii) cruise speed of transit is constant along the line in both directions,

iii) transit can achieve its cruise speed in any station interval†,

iv) the dwell times of transit are equal at all stations.

Assumption i) is made in order to avoid the complexity caused by the boundary condition. With this assumption the benefit surface for each station interval will have the same shape, that is, the shape that would be obtained if the benefit surface in Fig.3 were extended beyond $x=0$ to $x=-\infty$ and beyond $y=L$ to $y=+\infty$.

† Wirasinghe[15] has derived two formulas which give the local optimals of station spacings for the case when transit does and does not reach its cruise speed. He did not show clearly, however, how one can choose the global optimal from the results.

In accordance with this assumption, one may choose any point along the line as the origin of the co-ordinates for x,y and z. f(x, y) will be defined on the region $-\infty < x, y < \infty$, and in order to avoid unnecessary confusion, we assume that

$$\int_{-\infty}^{\infty}\int_{-\infty}^{\infty} f(x, y)\, dx\, dy < \infty . \tag{8}$$

Let D(z) denote the station spacing with center at z. With the assumptions ii), iii) and iv), eq.(1) can be re-written as

$$a(z) = \left(\frac{1}{v} - \frac{1}{V}\right) D(z) - S, \tag{9}$$

where V = cruise speed of transit

S = penalty caused by the transit stop, i.e., dwell time at the station at the beginning of this station interval plus the delay during acceleration and deceleration in this interval.

First we will compute the benefit gained in one station interval by people who have the trip length $0 < H (= y-x)$, i.e., the benefit for given H. Let

$$\lambda(z) = D(z) - \frac{v}{2}\, a(z),$$

which corresponds to the maximum trip length of an r trip, i.e. non-transit travel, in the interval D(z).

Fig. 6 shows a projection of a cut section of the benefit surface for $H \geqq D(z)$ along the line $y = x+H$ on the x - Benefit plane. The benefit for given H, B(H), will be computed by integrating this benefit weighted by the number of people who lie along the line $y=x+H$, $f(x, x+H)$, with respect to x.

Now let us assume that:

v) f(x, y) varies slowly enough so that f(x, y) could be replaced by the constant values of f(z, z+H) and f(z-H, z) within the region $\{(x,y) \mid z- D(z)/2 \leqq x \leqq z,\ y = x+H\}$ and $\{(x,y) \mid z \leqq y \leqq z+D(z)/2,\ x= y-H\}$ respectively for $H \geqq \lambda(z)$.

With this assumption, the trapezoid-shaped benefit can be approximated by a rectanglar one(hatched area) shown in Fig. 7 a.

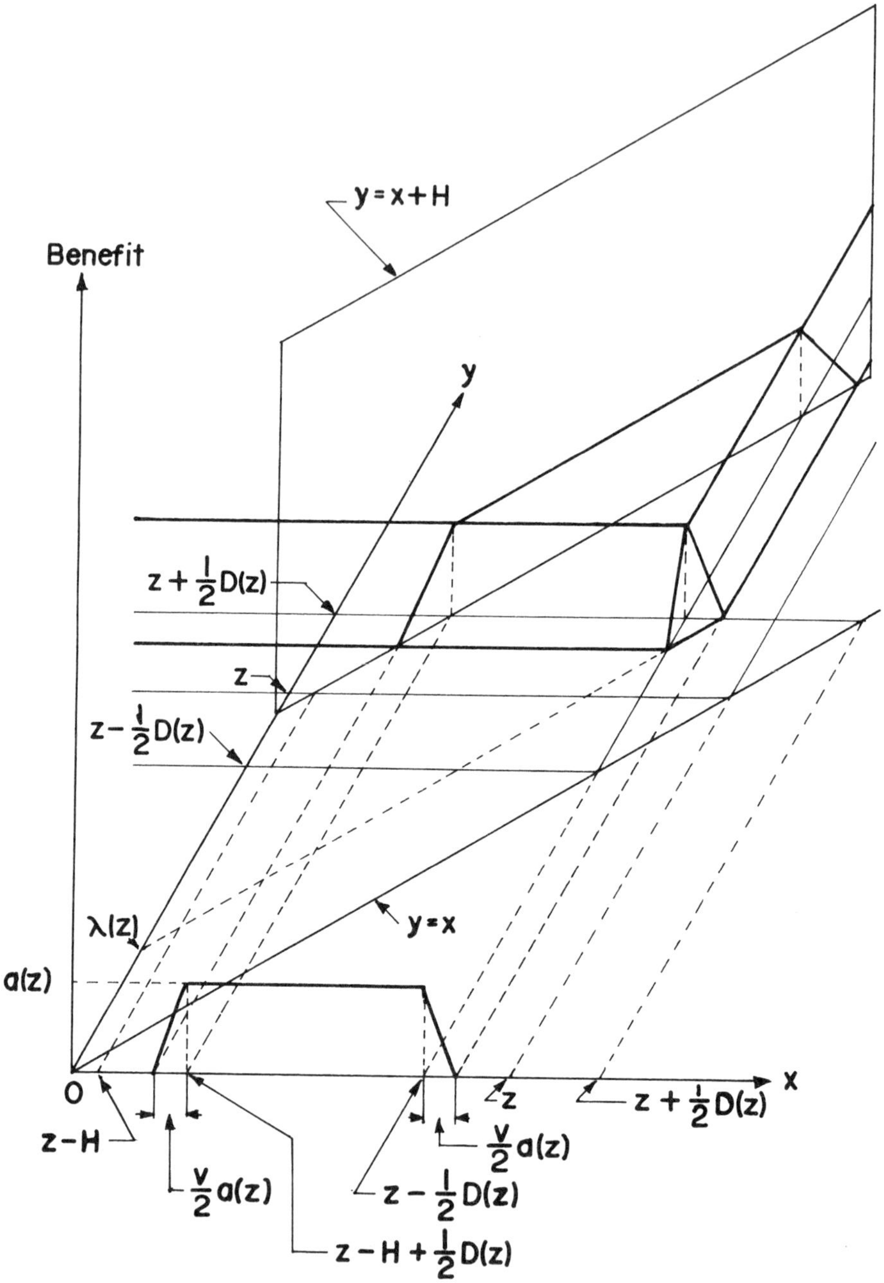

Fig. 6 Benefit surface and a cut section
($H \geqq D(z)$)

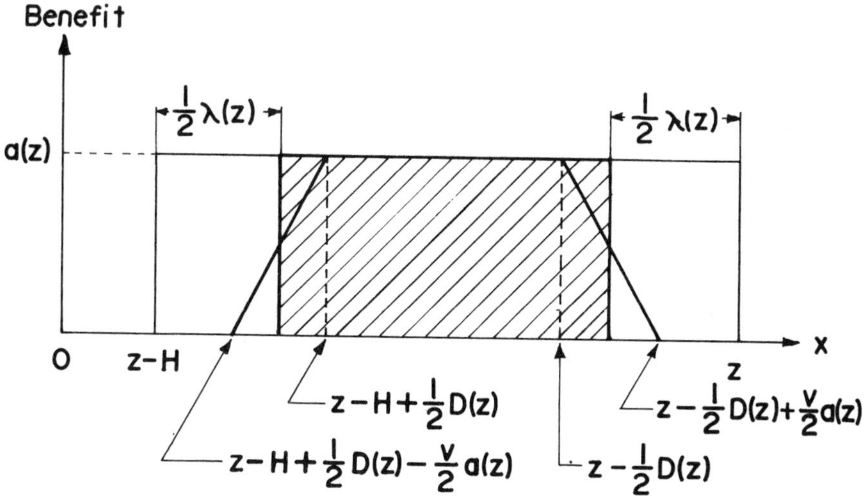

Fig. 7 a Projection of cut section for $H \geqq D(z)$

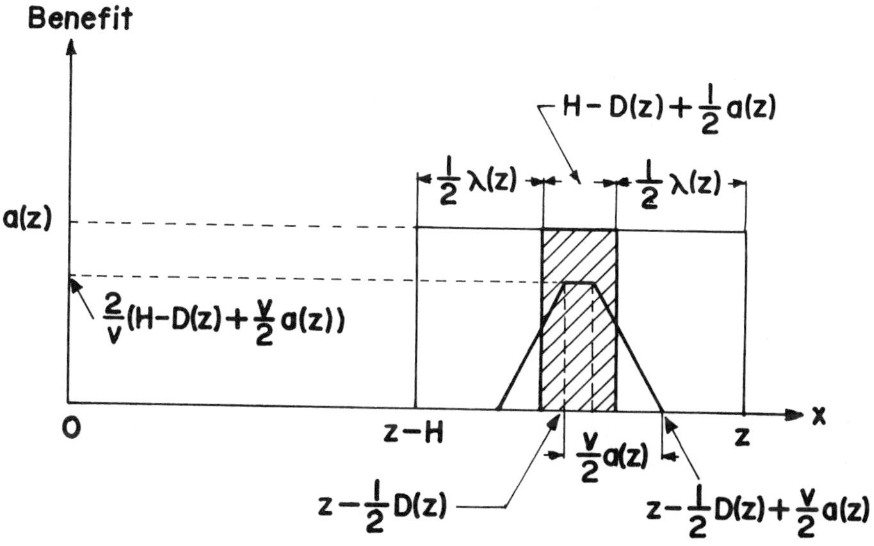

Fig. 7 b Projection of cut section for $\lambda(z) \leqq H < D(z)$

For $\lambda(z) \leqq H < D(z)$, the projection of the cut section is obtained as the trapezoid in Fig. 7 b. Since the area of the trapezoid is equal to that of the hatched rectangle with the assumption v, the benefit for this case also can be approximated by the rectangular shape.

From Fig. 7 a and b, one can construct the approximate benefit surface in three dimensional space as shown in Fig. 8.

Obviously $B(H) = 0$ for $0 \leqq H < \lambda(z)$. Referring to Figs. 7 a and b, the resultant formula of B(H) for all H can be obtained as

$$B(H) = \begin{cases} 0, & 0 \leqq H < \lambda(z) \\ a(z)\int_{z-H}^{z} f(x, x+H)dx - \{\frac{1}{2}a(z)\lambda(z)\} \\ \quad \times\{f(z-H, z) + f(z, z+H)\}, & \lambda(z) \leqq H \end{cases} \tag{10}$$

By integrating eqs. (10) with respect to H and then dividing by D(z), the average benefit per unit length at z, B(z), can be obtained as follows:

$$B(z) = \frac{a(z)}{D(z)} c_1(z) - \frac{1}{2}\frac{a(z)}{D(z)} \lambda(z)\{g_{1x}(z) + g_{1y}(z)\}, \tag{11}$$

where

$$c_1(z) = \int_{\lambda(z)}^{\infty}\int_{z-H}^{z} f(x, x+H)dx\, dH,$$
$$g_{1x}(z) = \int_{\lambda(z)}^{\infty} f(z, z+H)\, dH, \tag{12}$$
$$g_{1y}(z) = \int_{\lambda(z)}^{\infty} f(z-H, z)\, dH.$$

The $c_1(z)$ can be interpreted as the number of people who travel through the point z with trip length greater than $\lambda(z)$, and $g_{1x}(z)$ and $g_{1y}(z)$ as those who originate and terminate per unit length at point z with trip length greater than $\lambda(z)$. Note that people whose trip lengths are less than $\lambda(z)$ will not use transit. Thus, $\lambda(z)$ can be interpreted as a continuum approximation of the shed boundary between transit travel and non-transit travel.

Since the benefit figure for $x > y$ can be obtained simply by reflecting that for $x \leqq y$ along the line $y=x$, the benefit for $x > y$ can be obtained by changing the sign of H. Also, since H appears in only the terms in eqs.(12), the

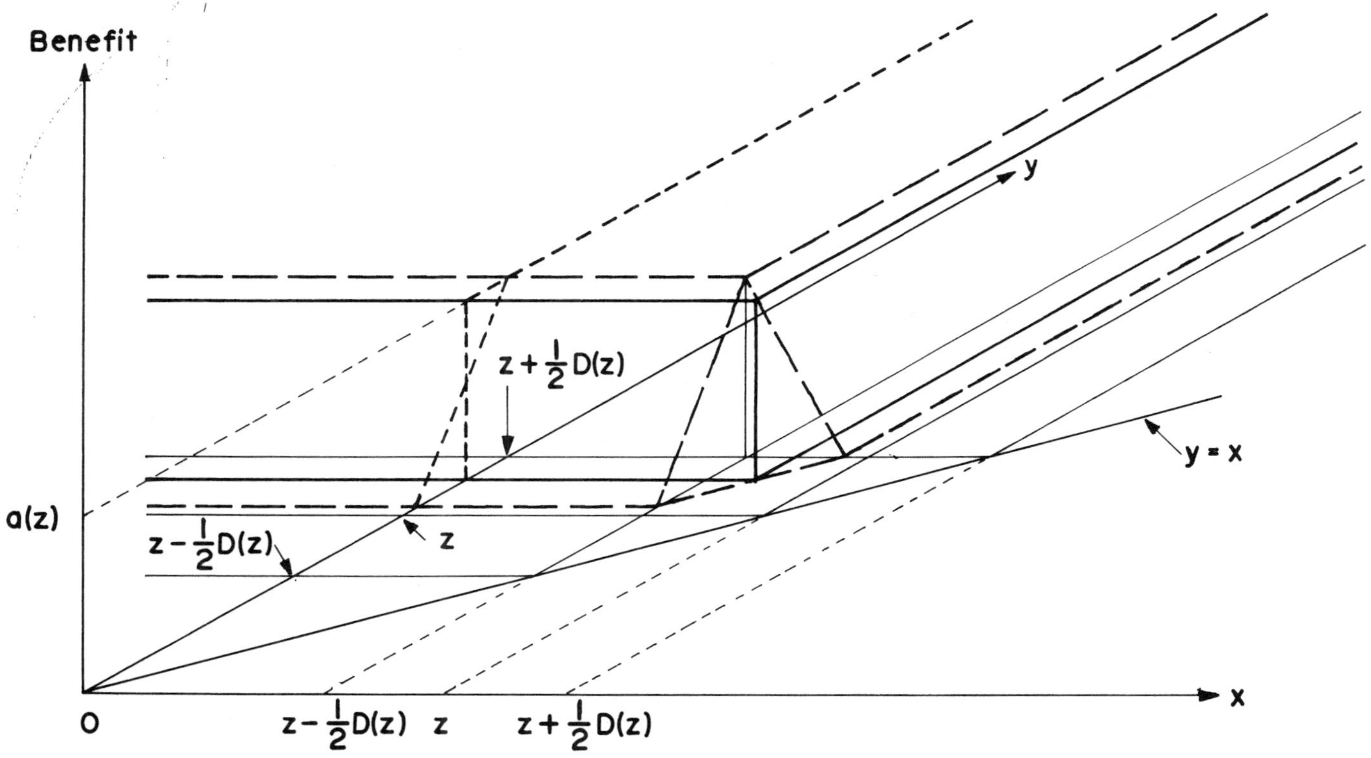

APPROXIMATE BENEFIT SURFACE

TRUE BENEFIT SURFACE

INTERSECTION OF THE TRUE AND APPROXIMATE BENEFIT SURFACES WITH THE Y - BENEFIT PLANE

Fig. 8 True and approximate benefit surfaces

average benefit of both directions will be obtained by substituting $c_1(z)$, $g_{1y}(z)$ and $g_{1x}(z)$ of both directions in eq. (11) for those of one direction. In the following we will continue to discuss the case with $x \leqq y$, but the discussion applies for both directions.

If f(x, y) satisfies the condition given by ineq. (8), the total benefit of all travellers is obtained by integrating the benefit per unit length with respect to z, which becomes

$$TB = \int_{-\infty}^{\infty} B(z)\, dz. \tag{13}$$

Now the objective is to find D(z), for all z, which maximizes the above integral, which is equivalent to maximizing the integrand, B(z), identically for each z†, that is, to find D(z) which maximizes B(z) for each z.

5. Optimal Station Spacing and Location

By taking derivatives of eq. (11) with respect to D(z) and by letting it be zero, we obtain the following formula for the station spacing:

$$D_1(z) = 2\left\{\frac{v\ S}{1 - \frac{v^2}{V^2}}\right\}^{1/2} \left\{\frac{c_1(z)}{g_{1x}(z)+g_{1y}(z)} - \frac{v\ S}{4}\right\}^{1/2}, \tag{14}$$

where it is assumed that $(g_{1x}(z) + g_{1y}(z))$ is positive for all z and $\lambda(z)$ in $c_1(z)$, $g_{1x}(z)$ and $g_{1y}(z)$ is evaluated at $D_1(z)$.

Since eq. (14) does not give the optimal station spacing explicitly one needs to solve it numerically for all z. In general this computation may not be easy. Practically, one can apply Dynamic Programming more generally and efficiently than eq. (14).

If, however, it can be assumed that:

vi) the trip lengths of people who travel through, originate or terminate at any point z are sufficiently large so that few of these travel a distance less than $\lambda(z)$ (or approximately one station interval),

† See Newell[7] p.349

then the optimal station spacing can be rewritten in the form

$$D_o(z) = 2\{\frac{v\ S}{1-\frac{v^2}{V^2}}\}^{1/2}\{\frac{c(z)}{g_x(z)+g_y(z)} - \frac{v\ S}{4}\}^{1/2}, \qquad (15)$$

where $c(z)$, $g_x(z)$ and $g_y(z)$ are obtained by substituting zero for $\lambda(z)$ in eqs. (12). Now $D_o(z)$ gives the explicit solution for the optimal station spacing. Further, since $c(z) = \int_{-\infty}^{z} g_x(x)dx - \int_{-\infty}^{z} g_y(y)dy$, $D_o(z)$ can be computed from only the trip origin density, $g_x(z)$, and destination density, $g_y(z)$.

Eq. (15) includes eq. (6.1) in Wirasinghe[11] and eq. (2.43) in Ghoneim[3] as special cases.

From $D_o(z)$ (or $D_1(z)$), one can determine the station location in the manner that Wirasinghe[11], Wirasinghe, Hurdle, and Newell[12] and Ghoneim[3] employed. The method can be summarized as follows. First, suppose that the stations are going to be allocated between two points, say between O and E, with a distance, L, between them and that the points O and E are specified as stations beforehand. Since $1/D_o(z)$ is considered as the density of stations at z, the number of stations between O and z ($0 \leqq z \leqq L$), $i(z)$, is given by $i(z) = \int_0^z \frac{dz}{D_o(z)}$. The $i(z)$ can be adjusted as $i^*(z) = \frac{N}{i(L)}\int_0^z \frac{dz}{D_o(z)}$ so that the boundary condition is satisfied. Here N is the total number of stations excluding the point O and is obtained by selecting the nearest integer to $i(L)$. Then the station locations may be approximately found as the points where $i^*(z)$ takes integer values. From eq. (9), $a(z)$ can be non-negative only if $D(z) \geqq vS/(1 - v/V)$, $(=D_m)$ where D_m can be interpreted as the theoretical lower boundary of $D(z)$. In the above computation, therefore, $D_o(z)$ should be replaced by D_m when $D_o(z) < D_m$.

Next let us consider the case when the station spacings are selected equally along the transit line. Referring to eqs. (11) and (13) and also employing the assumption vi), we can write the total benefit, TB', for equal station spacing,

D, as

$$TB' = \frac{a}{D}\int_{-\infty}^{\infty} c(z)dz - \frac{1}{2}\frac{a}{D}\lambda\{\int_{-\infty}^{\infty} g_x(z)dz + \int_{-\infty}^{\infty} g_y(z)dz\} \tag{16}$$

where a and λ are obtained by substituting D for D(z) in a(z) and $\lambda(z)$ respectively. Comparing eq.(16) with eq.(11), one can easily conclude that the optimal station spacing for this case can be obtained by substituting $\int_{-\infty}^{\infty} c(z)dz/\{\int_{-\infty}^{\infty} g_x(z)dz + \int_{-\infty}^{\infty} g_y(z)dz\}$ for $c(z)/(g_x(z) + g_y(z))$ in eq. (15). But it holds that

$$\int_{-\infty}^{\infty} c(z)dz = \int_{-\infty}^{\infty}\int_{-\infty}^{y} (y-x)\, f(x, y)dx\, dy$$

$$= \bar{H}\int_{-\infty}^{\infty} g_x(z)\, dz \qquad (\text{or } = \bar{H}\int_{-\infty}^{\infty} g_y(z)dz\ ) \tag{17}$$

where $\bar{H}$ is the average trip length. Therefore $\int_{-\infty}^{\infty} c(z)dz/\{\int_{-\infty}^{\infty} g_x(z)dz + \int_{-\infty}^{\infty} g_y(z)dz\} = \bar{H}/2$. By substituting $\bar{H}/2$ for $c(z)/(g_x(z) + g_y(z))$ in eq. (15), we obtain the following formula:

$$D_o = \left\{\frac{2vS\bar{H}\left(1 - \frac{vS}{2\bar{H}}\right)}{1 - \frac{v^2}{V^2}}\right\}^{1/2}. \tag{18}$$

This equation coincides with eq.(3)' in Fukuyama[2]. He derived the formula assuming a translationally symmetric trip distribution, but the above discussion showed that eq. (18) will also be true for more general trip distributions. Eq. (18) is also related to eq.(4.35) in Wirasinghe.[11] His formula, however, gives larger station spacing by a factor of $\sqrt{2}$ because of the assumption that people will only egress at the destination (CBD), i.e., the exclusion of the effect of $g_y(z)$.

Next we will investigate the condition in which the assumption vi) can be satisfied. From eqs.(17), we can derive the following relation:

$$\int_{-\infty}^{\infty} (g_x(z)+g_y(z))\left\{\frac{c(z)}{g_x(z) + g_y(z)} - \frac{\bar{H}}{2}\right\}dz = 0. \tag{19}$$

Since $(g_x(z)+ g_y(z))$ is positive, $c(z)/(g_x(z)+g_y(z))$ must vary around $\bar{H}/2$ in order that the integral be zero. Consequently $D_o(z)$ is also varying around D_o.

Now suppose that $D_o/\bar{H}$ is quite small, say $D_o/\bar{H} << 1$. This condition indicates that the trip concentration would stay far away from the line $y = x + D_o$ on the x-y plane. Obviously at the point where $D_o(z)$ is smaller than or comparable with D_o(even if it is larger than D_o), the effect of cutting off short trips from the integrals in $c_1(z)$, $g_{1x}(z)$ and $g_{1y}(z)$ becomes small. This follows also from the fact that the integral in eq. (19) is evaluated almost correctly at z where $c(z)/(g_x(z) + g_y(z))$ is less than $\bar{H}/2$ and, further, some portion of z where the former is larger than the latter. Therefore, where $c(z)/(g_x(z) + g_y(z))$ is much larger than $\bar{H}/2$, the former value should be approximately correct in order that the integral be exactly zero. Furthermore, since the concentration of trip origins and/or destinations must stay at the points where $c(z)/(g_x(z) + g_y(z))$ takes small values, the majority of travellers who pass through or originate(or terminate) at the points where $D_o(z)$ is large would terminate (or originate) at the points where $D_o(z)$ is small. That is, the length of these trips likely becomes large and consequently the effect of cutting off short trips also tends to be small. Thus, one can expect that assumption vi) will be satisfied and $D_o(z)$ applies (approximately)when $D_o/\bar{H}$ is small.

The value of $D_o/\bar{H}$, further, is a good index for judging the computational efficiency of the Dynamic Program. In order to provide sufficient accuracy for the computation, the number of proposed points for the stations must be much larger than the optimal number of stations.

Consider a corridor with the average trip length $\bar{H}$, and suppose that the length of the corridor is measured in units of $\bar{H}$. Since $D_o(z)$ varies around D_o, the value of $\bar{H}/D_o$ may be comparable with the optimum number of stations in an interval of length $\bar{H}$ or , in other words, in a unit length of the corridor. Therefore when $D_o/\bar{H}$ is large,i.e., the number of stations per unit length is small, the number of proposed points for the stations can be small and consequently the computation time becomes short. Conversely it requires a large number of proposed points for small values of $D_o/\overline{H}$.

In conclusion, the Dynamic Program can be applied efficiently when $D_o/\overline{H}$ is large, whereas $D_o(z)$ applies when $D_o/\overline{H}$ is small. Since $D_o(z)$ has the merit that the optimal station location can be determined from only the trip origin density and destination density, one should use $D_o(z)$ when $D_o/\overline{H}$ is small.

6. Numerical Example

We will consider the 10 km corridor (L = 10 km) with the OD distribution shown in Fig. 9. The trips are supposed to be distributed for $x \leqq y$. In the figure, c is an arbitrary constant. The dwelling time of transit at stations, S_o (assumed to be equal at all the stations), is 30 sec , and the acceleration and deceleration rates (A and B respectively) are 1 m/sec^2. Table 1 shows the cruise velocity of transit and the velocity of the access mode used in the example. The ratio v/V is also shown in the table.

In the Dynamic Programming computation, the length, L, is divided into 50 equal length intervals (0.2 km for each interval); that is, the total number of nodes is 51 including the first point, O, and the last point, E. The benefit a_{jk}, i.e., the benefit gained from a station interval between points j and k, is computed from

$$a_{jk} = \begin{cases} \dfrac{D_{jk}}{v} - \dfrac{D_{jk}}{V} - \{S_o + \dfrac{V}{2}(\dfrac{1}{A} + \dfrac{1}{B})\}, & \text{if } D_{jk} \geqq \dfrac{V^2}{2}(\dfrac{1}{A} + \dfrac{1}{B}) \\ \dfrac{D_{jk}}{v} - \{2\, D_{jk}(\dfrac{1}{A} + \dfrac{1}{B})\}^{1/2} - S_o, & \text{otherwise} \end{cases}$$

where D_{jk} is the distance between the points j and k. The first equation corresponds to the benefit when transit achieves its cruise speed and the second otherwise.

In Chart 1, the station locations determined from $D_o(z)$ are compared with those by the Dynamic Program. Table 2 shows the number of stations, the total benefit (the benefit computed by eq. (6) with c = 1 passenger/km^2/hour), difference in the benefit for the station location determined by

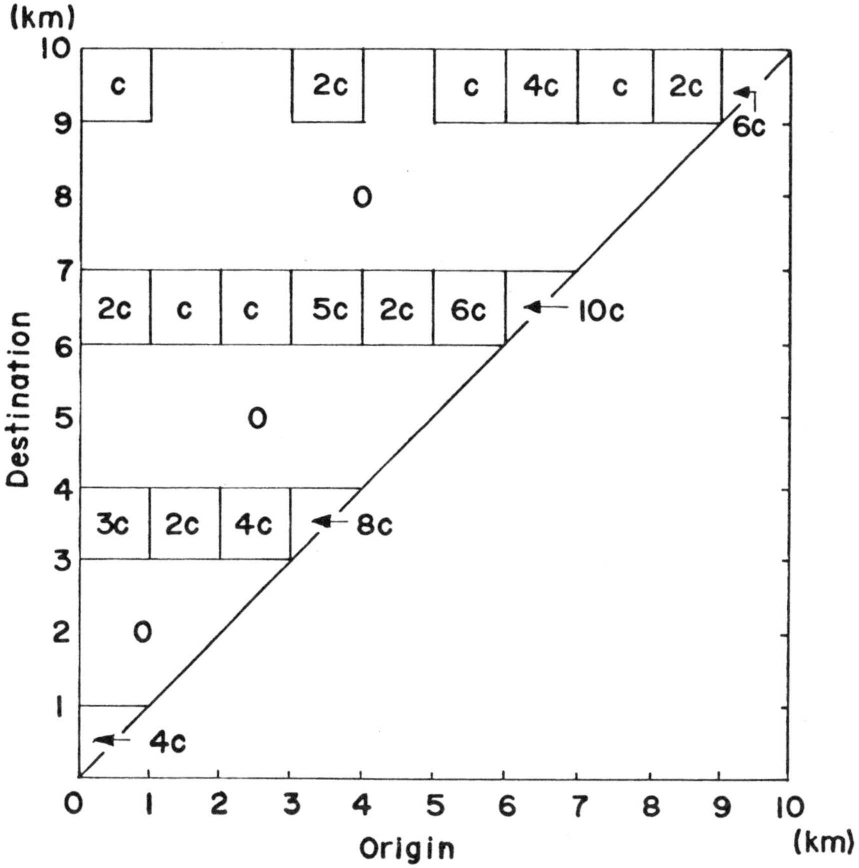

Fig. 9 OD Distribution

Table 1 Transit cruise velocity and velocity of the access mode

v	4	20	20
V	30	60	30
v / V	0.13	0.33	0.67

$D_o(z)$ and from the Dynamic Program, and the values of D_o and $D_o/\overline{H}$.

In Chart 1, the optimal station location obtained from both methods may not necessarily coincide. From Table 2, however, for v/V = 4/30 ($D_o/\overline{H}$ = 0.200), there is no perceptible difference in the total benefit. It appears that the benefit is quite insensitive to the exact location of the stations especially when $D_o/\overline{H}$ is small.

In the case with v/V = 20/60 and 20/30, one can see some errors in the results computed by $D_o(z)$.

7. Conclusion

The optimal station location which reflects the general trip distribution can be obtained either through eq.(15) or the Dynamic Programming computation. Eq.(15) can be applied when $D_o/\overline{H}$ takes small values. It only requires the trip origin density and destination density as trip data. The Dynamic Program, on the other hand, can be applied generally but it requires the OD distribution table.

The Dynamic Programming computation becomes especially efficient when $D_o/\overline{H}$ takes large values, that is, when there is a split of patronage between transit travel and non-transit travel. The split of the patronage, for example, may occur when express transit is introduced in order to increase the transportation capacity of the corridor. The express transit and existing system should share the patronage in such a way that the former is used for the longer trips and the latter the shorter trips. In such cases it may not be difficult to estimate the accurate OD distribution from the existing trip demand and thus the Dynamic Program can be effectively applied.

Acknowledgments

I would like to extend my cordial gratitude to Professor G.F. Newell who shared his time for discussion and gave me valuable suggestions in completing this paper.

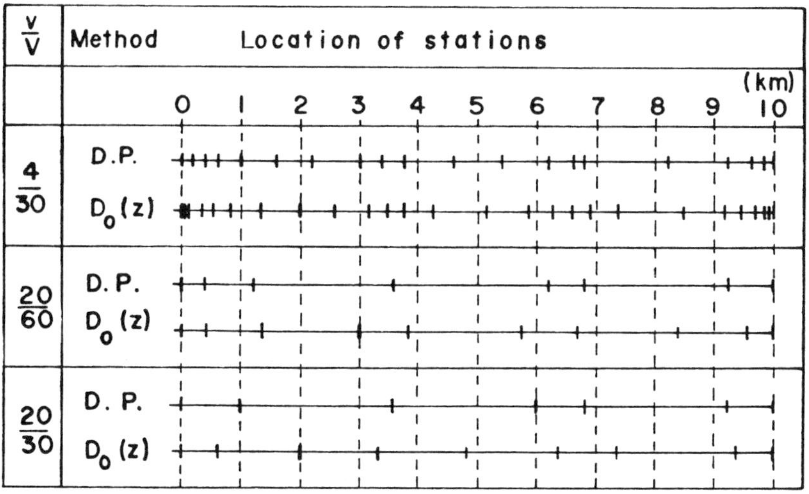

Chart 1 Location of stations

Table 2 Comparison of some results obtained by the Dynamic Programming computation and $D_o(z)$

$\frac{v}{V}$	Method	No. of stations	Total benefit (hour)	Comparison in the total benefit (percent)	D_o (km)	$\frac{D_o}{\bar{H}}$
4/30	D. P.	19	19.01	0.05	0.427	0.200
	$D_o(z)$	25	19,00			
20/60	D. P.	7	1.892	6.24	1.080	0.507
	$D_o(z)$	9	1.774			
20/30	D. P.	6	0.6977	15.02	1.245	0.584
	$D_o(z)$	8	0.5929			

References

1. Bendtsen,P.H.,"Urban and Surburban Railways.Study of Transport Problems and Network Design," Danmarks Nalurvidenskabelige Samfund, 1938, pp. 154-159.
2. Fukuyama,M., "Optimal Station Spacing for a Transit Line," Ph.D. dissertation, The Institute of Transportation Studies, University of California, Berkeley, July, 1977
3. Ghoneim, N.S.A., "Spacing and Location of Bus-Stops," Graduate Thesis of Master of Science in Engineering, Department of Civil Engineering, the University of Calgary, Calgary, Alberta, June, 1978
4. Hurdle, V.F.,"Minimum Cost Locations for Parallel Public Transit Lines," Trans. Sci. 7, 1973,pp.340-350.
5. Hurdle, V.F. and S.C.Wirasinghe, "Location of Rail Stations for Many to One Travel Demand and Several Feeder Modes," Journal of Advanced Transportation,Vol. 14, No. 1, Spring, 1980, pp. 29-45.
6. Landau, A.G., "Distance between Stops on Transport Service Minimizing Travel Times," Beitraege zur Theorie des Verkehrsflusses, Proceedings of the Fourth International Symposium on the Theory of Traffic Flow, Karlsruhe, 1968, pp.265-268.
7. Newell, G.F.,"Scheduling, Location, Transportation and Continuum Mechanics: Some Simple Approximations to Optimization Problems,"SIAM Journal of Applied Mathematics, special issue to honor Professor W.Prager, Vol. 25-3, 1973
8. Tomii,Y. and K. Kawachi,"The Optimization of Train-Running on Quadruple Line," Proc. of JSCE, No. 259, March, 1977, pp. 103-109 (in Japanese)
9. Vuchic, V.R.,"Interstation Spacing for Line-haul Passenger Transportation," Ph. D. dissertation-graduate Report, The Institute of Transportation and Traffic Engineering, University of California, Berkeley, December, 1966

10. Vuchic,V.R. and G.F.Newell,"Rapid Transit Interstation Spacings for Minimum Travel Time," Trans. Sci.2, 1968, pp. 303-339.

11. Wirasinghe,S.C.," A Model of a Coordinated Rail and Bus Transit System," Ph.D.dissertation, The Institute of Transportation and Traffic Emgineering, University of California, Berkeley, May, 1976

12. Wirasinghe, S.C., V.F. Hurdle and G.F.Newell,"Optimal Parameters for a Coordinated Rail and Bus Transit System," Trans. Sci. Vol. 11-4, November, 1977, pp. 359-374.

13. Wirasinghe, S.C.,"The Use of Analytical Models for Planning Public Transit Systems," Proceedings of the Second International Conference on Mathematical Modelling, Vol. 11, July, 1979, St. Louis, Missouri, U.S.A.

14. Wirasinghe, S.C.,"Nearly Optimal Parameters for a Rail/Feeder-Bus System on a Rectangular Grid," Transpn Res. A, Vol. 14 A, 1980, pp. 33-40.

15. Wirasinghe, S.C., "An Introduction to the Analysis of Public Transit Systems," Proceedings of the First Canadian Seminar on Systems Theory for the Civil Engineer, Department of Civil Engineering, The University of Calgary, Calgary, Alberta, October, 1979, pp. 169-180.

WAITING TO CROSS A MAJOR STREAM AT AN UNCONTROLLED ROAD JUNCTION

JOHN C. GOLIAS
Transport Studies Group, University College London, UK

An analytical model is suggested which estimates the statistics of delay to drivers who arrive randomly at an uncontrolled road junction and wait to cross a major stream. The complexity of the problem does not allow all the details of the real crossing process to be taken into account in the model. Some simplifying assumptions are therefore made in the analysis concerning the probability distribution of the 'first' gap encountered by the waiting driver and the independence between certain distributions.

Using thorough simulation, the effects of these assumptions on the estimated average delay have been investigated in the case of a cumulative Erlang distribution for the gap acceptance function, an Erlang distribution for the move-up time, and a gamma distribution for the major stream headways. The mean delay estimated by the analytical model was always well within 5 per cent of the value obtained by simulation, and the two values were usually almost equal. These results support the simplifying assumptions of the analytical model and suggest that it gives a satisfactory estimate of delay.

1. INTRODUCTION

Interaction between traffic streams takes place most frequently at priority-controlled intersections which are still the most numerous in both urban and rural areas. This paper presents a study of the delay to minor road drivers that are waiting to cross a major stream at such a junction.

There have been many theoretical treatments of this delay problem. Unfortunately though, there is no one single model to which one can attribute greatest generality since all of the models treat some features of the queueing situation in detail and either ignore others or approximate to them in a less than satisfactory way.

Before going through any of these models it is necessary to introduce the concept of the gap acceptance function which gives the probability that a driver will accept a gap of a certain length and that of the move-up time which is the time taken for a vehicle to move from second in queue to the head of it.

One of the first results is due to Tanner[1] who used a step acceptance function and a constant move-up time. A further restrictive assumption of Poisson traffic was made for both major and minor roads.

Yeo and Weesakul[2] and Evans, Herman and Weiss[3] have treated variants of Tanner's models in which the unsatisfactory assumption of a step gap acceptance function is kept. However, both of these models take into account the possibility of a distribution of critical gaps over the population.

A queueing model that appears to allow correctly for a general gap acceptance function $a(t)$ is one due to Hawkes[4]. The function $a(t)$ is assumed to satisfy the conditions

$$a(t) = 0 \qquad t < T_1$$
$$a(t) = 1 \qquad t > T_1 + \delta$$

where T_1 and δ are constants. It is further assumed that the move-up time M is a random variable which satisfies $\delta < M < T_1$. This inequality, however, is questionable in the light of experimental values.

Buckley and Blunden[5] proposed a queueing model that made use of known results for the M/G/1 queue. The authors assumed that a residual gap is independent of the total gap size when it is known that the gap has been accepted and suggested that this assumption would produce a negligible error on the final result. Unfortunately this is not true for the general case.

The simplicity, though, of this model led to an attempt to improve it to cope with its basic disadvantages, an attempt that led to the queueing model that this paper suggests.

2. THE ANALYTICAL QUEUEING MODEL

a. General

In the following analysis the headways on the main road are assumed to be identically distributed independent random variables with pdf $f(t)$. Then the initial headway that a driver arriving at a random instant faces has pdf $f_0(t)$ given[6] by:

$$f_0(t) = \int_t^{\infty} f(x)dx \Big/ \int_0^{\infty} xf(x)dx \quad .$$

The minor stream headways are assumed to have a negative exponential distribution.

The decision criterion for the acceptance of a headway by a driver is summarized by the gap acceptance function denoted by $a(t)$. To cover the case of a driver who tends to accept a shorter gap than if he had come to a complete stop a gap acceptance function denoted by $a_0(t)$ was used.

The move-up time, i.e. the time taken for a vehicle to move up from second in queue to the service point following the service of the previous vehicle, is assumed to have a pdf denoted by $n(t)$.

Before proceeding with the actual analysis it should be noted that in the sequel the i th moment about zero of a variable will be denoted by the letter which denotes the pdf of the variable and subscript i .

Let a minor road driver that arrives at the junction be considered. The driver arrives and joins the queue, maintaining strict queue discipline. If no queue exists his car becomes the lead vehicle. The service time is different for drivers joining an empty and those joining a non-empty queue.

The statistics of delay for the above problem were studied by Yeo[7]. He considered a generalisation of the queueing system M/G/1 where customers arriving at an empty queue have service time with pdf $E(t)$ while those arriving at a non-empty queue have service time with pdf $Q(t)$. The characteristic function (c.f.) $W_c(\theta)$ of the stationary queueing time distribution $w(t)$ was found to be

$$W_c(\theta) = P_0[i\theta - qQ_c(\theta) + qE_c(\theta)] / [i\theta + q - qQ_c(\theta)] \quad (1)$$

where P_0 is the probability that a customer arrives to find the queue empty and was found to be

$$P_0 = (1 - qQ_1) / (1 - qQ_1 + qE_1)$$

$Q_c(\theta)$, $E_c(\theta)$ are the c.f.'s of $Q(t)$ and $E(t)$ respectively, q is the arrival rate of the customers.

The queue size for the above system was also studied. Let P_n denote the probability that there are n customers in the queue. The probability generating function $P(z) = \sum_{n=0}^{\infty} P_n z^n$ then was found to be for $qQ_1 < 1$

$$P(z) = P_0[k(z) - zm(z)] / [k(z) - z] \quad (2)$$

where $k(z) = \sum_{n=0}^{\infty} k_n z^n \; , \quad k_n = \int_0^{\infty} e^{-qx} \frac{(qx)^n}{n!} \, dQ(x)$

$$m(z) = \sum_{n=0} m_n z^n \; , \quad m_n = \int_0^{\infty} e^{-qx} \frac{(qx)^n}{n!} \, dE(x)$$

The moments of the queueing time distribution may be found from Equation (1). The mean queueing time is

$$W_1 = qE_2 - q^2Q_1E_2 + q^2E_1Q_2 / 2(1-qQ_1)(1-qQ_1+qE_1) \ . \quad (3)$$

Therefore, the problem of delay to minor road drivers waiting to cross a major stream reduces to finding the service times of minor road drivers who arrive to find an empty and a non-empty queue. These two cases are going to be dealt with separately. In both cases the service time includes the move-up time. Any simplifying assumption made throughout the analysis is discussed at the end of it.

b. The case of a non-empty queue

Let $Q(t)$ be the pdf of the service time experienced by drivers arriving to find a queue. This service time consists of two parts: the move-up time with pdf $n(t)$ and the scanning time, which begins when the vehicle takes the lead position in queue and ends with the crossing of the minor road vehicle, with pdf $S(t)$. Therefore

$$Q_1 = n_1 + S_1 \quad (4)$$

$$Q_2 = n_2 + S_2 + 2n_1S_1 \quad (5)$$

For the derivation of Equation (5) the simplifying assumption of independence between the two parts was made.

The evaluation of the scanning time will be based on the delay to a single driver that waits to cross; its moments about zero are given[8] by:

$$D_1 = Y_{10} + \left(\frac{1-\bar{a}_0}{\bar{a}}\right) Y_1 \quad (6)$$

$$D_2 = Y_{20} + \left(\frac{1-\bar{a}_0}{\bar{a}}\right) Y_2 + \frac{2}{\bar{a}} D_1Y_1 \quad (7)$$

where

$$Y_i = \int_0^\infty t^i \ f(t)[1 - a(t)]\,dt$$

$$Y_{i0} = \int_0^\infty t^i f_0(t)[1 - a_0(t)]\,dt \tag{8}$$

$$\bar{a} = \int_0^\infty f(t)a(t)dt$$

$$\bar{a}_0 = \int_0^\infty f_0(t)a_0(t)dt\,. \tag{9}$$

The difference in dealing with the scanning time is that the 'first' gap encountered by the driver is very likely to be part of a gap which was accepted and therefore has a pdf $f_1(t)$ different from $f_0(t)$. Unfortunately, a general expression has not yet been developed for this distribution. The mathematical problem is made extremely difficult by the fact that $f_1(t)$ depends on the number of previous vehicles that have accepted the same gap, part of which is the 'first' gap encountered by the vehicle at the head of the queue now.

Let the case of a vehicle that is examining the remaining part of a gap accepted only by the previous vehicle be examined initially. Since the gap is acceptable by one vehicle at least, it has pdf given by

$$g_a(t) = f(t)a(t) / \bar{a}\,. \tag{10}$$

The first two moments S_1 and S_2 then of the scanning time of the waiting vehicle are given by Equations (6) and (7) where $\bar{a}_0$ and Y_{i0} must be replaced by $\bar{a}_Q$ and Y_{iQ} respectively as follows:

$$S_1 = Y_{1Q} + \left(\frac{1-\bar{a}_Q}{\bar{a}}\right) Y_1 \tag{11}$$

$$S_2 = Y_{2Q} + \left(\frac{1-\bar{a}_Q}{\bar{a}}\right) Y_2 + \frac{2}{\bar{a}} S_1 Y_1 \tag{12}$$

where

$$\bar{a}_Q = \int_0^\infty g_a(t)\left[\int_0^t n(x)a(t-x)dx\right]dt$$

$$+ \bar{a}_{0Q} \int_0^\infty g_a(t)\left[\int_t^\infty n(x)dx\right]dt \tag{13}$$

$$a_{0Q} = \int_0^\infty f_0(t)a(t)dt \tag{14}$$

$$Y_{i0} = \int_0^\infty g_a(t)\left[\int_0^t n(x)[1-a(t-x)](t-x)^i dx\right]dt$$

$$+ Y_{i0Q} \int_0^\infty g_a(t)\left[\int_t^\infty n(x)dx\right]dt \quad (15)$$

$$Y_{i0Q} = \int_0^\infty t^i f_0(t)[1 - a(t)]\,dt. \quad (16)$$

In each of Equations (13) and (15) the second term takes care of the possibility of a move-up time greater than an acceptable gap.

The form of Equations (10), (13) and (15) depends on the number of previous vehicles that have accepted the gap, the remaining part of which is the 'first' gap for the waiting driver. Even when this number is 2, these equations become very complicated. To overcome this difficulty the simplifying assumption that the scanning time moments are always given by Equations (11) and (12) was made. Therefore it was assumed that the 'first' gap encountered is part of the previously accepted gap which has pdf given by $g_a(t)$ for the general case.

Substituting the expressions for S_1 and S_2 in Equations (4) and (5), the first two moments of the service time are evaluated.

c. <u>The case of an empty queue</u>

Let $E(t)$ be the pdf of the service time of a driver who arrives to find an empty queue. This service time will consist in the general case of the remaining part of the move-up time period of the previous driver and of the scanning time; their i th moments will be denoted by n_{iE} and Z_i respectively. If the assumption of independence between the two parts of the service time is made its moments are given by

$$E_1 = n_{iE} + Z_1 \quad (17)$$

$$E_2 = n_{2E} + Z_2 + 2n_{iE}Z_1 \quad . \quad (18)$$

It is obvious that the service time will depend on whether the driver arrives sufficiently long after the completion of the service time of the previous vehicle to be unaffected by it as well as on whether he will seize the remainder of the gap accepted by the previous vehicle. Therefore it is evident to distiguish amongst five different cases for the arrival of the driver. For each case j, the probability $\bar{a}_{jE}$ of accepting the 'first' gap encountered, the ith moment Y_{iEj} of the 'first' rejected gap as well as the ith moment n_{iEj} are evaluated separately. These five cases are the following.

1. The driver arrives before the end of the gap accepted by the previous one but after the completion of his service time, i.e. after the move-up time period. Then

$$\bar{a}_{E1} = \int_0^\infty g_a(t)\left[\int_0^t qe^{-qz}a(t-z)\left[\int_0^z n(x)dx\right]dz\right]dt$$

$$Y_{iE1} = \int_0^\infty g_a(t)\left[\int_0^t qe^{-qz}[1-a(t-z)](t-z)^i\left[\int_0^z n(x)dx\right]\right]dz$$

$$n_{iE1} = 0$$

2. The driver arrives during the move-up time period which ends before the end of the gap accepted by the previous driver. For this case

$$\bar{a}_{E2} = \int_0^\infty g_a(t)\left[\int_0^t qe^{-qz}\left[\int_z^t n(x)a(t-x)dx\right]dz\right]dt$$

$$Y_{iE2} = \int_0^\infty g_a(t)\left[\int_0^t qe^{-qz}\left[\int_z^t n(x)[1-a(t-x)](t-x)^i dx\right]\right]dz$$

$$n_{iE2} = \int_0^\infty g_a(t)\left[\int_0^t qe^{-qz}\left[\int_z^t n(x)(x-z)^i dx\right]dz\right]dt$$

3. The driver arrives during the gap accepted by the previous driver and during the move-up time period which ends after the end of the accepted gap. The formulae for this case are given by

$$\bar{a}_{E3} = \bar{a}_{0Q}\int_0^\infty g_a(t)\left[\int_0^t qe^{-qz}\left[\int_t^\infty n(x)dx\right]dz\right]dt = \bar{a}_{0Q}\, P_3$$

$$Y_{iE3} = Y_{i0Q}\ P_3$$

$$n_{iE3} = \int_0^\infty g_a(t)\left[\int_0^t qe^{-qz}\left[\int_t^\infty n(x)(x-z)^i dx\right]dz\right]dt$$

4. The driver arrives after the end of the gap accepted by the previous driver and after his move-up time period. Then

$$\bar{a}_{E4} = \bar{a}_0 \int_0^\infty g_a(t)\left[\int_t^\infty qe^{-qz}\left[\int_0^z n(x)dx\right]dz\right]dt$$

$$= \bar{a}_0\ P_4$$

$$Y_{iE4} = Y_{i0}\ P_4$$

$$n_{iE4} = 0$$

5. The driver arrives during the move-up time period of the previous driver but after the end of the gap accepted by the latter. For this case

$$\bar{a}_{E5} = \bar{a}_{0Q} \int_0^\infty g_a(t)\left[\int_t^\infty qe^{-qz}\left[\int_z^\infty n(x)dx\right]dz\right]dt$$

$$= \bar{a}_{0Q}\ P5$$

$$Y_{iE5} = Y_{i0Q}\ P5$$

$$n_{iE5} = \int_0^\infty g_a(t)\left[\int_t^\infty qe^{-qz}\left[\int_z^\infty n(x)(x-z)^i dx\right]dz\right]dt$$

Therefore for the general case the moments of the scanning time of a driver that arrives to find an empty queue are given by

$$Z_1 = Y_{1E} + \left(\frac{1-\bar{a}_E}{\bar{a}}\right) Y_1 \tag{19}$$

$$Z_2 = Y_{2E} + \left(\frac{1-\bar{a}_E}{\bar{a}}\right)Y_2 + \frac{2}{\bar{a}} Z_1Y_1 \quad , \tag{20}$$

where $\bar{a}_E = \sum_j \bar{a}_{Ej}$

and $Y_{iE} = \sum_j Y_{iEj}$.

The ith moment of the remaining part of the move-up time is given in the general case by

$$n_{iE} = \sum_j n_{iEj} = \int_0^\infty qe^{-qz}\left[\int_z^\infty n(x)(x-z)^i dx\right]dz \quad . \tag{21}$$

Substituting in Equations (17) and (18) the first two moments about zero of the service time are evaluated.

d. Total delay and queue size

Using the expressions for the moments of the service time of the waiting drivers (Equations (4), (5), (17), (18)) and Yeo's queueing model (Equation (3)), and assuming that $Q(t)$ and $E(t)$ are individually and mutually independent, the mean queueing time W_1 of a driver waiting to cross a major stream can be calculated. The total delay is the convolution of the service time and the queueing time. In general these are not independent but the mean total delay d_1 can be evaluated by adding their means. Then

$$d_1 = W_1 + P_0 E_1 + (1-P_0)Q_1 \tag{22}$$

The probability P_n that there are n vehicles waiting in the queue can be found by differentiation from Equation (2)

$$P_n = \left.\frac{d^n P(z)}{n!\,dz^n}\right|_{z=0}$$

The average queue length $\bar{n}$ is found from Equation (2) to be $\bar{n} = P'(1)$. It can also be evaluated in a straightforward way by using $\bar{n} = qd_1$.

e. Assumptions

During the analysis some simplifying assumptions had to be made to reduce the complexity of the real crossing process and thus end up with a fairly realistic model which would retain a satisfactory accuracy.

The more important assumption made is that the pdf of the gap t which was accepted by a minor road driver and part of which is the 'first' gap t_r encountered by the next minor road driver is always given by $g_a(t)$. If $g_{ai}(t)$ denotes the pdf of t when the previous driver that crossed was the (i-1)th to cross with the same headway then the assumption made is expressed by

$$g_{ai}(t) = g_{a2}(t) = g_a(t).$$

The assumption implies that t_R has a pdf which is independent of the number of drivers that accepted so far the headway part of which it is.

This is true for the case of Poisson traffic in the major road, step gap acceptance function and constant move-up time. It is not true though for the general case. Nevertheless, the assumption is believed to have very little influence on the final results. Graphs of $g_{a2}(t)$ and other $g_{ai}(t)$ for the following combination of distributions:

$$f(t) = q^n e^{-qt} t^{n-1}/(n-1)! \;;\; a(t) = 1 - e^{-gt} \;;\; n(t) = \delta(t-M)$$

are given in Figure 1. It is clear from this that at least for the cases studied the assumption has little effect. Furthermore, the effect that it will have in certain cases depends on the product of the probability that a gap is acceptable by more than one driver and the probability that there is a queue of at least two vehicles. When one of these two probabilities increases the other decreases and the product is expected to be always low enough to give results with satisfactory accuracy for the general case.

To check this accuracy as well as the effects of the independence assumptions made throughout the analysis a simulation program was set up. The results of the simulation study supported the assumptions made. These results will be discussed, as contrasted with the above work which is theoretical, later in the present paper.

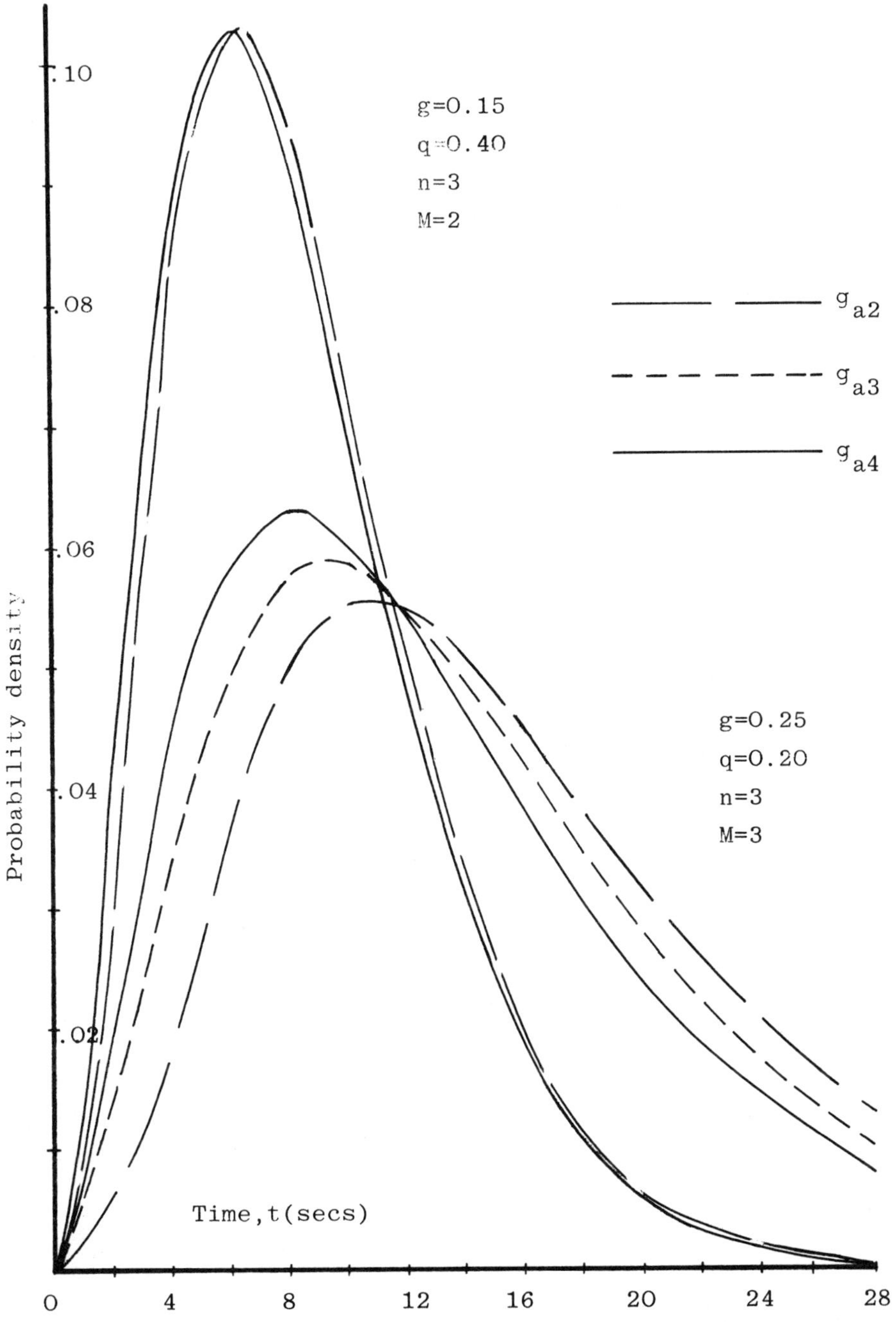

FIGURE 1 : GRAPHS OF $g_{\alpha i}(t)$

3. SIMULATION OF THE QUEUEING PROCESS

a. Introduction

It was quite clear from the start that the complexity of the crossing problem would make impossible the development of an analytical model that would take into account all the detailed real effects of the crossing process. To guide and complement the theoretical model, a simulation program was set up. The program has been set up in a rather general form on an IBM 360 digital computer and uses the G.P.S.S. simulation language[9]. For the purposes of this paper the simulation model has been developed for minor road vehicles arriving with exponentially distributed headways and waiting to cross one lane of priority traffic in which the headways have a gamma distribution. Each waiting driver is assumed to have a cumulative Erlang gap acceptance function and when a vehicle crosses there is a move-up time with an Erlang pdf before the next driver begins to consider gaps. The simulation results include histograms and statistics of time spent at the head of the queue and further back in the queue as well as total delay and queue length. The validity of the program has been verified by checking that the vehicles are following the desired paths, i.e. that the program performs the calculations and logical sequences intended.

b. Statistical analysis of output

The most important and critical part of the simulation is the analysis of the results. In this section the problems encountered in this case as well as the way they were treated are going to be discussed.

The simulation of the queueing process necessitated artificially starting and stopping a continuing system. It was thus important to decide when the starting conditions were no longer significantly influencing the results, i.e. when the system was considered to have arrived at a steady state. The procedure followed was to eliminate the transient effects by discounting any calculations up to that point. Prior to running the simulation for results, a few pilot runs provided an indication of the number of minor road drivers that had to be taken into account before the system could be considered to have reached the steady state. The mean delay was calculated successively for samples with increased size each time and when it appeared that the delay was becoming stable, the transient effects were considered no longer important. For each combination of parameters tried in the pilot runs, the procedure was repeated a few times changing the seeds of the random number generators and then a conservative size of sample was chosen. This size was found to range from 750 to 2500 minor road vehicles depending on the value of the mean delay in the system.

The problem of determining the length of a simulation run is now considered. It is known that if a very long run of the simulation model was made and the mean delay was computed this would converge to the true mean value. Unfortunately, in practice the limit can never be reached, so for a run of finite length an estimand of the real value can be obtained. The fact that for a single large sample this estimand is near the real value is comforting but does not help at all in the estimation of any kind of confidence interval.

We consider now the problem of determining the length of a simulation run. If we made a very long run of the simulation model and computed the mean delay,we know that this would converge to the true mean value. Unfortunately, in practice,we can never reach the limit, so for a run of finite length we end up with an estimand of the real value. The fact that for a single large sample this estimand is near the real value is comforting but gives us no chance to make any kind of confidence interval estimate.

The easiest way of overcoming this problem was found to be the repetition of simulation runs keeping all the design parameters constant, but using a different random number stream each time. Multiple runs of this nature are referred to as replications.

If n replications are run, and the random number streams used in each replication are not correlated with one another then the mean delay $\bar{w}$ and its variance s_w^2 are given by

$$\bar{w} = \left(\sum_{i=1}^{n} w_i \right) / n$$

$$s_w^2 = \left(\sum_{i=1}^{n} (w_i - \bar{w})^2 \right) / (n-1)$$

where w_i is the mean of the ith replication. Then using the central limit theorem the confidence interval estimate is obtained as

$$\bar{w} \pm t_a \, s_w / \sqrt{n}$$

where t_a is obtained from the student distribution tables and $1-a$ is the degree of confidence. For n sufficiently large ($n \geq 30$) one must replace t_a by Z_a, the corresponding value from the standard normal distribution.

Subjective judgements had still to be used in determining the run size of each replication for the longer the run length the smaller s_w will be. Some pilot runs were used to determine the trade-off between run length and number of replications, since increasing the latter also reduces s_w. The results of these runs as well as the restrictions imposed by computer running time due to the size of the jobs led to the decision that twenty replications would be generally enough for each combination of distribution parameters and that each replication should gather statistics on a number of minor road drivers ranging from 2000 to 4500 and depending on the value of the mean delay in the system.

It should be noted here that a number of variance reducing techniques[10][11] were also considered for the analysis of the simulation output. Nevertheless, the complexity of these techniques as well as the efficiency in computer time of the simulation program used made the simple procedure followed in the analysis preferable.

It should also be noted that any subjective judgement made in the analysis procedure has effect only on the width of the confidence interval of the mean delay. Therefore, a possible misjudgement in determining the beginning of the steady state or the total run length of the replications for certain extreme cases would not lead to an error but would necessitate more replications, i.e. more computational effort to get the required width for the confidence interval of the mean delay.

4. COMPARISON OF ANALYTICAL AND SIMULATION RESULTS

To check the assumptions made in the analytical model and the influence they have on the final results the analytical results obtained for a certain combination of distributions were compared to those obtained by simulation.

In making this comparison, the use of general distributions that would be suitable for a wide variety of cases was tried. Thus, for the pdf of main road headways the gamma distribution was chosen, which is of the form

$$f(t) = \beta^{a} e^{-\beta t} t^{a-1} / \Gamma(a).$$

With suitable adjustment of a and β, this distribution induces an element of regularity into arrival times and therefore can describe situations intermediate between regular and random arrivals. The minor road headways were assumed to have a pdf given by $f_m(t) = qe^{-qt}$.

For the gap acceptance function the cumulative Erlang distribution was used, which is of the form

$$a(t) = \int_0^t \left[g^k \, e^{-gx} \, x^{k-1} / (k-1)! \right] dx$$
$$= 1 - e^{-gt} \sum_{n=0}^{n=k-1} \frac{(gt)^n}{n!} \quad .$$

This distribution gives good agreement with experimentally derived results[12][13] and is also quite convenient for the mathematical calculations. For the gap acceptance function $a_0(t)$ which describes the behaviour of a driver that has not come to a complete stop, the same form of distribution as for $a(t)$ was used where parameter g was replaced by g_0 .

As far as the move-up time distribution is concerned, it was thought that an Erlang function of the form

$$n(t) = c^M \, e^{-ct} \, t^{M-1} / (M-1)!$$

is general enough to describe adequately the move-up time.

Making use of these distributions in the analysis gives a model that covers satisfactorily most of the cases as far as traffic flow and driver behaviour are concerned. On the other hand it complicates considerably the mathematical aspect of the problem. Nevertheless, after laborious calculations a general analytical expression for the mean delay was finally derived. The corresponding formulae are given in the Appendix.

In order to compare the analytical to the simulation results for the above combination of distributions the mean delay for different values of the parameters had to be evaluated. For each parameter values were chosen from a range which was considered to cover all the reasonable cases and a few values out of it on both sides to cover the extreme cases. Thus the mean major flow in the following analysis ranges from 0.10 to 0.50 with a ranging from 0.80 to 3.20. The mean minor flow ranges from 0.02 to 0.16. The mean move-up time takes values from 1.5 to 3.0 seconds with M ranging from 7 to 26. The mean acceptable gap and lag range from 3.5 to 5.5 seconds and from 3.0 to 4.5 seconds respectively with k taking values from 8 to 15.

A Fortran program was set up which using as input the values of the parameters evaluates the mean delay. The program was run for different combinations of the above parameters. Thus the mean delay w_a was evaluated analytically for 100 different parameter combinations. Then the simulation program was run for exactly the same parameter combinations and the mean delay w_s and the standard deviation S of the mean were evaluated.

Finally, the null hypothesis that the two means are equal was tested versus the alternative one that they are different. Given that the number n of simulation replications was generally small (less than 30) the t-statistic was used[14]. Assuming that the population from which samples were taken was roughly normal, the decision was based on the statistic

$$t = \frac{w_s - w_a}{S/\sqrt{n}} \quad .$$

Values of t were obtained from the student distribution tables with the number of degrees of freedom equaling $n-1$ and the level of significance being chosen equal to $a=0.05$.

The difference between the mean delay derived by simulation and the mean analytical delay can be seen in Figures 2 and 3 where the former is plotted together with its confidence interval $(a = 0.05)$ versus the latter for a variety of cases. The results did not reveal a specific trend for the difference between the two means or for the result of the statistical test made with respect to any of the distribution parameters. Thus no parameter values are given for each case in these two figures.

Using the t-test it was found that for about 85 per cent of the cases there was not enough evidence to reject the null hypothesis that the two means are equal. For the remaining

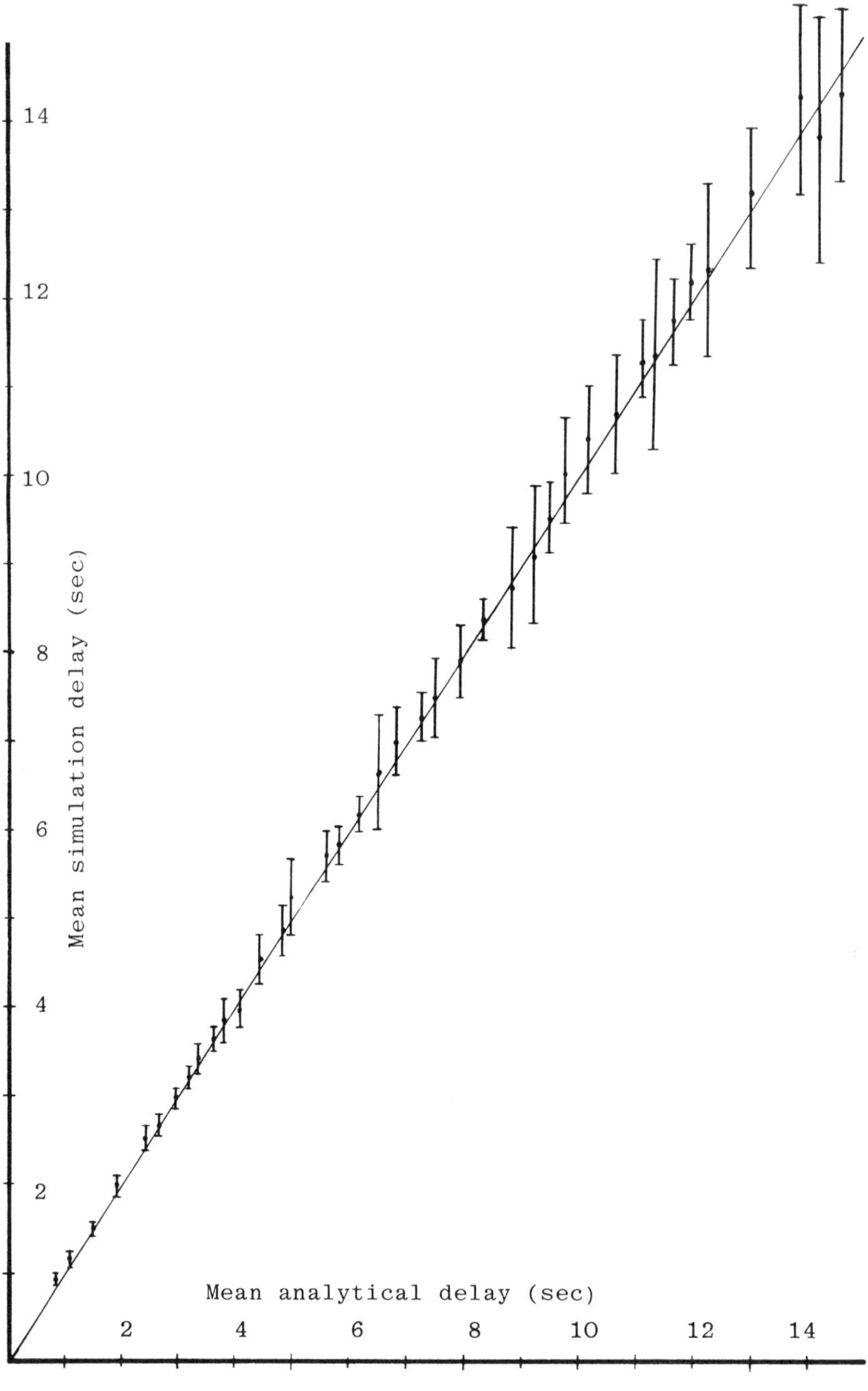

FIGURE 2 : MEAN ANALYTICAL VERSUS MEAN SIMULATION DELAY

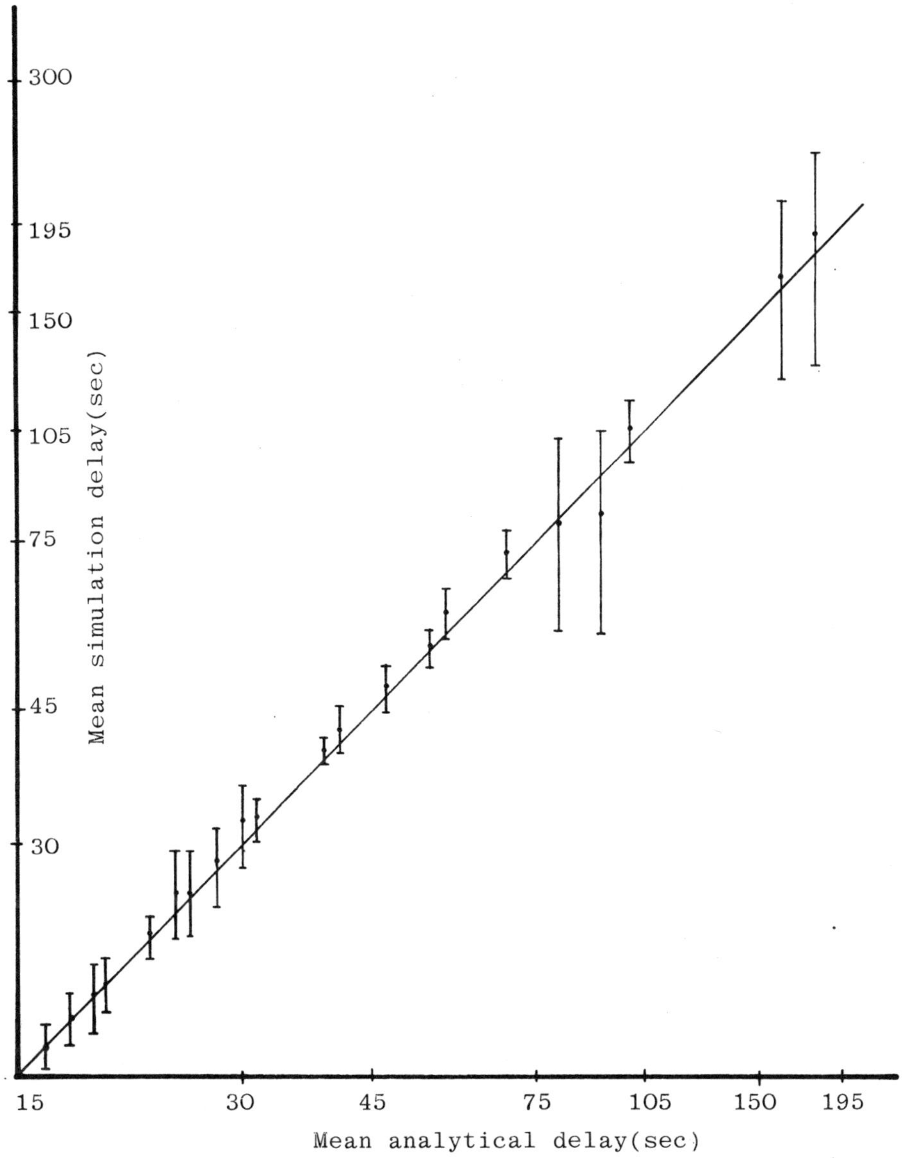

FIGURE 3 : MEAN ANALYTICAL VERSUS MEAN SIMULATION DELAY (Logarithmic axes)

15 per cent of the cases the above hypothesis was rejected; however, the hypothesis that the difference between the two means is only 5 per cent could not be rejected. These results supported the simplifying assumptions of the analytical model and suggested that it evaluates delay with a quite satisfactory accuracy.

5. GENERAL REMARKS AND CONCLUSIONS

The comparison between the mean delay evaluated analytically and that evaluated by simulation supports the view that the theoretical model presented in this paper provides a quite satisfactory solution to the queueing problem at least for the forms of distributions tried. These distributions are general enough to cover a quite wide variety of actual cases when their parameters are suitably adjusted. This fact alone would be enough to justify the use of the model. Nevertheless, it is believed that the analytical model behaves equally satisfactorily for any form of distribution concerning traffic flow and driver behaviour since no restriction concerning the forms was imposed during the analysis.

The model suggested can also be used to calculate delays to opposed turning traffic and uncontrolled intersections when there is no interference between the turning traffic and any other stream except the opposing one. An obvious example is the case of a driver turning right from a major road at an uncontrolled T-junction.

The preceding analysis can be modified to deal with the case of a major stream that has more than one lane of traffic travelling in the same direction. Using the same general idea and the statistics of the delay for a single driver waiting to cross a multi-lane major stream[15][16] the service time of a driver joining an empty and a non-empty queue can again be evaluated and fed into the model. Although this case has not been thoroughly examined it is believed that it may be as accurately represented by the analytical model.

ACKNOWLEDGEMENTS

The author is grateful to Mr. J.G.Wardrop for his constructive criticisms and suggestions. Thanks are also due to Professor R.E.Allsop for some helpful ideas and to members of the Transport Studies Group at University College London for interesting discussions on the subject.

REFERENCES

1. J.C.TANNER, "A theoretical analysis of delays at an uncontrolled intersection." Biometrika, 49, 163-170 (1962)

2. G.F.YEO AND WEESAKUL, "Delays to road traffic at an intersection." Journal of Applied Probability, 1, 297-310 (1964)

3. D.H.EVANS, R.HERMAN AND G.H.WEISS, "The highway merging and queuing problem." Operations Research, 12, 832-857 (1964)

4. A.G.HAWKES, "Gap acceptance in road traffic." Journal of Applied Probability, 5, 85-92 (1968)

5. D.J.BUCKLEY AND W.R.BLUNDEN, "Some delay flow characteristics for conflicting traffic streams." Proceedings of the 2nd International Symposium on the Theory of Traffic Flow, edited by J.Almond, Paris: OECD, 190-199 (1965)

6. D.R.COX AND W.L.SMITH, "On the superposition of renewal processes." Biometrika, 41, 91-99 (1954)

7. G.F.YEO, "Single-server queues with modified service mechanisms." Journal of Australian Mathematical Society, 2, 499-502 (1962)

8. G.H.WEISS AND A.A.MARADUDIN, "Some problems in traffic delay." Operations Research, 10, 74-104 (1962)

9. T.J.SCHRIBER, "Simulation Using GPSS." London: John Wiley ans Sons (1974)

10. J.M.HAMMERSLEY AND D.C.HANDSCOMB, "Monte Carlo Methods." London: Methuen and Co. Ltd. (1964)

11. M.FRANCKI, "The problem of confidence and three methods of variance reduction in the simulation of queueing." Proceedings of Australian Road Research Board, Volume 4, part 1, 567-580 (1968)

12. W.R.BLUNDEN, C.M.CLISSOLD AND R.B.FISHER, "Distribution of acceptance gaps for crossing and turning manoeuvres." Proceedings of Australian Road Research Board, Volume 1, part 1, 188-205 (1962)

13. R.ASHWORTH AND C.G.BOTTOM, "Driver gap acceptance behaviour at priority-type intersections." Final Report, University of Sheffield, Department of Civil and Municipal Engineering, Research Report No. R63 (1975)

14. J.E.FREUND, "Modern Elementary Statistics." 4th edition. London: Prentice/Hall International (1973)

15. J.C.GOLIAS, "The delay problem for a single driver crossing a two-lane stream. 12th Annual Conference of Universities Transport Study Group, Newcastle upon Tyne University (unpublished) (1980)

16. D.C.GAZIS, G.F.NEWELL, P.WARREN AND G.H.WEISS, "The delay problem for crossing an n - lane highway." Proceedings of 3rd International Symposium on Traffic Flow, New York: Elsevier, 267-279 (1967)

APPENDIX

The expressions for the various elements of the analytical model are given here when the following forms of distributions are used:

$$f(t) = \beta^a e^{-\beta t} t^{a-1} / \Gamma(a) \quad , \quad f_m(t) = qe^{-qt}$$

$$a(t) = 1-e^{-gt} \sum_{n=0}^{n=k-1} (gt)^n/n! \quad (g_0 \text{ instead of } g \text{ for } a_0(t))$$

$$n(t) = c^M e^{-ct} t^{M-1} / (M-1)!$$

To facilitate the manipulation of the analytical expressions the following are defined:

$$P(a_1+a_2,a_3) = (a_1+a_2-1)! / (a_1-1)! \, a_3^{a_1+a_2}$$

where a_2 can be a sum of terms.

$$b = \beta + c \quad , \quad d = \beta + g \quad , \quad f = \beta + c + g$$

The non-empty queue case will be dealt with at first. For the choice of distributions made we have

$$\bar{a} = 1 - \beta^a \sum_{N=0}^{N=k-1} g^N / N! \, P(a+N,d)$$

$$\bar{a}_{0Q}=1- \frac{k\beta}{ag} + \frac{k\beta^{a+1}}{agd^a} + \sum_{N=0}^{N=k-1} (k-N+1)\beta^{a+1}g^N P(a+N+1,d)/a(N+1)!$$

By replacing g by g_0 in the above formula $\bar{a}_0$ is obtained.

$$\bar{a}_Q = \bar{a}_{Q1}(\bar{a}_{0Q}-1)/\bar{a}_{0Q} + 1 - \bar{a}_{Q2}$$

where

$$\bar{a}_{Q1} = \frac{\beta^a}{\bar{a}} R_{Q1} = \frac{\beta^a}{\bar{a}} \sum_{m=0}^{m=M-1} \left[\frac{c^m}{m!} P(a+m,b) - \sum_{N=0}^{N=k-1} \frac{g^N P(a+N+m,f)}{N!} \right]$$

$$\bar{a}_{Q2} = (\beta^a c^M / \bar{a})\, R_{Q2}$$

$$R_{Q2} = \sum_{N1=0}^{N1=k-1} \sum_{N2=0}^{N2=k-1-N1} \frac{P(M+N2,c-g)g^{N1+N2}(-1)^{N2}}{N1!\ N2!} \Bigg[P(a+N1,d) - \sum_{N3=0}^{N3=M+N2-1} \frac{(c-g)^{N3} P(a+N1+N3,b)}{N3!} - \sum_{N=0}^{N=k-1} \frac{g^N}{N!} \Bigg[P(a+N+N1,d+g) - \sum_{L=0}^{L=M+N2-1} \frac{(c-g)^L P(a+N+N1+L,f)}{L!} \Bigg]\Bigg]$$

For the moments of the "first" gap encountered by the driver, given that it was rejected, we have

$$Y_i = \sum_{N=0}^{N=k-1} g^N \beta^a P(a+N+i,d) / N!$$

$$Y_{10Q} = A_k \left[\frac{\beta}{ag^2} - \frac{\beta^{a+1}}{ag^2 d^a} - \frac{\beta^{a+1}}{gd^{a+1}} \right] - \sum_{N=0}^{N=k-1} \frac{(A_k - A_N)\beta^{a+1} P(a+N+1,d)}{a(N+1)!}$$

where $A_L = L(L+1) / 2$

$$Y_{20Q} = E_k \left[\frac{\beta}{ag^3} - \frac{\beta^{a+1}}{ag^3 d^a} - \frac{\beta^{a+1}}{g^2 d^{a+1}} - \frac{\beta^{a+1}(a+1)}{2gd^{a+2}} \right] - \sum_{N=0}^{N=k-1} \frac{(E_k - E_N)\beta^{a+1} P(a+N+2,d)}{a(N+2)!}$$

where $E_L = \{L(L+1)(2L+1) / 6\} + A_L$.

By replacing g by g_0 in the formula for Y_{i0Q}, Y_{i0} is obtained.

$$Y_{1Q} = \bar{a}_{Q2}\{t\} - \bar{a}_{Q2}\{x\} + Y_{10Q}\,\bar{a}_{Q2} / \bar{a}_{0Q}$$

$$Y_{2Q} = \bar{a}_{Q2}\{t^2\} + \bar{a}_{Q2}\{x^2\} - 2\bar{a}_{Q2}\{tx\} + Y_{20Q}\,\bar{a}_{Q2} / \bar{a}_{0Q}$$

where

$\bar{a}_{Q2}\{t^r x^s\}$ is obtained from $\bar{a}_{Q2}$ by replacing a by a+r and M by M+s in the formula for R_{Q2}.

The moments S_i of the scanning time and n_i of the move-up time are given by

$$S_1 = Y_{1Q} + (1-\bar{a}_Q)\,Y_1 / \bar{a}$$

$$S_2 = Y_{2Q} + (1-\bar{a}_Q)\,Y_2 / \bar{a} + 2S_1Y_1 / \bar{a}$$

$$n_1 = M / c \quad , \quad n_2 = M(M+1) / c^2.$$

Then for the moments of the service time for the non-empty queue case we have

$$Q_1 = n_1 + S_1 \qquad Q_2 = n_2 + S_2 + 2n_2S_2$$

The case that a driver arrives to find an empty queue will now be considered.

$$\bar{a}_{E1} = \bar{a}_{E11} - \bar{a}_{E12}$$

where

$$\bar{a}_{E11} = \frac{1}{\bar{a}}\left[1 - \frac{\beta^a}{(\beta+q)^a}\right] - \frac{\beta^a}{\bar{a}} \sum_{N=0}^{N=k-1} \frac{g^N}{N!}\Big[P(a+N,d+q) - P(a+N,d)\Big]$$

$$+ \frac{q\beta^a}{\bar{a}} \sum_{L=0}^{L=M-1} \frac{c^L}{(q+c)^{L+1}} \left[\sum_{N=0}^{N=k-1} \frac{g^N}{N!}\left[P(a+N,d) - \sum_{m=0}^{m=L} \frac{(q+c)^m}{m!} P(a+N+m,f+q)\right]\right]$$

$$- \frac{q}{\bar{a}} \sum_{L=0}^{L=M-1} \left[\frac{c^L}{(q+c)^{L+1}} - \sum_{m=0}^{m=L} \frac{\beta^a (q+c)^m}{m!} P(a+m,b+q)\right]$$

$$\bar{a}_{E12} = \frac{q\beta^a}{\bar{a}} \sum_{N=0}^{N=k-1} \sum_{L=0}^{L=k-1-N} \frac{(-1)^L g^{N+L}}{N!L!} R_{E1}$$

$$R_{E1} = \frac{L!}{(q-g)^{L+1}} \Bigg[P(a+N,d) - \sum_{N3=0}^{N3=L} \frac{(q-g)^{N3}}{N3!} P(a+N+N3,\beta+q)$$
$$- \sum_{N1=0}^{N1=k-1} \Bigg[P(a+N+N1,d+g) - \sum_{N3=0}^{N3=L} \frac{(q-g)^{N3}}{N3!} P(a+N+N1+N3,d+q) \Bigg]\Bigg]$$
$$- \sum_{m=0}^{m=M-1} \frac{c^m (L+m)!}{m!(q+c-g)^{L+m}} \Bigg[P(a+N,d) - \sum_{I=0}^{I=L+m} \frac{(q+c-g)^I}{I!} P(a+N+I,b+q)$$
$$- \sum_{N1=0}^{N1=k-1} \Bigg[P(a+N+N1,d+g) - \sum_{I=0}^{I=L+m} \frac{(q+c-g)^I}{I!} P(a+N+N1+I,f+q) \Bigg]\Bigg]$$

$$Y_{1E1} = \bar{a}_{E12}\{t\} - \bar{a}_{E12}\{z\}$$

$$Y_{2E1} = \bar{a}_{E12}\{t^2\} + \bar{a}_{E12}\{z^2\} - 2\bar{a}_{E12}\{tz\}$$

where

$\bar{a}_{E12}\{t^r z^s\}$ is obtained from $\bar{a}_{E12}$ by replacing a by a+r and L by L+s in the formula for R_{E1}.

For the second sub-case of the empty queue case we have

$$\bar{a}_{E2} = \bar{a}_{E21} - \bar{a}_{E22} + \bar{a}_{E23}$$

$$\bar{a}_{E21} = \bar{a}_Q + \bar{a}_{Q2}\{\beta,\beta+q\} - \bar{a}_{Q1}\{\beta,\beta+q\}$$
$$+ \frac{\beta^a}{\bar{a}} \left[\frac{1}{(\beta+q)^a} - \sum_{N=0}^{N=k-1} \frac{g^N}{N!} P(a+N,d+q) \right]$$

where

$\bar{a}_{Qi}\{\beta,\beta+q\}$ is obtained from $\bar{a}_{Qi}$ by replacing β by $\beta+q$ in the formula for R_{Qi}.

$$\bar{a}_{E22}=\frac{1}{\bar{a}}\left[1-\frac{\beta^a}{(\beta+q)^a}-\sum_{N=0}^{N=M-1}\frac{qc^N}{(q+c)^{N+1}}\left[1-\beta^a\sum_{L=0}^{L=N}\frac{(q+c)^L}{L!}P(a+L,b+q)\right]\right]$$

$$-\frac{1}{\bar{a}}\sum_{m=0}^{m=k-1}\left[\frac{g^m\beta^a}{m!}P(a+m,d)-P(a+m,d+q)\right.$$

$$\left.-\sum_{L=0}^{L=k-1}\frac{qc^L}{(q+c)^{L+1}}\left[1-\sum_{N=0}^{N=L}\frac{(q+c)^N}{N!}P(a+m+N,f+q)\right]\right]$$

$$\bar{a}_{E23}=(\beta^a c^M/\bar{a})\,R_{E2}$$

where

$$R_{E2}=\sum_{N=0}^{N=k-1}\sum_{L=0}^{L=k-1-N}D\left[P(a+N,d)-P(a+N,d+q)\right.$$

$$-\sum_{m=0}^{m=L}E\left[P(a+N,d)-\sum_{I=0}^{I=m}\frac{(q+c-g)^I}{I!}P(a+N+I,b+q)\right]$$

$$-\sum_{J=0}^{J=k-1}\frac{g^J}{J!}\left[P(a+N+J,d+g)-P(a+N+J,d+q+g)\right.$$

$$\left.\left.-\sum_{m=0}^{m=M+L-1}E\left[P(a+N+J,d+g)-\sum_{I=0}^{I=m}\frac{(q+c-g)^I}{I!}P(a+N+I+J,f+q)\right]\right]\right]$$

$$D=(-1)^L\,g^{N+L}\,P(M+L,c-g)\,/\,(N!\,L!)$$

$$E=q\,(c-g)^m\,/\,(q+c-g)^{m+1}$$

$$Y_{1E2}=Y_{1Q}+\bar{a}_{Q2}\{(\beta,\beta+q),x\}-\bar{a}_{Q2}\{(\beta,\beta+q),t\}-\bar{a}_{E23}\{t\}+\bar{a}_{E23}\{x\}$$

$$Y_{2E2}=Y_{2Q}+2\bar{a}_{Q2}\{(\beta,\beta+q),tx\}-\bar{a}_{Q2}\{(\beta,\beta+q),x^2\}$$

$$-\bar{a}_{Q2}\{(\beta,\beta+q),t^2\}+2\bar{a}_{E23}\{tx\}-\bar{a}_{E23}\{t^2\}-\bar{a}_{E23}\{x^2\}$$

where $\bar{a}_{Q2}\{(\beta,\beta+q),t^r x^s\}$ is obtained from a_{Q2} by replacing β by $\beta+q$, a by a+r and M by M+s in the formula for R_{Q2}, and

$\bar{a}_{E23}\{t^r x^s\}$ is obtained from $\bar{a}_{E23}$ by replacing a by a+r and M by M+s in the formula for R_{E2}.

For the remaining cases when the driver arrives to find an empty queue we have the following formulae:

$$\bar{a}_{E3} = \bar{a}_{0Q}\, P_3 \quad , \quad Y_{iE3} = Y_{i0Q}\, P_3$$

where

$$P_3 = \frac{\beta^a}{\bar{a}} \sum_{N=0}^{N=M-1} \left[\frac{c^N}{N!} P(a+N,b) - P(a+N,b+q) - \sum_{L=0}^{L=k-1} \frac{g^L}{L!} \left[P(a+N+L,f) - P(a+N+L,f+q) \right] \right]$$

$$\bar{a}_{E4} = \bar{a}_0\, P_4 \quad , \quad Y_{iE4} = Y_{i0}\, P_4$$

$$\bar{a}_{E5} = \bar{a}_{0Q}\, P_5 \quad , \quad Y_{iE5} = Y_{i0Q}\, P_5$$

where

$$P_5 = \frac{\beta^a q}{\bar{a}} \sum_{m=0}^{m=M-1} c^m \sum_{L=0}^{L=m} \frac{(q+c)^{L-m-1}}{L!} \left[P(a+L,b+q) - \sum_{N=0}^{N=k-1} \frac{g^N}{N!} P(a+L+N,f+q) \right]$$

$$P_4 = \frac{\beta^a}{\bar{a}} \left[\frac{1}{(\beta+q)^a} - q \sum_{N=0}^{N=k-1} \frac{g^N}{N!} P(a+N,d+f) \right] - P_5 \, .$$

Then for the probability $\bar{a}_E$ of accepting the "first" gap and the moments Y_{iE} of the "first" gap given that it was rejected we have

$$\bar{a}_E = \sum_j \bar{a}_{Ej} \quad , \quad Y_{iE} = \sum_j Y_{iEj} \, .$$

The moments Z_i of the scanning time and n_{iE} of the remaining part of the move-up time are given by

$$Z_1 = Y_{1E} + (1 - \bar{a}_E)\, Y_1 \,/\, \bar{a}$$

$$Z_2 = Y_{2E} + (1 - \bar{a}_E)\, Y_2 \,/\, \bar{a} + 2Z_1 Y_1 \,/\, \bar{a}$$

$$n_{1E} = \sum_{N=0}^{N=k-1} \frac{qMc^{N-1}}{(q+c)^{N+1}} - \sum_{L=0}^{L=M-1} \frac{q(L+1)c^{L}}{(q+c)^{L+2}}$$

$$n_{2E} = \sum_{N=0}^{N=M+1} \frac{qc^{N-2}k(k+1)}{(q+c)^{N+1}} + \sum_{L=0}^{L=M-1} \frac{q(L+1)(L+2)c^{L}}{(q+c)^{L+3}} - 2\sum_{m=0}^{m=M} \frac{qMc^{m-1}(m+1)}{(q+c)^{m+2}} .$$

Then we have for the moments of the service time for the empty queue case

$$E_1 = n_{E1} + Z_1$$

$$E_2 = n_{E2} + Z_2 + 2\, n_{E1}\, Z_1$$

The mean queueing time w_1 and the mean delay d_1 are then given by

$$w_1 = \frac{qE_2 - q^2 Q_1 E_2 + q^2 E_1 Q_2}{2(1-qQ_1)\,(1-qQ_1+qE_1)}$$

$$d_1 = w_1 + P_0\, E_1 + (1 - P_0)\, Q_1$$

where

$P_0 = (1-qQ_1) \,/\, (1-qQ_1+qE_1)$ is the probability that a driver arrives to find the queue empty.

Characteristics of Travel Time and Dynamic User Equilibrium for Travel-to-Work

Chris Hendrickson
Daniel Nagin
Edward Plank
Carnegie-Mellon University, Pittsburgh, PA

Departure time decisions for travel-to-work are likely to vary in response to work start times and to variations in the travel cost over the course of the peak period of travel. One component of this cost is travel time. This paper summarizes some data on travel times at different departure times for five corridors in Pittsburgh, PA. Several corridors exhibit substantial variations in average travel time at different departure times. However, the variance of travel time at any departure time is small and displays no systematic relationship to departure time. The paper also considers the various elements of travel cost which may be important for the departure time decision problem, including congestion avoidance, reliability and schedule incompatibility. With the assumption that individuals minimize their own travel cost, several models of dynamic user equilibrium are developed, each corresponding to different user cost functions. These models incorporate the effects of travel time variability.

INTRODUCTION

Volumes and travel times are often assumed to be constant over the course of peak periods of travel. While this static assumption is analytically convenient, there can be significant variations in travel times and volumes over the course of a peak period of travel such as the morning commuting hours. Among other influences, these variations depend upon both departure time and route choice decisions of individual travelers. This paper investigates influences on departure time decisions in the context of a dynamic user equilibrium and presents some empirical evidence on these influences and their effects.

Understanding the reasons for variations in travel characteristics over the course of a peak period is important for a number of reasons. First, one of the major problems in providing urban transportation services is the expense associated with accommodating sharp peaks in volume. For mass transit services, increasing the peak period capacity may involve purchase of additional vehicles or employment of additional drivers. Expansion of roadway facilities to accommodate the peak period volume may result in over-investment in such facilities. Thus, policies which may affect time of departure decisions and thereby affect peaking patterns may have significant implications for the cost of providing urban transportation services. Examples of such policies are peak hour tolls, flexible work hours and staggered work hours. Second, the possibilities for ride-sharing (in the form of car-pools or van-pools) are limited not only by the location of origins and destinations but also by the time at which trips are made. Thus, if desired arrival times at destinations are not reasonably consistent, then it will be impossible to form ride sharing arrangements among a given group of travelers.

In this paper, we shall focus upon the limited case of travel to work over a single route. This restriction allows us to consider in relative isolation the effects of departure time decisions on travel characteristics. Evidence developed from measurement of travel times and a survey of travelers conducted for several areas in Pittsburgh, PA (USA) will be used to illustrate the discussion. First, we shall present some evidence of travel time variations. In the following two sections, we shall define dynamic user equilibrium and discuss various elements of user cost which influence travel decisions. In Section 4, we consider departure time decisions to avoid peak periods of congestion. In Section 5, we discuss the effects of travel time randomness or service reliability on departure decisions. In Section 6, delays due to schedule incompatibilities are considered. Finally, Section 7 summarizes some implications of the discussion and considers useful directions for future research.

1. SOME EVIDENCE OF TRAVEL TIME VARIABILITY FOR TRAVEL TO WORK

Several studies of travel time characteristics during the peak hour of travel have been made (such as Herman and Lam [1]). However, few studies report on changes in travel time characteristics during the course of the peak period. Accordingly, we have gathered data related to departure time decisions and travel time variations over the course of the peak for travel to a portion of the Central Business District (CBD) in Pittsburgh, PA. This data collection included a survey of work start time and work trip characteristics of 11,000 employees in the downtown study area. In addition, measurements were taken of travel times on several modes and routes into the CBD.

We consider here the travel characteristics from five postal zip code areas to the CBD. Travel from each depends upon a single route, and with the exception of the Shadyside zone, each route includes either a major tunnel or a bridge. Travel times to a point in the Pittsburgh CBD (corresponding to the headquarters building of the U.S. Steel Corporation) from a reference point in each of these five areas are shown in Table 1, along with the number of observations and the standard deviations of travel time. For Bethel Park, Shadyside, and Glenshaw, travel times were recorded from special vehicle trips in which all departures were within two minutes of each quarter hour. For Monroeville and Greentree, travel times were measured by bus travel times over an express portion of their route. During this express portion, the buses traveled on access-controlled freeways at the same speed as automobiles. For these two zones, departure times from the reference point were spread throughout each fifteen minute interval; travel times reported in Table 1 are for departures occurring within 7.5 minutes of the quarter hour.

There is a substantial amount of travel time peaking on some of the routes for which travel times were measured. The difference between the peak period travel time and the base or shortest travel time ranged from 3 to 16 minutes, representing between 25 and 52 percent of the minimum travel times (Table 2). For Glenshaw and Shadyside, there is no sharp peak travel time, but rather a plateau representing a peak period travel time (Figure 1). The routes to the CBD from these two areas seem to have sufficient capacity that no major bottlenecks exist. In contrast, Bethel Park, Greentree and

Table 1

Measured Travel Time Characteristics to the CBD from Five Points in Pittsburgh, PA.

Departure Time	Bethel Park	Glenshaw	Greentree (Bus)	Monroeville (Bus)	Shadyside
		Average Travel Time (min.)			
6:45	-	-	-	32.9	-
7:00	24.1	19.0	-	40.9	-
7:15	26.7	21.8	10.6	47.2	12.3
7:30	32.4	23.2	15.3	43.2	13.9
7:45	33.9	19.8	13.7	37.0*	14.8
8:00	32.4	21.0	13.6	32.0*	14.7
8:15	28.0	19.7	12.4	30.7	14.3
8:30	26.6	19.2	11.0	-	14.9
8:45	-	18.0	-	-	12.6
		Standard Deviation (min.)			
6:45	-	-	-	2.9	-
7:00	2.3	2.2	-	5.4	-
7:15	4.9	1.8	2.6	4.2	1.0
7:30	4.1	2.1	2.6	4.6	1.9
7:45	3.4	1.7	2.0	*	1.7
8:00	4.1	2.0	2.1	*	1.7
8:15	6.3	1.9	0.9	2.5	3.1
8:30	6.6	2.7	1.0	-	3.2
8:45	-	2.1	-	-	1.7
Number of Observations - (Average Number of Observations per Interval with Measured Times)					
	10	6	13	6	10

* Less than three observations

\- No observations available

Table 2

Summary of Measured Travel Time Characteristics

	Bethel Park	Glenshaw	Greentree (Bus)	Monroeville (Bus)	Shady-side
1. Uncongested Travel Time (min.)	24	18	11	31	12
2. Peak Travel Time (min.)	34	23	15	47	15
3. Extra Travel Time at Peak (min.) (2-1)	10	5	4	16	3
4. Extra Travel Time as % of Base (3/1 x 100)	42	28	36	52	25
5. Average Standard Deviation of Travel Time (min.)*	4.5	2.1	1.9	3.9	2.0

*Average of standard deviations for each departure time interval.

Monroeville exhibit definite peaking patterns.

The variability in travel times was relatively low, with the average ratio of the standard deviation to mean travel time (i.e. the coefficient of variation) equal to .13 over all routes and departure times. This variation is lower than that found by Herman and Lam [1], who found a coefficient of variation of less than .20 for travel in the vicinity of Detroit, Michigan. The difference in these estimated coefficients of variation may be attributed to the differences in the length of the departure time interval. Herman and Lam [1] included observations over a half hour departure time interval in calculating their coefficient of variation, whereas our coefficients are calculated for departures within a fifteen minute interval. There does not seem to be any pattern by which travel time variability is related to departure time in our observations.

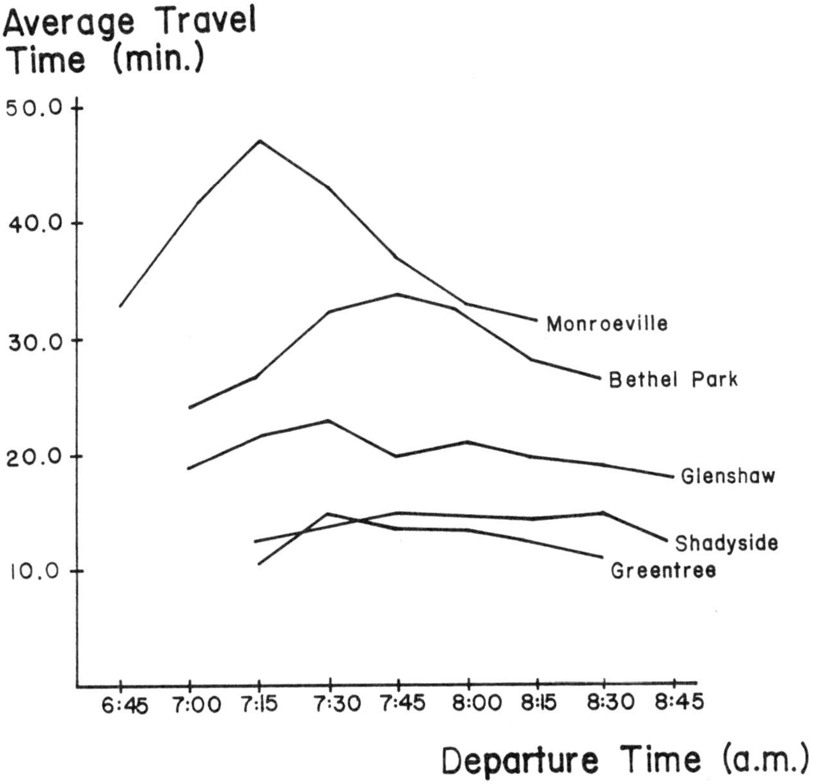

Figure 1
Average Travel Times by Departure Time
for Five Zones in Pittsburgh, PA.

In the next few sections, we shall consider possible user responses to this travel time peaking. Our discussion will assume that individuals attempt to minimize their own cost, based upon a knowledge of this pattern of travel time peaking; the aggregate result of this individual optimization will be called a dynamic user equilibrium, as defined below.

2. DYNAMIC USER EQUILIBRIUM

We characterize dynamic user equilibrium as a pattern of travel in which some characteristics of trip-making vary over time and in which no individual can unilaterally alter his trip-making so as to achieve a lower personal cost (or, alternatively, a higher level of utility, where utility reflects the costs of travel, including travel time). This definition is an extension of the familiar notion of user equilibrium, advanced by Pigou [2], Wardrop [3] and others, with the difference that various network and demand characteristics are expected to vary over time.

Two such characteristics may reasonably be expected to vary and have been discussed in the literature. The first postulates a situation in which origin/destination flows are time dependent but known. In this case, the choice of route, the travel cost (including time), and, possibly, the mode of travel are likely to vary over time. Preliminary studies of this dynamic equilibrium problem have been made by Hurdle [4] and Pearce [5]; a review and some extensions appear in Newell [6].

Another interesting situation is the case in which desired departure or desired arrival times are known, but in which actual departure (or arrival) times may differ from these desired times. In this case, it is reasonable to assume that costs are incurred for arrivals (or departures) which occur earlier or later than those desired. For example, travelers may wait and travel _after_ the peak period of congestion or, alternatively, depart before the peak, so as to avoid this congestion. In this case, the origin/destination flows at any given time depend on the distribution of desired arrival (or departure) times _and_ on the cost of travel at any given time. An example of this type of dynamic user equilibrium appears below.

While both these extensions are conceptually straightforward and quite likely to occur over the course of peak periods of travel, they create a number of analytical problems. For static or steady state user equilibrium, the problem of identifying equilibrium flows may be formulated as a mathematical programming problem with a convex constraint set and a non-linear objective function (see, for example, Nguyen [7]). A comparable formulation has not yet been developed for dynamic user equilibrium. One difficulty in formulating this problem is the specification of the set of node continuity constraints. For the static problem, total volume into a link equals the volume out, no matter what the travel time on the link. In the dynamic user equilibrium problem, the volume departing from a link at any given time depends upon the volume entering the link in previous periods and the travel time on the link. Consequently, node continuity constraints for the dynamic user equilibrium problem also depend upon these factors, thereby greatly increasing the complexity of the problem.

We might also note that several of the extensions or modifications of static user equilibrium which have received attention might be applied to the problem of dynamic user equilibrium. For example, one could characterize a stochastic, dynamic user equilibrium (following Daganzo and Sheffi [8]) or a dynamic user equilibrium with elastic demands. Again, we shall not pursue such extensions here, although we will analyze the effects of expectations of travel time randomness below.

3. USER COSTS OF TRAVEL TO WORK

For travel to work, we assume that user travel costs are a combination of travel time r, early arrival time s, lateness in arrival p, and any monetary charge f, all of which are functions of departure time t. For a linear combination of these components, we have:

$$UC_i(t) = a_0 + a_1 r(t) + a_2 s_i(t) + a_3 p_i(t) + f(t) \qquad (1)$$

where a_0, a_1, a_2 and a_3 are constant parameters and the subscript i refers to the ith traveler. For simplicity, we omit modal subscripts on the parameters (a_0, a_1, a_2, and a_3), but these parameters might vary among modes of travel. Each individual is assumed to have a work start time B_i, which may differ among individuals. Early arrival time is defined as the difference between B_i and the actual arrival time for early arrivals, so that $s_i(t) = \max \{B_i - r(t) - t;\ 0\}$, where t is the departure time. Similarly, late arrival time is defined as $p_i(t) = \max \{r(t) + t - B_i;\ 0\}$. We assume throughout that travel time is more onerous than early arrival time, so that $a_1 > a_2$ below. Several additional cost components might be added to our model (1), such as the variation in crowding on transit vehicles over time or the probability of securing a parking spot as a function of time, but we shall not consider these effects here.

We assume that individuals will choose the departure time which represents the minimum value of user cost as given by (1). Importantly, this departure time need not represent a minimum travel time or be equal to the time at which arrival would occur exactly at the work start time B_i. For example, suppose that an individual is faced with a constant toll and with the choice of arriving on time with a travel time of 30 minutes versus arriving 10 minutes early with a travel time of 25 minutes. Then, the early arrival will be chosen if:

$$a_0 + 25a_1 + 10a_2 < a_0 + 30a_1$$

or

$$a_2 < .5a_1$$

implying that early arrival would be chosen if two minutes spent at the destination are less onerous than a minute spent traveling.

We consider three major influences which may cause travelers to choose to arrive earlier or later than their work start time. First, reductions in travel time at different departure times may be sufficiently large to compensate for the increased costs of arriving early or late; the example above illustrates this point. The delay associated with early or late arrivals due to this influence can be termed "congestion avoiding" delay. Second, travel times may be uncertain and, if late time is more costly than early arrivals ($a_3 > a_2$ in (1)), it is less costly to arrive early (on average) so as to avoid large late penalties. We term this type of effect a "travel reliability" influence. Third, early or late arrivals may be required due to the schedules of shared vehicles, including scheduled transit services or car-pools. We shall call such influences delays due to "schedule incompatibilities." Finally, for completeness, monetary costs might vary by time of day due to, for example, peak/off-peak toll differentials. We shall not explore the influence of time variations in monetary cost here.

4. DYNAMIC USER EQUILIBRIUM WITH CONGESTION AVOIDING DELAYS

Hendrickson and Kocur [9] developed a simple model of dynamic user equilibrium for the case of flexible departure times and known desired arrival times. Their model assumes a deterministic queueing framework in which a bottleneck possesses a maximum service rate, g. Each traveler is expected to minimize his travel cost, given by (1). The user cost associated with early or late arrival is depicted as curve 1 in Figure 2; curves 2 and 3 represent different assumptions which are described below. For simplicity, the cost of travel to the bottleneck queue or after the bottleneck is assumed to be constant and is included in the constant term a_0, along with any toll. Also, the population of travelers is assumed to be homogeneous, so that the parameters a_0, a_1, a_2 and a_3 are constant for all travelers.

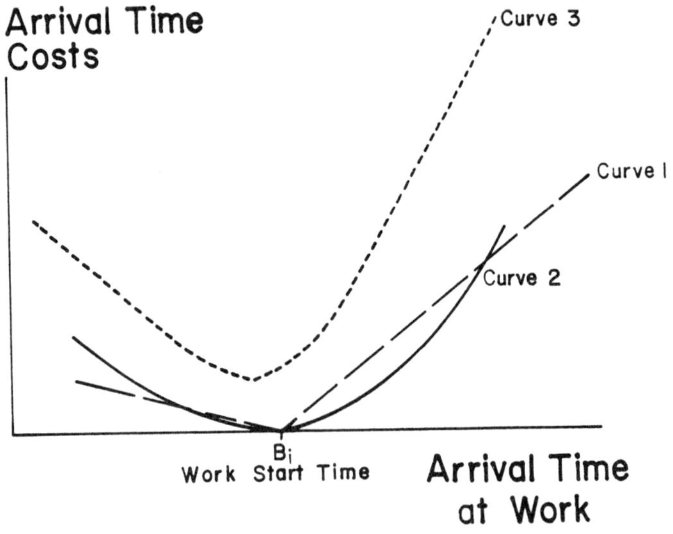

Curve 1: Linear Cost Model: $a_2 s(t) + a_3 p(t)$

Curve 2: Quadratic Cost Model: $a_2'(s(t))^2 + a_3'(p(t))^2$

Curve 3: Linear Cost Model with Random Travel Time: $a_2 E[s(t)] + a_3 E[p(t)]$

Figure 2

Illustrations of Early and Late Arrival Time Costs

Barring a discontinuity in the queueing time, a necessary condition for a dynamic user equilibrium for early arrivals is that the change in user cost with respect to the chosen arrival time at the bottleneck (or departure time from the residence) must be equal to zero for any individual. Otherwise, a change in departure time could be made to reduce user cost. Thus, for an early arrival, user cost is:

$$UC_i(t) = a_0+a_1r(t) + a_2(B_i-t-r(t)) \qquad (2)$$

and an equilibrium condition is:

$$\frac{\partial UC_i(t)}{\partial t} = 0 = a_1 \frac{\partial r(t)}{\partial t} - a_2-a_2 \frac{\partial r(t)}{\partial t} \qquad (3)$$

For a deterministic queue which begins at time t_0 and does not dissipate by time t, the waiting time in queue is equal to

$$r(t) = \frac{1}{g} \int_{t_0}^{t} (m(y)-g)\,dy \qquad (4)$$

where m(y) is the aggregate arrival rate at time y, so that (3) may be simplified to:

$$0 = a_1 \left(\frac{m(t)-g}{g}\right) - a_2-a_2 \left(\frac{m(t)-g}{g}\right) \qquad (5)$$

which can be re-written as:

$$m(t) = ga_1/(a_1-a_2) \qquad (6)$$

implying that the aggregate arrival rate, m(t), must be constant in equilibrium for early arrivals. A similar result is obtained for the equilibrium condition with late arrivals, although the arrival rate is different than for early arrivals. Details of the derivation, a discussion of several special cases, and consideration of different toll possibilities appear in Hendrickson and Kocur [9].

An example of an equilibrium pattern of departures for work, A(t), appears in Fig. 3, corresponding with the cumulative distribution of work start times, W(t), which was reported by survey respondents working in the Pittsburgh Central Business District. Fig. 3 illustrates a case with negligible travel time from the bottleneck to the work place, hence the bottleneck departure time is equivalent to the arrival time at work. The waiting time in queue, r(t), for an arrival at time t is given by the horizontal difference between the A and D functions, while early arrival time (or lateness) is the horizontal difference between the D and W functions. Costs of riding time, early arrival time, and late time as functions of desired arrival time at work appear in Fig. 4.

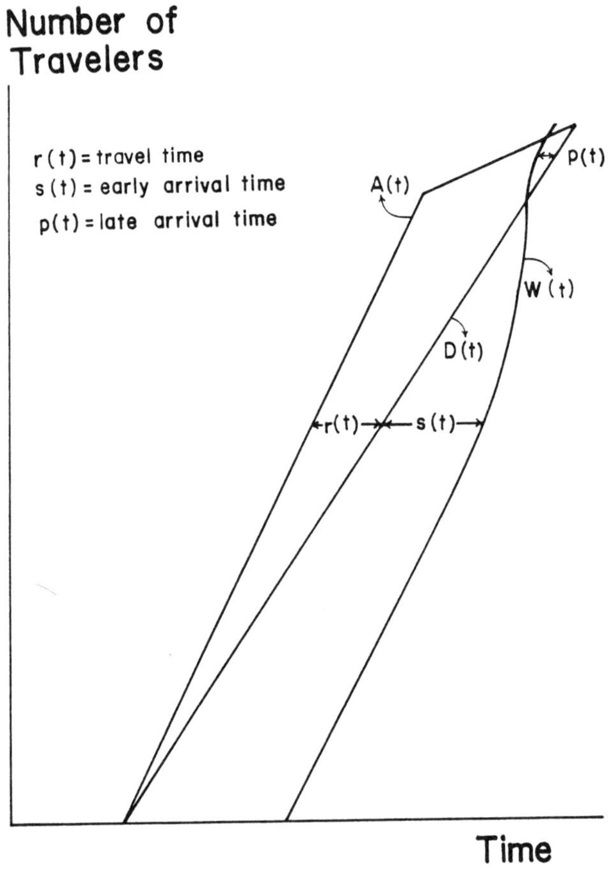

Figure 3
Costs of Travel Time, Early Arrival Time, and Late Arrival Time as Functions of Arrival Time at Work

Some implications of this simple model are worth noting. First, the travel time cost increases linearly to a peak value. This peak occurs at the time when the change-over from early to late arrivals occurs (Fig. 4). Secondly, the arrival rate of travelers at the bottleneck does not depend upon the shape of the work start time function W(t) (from Eq. 6), although the time at which the queue first forms is affected and thus the amount of early arrival time does depend upon W(t). Consequently, the value of user cost for any traveler depends upon W(t); altering work start times may change users' travel costs. Thirdly, a number of individuals choose to arrive early (or late) at work so as to reduce their overall cost. Finally, since the equilibrium aggregate arrival rate differs for early and late arrivals, the travelers are divided into a group of all early arrivals (with aggregate arrival rate given by (6)) and a group of all late arrivals. The transition between these two groups occurs at the point at which the equilibrium aggregate arrival rate ($\partial A(t)/\partial t$) changes, as illustrated in Fig. 3.

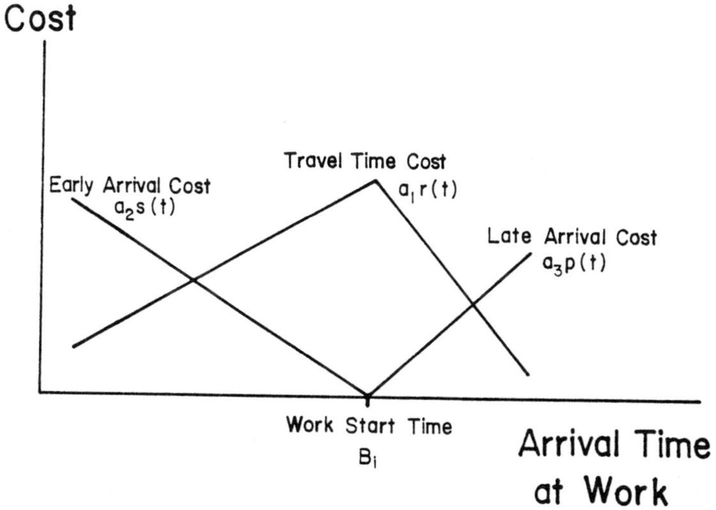

Figure 4
Costs of Travel Time, Early Arrival Time, and Late Arrival Time as Functions of Arrival Time at Work

Of course, this simple model omits some important facets of commuting-to-work. Some of these considerations, such as a non-linear user cost function, can be added. For example, suppose that the user cost function is a linear combination of waiting time, the square of early time and the square of late time:

$$UC_i(t) = a_0' + a_1'r(t) + a_2'(s_i(t))^2 + a_3'(p_i(t))^2 \tag{7}$$

where the parameters a_0', a_1', a_2' and a_3' may be different from those used in the linear user cost model (1). Costs of schedule delay or lateness are shown as curve 2 in Figure 2.

After making the substitution $s_i(t) = B_i-t-r(t)$, the equilibrium condition corresponding with (7) for early arrivals is:

$$\frac{\partial UC_i(t)}{\partial t} = 0 = a_1'\frac{\partial r(t)}{\partial t} + 2a_2's_i(t)\ (-1-\frac{\partial r(t)}{\partial t}) \tag{8}$$

which, using (4), simplifies to the arrival rate:

$$m(t) = ga_1'/(a_1' - 2a_2's_i(t)) \tag{9}$$

Similarly, the equilibrium arrival rate for late arrivals is:

$$m(t) = ga_1'/(a_1' + 2a_3'p_i(t)) \tag{10}$$

For some patterns of work start times, it is possible that an equilibrium pattern of arrivals corresponding with (9) and (10) does not exist; we shall not consider sufficient conditions for an equilibrium here.

Figure 5 illustrates the resulting dynamic user equilibrium with the non-linear user cost function (7). As can be seen from the figure, the general pattern of early and late arrivals as well as the peaking of the travel time is similar to the original example with a linear user cost function. However, the increase (and decrease) in travel time is not necessarily linear. Also, the arrival rate is not independent of the shape of the desired arrival time function, W(t).

The dynamic user equilibrium with a non-linear utility function exhibits an extreme form of traveler platooning. With the linear utility function (1), travelers are divided into groups of all early arrivals and all late arrivals. With the quadratic utility function (7), there will generally be only one departure time interval for which condition (8) is satisfied. Hence, travelers are not only grouped into all early and late arrivals, but also are ordered by their arrival times at the bottleneck in accordance with their work start times.

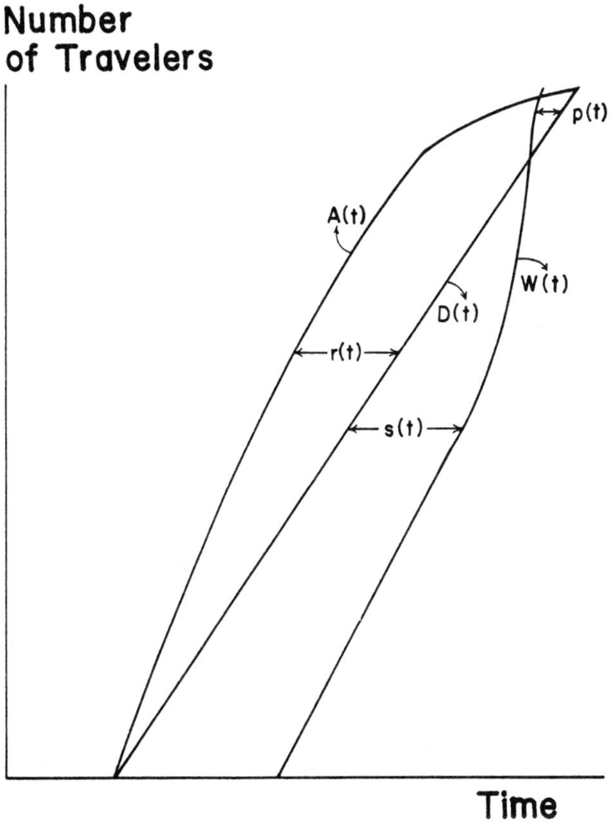

Figure 5

Cumulative Arrivals A(t), Departures D(t), and Work Starts W(t) as Functions of Time (Non-Linear User Cost Function)

The same type of travel patterns may be observed in the absence of a queueing model. For example, suppose the travel time over a roadway at any time t is a linear function of the arrival rate at the roadway, m(t):

$$r(t) = b_0 + b_1 m(t) \qquad (11)$$

This model presupposes that changes in m(t) are sufficiently slow or that some form of platooning exists such that arrivals during previous periods do not affect the travel time at time t. We shall explore this model using the linear user cost function (1).

For early arrivals, the necessary equilibrium condition is:

$$\frac{\partial UC_i(t)}{\partial t} = 0 = a_1 \frac{\partial r(t)}{\partial t} - a_2 - a_2 \frac{\partial r(t)}{\partial t} \qquad (12)$$

and substituting the derivative of r(t) defined by (11):

$$a_2 = (a_1 - a_2) b_1 \frac{\partial m(t)}{\partial t} \qquad (13a)$$

or:

$$\frac{\partial m(t)}{\partial t} = a_2/(b_1(a_1 - a_2)) \qquad (13b)$$

implying that the derivative of the arrival rate is constant, so that m(t) increases linearly with time and, from (11), travel time increases linearly with time. For late arrivals, the derivative of the arrival rate is also constant but negative, so that m(t) and r(t) decrease linearly with time. Thus, the resulting variation in travel time would be similar to that of Fig. 4, although the pattern of travel volumes would be quite different.

The departure time patterns observed in several of the sample zones are consistent with the prediction of these models. For example, the actual cumulative arrival at work and work start time functions for non-transit riders from Bethel Park closely resemble those shown in Fig. 5; as can be seen, most workers arrive substantially early for work.

The average early arrival times for all five postal zones are summarized in Table 3 for commuters who drive alone or car-pool. The correlation between average early time and travel time or travel time peaking is very strong for individuals who drive alone, as the simple models developed above would suggest. The correlation coefficient between average early time and peak travel time is .77, while the correlation is .72 between average early time and the difference between peak and base travel time. However, this strong correlation is not found for the average early time of travelers who share rides. For these travelers, the need to coordinate departure times is an important influence, as discussed below.

Table 3 Average Early Arrival Times for Five Postal Zones

	Bethel Park	Glenshaw	Greentree	Monroe-ville	Shady-side
Drive Alone					
Average Early Time (min.)	10	15	7	17	6
Number of Observations	36	61	22	19	25
Shared Ride					
Average Early Time (min.)	14	16	20	13	16
Number of Observations	63	86	30	27	17

NOTE: Early arrivals of greater than 50 minutes were omitted

5. DYNAMIC USER EQUILIBRIUM WITH STOCHASTIC TRAVEL TIMES

The previous section explored user equilibrium under conditions of deterministic travel times. Uncertainty about actual travel time may cause some travelers to depart earlier than they otherwise would. This is because of a presumed asymmetry in user cost for late versus early arrival, with the per unit time cost of late arrival greater than for early arrival (i.e., $a_3 > a_2$ in (1)). In this section we explore a simple model of dynamic user equilibrium with stochastic user cost. For simplicity, we consider the case in which all work start times are identically equal to B.

Sources of randomness in travel time are numerous and diffuse (e.g., accidents, weather, construction, irregularly scheduled events) and approaches to introducing uncertainty into models are correspondingly various. The formulation we will explore introduces uncertainty directly by treating travel time, $r(t)$, as a random variable. Alternative, more complex approaches could treat the capacity of the bottleneck, g, as a random variable, but the more direct and simple approach explored here is sufficient for our purposes. Specifically, we shall assume that $r(t)$ is uniformly distributed on the interval $[\bar{r}(t) - b, \bar{r}(t) + b]$, with a mean of $\bar{r}(t)$.

With the assumption of a uniform distribution of travel time conditional on time of departure, user equilibrium must be examined over three distinct time intervals: (1) The interval ending at t_e where early arrival occurs with certainty, (2) the interval beginning at t_l where late arrival is certain, and (3) the interval from t_e to t_l where either early or late arrival is possible.

The departure time marking the end of the period of certain early arrival, t_e, is the departure time such that $E[s(t_e)] = B - \bar{r}(t_e) - t_e = b$ where $E[s(t)]$ is expected early arrival time. Actual early arrival time for departures at t_e will range from 0 to 2b because travel time is uniformly distributed over a range of 2b. Similarly, the departure time marking the beginning of the period of certain late arrival, t_l, is the time such that $E[p(t_l)] = \bar{r}(t_l) + t_l - B = b$, where $E[p(t)]$ is expected late arrival time.

In this stochastic environment, user equilibrium is defined as before, except that tripmakers are assumed to minimize expected cost. The constituent elements of expected user cost are $E[r(t)]$, $E[s(t)]$, and $E[p(t)]$, where $E[r(t)] = \bar{r}(t)$. Expected early arrival time is equal to $B - \bar{r}(t) - t$ for certain early arrival ($t < t_e$) and 0 for certain late arrivals ($t > t_l$). For intermediate times ($t_e \leq t \leq t_l$), expected early arrival time is:

$$E[s(t)] = \int_{-b}^{B - t - \bar{r}(t)} (B - (t + \bar{r}(t)) + x) \frac{dx}{2b} = \frac{1}{4b} (b + B - t - \bar{r}(t))^2 \qquad (14)$$

where x is a uniformly distributed variable on the interval [-b,b]. In this expression, $t + \bar{r}(t)$ is the expected arrival time, $B - (t + \bar{r}(t))$ is expected early arrival time, and $B - (t + \bar{r}(t) + x)$ is actual early arrival time for a particular realization of x. Since travel time is stochastic, the arrival time could be as much as b time units earlier than the expected time. This defines the lower limit of the integral. The upper limit corresponds to an arrival at exactly B, with no early arrival time. Summarizing, expected early arrival time for the three intervals is:

$$E[s(t)] \begin{cases} = B - \bar{r}(t) - t & t < t_e \\ = \frac{1}{4b}(b + B - t - \bar{r}(t))^2 & t_e \leq t \leq t_1 \\ = 0 & t > t_1 \end{cases} \qquad (15)$$

Expected late time may be derived in a similar manner to expected early arrival time, with

$$E[(p(t)] \begin{cases} = 0 & t < t_e \\ = \frac{1}{4b}(b + t + \bar{r}(t) - B)^2 & t_e \leq t \leq t_1 \\ = t + \bar{r}(t) - B & t > t_1 \end{cases} \qquad (16)$$

With the assumption of normally or lognormally distributed travel times (as suggested in [1]), similar expressions for expected early or late arrival times may be obtained. These expressions do not have closed form solutions and make analysis of user equilibrium more difficult. Also, these distributions imply that there is no time, such as t_e, for which early arrival is guaranteed.

In this model of stochastic travel time, the expected cost of early plus late arrival time is equal to $a_2E[s(t)] + a_3E[p(t)]$. The expected cost of early or late arrival as a function of the <u>planned</u> arrival time, $t + r(t)$, is shown by curve 3 in Figure 2. Observe that unlike curves 1 and 2 in Fig. 2, where travel times are deterministic, the expected cost of schedule deviation is minimized at a planned arrival time prior to the work start. This is because of the assumption that the per unit time cost of late arrival exceeds that of early arrival. Thus, from the perspective of minimizing early plus late arrival time costs, a planned arrival prior to the actual start time is optimal.

User equilibrium is defined by a pattern of expected ride times, $\bar{r}(t)$, such that $E[UC(t)]$ is constant. For departure times ensuring either early or late arrival, derivation of $\bar{r}(t)$ is a straightforward application of the approaches demonstrated in the prior section. In the case of certain early arrival, $E[UC(t)] = a_0 + a_1\bar{r}(t) + a_2[B - \bar{r}(t) - t]$. For $E[UC(t)]$ to be minimized at the chosen departure time, $\partial E[UC(t)]/\partial t$ must equal zero. This implies that $\partial\bar{r}(t)/\partial t = a_2/(a_1 - a_2)$. Thus during the period of certain early arrival, expected travel time increases linearly with time. During the period of certain late arrival beginning at t_1, the assumption that travelers minimize expected user cost implies $\partial\bar{r}(t)/\partial t = -a_3/(a_1 + a_3)$, and therefore during this period expected travel time declines linearly with time.

Eqs. (17) and (18) below specify $\bar{r}(t)$ for these two periods. Included in each is a constant of integration. These constants will be determined by the free flow travel time. In the case of certain early arrival, this is the time at which congestion begins, and in the case of certain late arrival, the time at which congestion finally dissipates:

$$\bar{r}(t) = \frac{a_2}{a_1 - a_2} t + c_e \qquad t \leq t_e \tag{17}$$

$$\bar{r}(t) = \frac{-a_3}{a_1 + a_3} t + c_1 \qquad t \geq t_1 \tag{18}$$

Determining the equilibrium pattern of ride times for periods where either late or early arrival is possible is more complicated. The solution requires solving a differential equation of the form $dY(t)/dt(k_1 + k_2Y(t)) = 1$ where $Y(t) = B - t - \bar{r}(t)$. Two solutions are possible. These are:

$$\bar{r}(t) = B - t - c_1b \pm [(c_1b)^2 + 4c_2(t + c_0)b]^{\frac{1}{2}} \tag{19}$$

where

$$c_1 = \frac{2a_1 + a_3 - a_2}{a_2 + a_3}$$

$$c_2 = \frac{a_1}{a_2 + a_3}$$

c_0 = constant of integration which ensures continuity with the functions specifying travel times during the periods of either certain early or late arrival.

The negative root in (19) is probably not a stable equilibrium. The equilibrium characterized by the negative root requires that $\bar{r}(t)$ declines throughout the period of uncertainty. Prior to the period of uncertainty, when early arrival is certain, $\bar{r}(t)$ increases linearly. Thus, the negative root implies that upon entering the period of uncertainty, $\bar{r}(t)$ abruptly begins to decline. This is intuitively unappealing because it implies that at the moment there is any possibility of late arrival, travel times immediately begin to decline. Instead, we would expect that upon entering the period of uncertainty, the traveler would begin trading-off the costs of late and early arrival. In this case, the deivative of r(t) with respect to t, $\partial r(t)/\partial t$, should remain positive at the onset of the period of uncertainty but should continuously decline thereafter and ultimately reverse sign. This would imply that the maximum expected travel time, $\bar{r}(t)$, would occur during the period of uncertainty and would reflect travelers trading-off the costs of late and early arrival. For plausible parameter values, the equilibrium path of $\bar{r}(t)$ characterized by the positive root follows this quadratic-like behavior. Figure 6 illustrates the positive root solution for $\bar{r}(t)$ over the three time intervals of certain early arrival, the period of uncertainty, and the period of certain late arrival. Expected travel time increases linearly during the period of certain early arrival, declines linearly during the period of certain late arrival and reaches a maximum during the period of uncertainty.

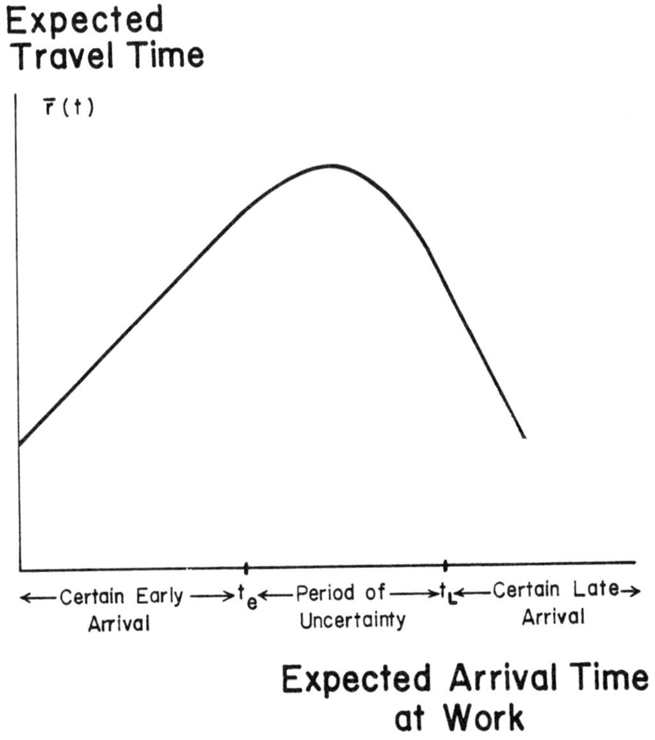

Figure 6

Expected Travel Time as a Function of the Expected Arrival Time with Random Travel Times

The effect of reliability on departure time decisions will be a function of the magnitude of the variations in travel time. Table 1 suggests that these variations are modest for travel by auto. The standard deviation of travel time is on average less than three minutes, and coefficients of variation are typically less than .13. This suggests that other considerations, such as variation in <u>mean</u> travel time over the course of the peak, are likely to predominate in the departure time decisions of auto users. We found no significant association by region between average early arrival delay and the average standard deviation of travel time controlling for other relevant factors.

However, uncertainty in travel times is likely to be both larger and more important for transit users. In addition to the variability in vehicle travel time, waiting time for vehicles may also vary substantially due to variations in inter-vehicle headways, lack of adherence to schedules, omitted vehicle runs, and vehicle overcrowding (which may require a wait for subsequent vehicles). Since waiting time is perceived to be more onerous than riding time, wait time variability may be quite significant in affecting mode choice decisions and departure time decisions for those choosing transit. The factors influencing variability in wait time are numerous, including transit operating policies, passenger volumes, and service frequencies. Hendrickson [10] discusses some simple models of such relationships and their effects. A dynamic user equilibrium analysis including transit service must also consider the effects of delays due to schedule incompatibilities. This issue is discussed in the next section.

6. DELAYS DUE TO SCHEDULE INCOMPATIBILITIES

Among the principal advantages of automobile travel is its flexibility. This flexibility typically includes the discretion to choose among alternative routes, to combine separate trips into tours, and to schedule trips at one's convenience. This departure time flexibility is reflected in our models by the assumption that travelers have complete discretion in choosing their departure times, and thus their preferred arrival time.

Effectively unlimited discretion in departure times is a unique advantage of solo tripmaking by private automobile. In contrast, departure and arrival times for transit trips are constrained by the transit service schedule. With thirty minute inter-vehicle headways, feasible arrival times are separated by thirty-minute intervals. Thus, a traveler might be faced with the options of, for example, arriving at work fifty minutes early, twenty minutes early or ten minutes late. This traveler is likely to incur a sizeable user cost associated with early (or late) arrival. With more frequent service, travelers would likely be able to arrive closer to their desired arrival time and thereby reduce this cost. In circumstances in which service frequencies impose large delay costs, we would expect that travelers would be more likely to choose auto travel.

As is the case with auto travelers, it is not necessary that all transit users choose the scheduled departure time that permits them to arrive at work as close as possible to their work start time. Transit travel times and crowding will increase with higher volumes, and some travelers will prefer to arrive early or late at work so as to avoid this congestion. Thus, the choice of departure time for transit service is a discrete analog to the continuous dynamic user equilibrium analyzed earlier.

Schedule delays also impinge upon the desirability of ride-sharing arrangements such as car-pooling and van-pooling. The feasibility of such arrangements depends not only on identifying a group of travelers who can travel to each of their origins and destinations as a group in an acceptable time, but also on the compatibility of their schedules. A car-pooling arrangement imposing only a minor travel time delay penalty on each potential member (in comparison to solo driving) will not be a feasible shared ride arrangement if the work start times of the potential participants differ greatly.

These observations suggest that schedule delay may be an important consideration in mode choice. The effect of schedule incompatibility considerations on this choice has received very limited attention in the demand modeling literature and has not been addressed in the context of a multiple mode equilibrium framework.

7. SUMMARY

In this paper we have examined the influence of travel cost variations during the peak period on departure time decisions for trips made during the peak period. One element of travel cost which varies over the course of the peak period is travel time. Measurements of travel time over the course of the peak period for trips to the CBD in Pittsburgh, PA, revealed that on several routes with a major bottleneck, variations in mean travel time were substantial. For these routes the ratio of the maximum to minimum average travel time in all cases exceeded 1.35. Interestingly, variations (or randomness) in travel time are modest at each departure time. The average coefficient of variation in mean travel time controlling for departure time is .13.

Among other factors, variations in mean travel time over the peak and randomness in travel time may cause a commuter to choose a planned arrival time either earlier or later than his/her work start time. Our survey of CBD workers revealed that nearly all workers arrive early, which suggests that the costs of late arrival are large. Early arrival time typically exceeds 10 minutes. We identified and highlighted three factors which may cause an individual to choose an actual or planned arrival time different from his work start time: congestion avoidance, reliability and schedule incompatibility. Models accounting for the former two factors were developed and the dynamic equilibrium pattern of travel times and volumes was explored. The patterns of departure time decisions implied by these factors are generally consistent with the observed patterns in Pittsburgh.

The results in this paper suggest several areas of further research. First, congestion avoidance, reliability considerations, and schedule incompatibilities will affect not only departure time decisions but also mode choice decisions. Incorporation of these effects in demand models of mode and time of departure decisions to assess their importance and to provide estimates of the various parameters of the user cost function would be extremely valuable. Some initial estimates are already available from the work of Abkowitz [11], McFadden, et al. [12], and Small [13]. Second, variation in travel cost over time will also affect route choices. Development of traffic assignment models which incorporate these variations would be a useful extension in the state-of-the-art of assignment models. Finally, analysis and evaluation of service changes should consider the impacts of the project on early and/or late arrival costs in addition to any effects on travel time cost. Development of approaches for taking these impacts into account would improve the urban transportation planning process.

ACKNOWLEDGEMENTS

Financial support in the preparation of this paper was received from the U.S. Urban Mass Transportation Administration, Program of University Research and Training, under Contract #PA-11-0024. We would like to thank several anonymous referees and Professor V. F. Hurdle for numerous comments which aided the presentation in this paper.

REFERENCES

1. Herman, R. and T. Lam, "Trip Time Characteristics of Journeys to and from Work," Proc. of the 6th Intl. Symp. on Trans. and Traffic Flow, Sydney, Australia, 57-85, 1974.

2. Pigou, A.C., The Economics of Welfare, 1st ed., New York: Macmillan, 1920; 4th ed., 1932.

3. Wardrop, J.G., "Some Theoretical Aspects of Road Traffic Research," Proc. of the Institute of Civil Engineering, Part II, 1, 325-378, 1952.

4. Hurdle, V.F., "The Effect of Queueing on Traffic Assignment in a Simple Road Network," Proc. of the 6th Intl. Symp. on Trans. and Traffic Flow, Sydney, Australia, 1974.

5. Pearce, C.E.M., "Time Dependence in Commuter Traffic Models," Trans. Sci. 9, 289-307, 1975.

6. Newell, G., Traffic Flow on Transportation Networks, Cambridge, MA: MIT Press, 1980.

7. Nguyen, S., "An Algorithm for the Traffic Assignment Problems," Trans. Sci. 8, 203-216, 1974.

8. Daganzo, C. and Y. Sheffi, "On Stochastic Models of Traffic Assignment," Trans. Sci. 11(3), 1977.

9. Hendrickson, C. and G. Kocur, "Schedule Delay and Departure Time Decisions in a Deterministic Model," Trans. Sci. 15(1), 62-77, 1981.

10. Hendrickson, C., "Travel Time and Volume Relationships in Scheduled, Fixed Route Public Transportation," Trans. Res. 15A(2), 173-182, 1981.

11. Abkowitz, M.D., The Impact of Service Reliability on Work Travel Behavior, Ph.D. Dissertation, Department of Civil Engineering, Massachusetts Institute of Technology, 1980.

12. McFadden, et. al., Demand Model Estimation and Validation, The Urban Travel Demand Forecasting Project, Volume V, University of California at Berkeley, 1977.

13. Small, K.A., "The Scheduling of Consumer Activities: Work Trips," Working Paper, Department of Economics, Princeton University, 1978.

GRAVITY MODEL FITTING WITH BOTH ORIGIN-DESTINATION DATA AND MODELLED TRIP-END ESTIMATES

HOWARD R. KIRBY
and
JOHN D. MURCHLAND

Institute for Transport Studies,
University of Leeds, Leeds, England

ABSTRACT

A new method of fitting gravity models of trip distribution is described. This combines, according to their accuracy, both zone to zone observations of trip numbers and estimates of trip-end totals from trip-end models. The main reason for developing this technique is to improve the reliability of estimates for the unobserved cells and for rows and columns with few observations. Estimation methods currently in use are outlined, together with some of the statistical and computational considerations involved in the development of the new method. The maximum likelihood principle is used. The trip observations are treated as effectively scaled-up Poisson samples, and the trip-end estimates as if they are independent data following the same distributional form. An iterative procedure is proposed, which is an adaptation of the customary calculational method. Data questions, such as the estimation of the variances of the estimates, are considered.

1. INTRODUCTION

Origin destination matrices obtained from survey data are often incomplete. For example, roadside interview stations sample only those movements that cross a cordon or screenline. Estimates of the movements not intercepted are however often required when forecasting, and indeed can be useful in the base year. The gravity model of trip distribution can be used for this purpose, after it has been fitted to the data for the intercepted movements. The extrapolation from observed to unobserved cells depends, of course, on faith in the model. There are two cases: one in which the base-year and future year trip matrices are to be similar gravity models, and another, in which the base-year gravity model is merely a convenient way of infilling the matrix, with no commitment to further use of the gravity model form. (The latter case divides into those uses in which a future matrix is produced by more or less complex growth factors and adjustments to the estimated base-year matrix, and those in which the matrix is only used directly, in particular after assignment, to produce estimated flows on numerous road links or collections of links.)

In addition to the observed trip matrix there may be further information in the form of a model for the total number of trips leaving each zone, and also for the inflow to each zone. Although these models, typically category analysis and some form of regression equation in United Kingdom practice, may be fitted to some or all of the trip data, they also include further truly independent variables, in particular zonal population, income and a measure of employment.

For practical purposes these trip-end estimates may be regarded as independent of the zonal totals of trips obtained with the distribution model for observed plus unobserved cells. Alternatively, the trip-end estimates could be incorporated in the method of estimation of the parameters of the distribution model. In this paper, a method is proposed which tries to make best use of both trip number estimates and modelled trip end estimates, combining these according to their accuracy: this is called the combined estimation procedure. Whilst this method is particularly appropriate where the trip matrix is only partially observed, it can also be used where the matrix is wholly observed. It appears that the limited accuracy of the modelled trip-end estimates means that they may provide little further information for reasonably well-observed rows or columns (provided of course that the distribution model holds good in both observed and unobserved cells).

The combined estimation procedure was developed in response to the difficulties encountered in fitting a nationwide trip distribution model in the Department of Transport's Regional Highway Traffic Model (RHTM) project. See Sacker and Harrison [24] and the RHTM Project Review [4] for details of the difficulties, the report of the Government's Standing Advisory Committee on Trunk Road Assessment [25] for a re-appraisal of the direction of the project (which in particular led to the abandonment of attempts to produce a nationwide distribution model at the level of a single 3613 by 3613 matrix), and Gunn, Kirby, Murchland and Whittaker [10,11] for reports on a technical investigation of some of the difficulties encountered with the distribution modelling.

The rest of the paper is structured as follows. Section 2 defines the model, data available, and the problem to be solved, and introduces the main notation. Section 3 outlines the main methods of estimation in use and the contribution of theory, giving the reasons for developing the combined estimation procedure. Section 4 derives the results for the chosen estimation procedure, giving brief details of the algorithm proposed in Section 5. Section 6 discusses the estimation of the variances of the trip and trip-end estimates, and suggests how to treat error in the separation estimates; and Section 7 concludes.

2. DEFINITIONS AND NOTATION

2.1 The gravity model

The most common model for the distribution of journeys from one zone, i, of a study area to another zone, j, is the gravity model. This has a multiplicative form: the number of journeys is the product of a 'generation factor' for zone i, an 'attraction factor' for zone j, and a factor representing the effect of the separation between zones i and j. The separation is usually measured in terms of journey time or a linear combination (known as generalised cost) of time taken, distance travelled, and money spent. In many studies an analytic form for the separation function is chosen at the outset, but for some purposes it is appropriate first to investigate the deterrent effect of separation without imposing an analytic form. This is done by supposing that the separation is split up into intervals, k, for each of which there is a separation factor. (In principle, one might wish to impose a constraint that the separation factors decrease monotonically with separation, but methods of doing this, and the question of the criterion to use in choosing the number and width of intervals, are outside the scope of this paper.) If there are several separation functions in the matrix, the k defines the function as well as the interval of separation. The model discussed in this paper thus has the form

$$T_{ijk} = a_i\, b_j\, c_k\, D_{ijk} \quad \text{where } D_{ijk} \text{ are non-negative constants.} \tag{1}$$

This is sometimes conveniently expressed in log-linear form

$$T_{ijk} = e^{(u_i + v_j + w_k)} \quad D_{ijk}$$

where

- i = generation zone ($i = 1, \ldots, I$)
- j = attraction zone ($j = 1, \ldots, J$)
- k = separation band ($k = 1, \ldots, K$)
- T_{ijk} = average (daily)number of trips generated at zone i and attracted to zone j for which the separation between i and j is in the kth band.

The i, j, and k will also be referred to as the row, column or level of a matrix such as (T_{ijk}).

a_i = exp u_i = generation factor for zone i

b_j = exp v_j = attraction factor for zone j

c_k = exp w_k = separation factor for band k

$$D_{ijk} = \begin{cases} 1 & \text{if the separation between zones i and j is in the kth band,} \\ 0 & \text{otherwise} \end{cases}$$

In accordance with usual practice, it is here assumed that there is only one separation band between a given pair of zones; and for such a situation this three-dimensional notation is not strictly necessary. However, it does simplify the expression of the various sums that arise and it has the advantage of enabling results to be extended to the situation where the use of more than one separation interval for a given zone-pair is appropriate.

The form of model given in (1) is known as a triproportional model (Evans and Kirby [7]). The quantities D_{ijk} are predetermined, being set by the definition of zones and separation intervals. The factors (a_i), (b_j) and (c_k) are to be estimated from the data - a process often referred to as model calibration.

It is convenient to abbreviate summations by using a '+' as subscript to denote the index over which a variable is summed.

2.2 The data

Three sets of data are considered.

i) For observed cells only, N_{ijk} = survey estimate of (daily) number of trips from i to j for band k

is used for estimating T_{ijk}. For those cells for which $D_{ijk} = 0$, $N_{ijk} = 0$. For unobserved cells, $N_{ijk} = 0$.

ii) For each row sum, X_i = trip end estimate of (daily) number of trips generated in zone i

is an estimate of $T_{i++} \equiv \Sigma_{jk}\, a_i\, b_j\, c_k\, D_{ijk}$

iii) For each column sum, Y_j = trip end estimate of (daily) number of trips attracted to zone j

is an estimate of $T_{+j+} \equiv \Sigma_{ik}\, a_i\, b_j\, c_k\, D_{ijk}$

Note that the last two sets are for complete row and column sums, not partial ones. Note too that, usually, $X_i \neq \Sigma_{jk} N_{ijk}$ and $Y_j \neq \Sigma_{ik} N_{ijk}$, not only because the N_{ijk} will not, in general, contain information on all possible ij movements (that is, the observed zone-to-zone matrix is incomplete, or partial), but also because the trip-end estimates are subject to sampling error and model mis-specification error, and may also contain planning information (such as estimates from a car ownership model) not contained in the trip information.

For each of these sets of data, the variances $V(N_{ijk})$, (X_i) and $V(Y_j)$ are assumed known, where $V(x)$ denotes the variance of a random variable x. Given information on the variances, it is not necessary to require that $\Sigma_i X_i = \Sigma_j Y_j$. (This is in contrast to the common practice of scaling trip attraction estimates in order to achieve such an equality.) However, it is necessary to assume that each of the sets of estimates (N_{ijk}), (X_i) and (Y_j) is an unbiased estimate of the true means given by (T_{ijk}), (T_{i++}) and (T_{+j+}) for the observed cells, and that the model (1) is true.

Under certain simplifying assumptions, the variance of the trip data N_{ijk} can be related to the sampling fractions used in the survey. Suppose that a sample count n_{ijk} is obtained when the sampling fraction is γ_{ijk}, then $N_{ijk} = n_{ijk}/\gamma_{ijk}$. Now the error distribution of the sample counts, e.g. at a roadside interview station, can, to a first approximation, conveniently be regarded as Poisson. If the expansion factors $1/\gamma_{ijk}$ are known without error, then we would have

$$V(N_{ijk}) = V(n_{ijk})/\gamma_{ijk}^2 = n_{ijk}/\gamma_{ijk}^2 = N_{ijk}/\gamma_{ijk}.$$

Of course, the expansion procedures are very much more complicated than this, and there are errors in the expansion factors. So the method of variance estimation for trips is more complicated: this is discussed in Section 6. But, if $V(N_{ijk})$ can be estimated, it is then possible to invert the argument and deduce an effective sampling fraction (or scaling factor) γ_{ijk}, defined as the reciprocal of the variance to mean ratio, and so estimated as

$$\gamma_{ijk} = N_{ijk}/V(N_{ijk}) \quad \text{for } N_{ijk} > 0 .$$

This is then such that the scaled down value

$$n_{ijk} \equiv \gamma_{ijk} N_{ijk}$$

has, approximately, a Poisson error distribution. Similarly, scaling factors defined by

$$\alpha_i = X_i/V(X_i) \quad \text{for } X_i > 0$$

and

$$\beta_j = Y_j/V(Y_j) \quad \text{for } Y_j > 0$$

enable variables with a Poisson-like distribution to be inferred for the trip ends. (Strictly speaking, of course, for a Poisson distribution of the effective sample counts, the variables N_{ijk} could take only the values 0, $1/\gamma_{ijk}$, $2/\gamma_{ijk}$, ... and similarly for the X_i and Y_j variables.)

It is convenient to extend the summation notation, such that

$$(\gamma T)_{i++} \equiv \Sigma_{jk} \gamma_{ijk} T_{ijk}.$$

In practice, zero values of the trip-end estimates X_i (or Y_j) are likely to be associated with rows (or columns) full of zero observed N_{ijk}, and for such circumstances the corresponding row or column of the matrix may be deleted. In the rest of the paper it will be assumed that, if a value is estimated, $X_i > 0$ for all i, and $Y_j > 0$ for all j. In cases where no independent trip-end estimate is available (as opposed to being available but with zero value), it is appropriate to set the corresponding $\alpha_i = 0$ or $\beta_j = 0$. There will, however, be many instances in which $N_{ijk} = 0$. Some - the structural zeroes - will correspond to cells for which no movements were observable by the surveys. For these, define $\gamma_{ijk} = 0$. Others - the sampling zeroes - will correspond to cells where a zero estimate was obtained by chance. For these, the appropriate value of γ_{ijk} is a weighted average of those obtained by the same surveys for non-zero cells. This is also discussed in section 6.

2.3 Problem statement

The problem is to estimate the parameters (a_i), (b_j) and (c_k) of the model (1) such as to optimise some function, L, say, of the departures, or probability of departures, of the data from the model's estimates, given the variability in the data, assuming the model to be well specified. The function L can be written as:

$$L\,\{a_i\, b_j\, c_k\, D_{ijk} \mid N_{ijk},\, V(\underline{N}_{ijk});\ \Sigma_{jk}\, a_i\, b_j\, c_k\, D_{ijk} \mid X_i,\, V(\underline{X}_i);\ \Sigma_{ik}\, a_i\, b_j\, c_k\, D_{ijk} \mid Y_j,\, V(\underline{Y}_j)\}$$

There are clearly a number of possible ways of defining the criterion to optimise. The method chosen is that of maximising the (log-) likelihood. Some reasons for this choice, and for the choice of error distribution, are presented in the following section, 3, after outlining other methods of estimation in use. The derivation of results is given in section 4.

3. METHODS OF ESTIMATION

Methods of fitting the triproportional model to data can be interpreted in three ways. The common sense, and most commonly used, approach is to fit so that the model reproduces important aggregate quantities, in particular, trip-end estimates. A second approach, which includes maximum likelihood estimation, is to minimise a function of the factors to be fitted. A third approach, more appropriate to forecasting, is to make the fitted matrix as similar as possible to a prior matrix, whilst satisfying required row, column and separation band sums.

A given minimisation problem to describe the second or third approach may be reformulated as an instance of the other approaches. Thus the three approaches may be called existence, dual and primal, from the familiar mathematical programming formulation of the standard problem.

Although the existence approach has the advantage of reproducing what are considered to be best estimates of important quantities, there is the disadvantage that not all existence approaches can be expressed as an equivalent minimisation problem. Unless this can be done, mathematical analysis may be very difficult. Other disadvantages of the existence approach are that it may be hard to choose just sufficient important quantities that the fitted model is neither over-determined nor under-determined, that the chosen quantities may be highly correlated or contradictory, and that the approach does not take into account the accuracy of the data.

The dual approach is the one most used in the theoretical literature, for it is more amenable to mathematical and statistical analysis. With this approach it is easier to take proper account of different levels of accuracy in the data, to show that the solution is unique, to devise an efficient algorithm which can be proved to converge, to determine something about the rate of convergence and remaining convergence error, and to assess how accurately the model parameters are determined by the data. If there was agreement on a single error measure by which the goodness of fit of the model could be assessed, then the estimation procedure would clearly seek to minimise that error. As there is no such agreement, a variety of objective functions could be chosen, such as maximum likelihood, least squares, least risk; and indeed the choice of basic variable could be either trips or travel (trips times travel time or cost), with a choice of the level of aggregation to be used, such as O-D flow or link-flow. (See Murchland [21].) There are, of course, different uses to which a gravity model may be put. For example, it may be used simply to economise in data, requiring only I + J + K - 2 independent factors instead of IJ data items. It may be required only to infill the unobserved cells in the base-year matrix. It may be required only to provide the base-year estimates of separation factors (c_k) that are to be used in forecasting, the base-year estimates (a_i), (b_j) being superseded in the forecast year by other values. It may be that these different uses of a gravity model deserve different methods of fitting. But it is generally felt for consistency that the whole base-year model should be produced and compared with the base-year and other data to assess its adequacy. This paper proceeds on that assumption.

The following methods have been used in transport planning practice in the United Kingdom for obtaining gravity model estimates, utilising O-D trip data and (in some cases) trip end estimates. The main differences are in the treatment of the unobserved cells.

a. Partial matrix technique. The factors (a_i), (b_j) and (c_k) are chosen such that the sums of T_{ijk} over observed cells only in each row, column or level are made to agree with those of the expanded survey data, namely (N_{i++}), (N_{+j+}), (N_{++k}). Consistency is assured because

$\Sigma_i N_{i++} = \Sigma_j N_{+j+} = \Sigma_k N_{++k}$.

The fitted parameters are then used to estimate trips in the unobserved cells. (See Kirby[16] for a description of the assumptions, theoretical basis, and problems.) Whilst it is well known that the fitted model's estimates of trips in the observed cells are unique (Evans and Kirby [7]), it is less well-known that the estimates in the unobserved cells may not be. The latter condition, known as non-identifiability, can be tested for (Day and Hawkins [3], Murchland [20]).

b. Base-year prediction method

The parameter estimates (a_i), (b_j) and (c_k) are obtained for the observed cells as in the partial matrix technique . Those for (c_k) are applied to all cells of the matrix (perhaps after smoothing them), and then the parameters (a_i), (b_j) recalculated using biproportional iterations to ensure that the row and column sums over all cells agree with the trip end estimates (X_i), (Y_j), where these have first been adjusted so that $\Sigma_j Y_j = \Sigma_i X_i$. (No literature on this method is known to the authors.)

c. Synthetic trip end method

Parameter estimates are desired such that the sums of T_{ijk} over all cells in each row and column agree with independent trip-end estimates (X_i) and (Y_j), where these have been scaled so that $\Sigma_j Y_j = \Sigma_i X_i$ and such that the sums over observed cells only of trips in each separation band agree with those of the survey, (N_{++k}). To avoid having an over-determined model, one additional factor needs to be introduced. This may be a scaling factor for the observed separation band sums, or a factor for the synthetic row (and hence also column) sums. (Experience with this method is described in Sacker and Harrison[24] and theoretical problems reviewed in Murchland[19].)

d. RHTM hybrid method

The parameter estimates (a_i), (b_j) and (c_k) obtained by the partial matrix method are retained for the observed cells. Those for (c_k) are then used in the unobserved cells and the factors (a_i), (b_i) recalculated for unobserved cells by estimates which ensure that the sums over unobserved cells in each row and column agree with the 'remaining' numbers of trips, $(X_i - N_{i++})$ and $(Y_j - N_{+j+})$. There is the problem that these 'remainders' can be negative, and that two generation factors would be required for each incompletely observed row and two attraction factors for each incompletely observed column, thereby making the fitted model(s) difficult to use as forecasting tools. This technique was developed in the RHTM project as a short-term response to the problems encountered in fitting the gravity model satisfactorily with the partial matrix technique or synthetic trip-end method at the level of a single 3613 by 3613 matrix. The variants on it that were tried are described in Alastair Dick & Associates[1].

e. 'Top L' method

Each row is treated either as a 'full' row or as a 'partial' row. For 'full' rows the model sum over all cells is made to agree with the trip-end model value X_i. For a 'partial' row the model sum over observed cells is required to equal the observed value N_{i++}. Similarly, and independently, for columns. Total modelled trips in each separation band are also made to agree with each observed sum N_{++k}. The method is typically used when a part of the matrix, for movements between external zones, is neither observed, nor modelled due to lack of separation and trip-end data. Hence the name. External rows and columns are necessarily treated as partial. This model is over-determined unless two additional factors are introduced. These may be taken to be a scaling factor for synthetic row sums, and for column sums, so that observed values are left unadjusted. The method is new to practice in the U.K., and is described in the Traffic Appraisal Manual[5].

It could be said that a major influence in devising fitting methods has been the ease with which the customary row, column and separation band balancing algorithm (known as the Furness method of iteration) operates over a great range in size of matrix, from a few zones to several thousand. Use of the algorithm long preceded any understanding of what it was doing, and all the practical methods listed above directly employ it.

Statistical estimation methods which have paralleled some of these methods have demonstrated the following:

(i) The constraints to observed row and column sums used in the partial matrix technique are appropriate if a maximum likelihood procedure is used with either a multinominal sample of trips (Evans [6]) or a Poisson sample of trips (Kirby [13]) but if the sampling fraction varies over the matrix, it is more appropriate for the marginal constraints to be between scaled-down model values ($\gamma_{ijk}\ T_{ijk}$) and effectively unexpanded trips ($\gamma_{ijk}\ N_{ijk}$). (Bexelius, Nimmerfjord, Nordqvist and Read [2], Kirby and Leese [17].)

(ii) If exact agreement to prescribed row and column sums is required, whilst utilising a (partially) observed matrix of trips, then the likelihood techniques of (i) should be modified by incorporating the required constraints with Lagrangean multipliers (Kirby[15] and Murchland[22].)

In many of the practical procedures described above, either one or the other kind of data (trips or trip ends) was used in some part of the estimation; and the relative accuracy of each set was not used. The combined estimation procedure described in the next section uses both, with their accuracy estimates. The use of a maximum likelihood criterion with an (effective) Poisson error distribution is regarded by the authors as being the most attractive. An advantage is that it generates the usual balancing algorithm. Other criteria were considered but rejected [10,Section 5.2].

4. THE COMBINED ESTIMATION PROCEDURE

The combined estimation procedure essentially maximises a likelihood function over an (I+1) x (J+1) matrix, in which the 'extra' row and column contain trip-end estimates. Gunn, Joliffe, Kirby and Redhouse all contributed to the initial development of ideas in 1977 - 78 in unpublished communications, but credit for the first full exposition of the appropriate theory is due to Redhouse[23]. An allied method of fitting, in which the model is fitted to grouped data - or, rather, in which data is available for general groups of model values - was described in Gunn[8] and this more general approach has recently been extended by Gunn in order to fit gravity models to independent estimates of not just trips and trip-ends, but also to traffic count and travel cost estimates as well[9].

The following distributional assumptions are made in developing the likelihood function.

1. For trips in cell ijk an effective sampling fraction γ_{ijk} was used, with an effective sample count of n_{ijk} such that this sample consists of independent Poisson observations for the cells which were observed. If the model is well specified, the mean of this sample (and also the variance) is $\gamma_{ijk} T_{ijk}$. With $\gamma_{ijk} = 0$ and $n_{ijk} = 0$ if cell ijk was not observed, the sample mean is $\gamma_{ijk} T_{ijk}$ for every cell. The probability of this sample is

$$\prod_{ijk} \exp(-\gamma_{ijk} T_{ijk}) (\gamma_{ijk} T_{ijk})^{n_{ijk}}/n_{ijk}! ,$$

interpreting the probability as 1 if $\gamma_{ijk} = 0$ (and so $n_{ijk} = 0$).
The negative logarithm of this is

$$\sum_{ijk} (\gamma_{ijk}T_{ijk} - n_{ijk} \ln(\gamma_{ijk}T_{ijk}) + \ln(n_{ijk}!)),$$

in which the only terms dependent on the T_{ijk} are

$$\sum_{ijk} (\gamma_{ijk}T_{ijk} - n_{ijk} \ln T_{ijk}). \qquad (2)$$

2. For trip ends, if the error distribution is known, then it would have to be used, but at this stage it seems reasonable to make a choice that is most convenient for analysis, provided that the mean and variance of the trip end estimates are respected. The choice of error distribution and its parameterisation affects how the variance of the trip end estimates changes as the mean changes (e.g. whether the variance should be proportional to the mean, or constant, for a given zone). If the choice results in a convex objective function to be minimised, the calculation of the solution should be robust, enabling it to be reached from poor starting points, and it will also follow that the solution is unique. For convenience, the distribution is, as in (1), taken to be scaled Poisson.

For row sums, the probability of obtaining X_i for row i with a scaling factor α_i is

$$\exp(-\alpha_i T_{i++}) \quad (\alpha_i T_{i++})^{\alpha_i X_i} / (\alpha_i X_i)!$$

where, strictly, X_i can only take the values $0, 1/\alpha_i$, $2/\alpha_i$, ...

That is, if the model is true, X_i is drawn from a distribution with mean T_{i++} and index of dispersion (variance to mean ratio) of $1/\alpha_i$, independent of T_{i++}. Note that as α_i tends to 0 the sample X_i becomes worthless, as α_i tends to ∞ the requirement that $T_{i++} = X_i$ becomes binding.

The negative logarithm of the probability of the sample of row sums is

$$\sum_i (\alpha_i T_{i++} - \alpha_i X_i \ln T_{i++}) \qquad (3)$$

after neglecting constant terms.

For column sums, the 'sample' is similar, with Y_j a scaled Poisson sample with mean T_{+j+} and index of dispersion $1/\beta_j$. The negative logarithm of the column sample probability is

$$\Sigma_j \ (\beta_j T_{+j+} - \beta_j Y_j \ln T_{+j+}) \tag{4}$$

after neglecting constant terms.

The negative logarithm of the combined probability of the three samples, supposed independent, is, from (2 - 4),

$$L = \sum_{ijk} (\gamma_{ijk} T_{ijk} - n_{ijk} \ln T_{ijk}) + \sum_i (\alpha_i T_{i++} - \alpha_i X_i \ln T_{i++})$$

$$+ \sum_j (\beta_j T_{+j+} - \beta_j Y_j \ln T_{+j+}),$$

apart from terms which do not depend on the T_{ijk}. Maximum likelihood estimates of the parameters in the model for T_{ijk} are obtained by minimising this with respect to those parameters.

The minimand L simplifies a little if

$$\gamma'_{ijk} = \gamma_{ijk} + \alpha_i + \beta_j$$

is introduced:

$$L = \sum_{ijk} \gamma'_{ijk} T_{ijk} - \sum_{ijk} n_{ijk} \ln T_{ijk} - \sum_i \alpha_i X_i \ln T_{i++}$$

$$- \sum_j \beta_j Y_j \ln T_{+j+} . \tag{5}$$

For the triproportional model

$$T_{ijk} = a_i b_j c_k D_{ijk} = \exp(u_i + v_j + w_k)\ D_{ijk},$$

the conditions that L be a minimum with respect to each u_i, v_j and w_k are

$$\frac{\partial L}{\partial u_i} = 0 = \sum_{jk} \gamma'_{ijk} T_{ijk} - \sum_{jk} n_{ijk} - \alpha_i X_i - \sum_j \beta_j Y_j \frac{T_{ij+}}{T_{+j+}}$$

for each i, (6)

$$\frac{\partial L}{\partial v_j} = 0 = \sum_{ik} \gamma'_{ijk} T_{ijk} - \sum_{ik} n_{ijk} - \sum_i \gamma_i X_i \frac{T_{ij+}}{T_{i++}} - \sum_j \beta_j Y_j$$

for each j, (7)

$$\frac{\partial L}{\partial w_k} = 0 = \sum_{ij} \gamma'_{ijk} T_{ijk} - \sum_{ij} n_{ijk} - \sum_i \gamma_i X_i \frac{T_{i+k}}{T_{i++}} - \sum_j \beta_j Y_j \frac{T_{+jk}}{T_{+j+}} \text{ for each } k. \quad (8)$$

It is assumed that for each i there is some $D_{ijk}>0$, and also for each j, so that no T_{i++} or T_{+j+} is inherently zero.

The conditions can be written more concisely as

$$(\gamma' T)_{i++} - n_{i++} - \alpha_i X_i - \sum_j \beta_j Y_j \frac{T_{ij+}}{T_{+j+}} = 0 \text{ each } i \quad (6a)$$

$$(\gamma' T)_{+j+} - n_{+j+} - \sum_i \alpha_i X_i \frac{T_{ij+}}{T_{i++}} - \beta_j Y_j = 0 \text{ each } j \quad (7a)$$

$$(\gamma' T)_{++k} - n_{++k} - \sum_i \alpha_i X_i \frac{T_{i+k}}{T_{i++}} - \sum_j \beta_j Y_j \frac{T_{+jk}}{T_{+j+}} = 0 \text{ each } k. \quad (8a)$$

Note that

$$\frac{T_{ij+}}{T_{+j+}} = \frac{\Sigma_k \exp(u_i+v_j+w_k)\, D_{ijk}}{\Sigma_{ik} \exp(u_i+v_j+w_k)\, D_{ijk}} = \frac{\Sigma_k \exp(u_i+w_k)\, D_{ijk}}{\Sigma_{ik} \exp(u_i+w_k)\, D_{ijk}}$$

does not depend on v_j, and similarly T_{ij+}/T_{i++} does not depend on u_i, T_{i+k}/T_{i++} does not depend on u_i, nor T_{+jk}/T_{+j+} on v_j. These ratios are also likely to be rather small quantities, since they are non-negative and their sum is unity.

The term $\Sigma_j (\beta_j Y_j T_{ij+}/T_{+j+})$ in the condition for row i is a partition of $\sum_j \beta_j Y_j$ amongst the rows.

Summing any set over its index gives the summary relation

$$(\gamma' T)_{+++} = n_{+++} + (\alpha X)_+ + (\beta Y)_+$$

or $$(\gamma T)_{+++} + (\alpha T)_{+++} + (\beta T)_{+++} = n_{+++} + (\alpha X)_+ + (\beta Y)_+ ;$$

thus the fitted model preserves the total of effective 'sample' data.

If all α_i and β_j are zero, the row and column samples have no effect and the problem reduces to the ordinary partial triproportional problem:

$$(\gamma T)_{i++} = n_{i++}, \quad (\gamma T)_{+j+}, = n_{+j+}, \quad (\gamma T)_{++k} = n_{++k}.$$

The conditions for row i can be solved for each $a_i = \exp u_i$:

$$a_i = \frac{n_{i++}}{(\gamma bcD)_{i++}} \tag{9}$$

and similarly

$$b_j = \frac{n_{+j+}}{(\gamma acD)_{+j+}} \tag{10}$$

$$c_k = \frac{n_{++k}}{(\gamma abD)_{++k}} \,. \tag{11}$$

5. CALCULATIONAL CONSIDERATIONS

The conditions (9 - 11) for the standard triproportional problem suggest an iterative calculational strategy. To accommodate problems with many rows and columns (and few non-zero D_{ijk} for each ij) so that all of the factors can be updated with one pass of the D array, accessed by rows, this can be arranged as follows. Initial estimates of the b_j and c_k are made. During one pass of the D array, each a_i is reset to

$$\frac{n_{i++}}{(\gamma bcD)_{i++}}$$

and then the contributions from the new a_i to $(\gamma acD)_{+j+}$ and to $(\gamma abD)_{++k}$ cumulated. At the end of the pass the complete sums are used to reset each b_j and each c_k. (Thus the new b_j use the new a_i but the old c_k, and the new c_k use the new a_i and the old b_j.)

This is the common method of solution. However, when the intrazonal flows T_{iik} are not small - and they are likely to be the largest in their row or column since their associated cost is least - it would seem better to pair a_i with b_i and solve for them jointly. At the cost of solving a quadratic equation accurately, this allows for the strong interaction which a_i and b_i will show in such circumstances. If intrazonal flows are absent from the model and data (D_{iik} is zero for each i), this becomes an algorithm in which a row and a column are balanced alternately (a_1, b_1, , ...).

The combined estimation problem may be solved by a small adaptation of the standard strategy for the ordinary triproportional problem. The same row-wise calculation is adopted, with the new values of the factors being calculated in the same order, and no more than one copy of each variable kept. Since the minimising value of a factor can no longer be found explicitly (as in 9, 10, 11 above), a quadratic approximation of the objective function is made around the current variable values, from which approximate minimising values are directly found. This is merely a single Newton iteration for each a_i in turn, then b_j, then c_k, if the standard algorithm is adapted; or each a_i, b_j pair, then each c_k, if the suggested variant is used. (Since the convergence of either algorithm for the ordinary problem is only linear, the quadratic approximation is more than accurate enough. If there were a more rapidly convergent algorithm for the combined problem, it would imply, of course, a better algorithm than is known for the ordinary problem.)

Convergence of the ordinary algorithm for the bi-proportional problem was analysed by Murchland (18), using a weighted, squared, relative change measure in each factor. Until a more specific analysis is available, the same measures can be adopted for the combined algorithm. As before, the important quantity is not the size of the total change over an iteration alone, but also the ratio of this total change to that of the previous iteration.

Compared with the standard algorithm, the combined algorithm needs to read more data for each row, the N_{ijk} and the γ_{ijk}, to do more arithmetic for each factor, for the second partial derivative as well as the first, and to store several further quantities per factor.

For the standard as well as the combined algorithm, it is computationally economical to obtain or pre-calculate all the fixed quantities for each row (including explicit or implicit D_{ijk} and γ_{ijk}) and to save these in what is often called a "C-file", to be read row by row as needed.

In passing, it may be worth mentioning that both algorithms allow for the incorporation of a more complex specification of the separation factors than the simple unrestricted form used throughout this paper. That is, to require that the factors for each function do not increase as separation increases or that values come from an exponential or gamma function, the algorithm is modified by separately minimising the objective function with respect to the separation function parameters after the iteration over the rows is completed, and the new a_i and b_j found. Again a quadratic approximation of the objective may be required, and whatever sums or partial derivatives are needed are built up during the pass over the rows. (It is important that this minimising should include an overall scaling factor.)

6. DATA CONSIDERATIONS

The combined method utilises measures of the accuracy of the trip cell estimates, and of the modelled trip-end totals. It is only very recently that it has begun to be appreciated that all survey estimates have errors, and that the size of these errors should be estimated, and the effects on modelled estimates determined. In this section, the factors involved in estimating these variances are discussed. The detailed calculations that would be necessary, particularly in the case of movements intercepted by roadside interview stations, are described in 6.1 and

6.2; rougher estimates may be based on an average scaling up factor and its variance. The calculation of variances of trip-end estimates is briefly discussed in 6.3, and the treatment of error in separation estimates in 6.4.

6.1 Variance calculations for observed trips

Trip estimates may arise from roadside interviews, household interviews or licence plate surveys. Assessing the accuracy of the estimate for one cell reduces to estimating the variance of the estimate by tracing the chain of factors successively applied to the original counts, and estimating the accuracy of each factor. Formally, the formula for the final trip estimate is written down, and the variance extracted, assuming that the factors and original count are all statistically independent. However, it is necessary to be very clear about the definition of the quantity finally estimated (is it average daily trips in a particular year, or in an average year?) and of each of the factors applied, and to identify carefully how many independent estimates of a factor are employed. For example, there may be only one estimate of a correction for a leaky cordon, or for trips uninterviewed during the night, applied to all trips for a particular origin-destination cell, or there may be separate estimates for each roadside interview station. For the same factor variance, the many estimates give a much lower contribution to the cell variance, of course.

It is certainly an inconvenience that each survey seems to need a separate treatment, owing to the variation in the kind of factors used, and in their incidence in the sense just described. It has to be assumed that there is no bias in any of the quantities, other than that represented as a random effect allowed for in the variances. Also, estimates for different cells are not independent, but positively correlated, due to the use of the same factors. (For example, the same seasonal correction factor may be applied to every cell in the matrix.)

It appears that the coefficient of variation of the overall expansion factor applied to survey counts can easily be as high as twenty per cent. This sets a limit to the number of counts it is worth obtaining for a cell, and has implications for the design of surveys.

Matrices from separate surveys may be combined to give a best joint estimate by using the estimated variance of each cell value. The best joint estimate is the one with the least coefficient of variation, and this is consistent with the way independent estimates are combined under the maximum likelihood assumptions of this paper: the effective sample counts add, and the effective sampling fractions add. This solves the old problem of how to combine household and roadside interview data.

6.2 Effective sampling fractions for unobserved cells

For cells of zero observed count there will be no difficulty in obtaining estimates of the applicable scaling-up factor, and hence of the variance, for household interview data, at least for home-based trips, since the average factor for the zone of generation can be used. But in the case of roadside interview data, the effective sampling fraction for such cells cannot be calculated simply, since the sites at which these movements may have been sampled are not known exactly. Two possible methods of calculating the average sampling fraction are as follows:

i) to determine the roadside interview station(s) which lies on the minimum time path for that movement, and take the average effective sampling fraction for this;

ii) to use the average effective sampling fraction for each of the cordons which intercept that movement.

The average should be taken only over those cordons or screenlines which intercept the given 'observed zero' movement. Of the above methods (ii) may be the most viable. The more accurate method, (i), would be more costly, since sites through which minimum time paths pass would have to be identified.

6.3 The variance of trip-end estimates

The trip-ends are not estimated by a sampling process, and thus there is no straightforward theoretical basis for estimating their variance. The typical sources of error that, in principle, need to be taken into account are as follows.

For the attraction model, errors in the dependent variable (the surveyed trip-ends) and in the independent variables (planning data) will lead to a fitted relationship which has itself a certain residual error, and which involves coefficients which are likewise estimated with error. The synthetic attraction estimates are then subject to errors in the independent variables on which they are based, errors in the coefficients which multiply these independent variables, and an additional component of residual error: three sources in all, excluding potential blunders and model mis-specification.

For the estimate of trips generated in a zone, a corresponding relationship is fitted at the level of the household; the resulting household model then also has errors in fitted parameters together with a characteristic residual error. A third source of error is introduced when the households are aggregated into zones, in that it is necessary to estimate the numbers of households in each of a number of categories chosen for the model, consistent with estimates of related planning variables. Errors at the aggregation stage are doubly complex in that the estimates of the numbers of households in the various categories are based on the output of yet another model, the car ownership model.

If there is appropriate information on the accuracy of the zonal population and employment data, and on the accuracy of the trip-end models, it should be possible to obtain, for each zonal estimate of the number of trip-ends, an estimate of its variance.

Because the calculation of trip end variances is not straightforward, very little is known of their accuracy at this time.

6.4 The treatment of error in the separation estimates

The method of fitting so far described takes into account the errors in trip and trip-end estimates, but assumes that the appropriate separation interval for each zone pair is known without error. It would be possible to devise a procedure which treated the separation as having an error distribution about some mean value, and modify the likelihood accordingly. However, a simpler procedure is possible: associate with each zone pair more than one estimate of separation. For example, if K_{ij} is the average separation between zones i and j and different proportions of travellers perceive the separation to lie in the intervals

$$k = K_{ij} - 1, K_{ij}, K_{ij} + 1,$$

the corresponding D_{ijk} will take the values associated with these proportions, and zero elsewhere, instead of the more usual definition. The theory in this paper thus embraces this extension.

One advantage of this method is that, because it links adjacent cost intervals, problems of non-identifiability are automatically avoided (Murchland[20]). Another is that it helps smooth the estimates of the separation factors.

7. CONCLUSIONS

Techniques for fitting a gravity model of trip distribution sometimes make use of estimates of row and column totals from trip-end models as well as the sample estimates for trips in cells. Most such techniques in U.K. practice use either one or the other of the two data values for row and column sums, with no account taken of their relative accuracy. The combined estimation method described uses all the data available, combining them, taking account of the errors in each, by use of a maximum likelihood approach. The algorithm developed for this is an extension of that ordinarily used, and therefore should work on problems of a similar size. The variances of all the data values are needed for this approach, and their derivation has been briefly indicated. Variances for all values are now becoming available in practice, though those for the trip-end model estimates deserve further study.

ACKNOWLEDGEMENTS

Much of the work described here was sponsored by the U.K. Department of Transport, as part of the RHTM project. J.D. Murchland was supported by a grant from the Science Research Council. The paper was prepared whilst Dr. Murchland was acting as a statistical consultant to the Assessment Policy and Methods Division of the Department of Transport.

Particular acknowledgement is made of the contribution of Hugh Gunn to the development of ideas, and to a number of transport consultants who have shared with us their experiences of calibration methods in practice.

No agreement of the sponsors to any of the statements made here is either expressed or implied.

REFERENCES

1. ALASTAIR DICK AND ASSOCIATES, "RHTM - Description of RHTM hybrid matrices, February/March 1979 version", Alastair Dick and Associates, Leatherhead.

2. S. BEXELIUS, G. NIMMERFJORD, S. NORDQVIST and E. READ, "Studies in traffic generetics", Document No. 2, National Swedish Institute for Building Research, Stockholm, 1969.

3. M.J.L. DAY and A.F. HAWKINS, "Partial matrices, empirical deterrence functions and ill-defined results", Traff. Engng. & Control 20 (8/9), 429-433 (1979).

4. DEPARTMENT OF TRANSPORT, Roads and Local Transport Directorate, "Regional Highway Traffic Model, Project Review", Alastair Dick and Associates, Leatherhead, September 1978.

5. DEPARTMENT OF TRANSPORT, Assessment Policy and Methods Division, "Traffic Appraisal Manual", Department of Transport, London, 1981.

6. A.W. EVANS, "The calibration of trip distribution models with exponential or similar cost functions", Transpn. Res. 5, 15-38 (1971).

7. S.P. EVANS and H.R. KIRBY, "A three-dimensional Furness procedure for calibrating gravity models", Transpn. Res. 8, 105-122 (1974).

8. H.F. GUNN, "Calibration of gravity models from grouped data", Proc. PTRC Summer Annual Meeting, Univ. Warwick, June 1977. Paper G23, PTRC Education and Research Services Ltd., London, 1977.

9. H.F. GUNN, "Spatial interaction models using aggregated information", Working Paper 148, Institute for Transport Studies, University of Leeds, Leeds, 1981.

10. H.F. GUNN, H.R. KIRBY, J.D. MURCHLAND and J.C. WHITTAKER, "Regional Highway Traffic Model. Trip Distribution Investigation", Department of Transport, London, 1980.

11. H.F. GUNN, H.R. KIRBY, J.D. MURCHLAND and J.C. WHITTAKER, "The RHTM Trip Distribution Investigation," Proc. PTRC Summer Annual Meeting, Univ. Warwick, July 1980, Paper P9, PTRC Education and Research Services Ltd., London, 1980.

12. J. JOLIFFE, Private communication referred to by Redhouse, 1978.

13. H.R. KIRBY, "Theoretical requirements for calibrating gravity models", Transpn. Res. 8, 97-104 (1974).

14. H.R. KIRBY, "Calibration of a gravity model from partial data given independent estimates of trip end totals", Note HRK/RN 15, Inst. Transp. Stud.. Univ. Leeds, Leeds, 1978.

15. H.R. KIRBY, "Calibration of a gravity model to obtain exact agreement with independent estimates of trip end totals", Note HRK/RN 22, Inst. Transp. Stud., Univ. Leeds, Leeds, 1979.

16. H.R. KIRBY, "Partial matrix techniques", Traff. Engng. & Control 20 (8/9), 422-427 (1979).

17. H.R. KIRBY and M.N. LEESE, "Trip distribution calculations and sampling error: some theoretical aspects", Environ. & Plann. A 10, 837-851, Part I, (1978).

18. J.D. MURCHLAND, "Convergence of gravity model balancing iterations", Proc. P153, PTRC Summer Annual Meeting, Univ. Warwick, July 1977, PTRC Education and Research Services Ltd., 1977, 305-312.

19. J.D. MURCHLAND, "Comments on Catling's examples of failures to calibrate the distribution model". RHTM Trip Distribution Investigation, Working Note 3, Inst. Transp. Stud., Univ. Leeds, Leeds, 1979.

20. J.D. MURCHLAND, "Non-identifiability in triproportional problems", Technical Note 23, Inst. Transp. Stud., Univ. Leeds, Leeds, 1979.

21. J.D. MURCHLAND, "What are the errors we should be seeking to minimise in the estimation process?" Working Paper 126, Inst. Transp. Stud., Univ. Leeds, Leeds, 1979.

22. J.D. MURCHLAND, "Maximum likelihood estimation of a trip distribution model from a partial data matrix when subject to given trip ends", JDM-325, Department of Transport Contract CON 31/2/028, London, 1980, unpublished.

23. A.M. REDHOUSE, "The RHTM distribution model calibration - an alternative approach", Privately circulated memorandum, North Eastern Road Construction Unit, Department of Transport, September 1978.

24. D. SACKER and D.A. HARRISON, "Regional Highway Traffic Model: the private vehicle distribution model", Proc. PTRC Summer Annual Meeting, Univ. Warwick, July 1978. PTRC Education and Research Services Ltd., London, Paper M10.

25. STANDING ADVISORY COMMITTEE ON TRUNK ROAD ASSESSMENT, "Forecasting traffic on trunk roads: a report on the Regional Highway Traffic Model Project. December 1979", HMSO, London, 1980.

A TIME DEPENDENT MARKOV RENEWAL MODEL OF TRIP CHAINING

Ryuichi Kitamura and Tenny N. Lam
Department of Civil Engineering
University of California, Davis

Previous Markovian trip chaining models are mostly time independent, representing identical travel and activity patterns throughout the day. The Markov renewal process of trip chaining is modified and extended in the present work to incorporate a time dependent transition probability and the sojourn time distribution. The new model allows a more realistic representation of the temporal variations in human activities and the possibility of considering trip coordination and scheduling under time budget constraints. A special case is studied for numerical solutions. In the example, the time dependent Markov renewal model is shown to be capable of offering qualitative features of trip chaining in agreement with previous empirical results. The present work further indicates that a time dependent Markov renewal process can be effectively employed as a unifying theoretical framework for describing trip chaining, activity scheduling, and spatial interactions important to travel behavior analysis.

1. INTRODUCTION

An important aspect of urban travel behavior is the way individuals coordinate and schedule their activities and, hence, trips to be made over a daily, or even a weekly, cycle. Previous urban travel behavior modeling and analysis have given attention largely to trip segments as independent units of analysis. Although such an approach deals with the fundamental aspect of travel, namely trips or trip segments, it lacks an ability to integrate temporal and spatial activity patterns with trip patterns. When there are few constraints on an individual's ability to satisfy his activity and travel needs, trip analyses are adequate for urban travel studies and useful for planning predictions. In recent years, however, urban travel behavior has become more complicated, especially for non-work trips. This trend is partially a result of the energy crises of 1974, 1975, and 1979, [1,2] and partially a result of increasing time and monetary budget constraints that affect the life style of many families and individuals. For multiple wage-earner households and smaller family units, their activity patterns and travel patterns are far more complex. [3,4] Recently, research interests have been attracted to the study of these complex time-space relationships of activities and trips.

Trip chaining considers the combination of trips for various purposes, starting times, destinations, travel modes, etc., into a coherent itinerary. This process has not been extensively modeled; despite that it has become increasingly important in urban travel behavior analysis and in urban transportation planning. The importance of trip chaining in travel demand analysis and forecasting is evident when one considers how an individual may effectively combine trips into chains to bring about a more efficient schedule and lesser transportation needs. The effect of trip chaining on total travel distance, spatial distribution of activities, and temporal variations of traffic flow can have a major impact on urban transportation. Focusing on the interdependencies of successive trips in a 24 hour time chain of activities offers a possibility for the development of a unified framework for studying the integrated nature of travel demand, spatial activity structure and transportation. A number of studies have been made to integrate some of these aspects.[5, 6, 7] In the past, most stochastic models of trip chaining were based on the Markovian process. The Markovian approach had offered the first comprehensive glimpse into the complex trip chaining behavior. Moreover, this approach "permits easy and economical reconstruction of journey patterns" [5] from observations, and the development of a systematic means of organizing travel behavior data for statistical analysis and for hypothesis testing. With the Markovian models, researchers were able to design meaningful observational and simulation experiments for studying trip chaining.

However, there are two major drawbacks in using the typical Markovian approach to describe trip chaining: the behavior is assumed to be independent of time of day and independent of its history. The first assumption implies that given the present state, the probability of making another trip is invariant of the time of day or the elapsed time since the beginning of the chain. The second assumption implies that future behavior is independent of the past trajectory given the present state (e.g., having already gone out for lunch does not preclude the possibility of going out to lunch again in the immediate future). The "forgetfulness" and temporal homogeneity of Markovian models contradict our intuition about travel behavior. Consequently, the validity and adequacy of the Markovian basis of describing trip chaining have often been questioned by researchers.[8, 9]

In light of the complexity of the behavior of trip chaining and the underlying decision process of activity scheduling, the stochastic approach to trip chaining seems to be, nevertheless, the best method for mathematical abstraction and data organization. An attempt to modify the Markovian approach would seem to be worthwhile. This paper presents such an attempt. A time dependent Markov renewal model for trip chaining has been studied as a simple extension of the Markovian trip chaining models. It is a modification of the Markov renewal process that takes into consideration time of day dependency of activities and time-budget constraints reflected through the elapsed time of the time chain of events. The model incorporates two unique components for this purpose. One is the time-dependent transition probability and the other is the time independent sojourn time distribution. The time dependent transition probability integrates two concepts in the spirit of Hemmens [5]: activity (trip purpose) linkage and time of activity occurrence. In this context, the sojourn time distribution may be considered as only activity dependent, since how much time one devotes to a particular activity is influenced by the activity involved.

The objective of the paper is to explore the properties of such a model for describing trip chaining. Although the model is simple, it may offer some indications as to whether or not this is a proper direction to extend the Markovian approach, and whether or not such an extension would offer profitable improvements. The model is kept intentionally simple, such that it may offer valuable insights into the desirability and limitations of the different aspects of the theoretical processes. It is obvious that even if a theoretical study such as this shows promise, further efforts are required to bring about a practical operational model. The focus of the present study is on comparing the theoretical properties of the time independent homogeneous Markovian process and the time dependent Markov renewal process in describing trip chaining behavior.

There are two motivations in the development of the model explored here, beyond the obvious one of getting a better replication of observations. The first is to study the effect of time constraints on travel patterns resulting from conscientious decisions of individuals to coordinate and to schedule travel and activities within a fixed time period. This leads to the development of the time dependent transition probability as a function of the time of day or the elapsed time since the beginning of the trip chain. The second motivation is to develop a general framework for integrating travel behavior and urban spatial structure of activities. This is done by eliminating the time homogeneity of previous Markovian models and, hence, allowing the incorporation of temporal aspects of activity choices, such as the temporal variations in activity opportunities made available to a trip decision maker. [10,11]

Although there may be much technical difficulty lying ahead, we are encouraged by our study of a simplified special case that shows the promise of the present modeling approach in depicting important aspects of observed trip chaining behavior. More exciting is the possibility that the present model may open new avenues in the development of a fundamental conceptual framework for trip chaining analysis. In addition, certain planning and policy issues may be entertained, even with such a simple model presented here. Since the model is formulated on the basis of a general discrete state space, by appropriately defining the state space the model may be applied to many relevant aspects of urban travel behavior. For example, the model serves as a basis for analyzing the dynamic spatial-interactions between activities when the state space is defined in terms of geographical locations. Activity allocation, or time budget analysis, may be explored within the model framework by defining the state space in terms of activity categories. It is worthy to note that when trip chaining, spatial interaction, and activity allocation are integrated into a single process, the complete human activity pattern and travel behavior are described.

2. MODEL FRAMEWORK

In a Markovian analysis of trip chaining, the state space may be defined by trip purposes, destination locations, activities or a combination of them. A Markov chain model describes the transition between the states by specifying the transition probability without any concern for the amount of time spent in each state or the time at which the transition takes place. We shall define the state space in terms of trip purposes, without loss of generality, in the model development discussed here. If X_n, $n \varepsilon N = \{0, 1, 2. . .\}$, is the purpose of the n-th trip in a chain, a transition takes place when the trip chain is extended by an additional trip, the (n+1)th trip, of purpose X_{n+1}. The Markov chain model describes the stochastic trip chaining process through the transition probability,

$$\Pr[X_{n+1} = j \mid X_0, X_1, \ldots, X_n]$$

for all $n \varepsilon N$ and all $j \varepsilon E$, where E is the state space of the process. In general, only Markov chains of the first order are used to describe trip chaining, namely,

$$\Pr[X_{n+1} = j \mid X_0, X_1, \ldots, X_n] = \Pr[X_{n+1} = j \mid X_n = i] \qquad (1)$$

For a homogeneous, time stationary, Markov chain of the first order, the process is defined by the transition matrix

$$P = \begin{bmatrix} p_{11} & p_{12} & p_{13} & \cdots \\ p_{21} & p_{22} & p_{23} & \cdots \\ \cdot & \cdot & \cdot & \cdots \\ \cdot & \cdot & \cdot & \cdots \\ p_{i1} & p_{i2} & p_{i3} & \cdots \\ \cdot & \cdot & \cdot & \cdots \\ \cdot & \cdot & \cdot & \cdots \\ \cdot & \cdot & \cdot & \cdots \end{bmatrix}$$

where

$$p_{ij} = \Pr[X_{n+1} = j \mid X_n = i]$$

for all i, j ε E. The stationarity assumption of the Markov chain implies that the transition matrix is invariant for all transitions.

Horton and Shuldiner [12] applied this Markov-chain structure to describe the succession of land use types visited in a series of trips. Following this, Horton and Wagner [13] analyzed trip purpose transitions and obtained factors characterizing trip chain pattern differences. Sasaki [14] also applied a Markov chain model to trip purpose chains. The work by Sasaki was extended to incorporate travel mode and empirically examined by Kondo. [15]

Trips in a trip chain not only evolve from state to state in the state space E, but also evolve along the time dimension. The discrete state space Markov chain process, and/or even the continuous state space Markov process in discrete time, has clearly limited capabilities in describing the evolution of the trip chaining process along a continuous time domain with randomly distributed sojourn times. The sojourn time is the time interval between the transition to the n-th trip purpose and the transition to the (n+1)th trip purpose. This time is made up of two components, the trip time and the time spent on the activity associated with the n-th trip purpose. Although the renewal process can be used to describe the stochastic process of transitions, it is not capable of simultaneously describing the evolution of the states.

The Markov renewal process allows the incorporation of time into the Markov chain process and allows the incorporation of state transitions into the renewal process. This is done by defining the transition probability as

$$\Pr[X_{n+1} = j,\ T_{n+1} - T_n \le z \mid X_o, X_1, \ldots, X_n;\ T_0, T_1, \ldots, T_n]$$

$$= \Pr[X_{n+1} = j,\ T_{n+1} - T_n \le z \mid X_n, T_n],$$

$$\text{for } 0 = T_0 \le T_1 \le T_2 \le \ldots \le T_{n+1} \qquad (2)$$

where T_n is the time of occurrence of the n-th transition; T_{n+1} the time of the (n+1)th transition; and $X_k \varepsilon E$ is the state of the k-th trip in the chain. The time interval $T_{n+1} - T_n$ is the sojourn time. Again, this process is generally history and time independent. Given that one knows the present state and the time of transition to the present state, the future process is described by the time invariant state transition probability and the sojourn time distribution.

Application of the Markov renewal model to trip chaining behavior was discussed by Gilbert et. al. [16] who carefully discussed the assumptions involved and the practical applicability of the model. No empirical data were analyzed in Gilbert's paper. Lerman [17] estimated parameters of a semi-Markov trip chaining model (which describes essentially the same process as the Markov renewal process), where the transition probability was estimated using discrete travel choice models.

In order to incorporate the two features that we feel are important in describing the trip chaining process, several extensions and modifications are made to the basic Markov-renewal model here. The first element involves the concept of time-dependent state transitions. Consider the following formulation of the conditional transition probability:

$$\Pr[X_{n+1} = j \mid X_0, X_1, \ldots, X_n;\ T_0, T_1, \ldots, T_n, T_{n+1}]$$

$$= \Pr[X_{n+1} = j \mid X_n = i,\ T_n = s,\ T_{n+1} = t], \quad \text{for } i, j \varepsilon E \text{ and } 0 \le s \le t, \qquad (3)$$

where, as before, X_n is the state of the n-th trip and T_n is the time when the n-th trip was started. Similar to the Markov-renewal process, the transition probability is assumed to be independent of the past history of the process; only the present state, X_n, and when this present state was entered, $T_n = s$, are of relevance. However, the probability of the state of the next trip, $X_{n+1} = j$, is dependent not only on the present state X_n and its entry time T_n, but also on the time when the transition to the (n+1)th trip occurs, namely $T_{n+1} = t$. In the context of trip chaining, the conditional dependence of the state transition on the time when the transition takes place represents the time dependence of the activity pattern of the individuals over a daily routine. For example, the probability that one is heading home, state j, given that this person has been working at the office, state i, since 8:00 a.m. ($T_n = s$) is a function of when he will be leaving the office ($T_{n+1} = t$). This transition will, of course, be far more likely if t is 5:00 p.m. than if it is 9:30 a.m.

It is, furthermore, not unreasonable to assume that an individual leaves his activity site when that activity is completed and moves immediately on to the next activity and its location. The time of the completion of the present activity and, hence, the time for the beginning of the trip to the next activity would presumably dominate the transition probability from the present trip purpose to the next trip purpose. The time when the present activity commenced (T_n) is likely to be only remotely related to the probability of the next trip purpose, although T_n and T_{n+1} are interrelated through the sojourn time. The structure of the sojourn time distributions is, yet, another major element of the present model. What is important is that the transition probability may be decoupled from the sojourn time, such that T_n may be removed from the conditional transition probability, so that we may assume

$$\begin{aligned} & \Pr[X_{n+1} = j \mid X_n = i, T_n = s, T_{n+1} = t] \\ &= \Pr[X_{n+1} = j \mid X_n = i, T_{n+1} = t] \qquad (4) \\ &= p(i,j,t), \text{ for } t \geq 0, \end{aligned}$$

where p(i,j,t) will be called the "time dependent transition probability". This time-of-day dependency of the trip purpose transitions seems reasonable and adequate for describing the daily activity pattern of typical individuals. Moreover, depending on the definition of the time origin, one may also treat the dependency in terms of the elapsed time since the beginning of the trip chain. By doing so, the possible dependency of the transition probability on the number of previous trips may not be necessary for the model. As a result, the time dependent transition probability needs only to be represented as a function of time and not as a function of the number of previous trips up to the present, n.

Given the state-to-state transition probability

$$\Pr[X_{n+1} = j \mid X_n = i, T_n = s, T_{n+1} = t] \quad ,$$

the time dependent Markov renewal process can be expressed as

$$\Pr[X_{n+1} = j, T_{n+1} \le t \mid X_0, X_1, \ldots, X_n; T_0, T_1, \ldots, T_n]$$

$$= \Pr[X_{n+1} = j, T_{n+1} \le t \mid X_n = i, T_n = s] \tag{5}$$

$$= \int_{\xi = s}^{t} \Pr[X_{n+1} = j \mid X_n = i, T_n = s, T_{n+1} = \xi] \; d\Pr[T_{n+1} \le \xi \mid X_n = i, T_n = s]$$

In this general formulation, the first term in the integral is the time dependent transition probability from state i at the n-th epoch to state j at the (n+1)th epoch, as previously defined in Eq. 4. The second term in the integral is the sojourn time distribution. In Eq. 5, both terms are expressed in forms that show the dependence on T_n. In the second term, the sojourn time $T_{n+1} - T_n$ is $(\xi - s)$. The model, represented by Eq. 5, follows the common assumption of Markovian history independence.

In the present model, the sojourn time is defined as the time interval between the transition to the n-th state and the transition to the (n+1)th state. When the state space is defined in terms of trip purposes, this time is made up of two components, the trip time and the time spent on the activity associated with the n-th trip purpose. Alternatively, one may define two sets of states, one representing traveling to and from activities and the other representing a stay at a location for activity pursuit. Then, the sojourn time between an activity state and a traveling state is the trip time and the sojourn time between a traveling state and an activity state is the activity time. When the process is modeled this way, the probability of transition among traveling states and among activity states may be set to zero in order to avoid any mathematical difficulties. However, given the transition probability and the sojourn time distributions, it is possible to determine mathematically the distribution of the combined sojourn time for a pair of successive traveling and activity states.

In considering the sojourn time distribution, one may further assume that, in trip chaining, the duration required to complete the trip and its destination trip purpose is only a function of the activity and not a function of the time of day. In other words, a shopping trip takes as much time (trip time and activity time) at 3:00 p.m. as at 7:00 p.m. Furthermore, it is assumed that the sojourn time of the present activity is independent of the previous activity or the next activity. These assumptions seem intuitively plausible, although there are obvious exceptions. For example, queueing and traffic congestion effects may prolong a particular sojourn during peak business and rush hours. There may also be a difference in the sojourn time depending on the sequence of activities; for example, a trip from home to shop may take longer than a trip from work to shop. Such differences in the sojourn time may also reflect travel distance, trip time, and travel mode differences. However, in the present work, these effects are ignored pending further observational evidences.* For the model discussed here, the sojourn time distribution is assumed to be time of day or elapsed time independent. The distribution is assumed to be dependent only on the trip purpose or activity type, giving

$$\begin{aligned} & \Pr[T_{n+1} \leq t \mid X_n = i,\ T_n = s] \\ & = \Pr[T_{n+1} - T_n \leq t - s \mid X_n = i] \qquad (6) \\ & = F_i(t - s), \qquad \text{where } 0 \leq s \leq t \end{aligned}$$

Although the sojourn time distribution is assumed to be time of day independent for mathematical convenience, a more elaborate model may take into consideration the time dependence of the sojourn time.

The time dependent state-to-state transition probability, Eq. 4, and the time independent sojourn time distribution, Eq. 6, may now be combined to describe the time dependent trip chaining process modeled within the Markov renewal framework given in Eqs. 2 and 5,

---

*There are, however, somewhat contradicting empirical observations. Hemmens [5] tabulated the average activity durations for 42 trip purposes using about 9,000 trip records from the Buffalo, New York area. The tabulation was made for three periods of the day. Some activities were found to have the longest average durations when they started in the morning period (till noon), while others had longest averages when started in other periods. No statistical tests were conducted to examine the differences in the averages across the time periods. An examination of the results together with additional tables presented in the paper, however, indicated that the number of observations used are extremely small (around 10 to 20) for those activity categories that exhibited larger temporal variations in their average durations. Thus, Hemmens' result is statistically inconclusive.

$$dPr[X_{n+1} = j, T_{n+1} \leq t \mid X_n = i, T_n = s]$$

$$= Pr[X_{n+1} = j \mid X_n = i, T_n = s, T_{n+1} = t] dPr[T_{n+1} \leq t \mid X_n = i, T_n = s]$$

$$= p(i,j,t) \; dF_i(t - s), \quad \text{for } t \geq s \geq 0 \quad , \tag{7}$$

where $F_i(\xi)$ is the cumulative distribution of the sojourn time for the present activity i, as defined in Eq. 6. Equation 7 gives the probability that the transition is next to occur at time t from the present state i to a new state j, if, by time t, the present activity i has had a duration $\xi = t - s$. By integrating Eq. 7 over [0, t], one obtains

$$Pr[X_{n+1} = j, T_{n+1} \leq t \mid X_n = i, T_n = s]$$

$$= \int_{\xi = s}^{t} p(i, j, \xi) \; dF_i(\xi - s), \quad t \geq s \geq 0 \quad , \tag{8}$$

which is the probabilistic formulation of the transition mechanism described by the model. Note that while the transition probability and sojourn time distribution have been structurally decoupled, they both include the parameter that represents the time of transition. Because of this, the process depicted by the model exhibits important interactions between the trip purpose successions and sojourn times as is shown later with a numerical example. Also note the similarity between Eq. 8 and the Markov renewal model given by

$$Pr[X_{n+1} = j, T_{n+1} - T_n \leq z \mid X_0, X_1, \ldots, X_n; T_0, T_1, \ldots, T_n]$$

$$= Pr[X_{n+1} = j, T_{n+1} - T_n \leq z \mid X_n = i], \quad t \geq 0$$

$$= p_{ij} \; F_i(z) \tag{9}$$

Both share the same Markovian assumption that, given the present state, the conditional transition probability is independent of the past. Also assumed is the independence of the sojourn time distribution on the time when the present state was entered.

There is, however, a critical difference between the two. The transition probability is defined in Eq. 9 for the relative time difference between the two successive transitions ($T_{n+1} - T_n$). This, however, is exactly the definition of the sojourn time. Thus the time element is represented in Eq. 9 only in terms of the sojourn time, whose distribution is assumed to be time independent. The remaining component of the model in Eq. 9, which is the counterpart of p(i,j,t), is atemporal, i.e., activity linkage is time independent. The formulation of the transition mechanism in Eq. 8, on the other hand, results in time dependent activity linkage, since it is formulated for the time <u>of</u> the next transition, not for the time <u>until</u> the transition. The decoupling of the Markovian aspect and the renewal aspect in Eq. 8 is also very apparent, hence, offering a simple approach to incorporating the new features into the Markov renewal model of trip chaining.

The time dependent transition probability, p(i,j,t), is the term where activity choice and the environment interface. At time periods when a large number of activity opportunities are available, the choice probability of those activities will naturally be large. On the other hand, certain activities are not available during other time periods. The term p(i,j,t) offers a mechanism where this interface can be incorporated. In this sense, the model can serve as a methodology for the analysis of trip chaining and its interfaces with the environment along the time dimension. The term also offers the possibility of representing the time dependent nature inherent in human behavior. There are thus many possibilities in modeling p(i,j,t). In previous analyses, the concept of activity linkage has hardly been related to any exogenous factors. The formulation of the time dependent linkage now makes it possible to relate the linkage to temporal variations in the supply of activity opportunities, or other temporal factors. Such modeling effort will provide a useful means in predicting activity and travel characteristics in response to changes in the urban travel environment. In addition, the model can serve as a logical approach to activity chaining.

Given the formulation of the transition process, the next step of model development is to derive an explicit probabilistic expression for the state of the process at an arbitrary time t. This requires the functional specification of p(i,j,t). The discussion in the next section is limited to a special case that is used simply as a convenient means to obtain analytical and tractable solutions. Although the specification is based on intuitive and plausible assumptions, it, nonetheless, leads to some useful numerical results which show the important properties of the time dependent model in contrast with those of the atemporal models.

3. A SPECIAL CASE

Suppose the state space, E, is defined as a collection of out-of-home activities and two "home" states, one being an absorbing "home" state (denoted by h) and the other being a transient "home" state, treated like any other state. The visit to the absorbing home state implies that no additional out-of-home activities will be pursued during the observational period. On the other hand, the visit to the transient home state indicates that a series of out-of-home activities (or, a trip chain) is terminated, and that another series will be undertaken. Thus, the process defined with E represents a travel history that may involve more than one trip chain.

Now consider the conditional transition probability to the absorbing home state given the time of transition, t, to the state, i.e., the probability that an activity just completed is the last one to be pursued outside the home for the day. One can reasonably assume that this probability is an increasing function of time. In other words, the probability of permanently returning home is higher and higher towards the end of the day. This assumption is compatible with the observation that travel behavior is subject to the time constraints that one must, and does, return home eventually. Further, suppose the rate of increase can be expressed as an exponential function of time, viz.

$$\Pr[X_{n+1} = h \mid T_{n+1} = t] = q(h, t)$$
$$= 1 - \alpha e^{-\eta t}, \quad t \geq 0 \quad , \qquad (10)$$

where α and η are constants ($0 \leq \alpha \leq 1$; $\eta \geq 0$). Since

$$\sum_{j \varepsilon E} q(j, t) = 1 \quad , \qquad (11)$$

it follows from Eq. 10 that

$$\sum_{j \neq h} q(j, t) = \alpha e^{-\eta t} \qquad (12)$$

A possible form of q(j,t) that satisfies the above condition is

$$q(j,t) = \alpha_j e^{-\eta t},$$

where
$$\sum_{j \neq h} \alpha_j = \alpha; \quad \alpha_j > 0, \quad j \neq h \tag{13}$$

Thus,
$$q(j,t) = \begin{cases} 1 - \alpha e^{-\eta t}, & j = h \\ \alpha_j e^{-\eta t}, & \text{otherwise} \end{cases} \tag{14}$$

The conditional transition probability for state $j \neq h$, therefore, decreases exponentially as time proceeds; while for the absorbing home state, h, the probability increases with time.

The exponential formulation of q(j,t) suggests the following specification for p(i,j,t):

$$p(i,j,t) = \begin{cases} 1 - \sum_{k \neq h} p_{ik} e^{-\mu_k t}, & j = h \\ p_{ij} e^{-\mu_j t}, & \text{otherwise} \end{cases} \quad , \tag{15}$$

where p_{ij} and μ_j are constants. The parameter μ_j (which replaces η) is different for different state j. The parameter α_j is replaced by p_{ij}, assuming that there is a certain linkage pattern between pairs of activities.

The time dependent transition probability p(i,j,t) as presented here, however, is formulated mainly for mathematical convenience and, obviously, is not the only possible formulation. For example, Kitamura et. al. [18] showed that by using an abstract model of travel pattern, the conditional probability of returning home is a positive exponential function of time. The model used is an integration of the Hagerstrand prism, [19] the trip-linkage concept and the intervening opportunities model of trip distribution, and is for a highly idealized representation of a one-dimensional linear city. This result suggests a possible formulation of p(i,j,t) as

$$p(i,j,t) = \begin{cases} 1 - \sum_{j \neq h} (p_{ij} - q_{ij} e^{\mu_j t}), & j = h \\ & 0 \leq t \leq T , \\ p_{ij} - q_{ij} e^{\mu_j t}, & \text{otherwise} \end{cases} \tag{16}$$

where p_{ij}, q_{ij}, μ_j and T are all positive parameters. Although the positive exponential formulation of Eq. 16 has more realism, the present paper employs the formulation in Eq. 15 for mathematical simplicity. Since the analysis is essentially the same for both models, for illustrative purposes, the model in Eq. 15 is preferred over a formulation based on Eq. 16 which is by far messier.

Using Eq. 15 for p(i,j,t) we have

$$dPr[X_{n+1} = j,\ T_{n+1} \le t \mid X_n = i,\ T_n = s]$$

$$= \begin{cases} p_{ij}\, e^{-\mu_j t}\, dF_i(t - s), & t \ge s \ge 0,\ j \ne h \\ \{1 - \sum_{k \ne h} p_{ik}\, e^{-\mu_k t}\}\, dF_i(t - s), & t \ge s \ge 0,\ j = h, \end{cases} \tag{17}$$

where F_i (ξ) is the sojourn time distribution for state i. One may use the notation

$$U^n(i,j,t) = Pr[X_n = j,\ T_n \le t \mid X_0 = i] \tag{18}$$

with $T_0 = 0$, to represent the semi-Markov kernel of the time dependent Markov renewal process. The explicit functional expression of $U^n(i,j,t)$ can now be derived as follows.

For n = 1, one immediately obtains

$$U^1(i,j,t) = \begin{cases} \int_{\xi=0}^{t} p_{ij}\, e^{-\mu_j \xi}\, dF_i(\xi), & j \ne h \\ \int_{\xi=0}^{t} \{1 - \sum_{k \ne h} p_{ik}\, e^{-\mu_k \xi}\}\, dF_i(\xi), & j = h\,. \end{cases} \tag{19}$$

For n = 1, 2, . . . ,

$$dU^{n+1}(i,k,t) = \sum_{j \ne h} \int_{\xi=0}^{t} d\,U^n(i,\ j,\ \xi)\ p_{jk}\, e^{-\mu_k t}\ d\,F_j\ (t - \xi),$$

$$k \ne h,\ n = 1,\ 2,\ \ldots, \tag{20a}$$

and for the absorbing home state,

$$dU^{n+1}(i,h,t) = \sum_{j \ne h} \int_{\xi=0}^{t} dU^n(i,j,\xi) \left\{1 - \sum_{k \ne h} p_{jk}\, e^{-\mu_k t}\right\} d\,F_j\ (t - \xi)$$

$$= \sum_{j \ne h} \int_{\xi=0}^{t} dU^n(i,j,\xi)\ d\,F_j\ (t - \xi) - \sum_{k \ne h} dU^{n+1}(i,k,t),\ n = 1,\ 2,\ \ldots \tag{20b}$$

The recursive relationship of U^n can be solved to yield

$$\tilde{U}^{n+1}_{ik}(\alpha) = \sum_{j_1}\sum_{j_2}\cdots\sum_{j_n} p_{ij_1}p_{j_1j_2}\cdots p_{j_nk}\tilde{F}_i(\alpha+\lambda_{1,k})\tilde{F}_{j_1}(\alpha+\lambda_{2,k})$$
$$\cdots\tilde{F}_{j_{n-1}}(\alpha+\lambda_{n,k})\tilde{F}_{j_n}(\alpha+\lambda_{n+1,k}),\ \text{for } k \neq h,\ n = 1, 2, \ldots \tag{21a}$$

and for the absorbing home state,

$$\tilde{U}^{n+1}_{ih}(\alpha) = \sum_{j \neq h}\tilde{U}^n_{ij}(\alpha)\tilde{F}_j(\alpha) - \sum_{k \neq h}\tilde{U}^{n+1}_{ik}(\alpha),\quad n = 1, 2, \ldots, \tag{21b}$$

where $\tilde{U}^{n+1}_{ik}(\alpha)$ and $\tilde{F}_j(\alpha)$ are respectively the Laplace-Stieltjes transforms of $U^{n+1}(i,k,t)$ and $F_j(t)$. The parameters $\lambda_{\ell,k}$ are given by

$$\lambda_{\ell,k} = \mu_{j_\ell} + \mu_{j_{\ell+1}} + \cdots + \mu_{j_n} + \mu_k,\qquad \ell = 1, 2, \ldots, n,$$

and

$$\lambda_{n+1,k} = \mu_k \tag{22}$$

Given the form of the distribution F_j (t) and the inverse transformation of $\tilde{U}$, the integration over [0, t] provides U^{n+1} (i,k,t). For example, let the sojourn time be negative exponentially distributed with parameter ρ_j:

$$F_j(t) = 1 - e^{-\rho_j t},\qquad t \geq 0,\ j \neq h$$

Then for the non-absorbing states,

$$\tilde{U}^{n+1}_{ik}(\alpha) = \sum_{j_1}\sum_{j_2}\cdots\sum_{j_n}\left\{p_{ij_1}p_{j_1j_2}\cdots p_{j_nk}\,\rho_i\,\rho_{j_1}\,\rho_{j_2}\cdots\rho_{j_n}\right\}$$
$$\left\{(\alpha+\gamma_{1,k})(\alpha+\gamma_{2,k})\cdots(\alpha+\gamma_{n+1,k})\right\}^{-1},\ k \neq h,\ n = 1, 2, \ldots, \tag{23}$$

where $\gamma_{1,k} = \lambda_{1,k} + \rho_i$

and $\gamma_{\ell,k} = \lambda_{\ell,k} + \rho_{j_{\ell-1}},\qquad \ell = 2, 3, \ldots, n+1.$

On the condition that $\gamma_{\ell,k} \neq \gamma_{m,k}$ for ℓ, m = 1, 2, ..., n + 1, and $\ell \neq m$,

$$\prod_{\ell=1}^{n+1} (\alpha + \gamma_{\ell,k})^{-1} = \sum_{\ell=1}^{n+1} A_{\ell,k}^{(n+1)} / (\alpha + \gamma_{\ell,k})$$

where $A_{\ell,k}^{(n+1)}$ is defined as

$$A_{\ell,k}^{(n+1)} = (-1)^n \Big/ \prod_{\substack{m=1 \\ m \neq \ell}}^{n+1} (\gamma_{\ell,k} - \gamma_{m,k}), \ell = 1,2,\ldots,n+1;\ n = 1,2,\ldots$$

(see reference 20, Formula 20, page 232). Applying this to Eq. 23, and integrating over [0, t], the desired expression is obtained as

$$U^{n+1}(i,k,t) = \sum_{j_1} \sum_{j_2} \cdots \sum_{j_n} \Big\{ p_{ij_1} p_{j_1 j_2} \cdots p_{j_n k}\, \rho_i\, \rho_{j_1}\, \rho_{j_2} \cdots \rho_{j_n}$$

$$\sum_{\ell=1}^{n+1} A_{\ell,k}^{(n+1)} (1 - e^{-\gamma_{\ell,k} t}) / \gamma_{\ell,k} \Big\},\quad k \neq h;\ n = 1, 2, \ldots \tag{24}$$

The expression for the absorbing home state can be obtained in a similar manner.

With a slight modification, the result also applies to the case where the sojourn time distribution is Erlang, which may be appropriately used to represent many observed distributions. Suppose, for example, the distribution is Erlang type II, viz.

$$F_j(t) = 1 - e^{-\rho_j t} - \rho_j\, t\, e^{-\rho_j t},\quad t \geq 0 \tag{25}$$

Then the expression corresponding to Eq. 23 is

$$\tilde{U}_{ik}^{n+1}(\alpha) = \sum_{j_1} \sum_{j_2} \cdots \sum_{j_n} \Big\{ p_{ij_1} p_{j_1 j_2} \cdots p_{j_n k}\, \rho_i^2\, \rho_{j_1}^2 \cdots \rho_{j_n}^2 \Big\}$$

$$\Big\{ (\alpha + \gamma_{1,k})(\alpha + \gamma_{2,k}) \cdots (\alpha + \gamma_{n+1,k}) \Big\}^{-2},\quad k \neq h;\ n = 1, 2, \ldots \tag{26}$$

Noting that

$$\prod_{\ell=1}^{n+1} (\alpha + \gamma_{\ell,k})^{-2} = \{ \sum_{\ell=1}^{n+1} A_{\ell,k}^{(n+1)} / (\alpha + \gamma_{\ell,k}) \}^2 \tag{27}$$

the inverse transformation can be carried out in the same manner as before.

Given the explicit functional form of U, useful information can be obtained, including:

(a) The expected number of visits to j by time t, given $X_0 = i$:

$$E\left(N_j(t) \mid X_0\right) = \sum_{n=1}^{\infty} U^n(i,j,t), \quad j \neq h , \tag{28}$$

where $N_j(t)$ is the number of visits to state j by time t.

(b) The conditional distribution of chain duration, given chain length n and X_0:

$$\Pr\left(T_n \leq t \mid X_n = h, X_0 = i\right) = U^n(i,h,t)/U^n(i,h,\infty) \quad , \tag{29}$$

where the chain duration is defined as the time since the chain started until the transition to the absorbing home state.

(c) The conditional probability of being at state j at time t, given X_0:

$$\begin{aligned} \Pr\left(X(t) = j \mid X_0\right) &= \sum_{n=1}^{\infty} \int_{\xi=0}^{t} dU^n (i,j,\xi) \{1 - F_j (t - \xi) \} \\ &= \sum_{n=1}^{\infty} \{ U^n (i,j,t) - V^n (i,j,t)\} , \quad j \neq X_0 , \end{aligned} \tag{30}$$

where

$$V^n(i,j,t) = \int_{\xi=0}^{t} dU^n (i,j,\xi) F_j (t - \xi) \tag{31}$$

4. NUMERICAL RESULTS

In order to illustrate some important features of the model, this section presents a numerical example using a hypothetical set of parameter values. The model is applied to a three-state case, {1, 2, h}, with the home state, h, being absorbing. This may be interpreted as representing trip chaining behavior with a binary classification of trip purposes. The sojourn time distribution employed is negative exponential. Time is defined in the model as the time since the chain started, i.e., $T_0 = 0$.

Table 1 summarizes the parameter values used. The transition probabilities at time 0, (p_{ij}), indicate that the transitions from state 1 to 2 and from state 2 to home have relatively high probability. The time rates of change in the transition probabilities to states 1 and 2 are assumed to be, respectively, -0.2 and -0.5 for every two hour calculation interval. Correspondingly, $\mu_1 = -\ell n\ (0.8)/2$ and $\mu_2 = -\ell n\ (0.5)/2$. In this example, the transition probability to state 2 rapidly decreases, as shown in Figure 1. Figure 1 shows the time dependent transition probabilities in the example. The mean sojourn times are 4/3 hours and 3/4 hour at states 1 and 2, respectively. Computation is carried out with the initial state being state 1, and truncated beyond n = 10.

Figure 2 presents the distribution of the chain length, which is defined as the number of sojourns per chain inclusive of the sojourn at the initial state, but exclusive of the sojourn at the absorbing home state. The distributions for both time dependent and time independent cases are presented. The latter employs the (p_{ij}) at t = 0, given in Table 1, as its stationary transition probability.

TABLE 1

PARAMETER VALUES USED IN THE NUMERICAL ANALYSIS

Transition Probabilities, p_{ij}, at t = 0

	Home*	State 1	State 2
Home*	1.0	0.0	0.0
State 1	0.3	0.3	0.4
State 2	0.5	0.2	0.3

* an absorbing state

Rate Of Change Of The Transition Probability

State	μ_j	Time Rate of Change in p(i,j,t) per 2 hours
1	.1116	- .2
2	.3466	- .5

Sojourn Time Distribution

State	ρ_i	Mean Sojourn Time (hr)
1	.75	1.33
2	1.3	0.75

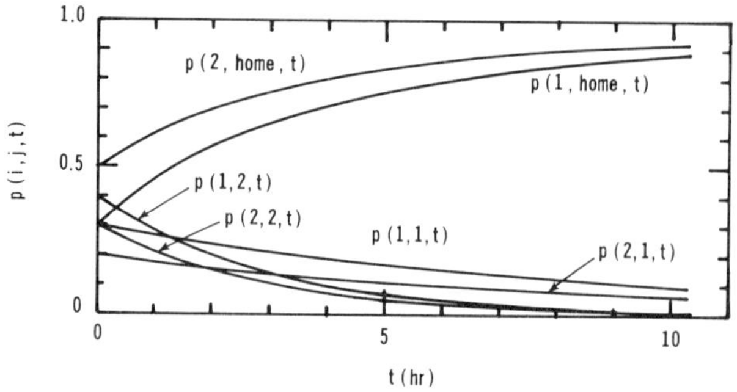

FIGURE 1. Time dependent Transition Probabilities Used in the Numerical Analysis

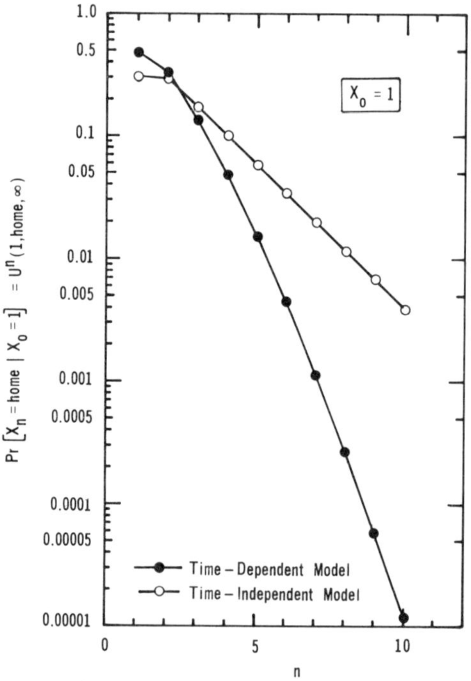

FIGURE 2. The Probability of a Trip Chain of Length n

The result of the time independent model is presented here to evaluate how the time dependent transition probability affects the behavior of the process. In terms of such measures as the expected chain length, the parameter values used are not comparable between the two cases. However, substantial differences in the chain length distribution between these two cases are evident from the figure. The time dependent model yields larger probabilities for shorter chains ($n = 1$ and 2). On the log scale, the distribution is almost linear beyond $n = 1$ for the time independent case, but declines with a slightly increasing rate for the time dependent case. The expected chain length, given $X_0 = 1$, is 1.84 for the time dependent case. For the particular set of parameter values employed, the probability that the chain length exceeds 10 is less than 0.3×10^{-5}, which eventually imposes an upper limit to the chain length.

Additional computation was done using the time independent model with a stationary transition matrix that yields identical expectations for the chain length and the number of visits to each state as in the time dependent example. The probability that the chain length exceeds 10 was found, for this case, to be greater than 0.25×10^{-3}. The chain length distribution from the time independent model is essentially geometric and the probability of terminating the chain is invariant, regardless of n, after the effect of the initial state has diminished. Although some observations [21] appear to support this result for relatively small n, its validity for larger n is questionable. Another empirical result, [15] in fact, indicates that the chain length distribution is not geometric; hence, shorter chains are substantially overrepresented. The time dependent model, on the other hand, demonstrates an increasing probability of terminating the chain with increasing n. A large number of trips in a chain is practically impossible because of time constraints, and the time dependent model appears to represent this situation very well. The numerical results also show little error resulting from the truncation rule employed.

As before, we define the "chain duration" as the elapsed time until the transition to the home state takes place. The conditional expectation of the chain duration, given the chain length n and X_0, is given by

$$\bar{t}^{(n)}_{i,h} = E(T_n \mid X_0 = i, X_n = h)$$

$$= \int_{t=0}^{\infty} t \, dU^n (i,h,t)/U^n (i,h,\infty)$$

This expected chain duration is plotted in Figure 3 for both the time dependent and time independent cases (again using the p_{ij} of Table 1 as the stationary transition probability). The figure also shows the mean sojourn time per state, defined as $\bar{t}_{i,h}^{(n)}/n$. For the time independent case, the expected chain duration increases almost linearly with the chain length beyond n = 1. Accordingly, the mean sojourn time is almost constant, regardless of the chain length. The expected chain duration of the time dependent process, on the other hand, increases with the chain length at a decreasing rate. As a result, the average sojourn time decreases with increasing number of sojourns in the chain.

The negative correlation between the mean sojourn duration and the number of sojourns has been empirically observed. [18, 22] This observation and our numerical results agree with intuition. One obvious reason for this correlation is the influence of the time constraint imposed upon the daily activity pattern, by the fact that one must return home within a certain time period after leaving home. Because of this constraint, the average sojourn duration must be shorter when a larger number of sojourns are pursued; or the number of sojourns must be cut back when longer sojourns are involved. With the introduction of a time constraint into the model structure through the time dependent transition probability, the model studied produces the negative correlation between average sojourn time and the number of sojourns as previously observed.

Another interesting feature observed in the example is the dependence of the sojourn time distribution upon the state of the next visit for the time dependent process. As an example, Figure 4 shows the conditional distribution of sojourn times at the initial state (X_0), given the state of the next visit (X_1). The conditional expectation ranges from 0.912 (given X_1 = 2) to 1.678 (given X_1 = h). The distribution for X_1 = 1 or 2 is exponential, although the distribution for X_1 = h is not. The implication is that, even when a sojourn is at the same state (for the same trip purpose), the time spent there is dependent upon what the tripmaker is going to do next, viz., whether go back home or make another trip. The time independent model, on the other hand, would show identical sojourn time distributions, regardless of the state of the next visit. Again, it appears that the time dependent model is capable of reflecting some aspects of a tripmaker's activity scheduling under time constraint.

Figure 5 presents the probability of being at various states, given X_0, computed according to Eq. 30 with truncation beyond n = 10. For the parameter values used here, the tripmaker returns home within 10 hours with a probability of 0.997, and within 16 hours with a probability exceeding 0.9999. Hence, the model assures that, eventually, everybody is back home within a one-day period.

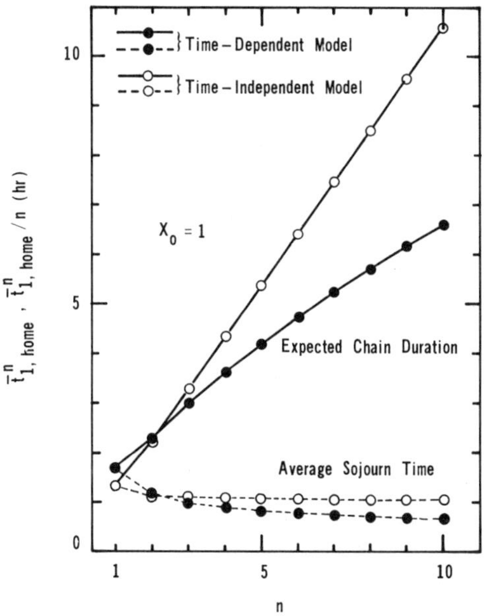

FIGURE 3. Expected Chain Duration, $\bar{t}^n_{1,\text{home}}$, and Average Sojourn Time, $\bar{t}^n_{1,\text{home}}/n$, versus the Number of Sojourns, n, in the Chain

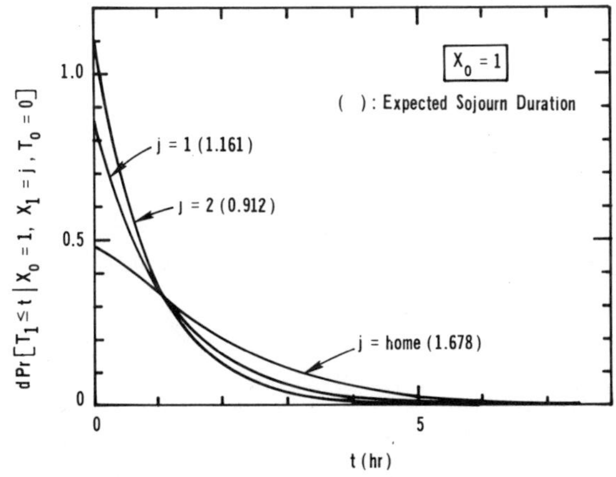

FIGURE 4. The Conditional Probability of the Sojourn Time, t, at the Initial State $X_0 = 1$ for Various States of the Next Visit, $X_1 = j$. The Values in Parentheses are the Conditional Expectations of the Sojourn Time

The probability of being at state 2 becomes a maximum at about 1 hour after the chain started, while the maximum for state 1 occurs at about 1.5 hours. This reflects, among others, the different rate of change in the time dependent transition probability. For the same reason, and also because of its shorter expected sojourn time, the probability of being at state 2 is relatively small. However, the expected number of visits for state 1, given $X_0 = 1$, is 0.423; while for state 2 the expectation is 0.417, very comparable to that for state 1.

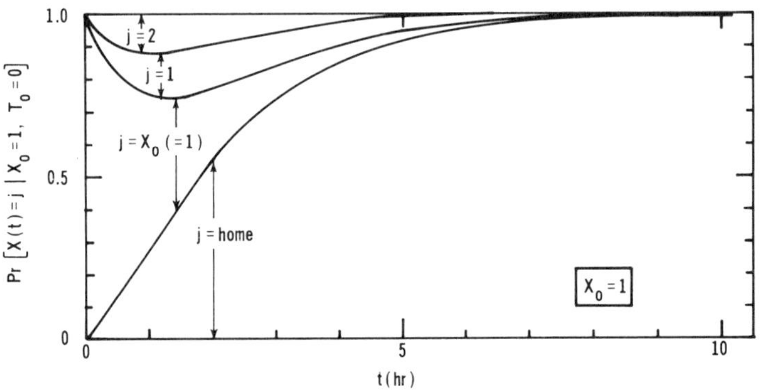

FIGURE 5. Probability of Being at Various States j at Time t Given the Initial State $X_0 = 1$

Although the parameter values used are hypothetical, this numerical example illustrates, nonetheless, some important features of the model which were not found in previous Markovian models. These features agree with intuition and with empirical observations published in the literature. Although the example has only shown a capability of the time dependent model to produce some qualitative aspects of trip chaining behavior under time constraint, the model is sufficiently flexible for other relevant aspects of trip chaining behavior to be studied.

5. DISCUSSION

The subject of urban travel behavior analysis is about the movement of people in space and time in pursuit of their daily activities. The movements are represented by trajectories consisting of individual sequences of activities by time and location over a period of time. Depending upon the domain of the trajectories thus defined, studies of travel behavior may be termed "trip chaining", "activity scheduling", or "time budgeting" analysis. This paper presents a stochastic model that can serve as a framework for trip chaining or activity scheduling analysis. The present model emphasizes temporal inhomogeneity and time budget constraint in a Markov renewal process. In previous Markovian models, time independence is assumed, although trip chaining or travel behavior is most unlikely to be homogeneous in time. The effort that has been made in the paper is an attempt to bring about a more realistic stochastic representation of the trip chaining behavior. Through a simple example of the model, it is shown that a time dependent Markovian process is capable of providing many qualitative features of travel behavior observed in previous empirical studies.

The improvements offered by the present model in representing trip chaining behavior came from two components incorporated into the Markov renewal process. These are the time dependent transition probability and the sojourn time distribution. This model, as a result, has several distinctive features in comparison with previous time independent models. For example, the time dependent model shows a negative correlation between the average sojourn duration and the number of sojourns in a chain, an empirically observed property of trip chaining [18,22] that previous models failed to produce. The model also shows that the conditional distribution of sojourn times at a state is dependent upon the state of the next visit. More importantly,the spatial movements of people depicted in the time dependent model would show a clear orientation. It is no longer the movement of a "frog jumping between lily pads".

Despite the simple-minded assumptions made in the special case example, the time dependent model demonstrates certain properties that suggest its usefulness. For example, suppose that the model is specified with the sojourn time as the sum of travel time and activity duration, and is applied to estimate the impact of a service level improvement upon the travel pattern. The service improvement is reflected in a shift in the sojourn time distribution due to reduced travel time (it is assumed that the p_{ij}'s remain the same). A time independent model, if applied, would give the same trip pattern after the change. The time dependent model, on the other hand, would show that the expected number of trips would increase as a result. The latter result is more plausible, especially if travel behavior is known to be influenced by accessibility.

Consider the application of the model with the state space being defined as a set of geographical zones. In this case, the process describes tripmakers' spatial movements over time. It may be possible, then, to interpret the μ_j's in the model as parameters related to spatial, temporal, and transportation constraints in trip chaining. Consider the tripmakers who reside in a particular traffic zone, and define the μ_j for this group of tripmakers. From the principles of spatial interaction, one can expect that a destination zone farther away (say zone j) would have a smaller probability of being visited at t = 0. One can also expect that the probability of visiting zone j further decreases as time proceeds, since the time budget constraints will get tighter and visiting the zone will become less and less desirable. This situation can be represented by a large parameter value assigned to μ_j (the probability of visiting zone j may be larger when the present state, zone k, say, is closer to zone j; but, in this case, p_{kj} will have a larger value). The model appears to have the characteristics to approximate this concept of the Hagerstrand "prism"[19] in a stochastic framework, when appropriate modifications are made. For example, since the model includes travel time in the sojourn time, the sojourn time distribution for j must be made dependent on both k and j, not only on j. Parameter μ_j, hence, provides an orientation to the tripmakers' movements, which the model depicts.

There are, however, many limitations to the present model. At this stage of development, it is not possible to exactly incorporate the availability of activities (e.g., store hours) into the model. It is rather doubtful whether the model, in its present form, applies to trip chains that contain work trips. The exponential formulation of transition probability still remains to be validated. In spite of these limitations, it is believed that the model developed here is an improvement towards a more plausible representation of trip chaining, and will serve as a more useful tool for the analysis and understanding of trip chaining.

Many other issues remain to be addressed in further empirical study of the model. For example, little is known at present about the significance of the independence between the sojourn time distribution and time of day as assumed in the present model on the statistical characteristics of trip chaining. It may also be the case that travel time and sojourn time are interrelated for certain definitions of the activity categories that form the state space. All these issues must be resolved in the context of the statistical model that is proposed here. Within the statistical model, assumptions were made, not so much on the behavioral characteristics of the trip makers as individuals, but mostly on the basis of the distributional characteristics of all the travelers as a collective statistical ensemble. Within the stochastic structure of random events, the specific trajectory or realization by an individual is not the subject of analysis. The results of the model are meaningful only as statistics. In this sense, the intent of the present work is to develop the mathematical structures of the collective trip chaining characteristics observed over repeated trials involving large samples of trip chaining events (not trip makers).

ACKNOWLEDGEMENT

A part of this research by one of the authors [RK] is supported by National Science Foundation Grant CME8006293.

REFERENCES

1. Hartgen, D. T., A. J. Neven, J. M. Brunso, J. S. Banas and J. Miller, Changes in Travel Response to the 1979 Energy Crisis, Preliminary Research Report No. 171, Planning Research Unit, New York State Department of Transportation, Albany, New York, 1980.

2. Peskin, R. L., "Policy Implications of Urban Travel Response to Recent Gasoline Shortages", Paper presented at Energy Users Conference, San Antonio, Texas, April 1980.

3. Damm, D., "Interdependencies in Activity Behavior", Transportation Research Record, No. 750, 1980, pp. 33-40.

4. Hummon, N. P., "An Analysis of Household Travel Patterns Based on Time Budget Data", Paper presented at the Joint National Meeting of Operations Research Society of America and The Institute of Management Sciences, Milwaukee, Wisconsin, October 1979.

5. Hemmens, G. C., "Analysis and Simulation of Urban Activity Patterns", Socio-Economic Planning Sciences, Vol. 4, 1970, pp. 53-66.

6. Lenntorp, B., Paths in Space-Time Environments: A Time-Geographic Study of Movement Possibilities of Individuals, Lund Studies in Geography, Series B, No. 44, CWK Gleerup, Lund, Sweden, 1976.

7. Chapin, F. S., Jr., Human Activity Patterns in the City, Wiley, New York, New York, 1974.

8. Vidakovic, V., "Modal Mixture of the Serial Transportation Demand: An Explorative Study", Proceedings of the International Conference on Transportation Research, Transportation Research Forum, Bruges, Belgium, 1973, pp. 679-683.

9. Hanson, S., "Urban Travel Linkages: A Review", in Hensher, D. and P. L. Stopher, eds., Behavioral Travel Modelling, Croom Helm, London, 1979, pp. 81-100.

10. Tomlinson, J., N. Bullock, P. Dickens, P. Steadman, and E. Taylor, "A Model of Students' Daily Activity Patterns", Environment and Planning, Vol. 5, 1973, pp. 231-266.

11. Shapcott, M. and P. Steadman, "Rhythms of Urban Activity", In Carlstein, T., et.al., eds., Timing Space and Spacing Time Volume 2: Human Activity and Time Geography, Edward Arnold, London, 1978, pp. 49-74.

12. Horton, F. E. and P. W. Shuldiner, "The Analysis of Land-Use Linkages", Highway Research Record, No. 165, 1967, pp. 96-107.

13. Horton, F. E. and W. E. Wagner, "A Markovian Analysis of Urban Travel Behavior: Pattern Responses by Socioeconomic-Occupational Groups", Highway Research Record, No. 283, 1969, pp. 19-29.

14. Sasaki, T., "Estimation of Person Trip Patterns Through Markov Chains", In Newell, G. F., ed., Traffic Flow and Transportation, American Elsevier, New York, New York, 1972, pp. 119-130.

15. Kondo, K., "Estimation of Person Trip Patterns and Model Split", In Buckley, D. J., ed., Transportation and Traffic Theory, Elsevier, New York, New York, 1974, pp. 715-742.

16. Gilbert, G., G. L. Peterson and J. L. Schofer, "Markov Renewal Model of Linked Trip Travel Behavior", Transportation Engineering Journal, Vol. 98, No. TE3, 1972, pp. 691-704.

17. Lerman, S. R., "The Use of Disaggregate Choice Models in Semi-Markov Process Models of Trip Chaining Behavior", Transportation Science, Vol. 13, 1979, pp. 273-291.

18. Kitamura, R., L. P. Kostyniuk and M. J. Uyeno, "Basic Properties of Urban Time-Space Paths: Empirical Tests", Paper presented at the 60th Annual Meeting of the Transportation Research Board, Washington, D.C., January 1981, (forthcoming Transportation Research Record).

19. Hagerstrand, T., "What About People in Regional Science?", Regional Science Association, Papers, Vol. 24, 1970, pp. 7-21.

20. Erdelyi, A., ed., Table of Integral Transforms, Vol. 1, McGraw-Hill, New York, New York, 1954.

21. Bentley, G. A., A. Bruce and D. R. Jones, "Intra-Urban Journeys and Activity Linkages", Socio-Economic Planning Sciences, Vol. 11, 1977, pp. 213-220.

22. Vidakovic, V., "A Study of Individual Journey Series: An Integrated Interpretation of the Transportation Process", In Newell, G. F., ed., Traffic Flow and Transportation, American Elsevier, New York, New York, 1972, pp. 57-70.

SOME FINDINGS AND AN OVERVIEW ON VEHICULAR FLOW CHARACTERISTICS

M. KOSHI
University of Tokyo, Japan

M. IWASAKI
Musashi Institute of Technology, Tokyo, Japan

I. OHKURA
Yokohama National University, Yokohama, Japan

Vehicular flow characteristics, especially those in a congested region, are dealt with in the paper, from both macroscopic and microscopic points of view. Data on traffic flows on Tokyo Expressways have been collected by various methods such as vehicle detectors, cine cameras and aerial photographs. Car-following spacings were measured by a test car method and roadside cine cameras on Tokyo Expressways. A car-following experiment of 29 cars was also conducted on a test ground. Various features of vehicular flow were revealed such as the oscillation characteristics in the congested region and the discrepancy in volume-density as well as in speed-spacing relationships between the free flow and the congested flow regions. A qualitative hypothesis is presented to explain why and how oscillations are generated.

INTRODUCTION

This paper deals with vehicular flow characteristics especially in a congested region, and attempts to describe what really happens and why.

It has been pointed out that vehicular flows oscillate in congested conditions and that there is a discrepancy in the speed-density-volume relationships between free flow and congested flow regions [1]-[6]. The phenomena, however, seem not to have been explored thoroughly enough to understand the total picture of vehicular flow characteristics.

The authors have been faced with difficulties in accurate congestion detection and travel time estimation in the course of developing motorway traffic surveillance and control systems, largely because of the disturbances due to the oscillations of congested flows. This experience motivated the authors to collect and analyze data on the phenomena for better understanding of congested flows.

In the first part of this paper, the macroscopic features of vehicular flows such as volume-density relationships and oscillations are described based on the data collected on Tokyo Expressways. In the middle part, car-following behavior as a microscopic aspect of traffic flow is analyzed for both real traffic observed on Tokyo Expressways and an idealized car-following experiment conducted on a test ground. In the last part, a hypothetical explanation for the traffic flow phenomena observed on Tokyo Expressways is proposed, based on the analyses presented earlier.

VOLUME-DENSITY RELATIONSHIPS

Data on vehicular traffic flows have been collected on Tokyo Expressways through various methods: 16 mm cine cameras, aerial photographs and vehicle detectors of the surveillance and control system. Table 1 summarizes the data on which this paper is based. Some of the vehicle detectors of the system are of the twin-loop type and others are of the single-loop type. Speed can be detected by the former and density then can be calculated, but only volume and occupancy are measured by the latter. Occupancy-density (calculated) relationships from the data of the former were plotted and a regression curve was obtained as shown in Fig. 1. The occupancy values measured by the single loop detectors were converted to density values using the regression curve.

Table 1 The List of Data

DATA #	METHODS	SECTIONS	LENGTH OF SECTIONS	MEASUREMENT INTERVAL	MEASUREMENT PERIOD	REFERENCES
I.1	16mm Cine Camera	Loop, Clockwise Entrance to Shiodome Tunnel	60 m	0.25 sec.	2 hrs.	(7)
I.2		Radial #7 Inbound Ohshimadanchi	120 m	0.5 sec.	10 hrs.	(8),(9)
I.3		Radial #5 Inbound	1800 m	0.25 sec.	8 hrs.	
I.4		Radial #4 Inbound Sendagaya	100 m	0.5 sec.	20 hrs.	(10)
II.1	Aerial Photograph	Radial #3 Inbound Tanimachi	1400 m	10 & 20 sec.	2.8 hrs.	(11)
II.2		Loop, Clockwise Kandabashi	800 m	10 sec.	36 min.	
III.1	Vehicle Detector	Radial #7 In- and Outbound	10.6 km	1 min.	74 hrs.	(8),(9)
III.2		Radial #1 Inbound	9.0 km	1 min.	168 hrs.	
III.3		Radial #3 In- and Outbound	11.8 km	1 min.	168 hrs.	
III.4		All of the Network	109 km	1 min.	38 hrs.	

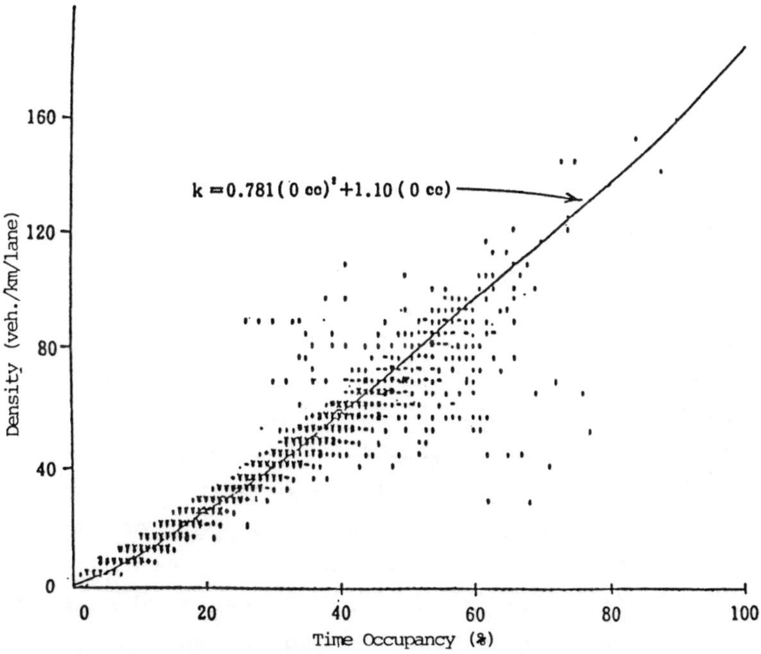

Fig. 1 One Min. Occupancy vs. Calculated Density Measured by Twin-Loop Vehicle Detectors (Radial #3 Inbound, Outer Lane)

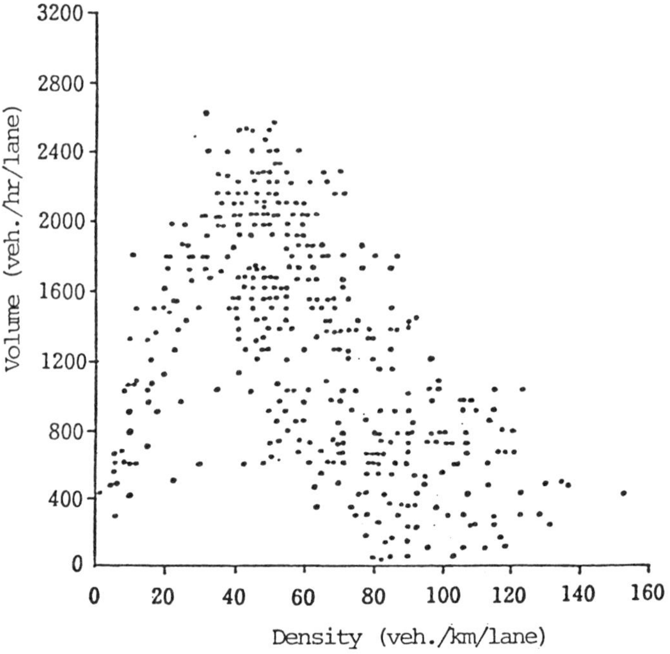

Fig. 2 One Min. Volume vs. Density (Measured by 16 mm Cameras, Data #1.4, Inner Lane)

Fig. 2 shows an example of volume-density relationships obtained from 16 mm cine photographs. It is seen that the volume-density plots are more dispersed in the congested region than in the free flow region. It is not quite clear whether or not there is a discontinuity between the two regions, perhaps because of insufficient data. Figs. 3 and 4 show the volume-density relationships obtained through the vehicle detectors for two different locations on the same route, 0.4 km and 5.7 km upstream of the bottleneck, respectively. In comparing these two figures, it can be seen that the discontinuity between the two regions becomes obvious at some distance upstream of the bottleneck, while the boundary between the regions is not clear immediately upstream of the bottleneck. It is also seen in Fig. 4 that the two regions of flow form not a single downward concave curve on the volume-density plane but a shape like a mirror image of the Greek letter lamda (λ). The other data listed in Table 1

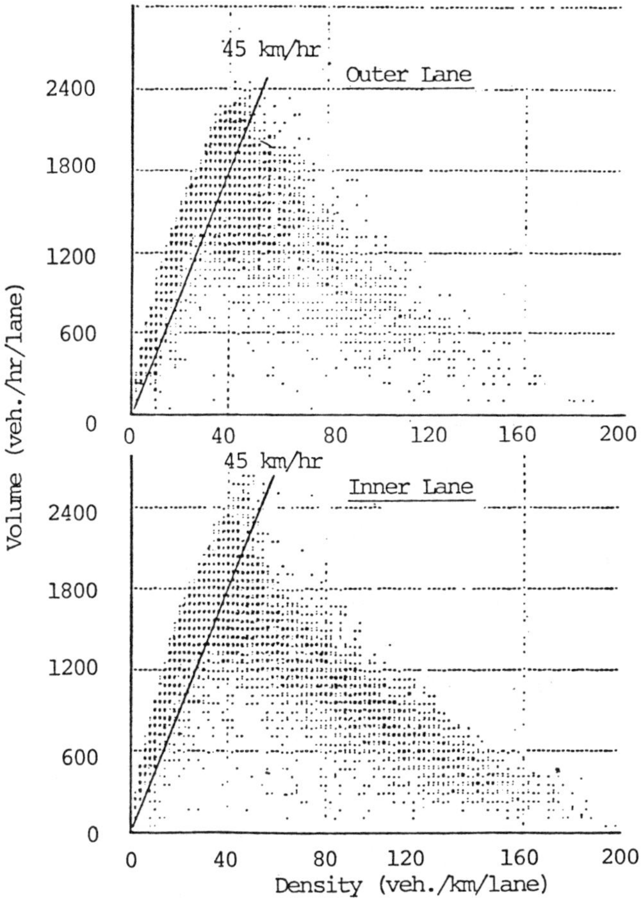

Fig. 3 One Min. Volume vs. Density at 0.4 km Upstream From Merging Point (Measured by Vehicle Detectors, Data #111.2)

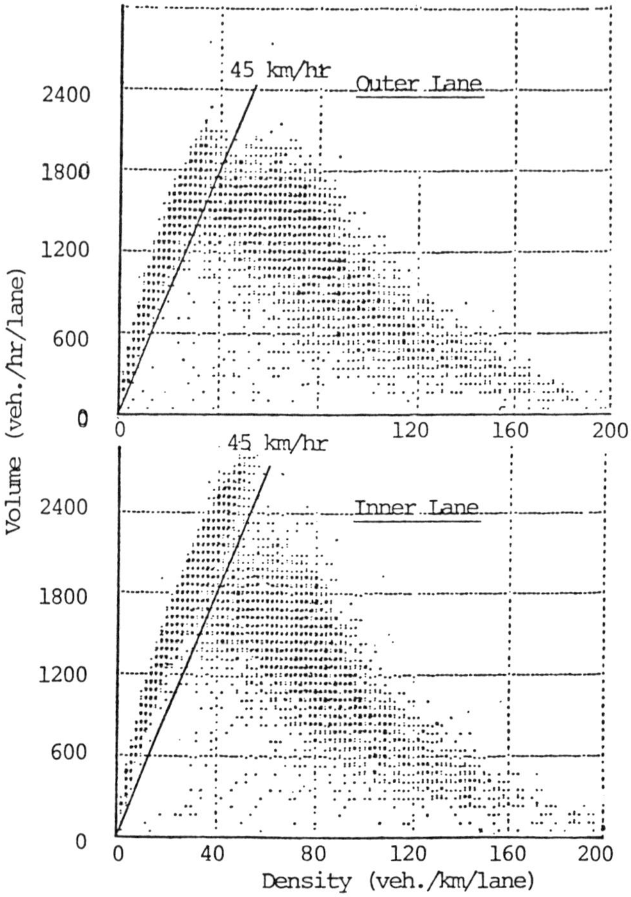

Fig. 4 One Min. Volume vs. Density at 5.7 km Upstream From Merging Point (Measured by Vehicle Detectors, Data #111.2)

also showed similar results. This implies that the speed-spacing relationships in car-following should differ between the two regions. This subject will be discussed later in this paper.

Another important feature to be pointed out in Figs. 2, 3 and 4 is that the volume-density curve in the congested region is not downward concave as represented by Greenshields' and Greenberg's models, but upward concave, which implies that oscillations with higher speed flow propagate faster than lower speed ripples of flow.

OSCILLATIONS IN CONGESTED FLOW

Fig. 5 illustrates a typical example of oscillations in congested flow on a 9 km section upstream of a merging point

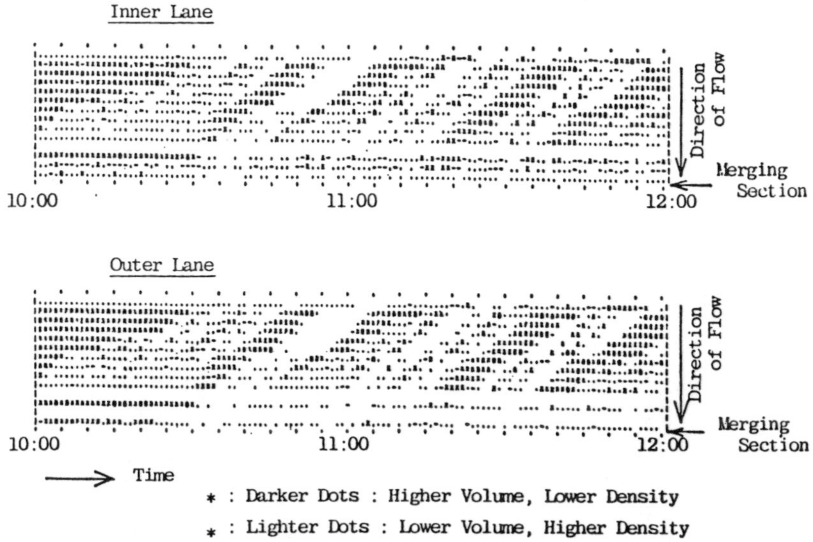

Fig. 5 Oscillation in Congested Conditions on Time-Space Plane (Measured by Vehicle Detectors, Data #111.2)

(Data III. 2 in Table 1). In the figure the values of products of 1 min. volume and mean speed are classified and indicated by means of darkness of the dots. The darker the dots, the higher are the flow rates and the speeds. It is seen that the predominant waves of flow on the two lanes of the roadway are synchronized and propagate upstream at the speed of 18 to 20 km/hr. Similar results were obtained from other sets of data.

Fig. 6 shows the fluctuations of 1 min. speeds of congested flow on another section. It is seen in the figure that (a) the oscillations are not very large in amplitude immediately upstream of the bottleneck but they are amplified as they propagate upstream until the speed of the lower density (higher speed) part of the waves reaches about 40 km/hr and (b) the oscillations of the two neighbouring lanes become more synchronous as they propagate upstream.

Fig. 7 shows a similar aspect of oscillations on the same section as Fig. 6 but for two different volume levels. The capacity of the toll gates which form the bottleneck was raised from 560 vph to 1400 vph at 12:00 noon because the number of toll collecting lanes was increased. (The lanes are controlled for the purpose of protecting the downstream sections from heavy congestion.) Fig. 7 and Fig. 6, which is for a volume level of 1600 vph, show that the shape (height and length) of the individual predominant high speed waves does not significantly change depending on the volume level whereas the frequency does. The high speed waves of similar shape appear less frequently at lower volume levels.

Fig. 8 compares the power spectra of 1 min. volume x speed of the two different volume levels, 560 vph and 1600 vph, in order to examine the frequency characteristics. It can be seen that there exists no common predominant frequency in the oscillations.

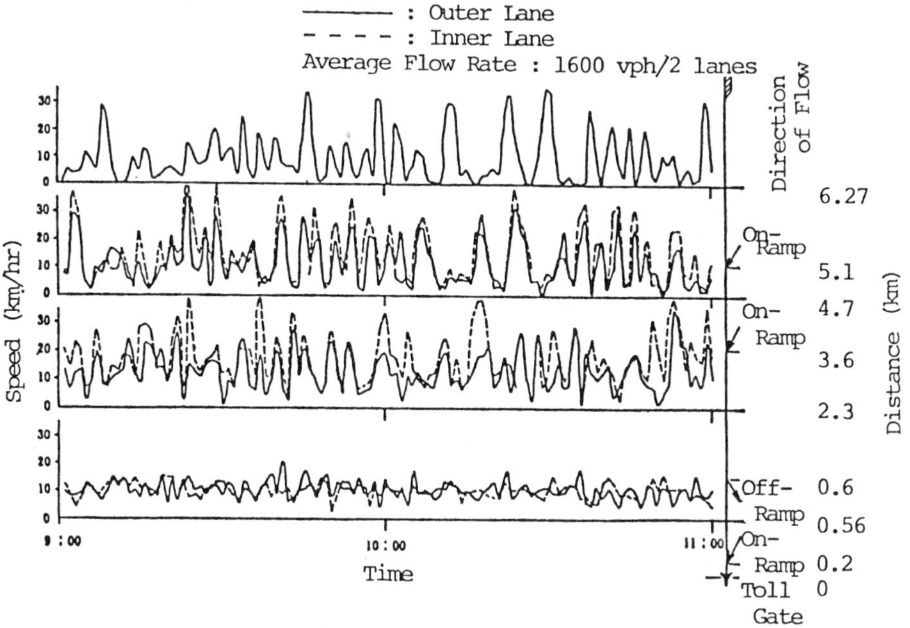

Fig. 6 Fluctuations of Speed in Congested Conditions (Measured by Vehicle Detectors, Data #111.1)

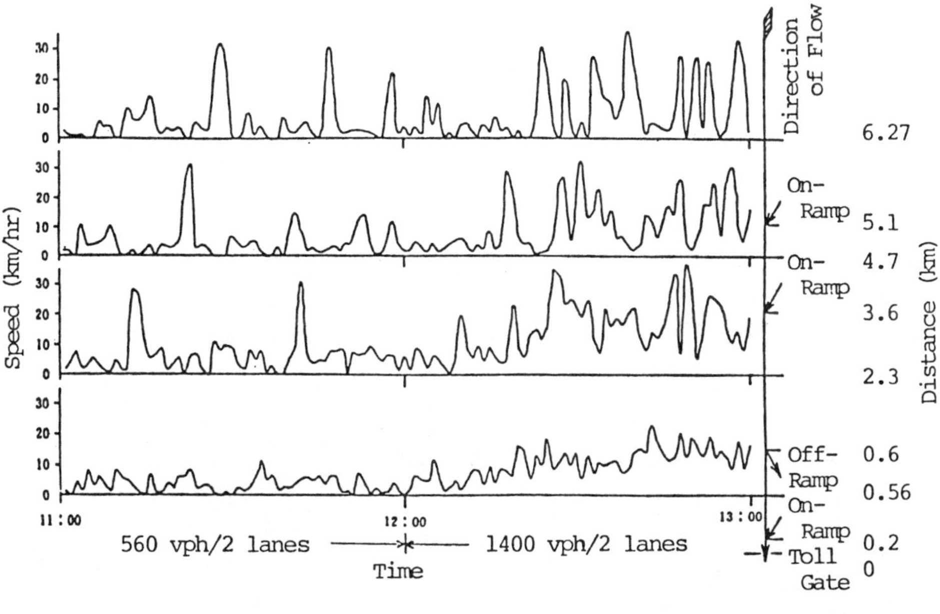

Fig. 7 Fluctuations of Speed in Congested Conditions (Measured by Vehicle Detectors, Data #111.1)

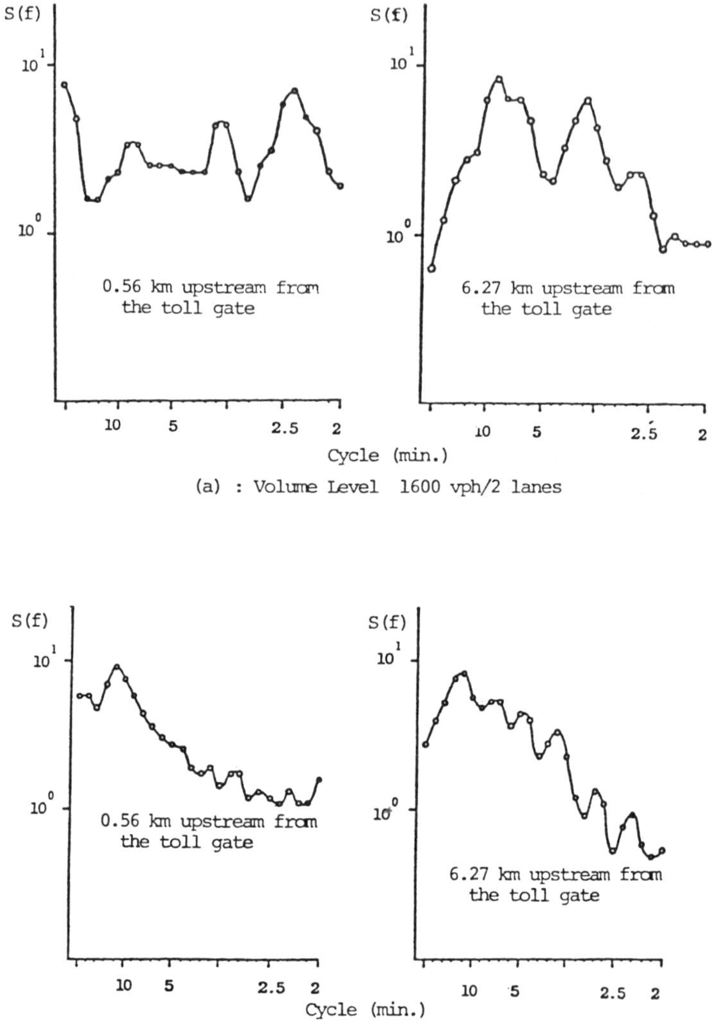

Fig. 8 Power Spectra of 1 Min. Volume X Mean Speed (Data #111.1)

CAR-FOLLOWING IN REAL FLOW

A test car equipped with a 16 mm cine camera pointing out the back window was driven on Tokyo Expressways and the longitudinal clearance between the test car and the car running behind was measured(12). Speed of the test car was also recorded on the films.

Fig. 9 shows speed-clearance plots of one of the cars which ran behind the test car. "A", "D" and the other marks show acceleration ($\geq$0.1g), deceleration ($\leq -$0.1g) and steady run (within $\pm$0.1g). It is seen that the speed-clearance relationship of even a single driver is not consistent on a single curve and the clearance varies in a wide range for given speeds. It is also seen that clearances are longer when accelerating than when decelerating.

Fig. 10 summarizes speed-clearance relationships of steady runs (within $\pm$0.1g) of all the 19 vehicles which followed the test car. There are very few data points at around 45 km/hr speed, which means that steady runs did not occur so often at and around this speed and suggests the discontinuity between free flow and congested flow. The relationship of speed vs. clearance may be regarded either as following a single monotonic curve as shown by the broken line in Fig. 10 or as made up of two different curves, one for congested flow and the other for free flow, as shown by the solid lines in Fig. 10. To the authors, however, the latter interpretation seems less biased. If this can be accepted, we can see a rather large discrepancy in clearance between the two curves at the boundary speed of about 45 km/hr.

The roadside observations by 16 mm cine cameras listed in Table 1 also gave speed-spacing data of real traffic.

Fig. 11 shows 1 min. average spacing vs. 1 min. average speed observed by a roadside 16 mm cine camera. The average speeds here are 1 min. averages of the travel speeds for a 60 m section and the average spacings are 1 min. averages of

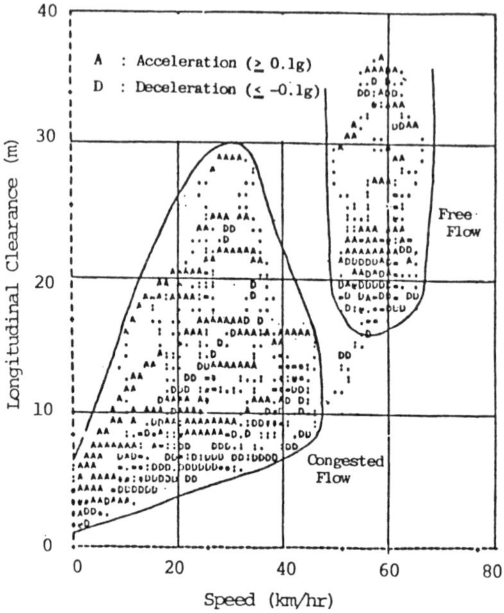

Fig. 9 Longitudinal Clearance of a Car Which Ran Behind the Test Car

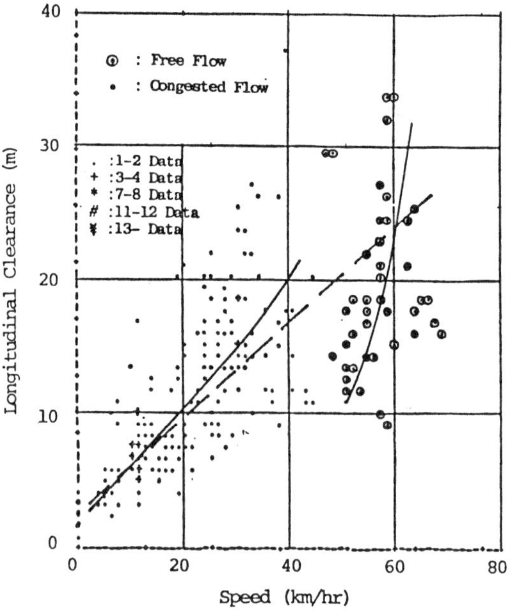

Fig. 10 Longitudinal Clearance vs. Speed of all 19 Vehicles Which Ran Behind the Test Car (-0.1 g Acceleration 0.1 g)

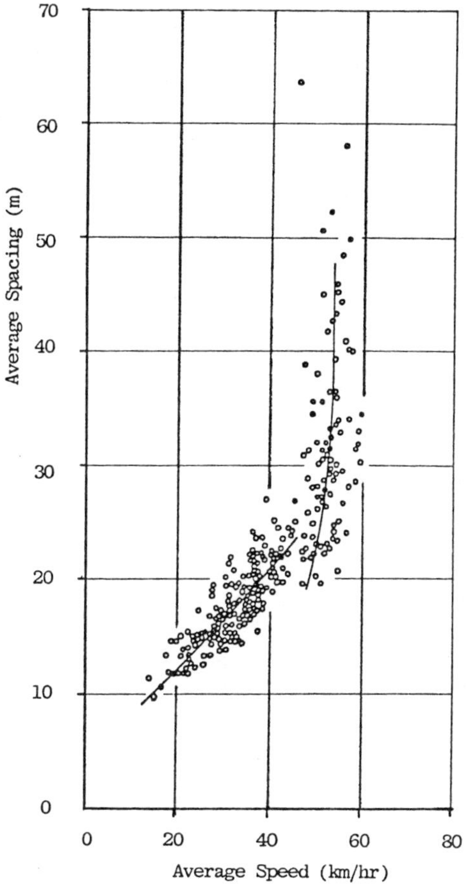

Fig. 11 1 Min. Average Spacing vs. Average Speed
(Measured by 16 mm Cameras, Data #1.1)

spacings observed at 10 second intervals. A discussion similar to that for Fig. 10 applies here, and there seem to the authors to exist two different curves of speed vs. spacing, one below 45 km/hr and the other above that speed.

It should be noted that the plots of spacings and clearances in a free flow region in both Figs. 10 and 11 do not necessarily indicate those of "car-following" because not all the vehicles are "followers" in free flow conditions. The average spacings of car-following in a free-flow region must be somewhat shorter than the average values in Figs. 10 and 11.

Fig. 12 is an image sketch of speed-spacing relationships. The volume-density expression of Fig. 12 would then be somewhat similar to Fig. 13, which resembles the observed phenomena shown in Figs. 3 and 4.

CAR-FOLLOWING IN EXPERIMENTAL CONDITIONS

A car-following experiment was carried out in December 1979 in order to make an in-depth examination of the car-following behaviour of drivers. A total of 29 passenger cars, one leader and 28 followers who were instructed to follow the vehicle ahead, were driven by student drivers on the test circuit of the Public Works Research Institute in Tsukuba. The circuit is level and 6,150 m in perimeter, with two curves 455 m and 295 m in radius. Cine pictures were taken by a 35 mm cine camera mounted on a helicopter which hovered over the circuit following the platoon.

Fig. 14 shows speed-spacing plots obtained from the experiment. In this figure, no discrepancy in the speed-spacing relationship is found. This point is quite different from in Figs. 10 and 11, which are the speed-spacing relationships observed in the real traffic flow on Tokyo Expressways.

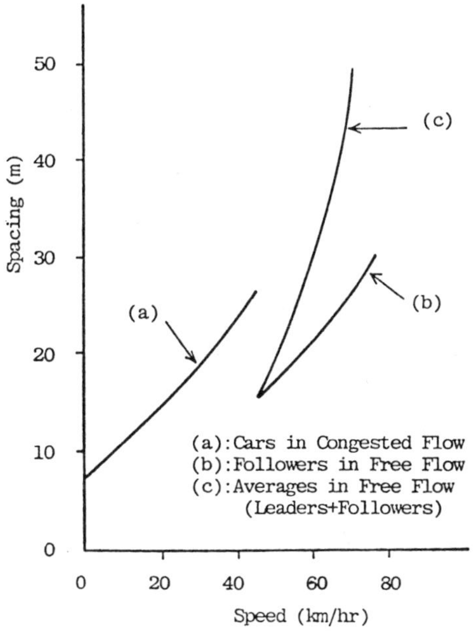

Fig. 12 Conceptual Sketch of Speed vs. Spacing Relationship

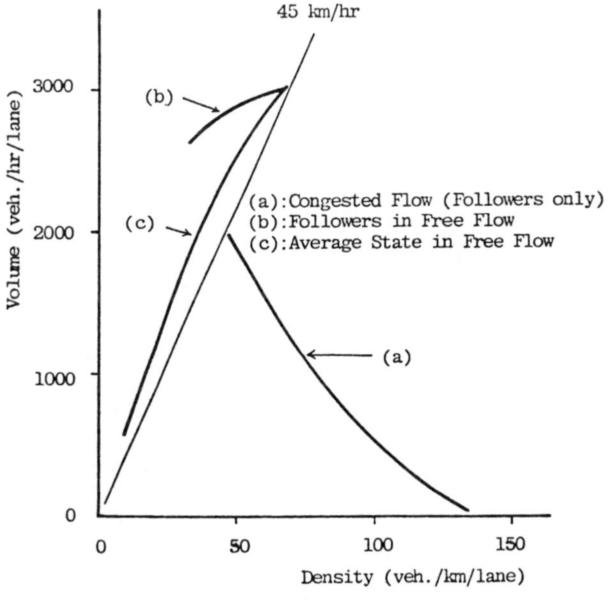

Fig. 13 Volume vs. Density Expression of Fig. 12

Fig. 15 compares the results from the Tsukuba experiment with those obtained from the real traffic measurements already shown in Fig. 10. The average car length of 4.2 m in the Tsukuba experiment was added to the clearance values shown in Fig. 10 in order to convert them to spacing values. The spacings in the Tsukuba experiment are found to lie at the lower limit of real flow spacings in the region below the speed of about 45 km/hr. In the region above that speed, however, the Tsukuba spacings seem to be close to the average car-following spacings of real traffic if one keeps in mind that not all vehicles are followers in free flow conditions in real traffic and some of the real flow spacings are not those of followers.

The only interpretation of the difference in spacing between the experimental and real flows in the congested region is that in congested conditions drivers do not seriously "follow" the cars ahead as closely as they do in steady flow conditions, perhaps because they know that they are in an oscillating flow and will be forced to decelerate again sooner or later. Because of this behavior of drivers average spacings are longer in congested conditions than in steady flow conditions, which is represented in Fig. 14.

It is also shown in Fig. 14 that the tangent from the origin touches the speed-spacing curve at the speed of about 45 km/hr, which implies that the maximum flow rate is obtained at this speed. Above this speed, flow rate increases as speed decreases, and below this speed flow rate decreases as speed decreases.

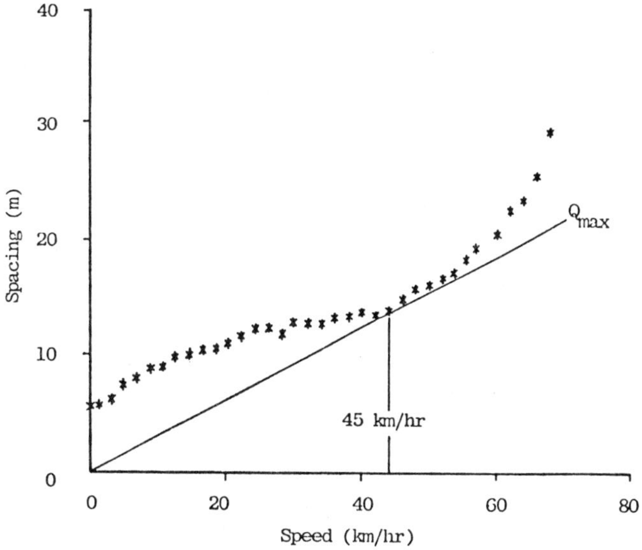

Fig. 14 Speed vs. Spacing Plots of the Car-Following Experiment in Tsukuba

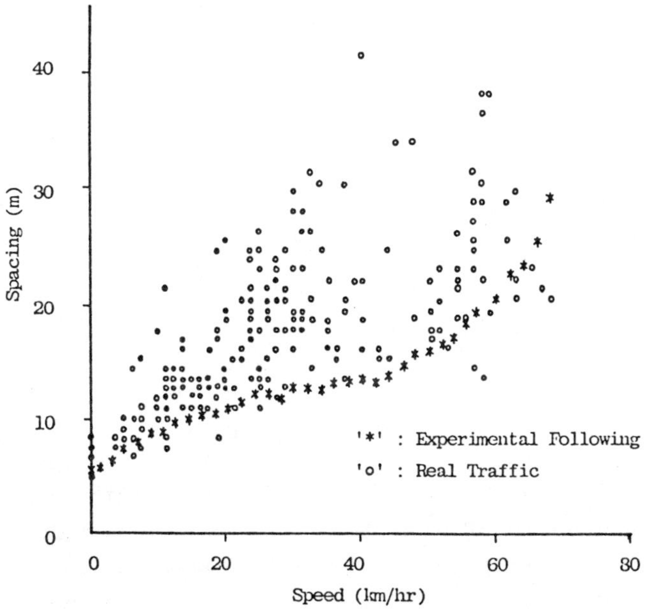

Fig. 15 Comparison Between Real and Experimental Spacings

HYPOTHETICAL EXPLANATION OF OSCILLATION IN THE CONGESTED REGION

The explanation of the cause and the process of oscillation in the congested region that will be presented here is still no more than a qualitative hypothesis. Further analyses of the traffic flow observation data and the car-following experiment results are now being made in order to obtain quantitative confirmation.

It is obvious that shock waves, once they are generated, propagate backward to the upstream traffic in congested regions where flow rate decreases as speed decreases or as density increases. Also there are probably slight fluctuations or "ripples" of speed at bottlenecks due to mere speed control errors of drivers, merging of ramp vehicles, stops at toll plazas or other reasons. The problems left to be answered, therefore, are as follows:

1) Why do the ripples of speed grow larger in terms of both height and length of the waves as they propagate upstream?

2) Why do the waves stop growing when they reach a certain size in height and length?

3) Why do the waves on the neighbouring lanes become synchronous as they propagate?

1. Why do the ripples grow?

The volume-density curve is upward concave in the congested region as shown in Figs. 2, 3 and 4. This implies that shock waves with higher speed propagate faster and higher speed ripples "swallow" those of lower speed. Fig. 16 illustrates this process. The shock wave S_{12} propagates

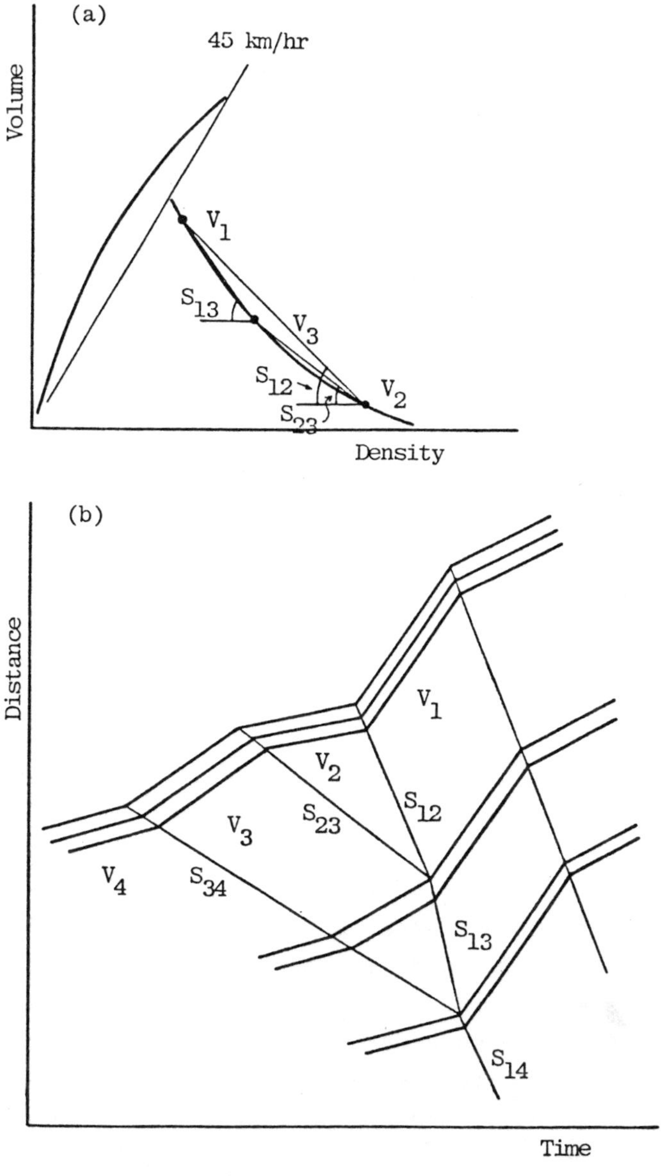

Fig. 16 Speeds of Shock Waves

faster than S_{23} as shown in Fig. 16 (a) and therefore takes over S_{23}. Then a new shock wave S_{13} is born which propagates even faster than S_{12} and takes over S_{34}. Ripples of lower speed are thus filtered out and are swallowed by those of higher speed resulting in fewer waves of larger length and higher speed.

Time lag in car-following is also a major factor which amplifies the ripples and makes the height of them grow.

2. Why do the waves stop growing at a certain size?

As mentioned earlier, the actual spacings in congested conditions are longer than those of pure car-following observed in the experiment, which are continuous with those in free flow car-following as shown in Fig. 15. The point here is the transition from the longer spacing state to the shorter. How do drivers transfer from the longer spacing state to the shorter one?

The authors' assumption is that this transition takes place at a certain level of speed which is close to the critical speed (about 45 km/hr), as shown in Fig. 17, after the driver has run at that speed for a certain period of time (perhaps of the order of several to ten seconds). If this assumption is correct, the size of waves (height and length) is kept within a certain limit such as shown in Figs. 6 and 7. When a driver starts to transfer from the longer to the shorter spacing state, the flow rate immediately increases and the propagation speed of the deceleration shock wave which lies ahead rises so that the next driver has to decelerate before he starts to transfer.

3. Why do the waves on the neighbouring lanes synchronize?

Two reasons can be thought of. One is a psychological

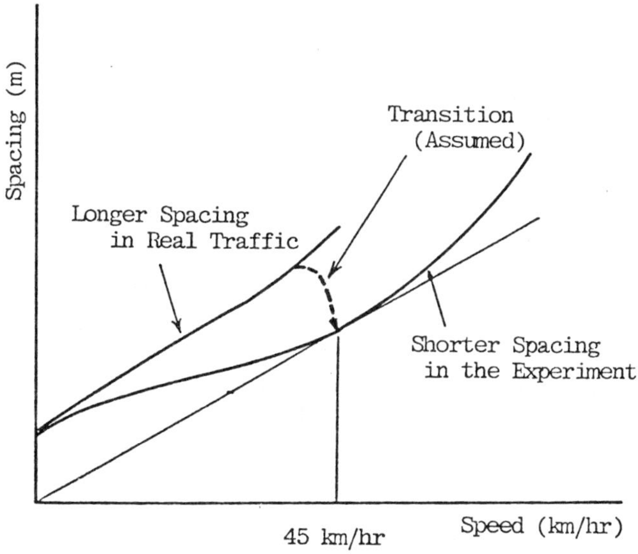

Fig. 17 Transition From the Longer Car-Following State To the Shorter

factor that drivers are motivated to move when the cars in the neighbouring lane start moving. The other is an effect of lane changing. Most lane changes in oscillating conditions are moves from the lower to the higher speed lanes, and therefore also act as a synchronizing factor.

SUMMARY AND CONCLUSIONS

1) The volume-density relationship of vehicular flow was found not to follow a single and continuous downward concave curve as has often been used but to have a shape like a mirror image of the Greek letter λ, in which the free flow and congested flow regions have separate volume-density relationships. Free flow has a downward concave curve, whereas congested flow has an upward concave curve.

(2) Correspondingly, speed-spacing relationships of real flow seem to be comprised of separate curves which are discrepant from one another.

(3) The boundary of these two regions of flow lies at the speed of about 45 km/hr.

(4) In the experimental conditions, however, no discrepancy in the speed-spacing relationship was found between the two regions. Drivers seem to adopt longer spacings in real congested flow than they do in the experimental car-following conditions.

(5) Oscillations in congested flows are not uniform over the congested road section. There are only ripples of small amplitude immediately upstream of the bottleneck. As they propagate upstream, the ripples grow to waves of larger amplitude, but become fewer in number.

(6) There seems to exist a limit on the size of high speed parts of oscillation waves in terms of height (maximum speed) and length (time duration). The waves stop growing when they reach a certain size which is seemingly the limit.

(7) The above mentioned limit of the size of high speed parts does not vary very much depending on volumes, but their frequency does.

(8) The high speed parts of oscillation waves in neighbouring lanes tend to synchronize as they propagate upstream.

(9) A qualitative hypothesis on the reason and the process of the oscillation phenomena is presented in this paper. Further analyses are necessary for a quantitative confirmation.

REFERENCES

1. H.S. Mika, J.B. Kreer and L.S. Yuan, "Dual Mode Behavior of Freeway Traffic," Highw. Res. Rec., No. 279, pp. 1-13, 1969.

2. T. Lam and R.W. Rothery, "The Spectral Analysis of Speed Fluctuations on a Freeway," Transp. Science, Vol. 4, No. 3, pp. 293-310, 1970.

3. B.D. Hillegas, D.G. Houghton and P.J. Athol, "Investigation of Flow-Density Discontinuity and Dual Mode Behavior," Transp. Res. Rec., No.495, pp.53-63, 1974.

4. L.C. Edie, "Car-Following and Steady State Theory for Noncongested Traffic," Opers. Res., Vol. 9, No. 1, pp. 66-76, 1961.

5. A. Ceder and A.D. May, "Further Evaluation of Single- and Two-Regime Traffic Flow Models," Transp. Res. Rec., No. 567, pp. 1-15, 1976.

6. A. Ceder, "A Deterministic Traffic Flow Model for the Two-Regime Approach," Transp. Res. Rec., No. 567, pp.16-30, 1976.

7. Japan Soc. of Traffic Engs., "A Study for Smoother and Safer Traffic on Tokyo Expressways" (in Japanese), pp.78-101, March, 1978.

8. Japan Soc. of Traffic Engs., "A Study for Future Traffic Control System on Tokyo Expressways" (in Japanese), pp. 76-120, March, 1977.

9. Japan Soc. of Traffic Engs., "New Traffic Control Methods on Tokyo Expressways" (in Japanese), March,

1976.

10. Metropolitan Expressway Public Cooporation, "An Analysis of Traffic Phenomena" (in Japanese), Jan., 1977.

11. Metropolitan Expressway Public Cooporation, "An Analysis of Congested Traffic Flows by Aerial Photogrammetry" (in Japanese), March, 1977.

12. K. Kuwata, "The Car-Following Characteristics in the Congested Traffic Flows" (in Japanese), M. Eng., Master Thesis at University of Tokyo, Feb., 1976.

ON A SIMULATION MODEL FOR THE TRAFFIC STREAM IN A FREEWAY MERGING AREA

Yasuji Makigami, Tsunehiko Nakanishi
Masahito Toyama & Ryoichi Mizote

Department of Engineering and Science
Ritsumeikan University

This report describes a merging traffic simulation model developed based on a series of surveys conducted on the merging areas between the Central Ring Road and radial routes of the Hanshin Expressway in Osaka. Applying the theory of the Markov process to lane changes in the merging area, the model simulates the traffic stream characteristics in each lane of the merging area. The model also traces the propagation of congested areas using the compressible fluid theory. After simulating the actual traffic conditions, the model is used to estimate an allowable input demand as well as to evaluate the effect of control schemes on the merging traffic flow.

1. INTRODUCTION

As is often the case with expressway systems in a metropolitan area, urban expressway systems form complicated networks which consist of radial, ring and circumferential routes. The nodes of the network, where traffic merges and diverges, often become the bottlenecks of the system. In the case of an expressway system which consists of only one central ring road and several radial routes merging into it, the function of the total system might be damaged if a traffic jam, started from any node, is extended upstream toward the central ring road. The objective of the study is to develop a methematical model which simulates traffic flow characteristics in the merging zones of urban freeway junctions, in order to evaluate traffic operation techniques related to the freeway merging zone. The freeway merging model was developed based on the results of traffic surveys, with time lapse cameras and video television cameras, of the merging areas in the junctions connecting radial routes and the central belt of the Hanshin Expressway in Osaka.

2. THE PROCEDURE OF THE STUDY AND ASSUMPTIONS FOR THE MODEL

a. The Procedure of the Study and Outline of the Model

The study started in August 1978 with activities of the Committee of Traffic Control for the Hanshin Expressway. The committee conducted a survey to evaluate the effectiveness of merging traffic control techniques, such as the control of the use of lanes. The location of the traffic survey was the merging zone of the Moriguchi-Line Junction and the Airport-Line Junction, both on the Central Ring of the Hanshin Expressway network as shown in Fig. 1. Based on the results of the survey, the study proceeded towards developing a computerized mathematical model which simulates the traffic stream on the merging zone of a freeway junction. Sponsored by Japan Highway Public Corporation from September 1980, the study was continued to the status reported in this paper, with a second traffic survey for confirming merging traffic stream characteristics.

The freeway merging model consists of two submodels: 1) a free-flow model and 2) a congested-flow model. The free-flow model computes the traffic volume on each point of each lane through the merging area for a given traffic demand, and also the traffic density and running speed for the corresponding point for each lane. The congested-flow model simulates the movement of a shock-wave in the traffic stream.

b. The Condition of the Roadway and Traffic, and the Assumptions for the Model

It is necessary to make the roadway and traffic conditions clear in the first place, and then to formulate them

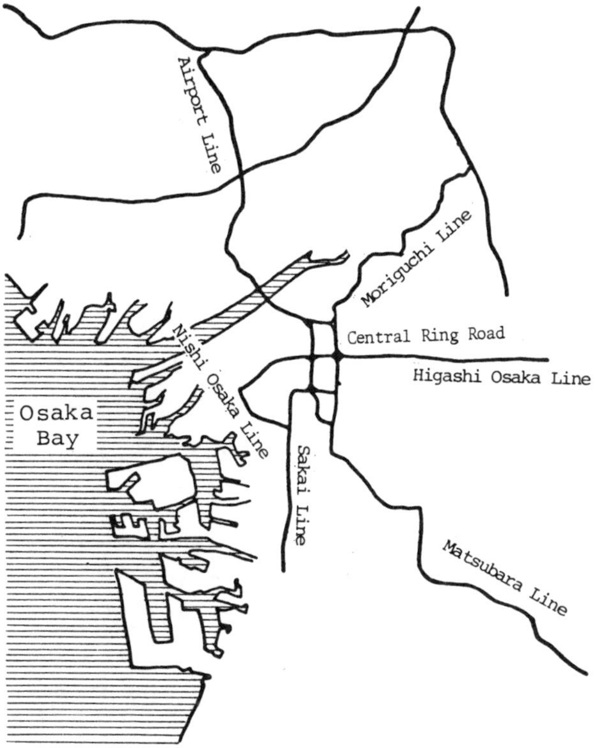

Fig. 1 Network of the Hanshin Expressway

numerically in order to make a simulation for merging traffic. In order to introduce the roadway conditions (mainly the geometric properties of the merging area) into the computation process of simulation, the section of the expressway which contains the merging zone of a junction is divided and marked as shown in Fig. 2. Each lane is marked A, B or C, except the taper-shaped merging lane, and each lane is divided every 30 to 40 meters, starting from the tapered end and ending up at the nose-end of the zebra zone. Stations are numbered from "1" to "n" as shown in Fig. 2.

Following the numbering system, variables describing traffic stream characteristics, such as capacities and lane traffic volume for each lane, are defined as follows:

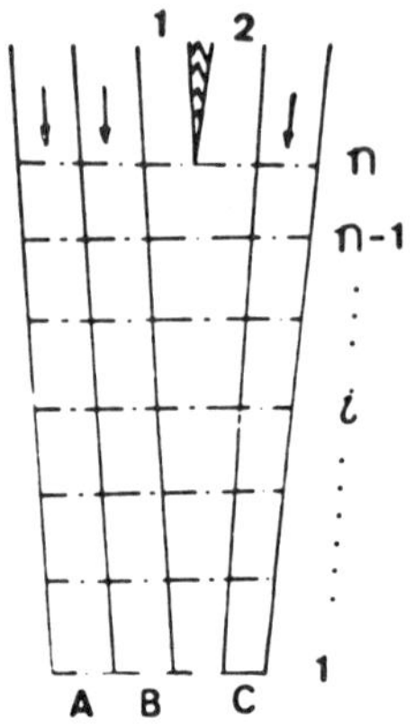

Fig. 2 Coding System of the Merging Zone

$C_x(i)$ = Capacity for lane x at station i (vehicles per hour)

$k_x(i)$ = Traffic density for lane x at station i (vehicles) per kilometer),

$q_x(i)$ = Traffic volume for lane x at station i (vehicles per hour),

$v_x(i)$ = Average running speed for lane x at station i (kilometers per hour),

where x is A, B, C, or blank corresponding to the lane marking system. All variables except capacities are marked with a prime, $k'_x(i)$, $q'_x(i)$ and $v'_x(i)$, if the lane x is congested at station i.

The assumptions for the model structure are as follows:
(i) Traffic is treated as a compressible fluid. Each vehicle is regarded as an integral part of the fluid and is not considered individually. Furthermore, a linear relationship is assumed between the average running speed and the traffic density in that compressible fluid.
(ii) Within a given time interval, traffic flow remains constant and does not fluctuate over that time interval. In order to simulate traffic demand fluctuation, traffic

demands are input to the merging zone as a step function over the entire time period under consideration.

(iii) Unless the merging area is congested, the lane changing maneuvers between neighboring lanes in the merging zone occur according to the transition probabilities determined for each section from the collected data. That is, the lane changing phenomenon on the merging area is assumed to be a Markov process.

(iv) When congestion develops in a lane in the merging area, the number of lane changes between the congested lane and the neighboring lanes is assumed to be a function of the difference of the traffic densities of the two neighboring lanes.

3. LOGIC IN THE COMPUTATION PROCESS

a. Free-Flow Model

(i) Computation of lane traffic volumes

If the collected data for a merging zone gives the lane distribution rate of traffic volume at the upstream end of the merging zone, which is station n in the numbering system shown in Fig. 2, then the lane traffic volume for each lane is obtained by multiplying the in-put demand by the lane distribution rate. Let $\mathbf{Q}(n) = [q_A(n), q_B(n), q(n), q_C(n)]$ be the lane traffic volumes at station n for lane A, lane B, merging lane, and lane C respectively, and furthermore, let $\mathbf{P}(i)$ be the transition probability matrix for lane changes between stations (i) and $(i-1)$ as shown below:

$$\mathbf{P}(i) = \begin{array}{c} \\ \text{Lane-A} \\ \text{Lane-B} \\ \text{Merging} \\ \text{Lane-C} \end{array} \begin{array}{c} \begin{array}{cccc} \text{A} & \text{B} & \text{Merging} & \text{C} \end{array} \\ \begin{bmatrix} {}_AP_A(i) & {}_AP_B(i) & 0 & 0 \\ {}_BP_A(i) & {}_BP_B(i) & {}_BP(i) & 0 \\ 0 & P_B(i) & P(i) & P_C(i) \\ 0 & 0 & {}_CP(i) & {}_CP(i) \end{bmatrix} \end{array} \qquad (1)$$

where ${}_XP_Y(i)$; the probability of lane change from lane X to Y between station (i) and $(i-1)$.

Then, the relationship between the lane traffic volume at station i and $(i-1)$ can be described as follows:

$$\mathbf{Q}(i-1) = \{q_A(i-1),\ q_B(i-1),\ q(i-1),\ q_C(i-1) = \mathbf{Q}(i)\ \mathbf{P}(i).$$

Therefore $\mathbf{Q}(i) = \mathbf{Q}(n) \cdot \prod_{j=i+1}^{n} \mathbf{P}(j)$,

where $\mathbf{Q}(i)$ = the lane traffic volume at station i.

(ii) Calculation of the average speed and traffic densities

Let k_j and v_f be the jam density and the free running speed respectively; then the relationship between the traffic density and the average running speed for the normal lanes and the merging lane is expressed as the Greenshield linear relation following the assumption made before. Therefore,

for normal lanes, $V_x(i) = V_f - \frac{V_f}{k_j} \cdot k_x(i)$

for the merging lane, $V(i) = V_f - \frac{V_f}{k_j\ (1+\frac{i-1}{n-1})}\ k(i)$ (3)

where x = A or B or C

In the second part of equation (3), it is assumed that the jam density varies with the station number in proportion to the merging lane width.

Once the input demand from station is given, the lane traffic volume for each station can be computed by Equation (2); then the traffic density at each point can be derived from the continuity equation as follows:

for normal lanes

$$k_x(i) = \frac{k_j}{2v_f}\left\{v_f \pm \left(v_f^2 - \sqrt{\frac{4q_x(i)\cdot v_f}{k_j}}\right)\right\} = \begin{cases} k_{frex}\{q_x(i)\} \\ k_{jamx}\{q_x(i)\} \end{cases}$$

for the merging lane

$$k(i) = \frac{k_j(1+\frac{i-1}{n-1})}{2v_f}\left\{v_f \pm \sqrt{v_f^2 - \frac{4q(i)\cdot v_f}{k_j(1+\frac{i-1}{n-1})}}\right\} = \begin{cases} k_{free}\{q(i)\} \\ k_{jam}\{q(i)\} \end{cases}$$

where $k_{free}\{q(i)\}$ and $k_{frex}\{q(i)\}$ indicate the traffic density as a function of lane traffic volumes for the merging lane and the other lanes in free traffic flow conditions, and $k_{jam}\{q(i)\}$ and $k_{jamx}\{q_x(i)\}$ indicate the traffic density in congested traffic flow conditions.

b. Congested Flow Model

(i) Formulation of shock wave behavior

The free flow model simply computes the lane volumes for each station of the merging zone for a given in-put traffic demand according to the transition probability matrix for lane changes. However, once the calculated lane volume exceeds the lane capacity at any station of a lane (or lanes) in the merging area, the congestion starts extending up-stream from the excessive demand section. The Congested Flow Model checks the demands and capacities for every station of each lane first; then, if the demand exceeds the capacity at any station in the merging zone, the model pursues the behavior of the shock wave.

The congested model first compares the lane capacities with the lane traffic volumes, which are given as the output of the Free Flow Model for every station of each lane, in order to find the stations with excess demand. If two or more excess demand stations are found, it is assumed that the congestion starts up-stream from the station with the minimum capacity among those excess demand stations. Following this assumption, the congestion usually starts forming at the downstream end of the merging lane where the lane capacity is the lowest; this phenomenon is often observed in actual merging traffic. For a brief explanation, let's consider a level of in-put demand in which the demands exceed the capacities in the merging lane, but there is no excess demand in the other lanes. Then the congestion starts at the downstream end of the merging lane, that is, in station 1 . Furthermore let's consider the instant in which the shock wave passes through the station i of the merging lane.

The traffic volume in the congested region of a merging lane at section i is calculated by

$$q'(i) = c(1) + \sum_{j=1}^{i-1} \left\{ d_B(j) + d_C(j) \right\}, \qquad (5)$$

where $d_B(j)$ and $d_C(j)$ are the hourly number of vehicles that change lane from the congested merging lane to adjacent lanes B and C, respectively, between station $(j-1)$ and (j). Both are calculated from a function of the difference of traffic densities of adjacent lanes as shown below:

$$d_x(j) = \alpha \left\{ \frac{k'(j) + k'(j-1)}{2} - \frac{k_x(j) + k_x(j)}{2} \right\}^{\beta} \quad (x = B \text{ or } C) \qquad (6)$$

The density of a merging lane can be computed from Eq(4) and Eq(5) as $k'(j) = k_{jam}\{q'(j)\}$.

The traffic density of lanes B and C at station i can be computed by applying lane traffic volumes obtained from the Free Flow model to $k_{frex}\{q_x(i)\}$ of Eq(4). However, for the stations down-stream of station i, lane traffic volumes for lanes B and C should be modified by replacing the number of lane changes according to the transition matrix in free flow stage by the number of lane changes from the congested lane to the adjacent lanes.

Let $\mathbf{Q}^{(0)}(i) = (q_A^{(0)}(i), q_B^{(0)}(i), q^{(0)}(i), q_C^{(0)}(i))$

be the lane traffic volume at station (i) that is obtained from the Free Flow Model. Leaving aside the lane change from the congested merging lane, the transition matrix $\mathbf{P}(j)$ between station j and station j-1 should be replaced by $\mathbf{P}'(j)$ which is

$$\mathbf{P}'(j)=\begin{array}{c} \begin{matrix} \text{Lane-A} & \text{Lane-B} & \text{Merging} & \text{Lane-C} \end{matrix} \\ \begin{bmatrix} {}_AP_A(j) & {}_AP_B(j) & 0 & 0 \\ {}_BP_A(j) & {}_BP_B(j)+{}_BP(j) & 0 & 0 \\ 0 & 0 & P_B(j)+P(j)+P_C(j) & 0 \\ 0 & 0 & 0 & {}_CP(j)+{}_CP_C(j) \end{bmatrix} \end{array}$$

$$=\begin{bmatrix} {}_AP_A(j) & {}_AP_B(j) & 0 & 0 \\ {}_BP_A(j) & 1-{}_BP_A(j) & 0 & \\ 0 & 0 & 1 & 0 \\ 0 & 0 & 0 & 1 \end{bmatrix} \tag{7}$$

Then the lane traffic volume without lane changes between the congested merging lane and adjacent lanes $\mathbf{Q}^{(0)}(j)$ is calculated as follows:

$$\begin{aligned} \mathbf{Q}^{(0)}(j) &= \mathbf{Q}^{(0)}(i)\cdot \mathbf{P}'(i-1)\cdot \mathbf{P}'(i-2)\ldots\ \mathbf{P}'(j) \\ &=(q_A{}^{(0)}(j),\ q_B{}^{(0)}(j),\ q^{(0)}(j),q_C{}^{(0)}(j)). \end{aligned}$$

Then the traffic density for lane B and C both in free flow conditions in Eq(4) is given by:

$$\left.\begin{aligned} &k_x(j) = k_{frex}\{q_x(j)\} \\ &\qquad \text{where } q_x(j) = q_x{}^{(0)}(j) + \sum_{k=j}^{i-1} d_x(k) \\ &\qquad\qquad (x=B \text{ or } C,\ j=1,\ 2,\ldots i-1) \end{aligned}\right\} \tag{8}$$

Actually $d_x(j)$, that is the hourly number of vehicles which change lanes from the congested merging lane to adjacent lanes, cannot be computed without related lane traffic densities. However, as shown by Eq(8), traffic density for lanes B and C varies with $d_x(j)$ itself. Therefore, it is necessary to iterate computations between Eq(5) and (8) starting from the condition $d_x(j)=0$ in order to fix lane traffic volumes and traffic densities as well as $d_x(j)$.

The velocity of the shock wave $S(i)$ at the moment when it crosses station i is calculated as follows:

$$S(i) = \frac{q^{(o)}(i) - q'(i)}{k_{free}\left\{q^{(o)}(i)\right\} - k'(i)} \tag{9}$$

Therefore, the period of time T_i necessary for the shock wave to traverse the merging lane from station i to station $(j-1)$ is:

$$T_i = \sum_{j=2}^{i-1} [2\ell / \left\{S(j-1)+S(j)\right\}]$$

where ℓ is the distance between stations and $S(1) = S(2)$.

Thus, it is possible to follow the behavior of the shock wave for given levels of input demand. Furthermore, as the shock wave moves upstream, the traffic volumes of the adjacent lanes are increased by the lane changing vehicles from the congested merging lane as indicated from Eq(8), causing congestion in the adjacent lanes. That situation can also be simulated with a method similar to that described above.

4. OUTLINE OF THE TRAFFIC SURVEY AND THE VALUES OF TRAFFIC STREAM CHARACTERISTICS FOR THE MODEL

a. Outline of the Traffic Survey

In order to develop the merging model, traffic surveys were conducted twice on the merging area between the Radial Routes and the Central Ring of the Expressway system in Osaka.

The survey was conducted at the merging points of the Central Ring with the two radial routes; these are the airport line and the Moriguchi Lane. Two time lapse cameras and two television cameras as well as the manual count crew were placed on the rooftop of tall buildings near the merging zone. The survey was conducted on a weekday in September 1980 from 6:00 AM to 11:00 AM for the Moriguchi Junction and from 1:00 PM to 4:00 PM for the Airport line. Fig.3 shows the location of the survey station at the Moriguchi merging zone. In Table 1, the objectives, methods and main instruments for the traffic survey are summarized briefly.

b. <u>The Values of the Traffic Stream Characteristics for the Model</u>

(i) The speed-density relationship

The merging model is based on the assumption of a linear relationship between the speed and the traffic density. The speed-density relationships estimated from the observed data are shown in Fig.4. All the points in Fig.4 are plots of 5 minute average density versus 5 minute average running

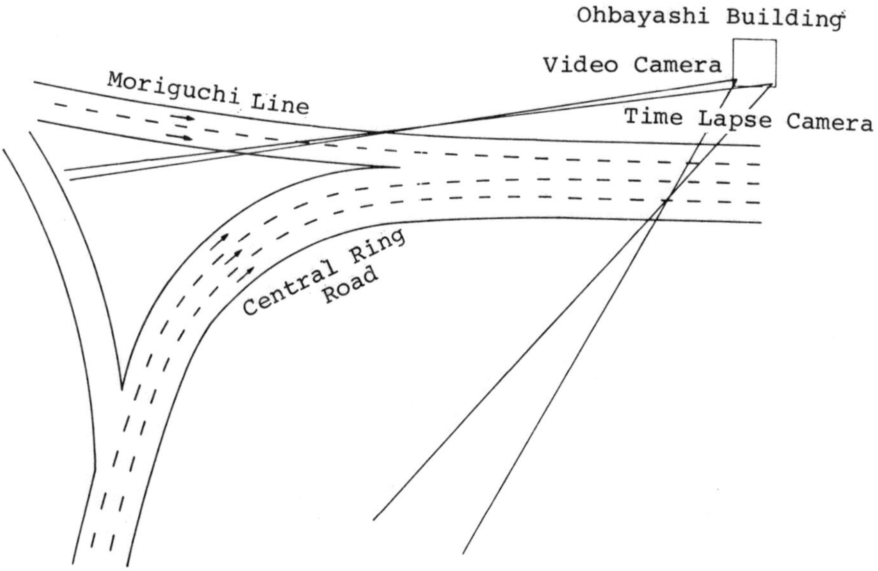

Fig. 3 Location of the Survey Station At the Moriguchi Junction

Table 1
The Objectives, Methods & Main Instruments of the Traffic Survey

Item	Instruments	Method	Objectives	Remarks
Time lapse Camera	-Bolex .H16 Reflex -Fujica Single 8 Momo Motion Camera	-Time lapse shooting of one Frame per second using 16 mm and 8 mm Color Film for 50 min. with 10 min. pause	-Measurement of Traffic Density -Measurement of Lane Changes -Measurement of Lane Changes from Congested Lane	-Supplemental use of 8 mm camera
Video Camera	Video Camera Type -Sony AVC 3500 -Pulse Height Analizer E551 NAIG	-Record on Black and White Video Tape for 25 min. with 5 min. pause	-Measurement of Average Running Speed	-One Camera for Up-stream Merging Zone and One for down-stream Merging Zone
Manual Count	Manual Counter one for each Lane	-5 min. Manual Lane Volume Count	-Measurement of continuous vol-ume count to record In-put Demand Varia-tion with time	-data also collect-ed from Traffic Control and Surveillance System

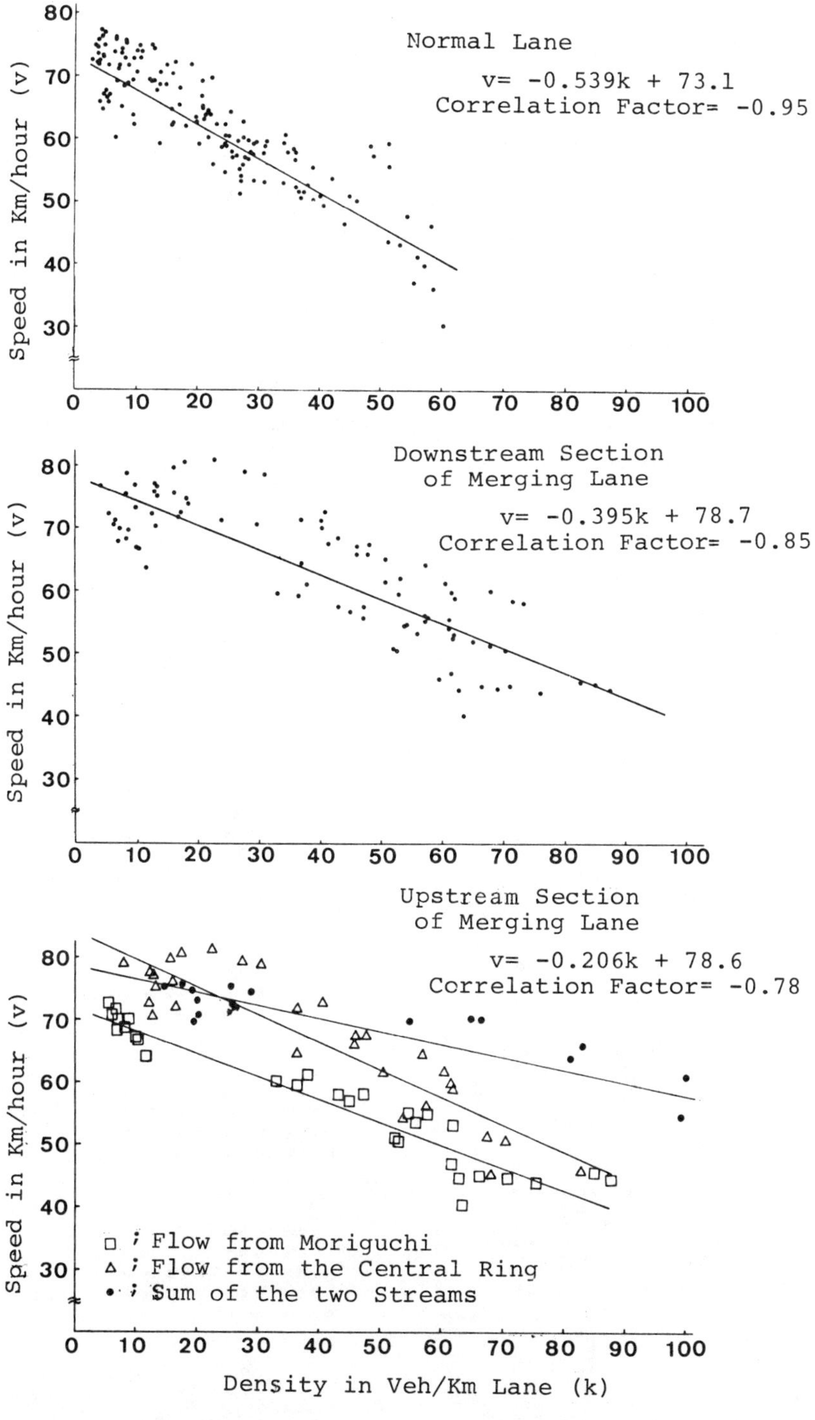

Fig. 4 Relationship between Speed and Density

speed for the upstream half and down-stream half of the merging lane as well as for the normal lane.

The 5 minute average density is the average of the density measurement made every ten seconds from a time lapse film. The 5 minute average running speed is measured from the television monitor by placing two photodetectors and measuring the time interval for each vehicle to traverse the two points. Judging from Fig. 4, the value of 75 kilometers per hour for the free flow speed and the value of 147 vehicles per kilometer per lane for the jam density were selected for the model.

(ii) The transition probability matrix for lane changing

Fig. 5 shows the results of film reading of the lane use of vehicles traversing the study area. The volume distribution rates over lanes are shown for each station as a percent, and the lane changing rates are the ratios of the

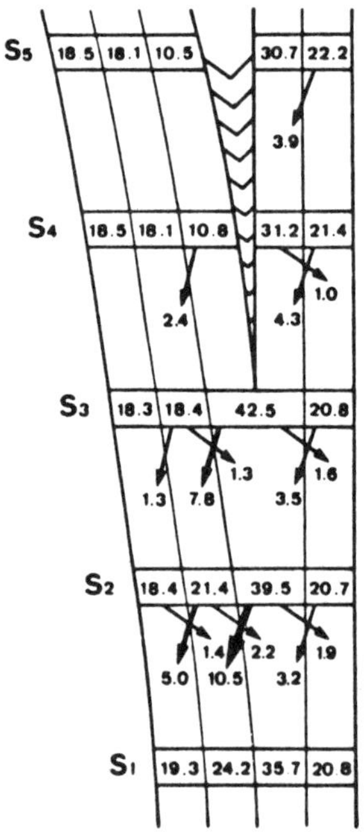

$$\mathbb{P}_{(4)} = \begin{matrix} & A & B & 1 & 2 & C \\ A & 0.995 & 0.005 & 0.000 & 0.000 & 0.000 \\ B & 0.001 & 0.995 & 0.004 & 0.000 & 0.000 \\ 1 & 0.000 & 0.004 & 0.996 & 0.000 & 0.000 \\ 2 & 0.000 & 0.000 & 0.009 & 0.989 & 0.002 \\ C & 0.000 & 0.000 & 0.000 & 0.039 & 0.961 \end{matrix}$$

$$\mathbb{P}_{(3)} = \begin{pmatrix} 0.991 & 0.009 & 0.000 & 0.000 \\ 0.002 & 0.992 & 0.006 & 0.000 \\ 0.000 & 0.024 & 0.976 & 0.000 \\ 0.000 & 0.000 & 0.990 & 0.010 \\ 0.000 & 0.000 & 0.043 & 0.957 \end{pmatrix}$$

$$\mathbb{P}_{(2)} = \begin{pmatrix} 0.991 & 0.009 & 0.000 & 0.000 \\ 0.013 & 0.974 & 0.013 & 0.000 \\ 0.000 & 0.078 & 0.906 & 0.016 \\ 0.000 & 0.000 & 0.035 & 0.965 \end{pmatrix}$$

$$\mathbb{P}_{(1)} = \begin{pmatrix} 0.986 & 0.014 & 0.000 & 0.000 \\ 0.050 & 0.928 & 0.022 & 0.000 \\ 0.000 & 0.105 & 0.876 & 0.019 \\ 0.000 & 0.000 & 0.032 & 0.968 \end{pmatrix}$$

Fig. 5 Volume Distribution Rate and Lane Changing Rate

number of lane changes between the two stations to the lane volume at the upstream station as a percent. The lane changing rate does not vary so much with the level of total in-put volume, at least during the peak two hours. The transition probability matrix is then derived for each station directly as shown in Fig. 5.

(iii) The Lane Change from the Congested Area

In order to identify the congestion on the merging area, the minute average running speed was computed for each lane in the study area. Then for the time periods with a very low value for the one minutes average running speed in the merging lane, measurements were made for the lane changings from the merging lane to the adjacent lane as well as the difference of traffic density for both lanes. Fig. 6 shows the results of that analysis.

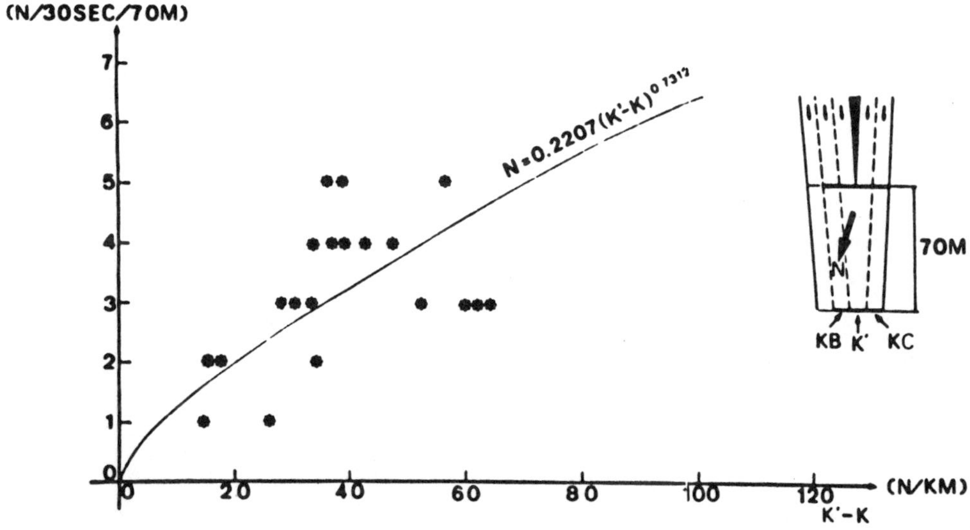

Fig. 6 Lane Change from the Congested Area

5. PROCEDURE FOR COMPUTER CALCULATION AND RESULTS OF THE SIMULATION

a. Procedure for Computer Calculation

Input data for the model are as follows:

i) In put for traffic stream characteristics: 1) free speed (v_f); 2) jam density (k_j); and 3) coefficient of lane changes from congestion.

ii) Volume distribution over lanes at the upstream input station-------------------------- **Q** (n)

iii) Transition probability matrix for lane changes for non-congestion stage--------------- **P**(*i*)

iv) Number of stations and lanes for the study area as well as the length of one time interval.

v) In-put demand.

The flow diagram of the Freeway Merging Model is shown in Fig.7. After reading the input data, the lane traffic volume for each station for each lane is calculated from the Free Flow Model. If the lane traffic volume exceeds the lane capacity or congestion remains from the previous time interval, the iteration process starts from station 1 to pursue the behaviour of the shock wave. The iteration process continues until the simulation times exceeds the time interval; then the input demand for the next time interval is fed into the model. These procedures are repeated until computations are completed for all time intervals or the shock wave passes through the study area.

An example of output from the Freeway Merging Model is shown in Fig. 8. The lane traffic volume, traffic density and average running speed for each section of each lane in the study area are typed out in the format as shown in figure. In addition to that, the lane traffic demand from the Free Flow Model, and the time when the shock wave passes through each station, as well as the number of lane changes are also included in the format.

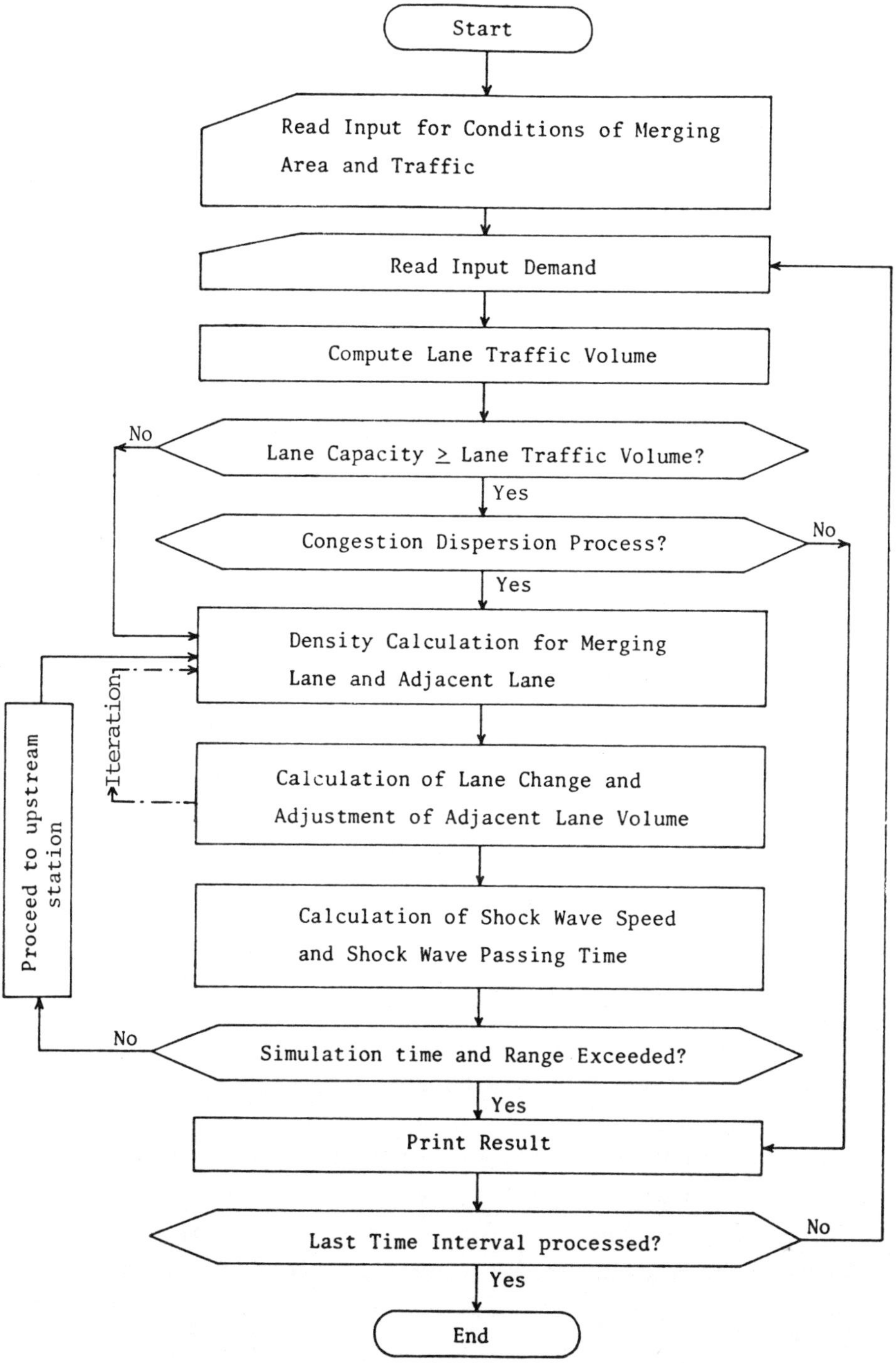

Fig. 7 Flow Diagram of Freeway Merging Model

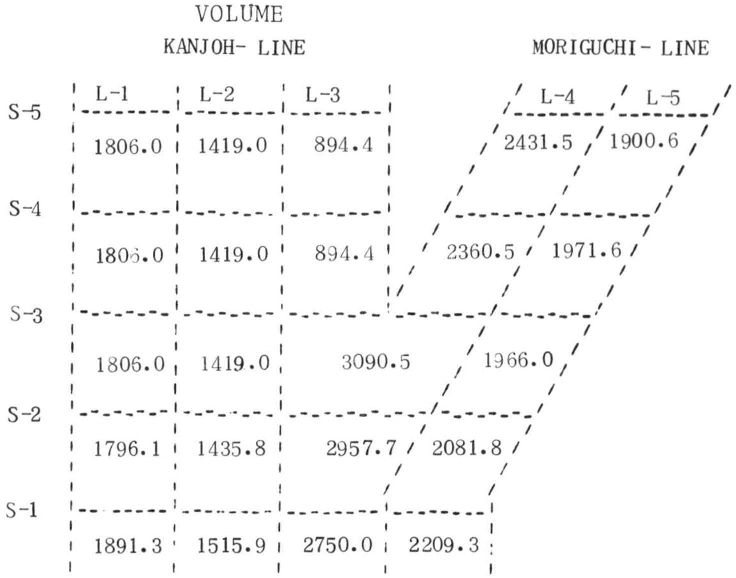

SHOCK WAVE PASSING TIME SEC.	SPEED OF SHOCK WAVE M/SEC
T5 = 209.02	S5 = -0.95525
T4 = 178.45	S4 = -1.33448
T3 = 145.80	S3 = -0.80915
T2 = 87.83	S2 = -0.39848
T1 = 0.00	S1 = -0.00000

Fig. 8 Output from the Freeway Merging Model

b. Results of the Simulation

(i) Simulation of actual conditions

Fig. 9 shows the speed and density contour diagrams for twelve minutes on the merging lane of the Moriguchi Junction. The input demand is also included in the figure for a three minute time interval. The simulation interval of three minutes is simply selected by judging from the density and speed contour diagram. Figure 10 shows the results of the simulation for the twelve minute periods with the input demands shown in Fig. 9. The density and speed diagram from the computer output simulate the propagation of the con-

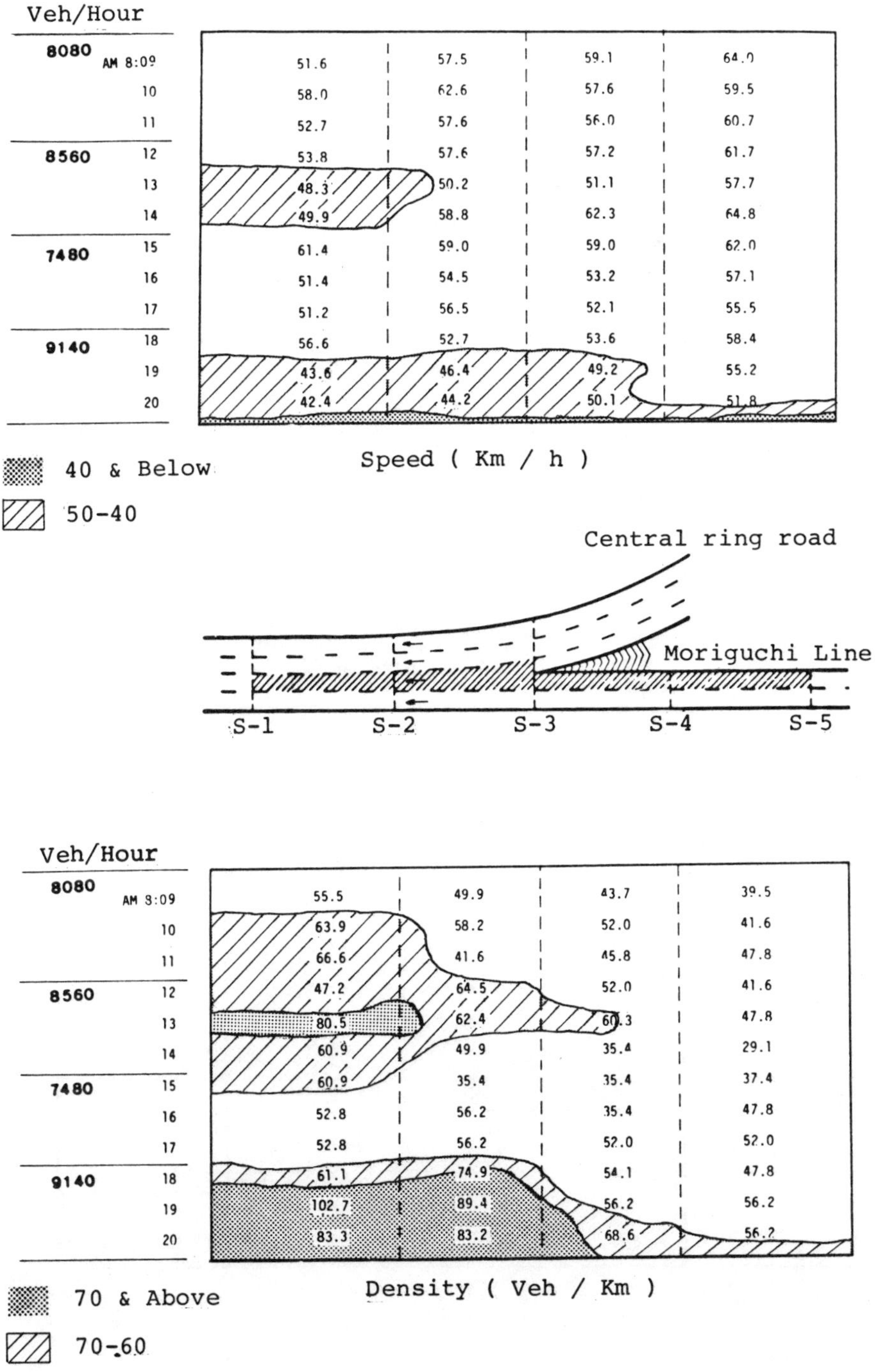

Fig. 9 Speed Density Contour Diagram from Observed Data

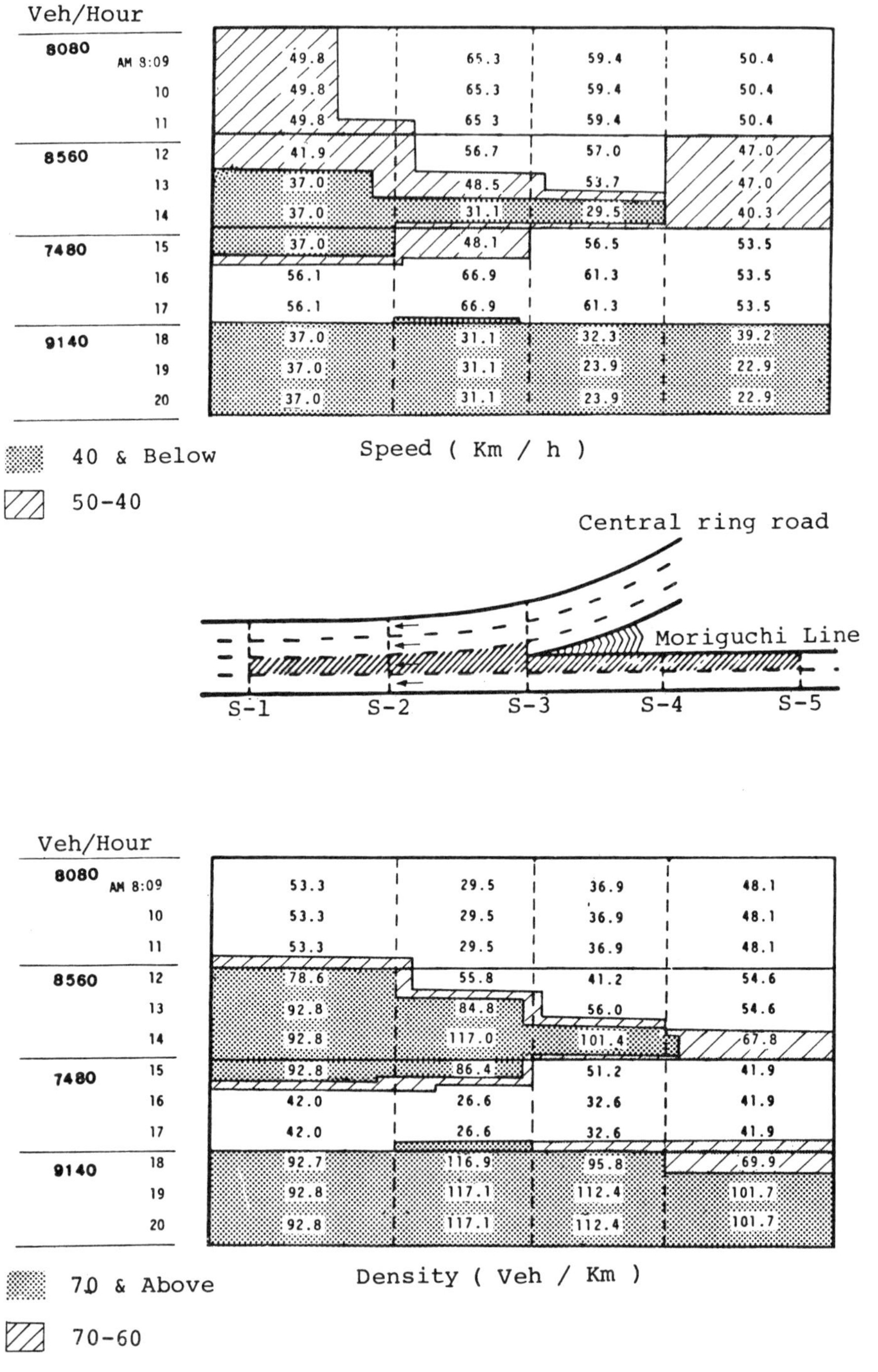

Fig. 10 Speed Density Contour Diagram from Computer Output

gested area and density variation in the merging area reasonably well as a whole, although the model cannot locate delicate fluctuations in the density and speed contour diagram.

(ii) Input demand analysis

Fig. 11 shows the results of input demand analysis for the Moriguchi Junction making use of the Freeway Merging Model. In this analysis, the input demands are increased by 20 vehicle per hour increments. For each level of input demand, the congestion propagation is pursued until the shock wave stops extending upstream or gets through the study area. Fig. 11 shows the lane traffic volume, as well as the time of the shock wave passage at each station of the merging lane. The figure indicates that the congestion starts building up at the input volume level of 8250 vph and passes through station 5 at the volume of 8600 vph. Therefore it is concluded that the congestion will extend upstream toward the Moriguchi Line if the input demand exceeds the level of 8600 vph.

iii) Evaluation of operational schemes

The Freeway Merging Model can be applied to evaluate operational schemes such as ramp metering or closure. In order to show how the model reacts for different traffic conditions, one simple operational scheme was tested, having a ramp closed or metered at the on-ramp which is the nearest upstream on-ramp along the Moriguchi line.

The result of the model for the on-ramp control plan is shown in Fig. 12. The behavior of the shock wave is depicted for ramp metering rates of 1/5, 1/2, and 1/1. In this case a ramp metering rate of 1/5, for example, corresponds to decreasing the on-ramp volume by one fifth by ramp metering; therefore, a ramp metering rate of 1/1 corresponds to a ramp closure. It is suggested that the ramp metering alone

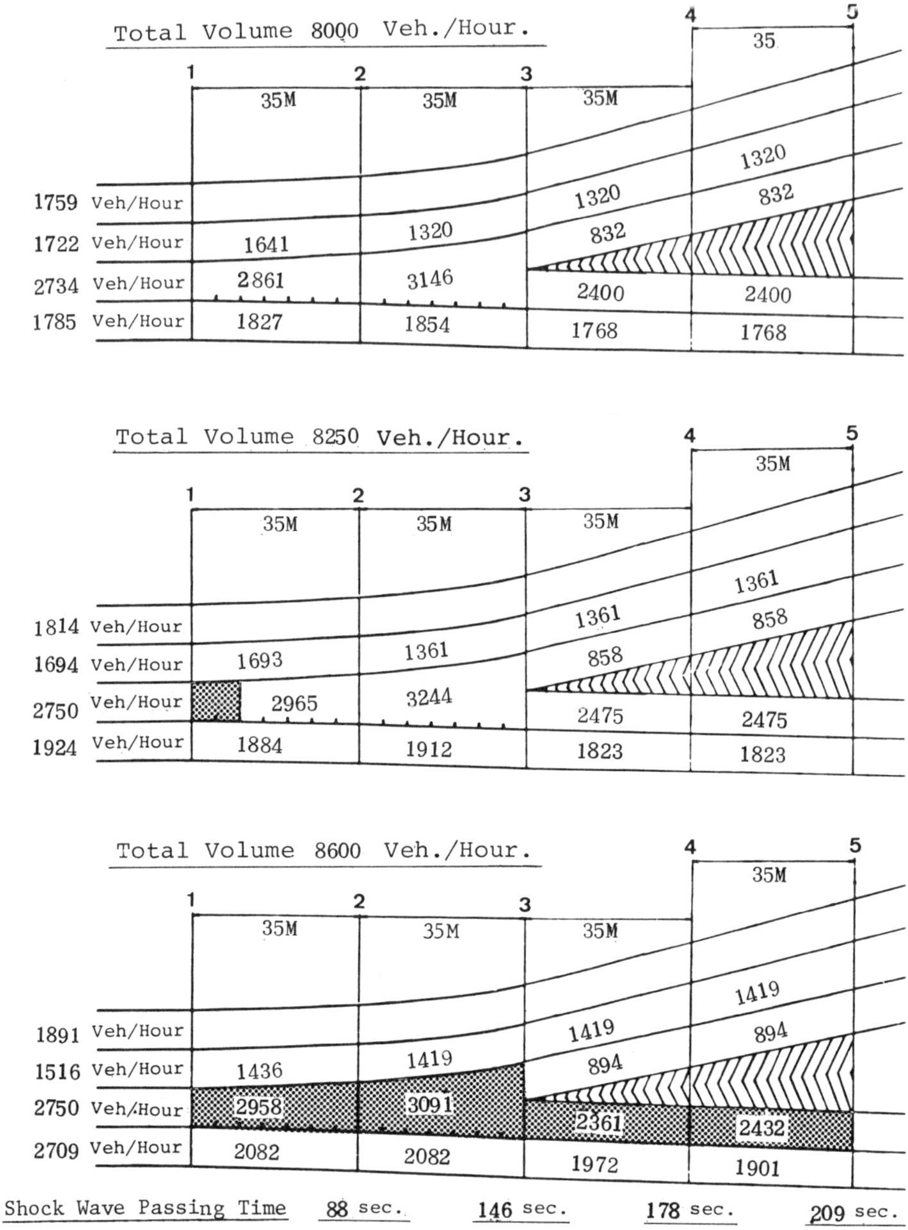

Fig. 11 Results of Input Demand Analysis

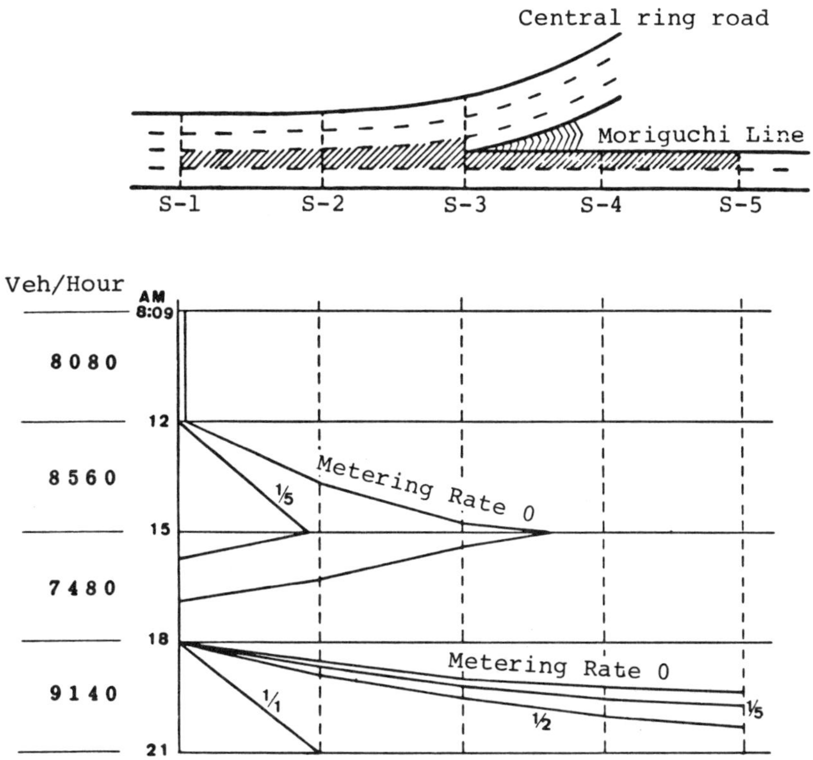

Fig. 12 Effect of Ramp Closure on the Merging Zone Traffic

is not able to effectively eliminate the congestion on the merging lane.

6. CONCLUSIONS

The experience gained in working with the development of the Freeway Merging Model led to the following conclusions:

(i) The Freeway Merging Model, based on the transition probability of lane changes and on the shock wave theory, simulates the traffic stream in the merging area both for congested and non-congested situations. The model can be used to evaluate the effectiveness of operational improvement plans.

(2) The following are possible future research topics which might further develop the model.

- i) To establish a systematic and reasonable procedure to estimate the transition probability,
- ii) To find an appropriate time interval of input demand.

REFERENCES

(1) M.J. Lighthill, "On Kinematic Wages; A theory of traffic flow on long crowded road", Record Print of Royal Society of London, Proceedings, Part A, V.229, No.1178, May 1955.

(2) E. Kometani, et al., Traffic Engineering, Kokuminkagakusha, Tokyo, 1977.

(3) Yasuji Makigami and William I. Woodie, "Freeway Travel Time Evaluation Technique" Highway Research Record No.321, 1970.

(4) P.K. Munjal, Yuan-Shih Hsu and R.L. Lawrence, "Analysis and Validation of Lane-drop Effects on Multilane Freeways", Vol. 5, Transpn. Res., pp. 241-255.

(5) P.K. Munjal and L.A. Pipes, "Propagation of On-Ramp Density Perturbations on Undirectional Two-and Three-Lane Freeways", Vol.5, Transpn. Res., pp. 257-266, 1971.

(6) Report of Committee of Traffic Control for the Hanshin Expressway, Traffic Engineering Association in Japan, March, 1976.

(7) Report of Committee of Traffic Control for the Hanshin Expressway, Traffic Engineering Association in Japan, March, 1980.

AN ANALYSIS OF HEURISTICS FOR THE CONTINUOUS NETWORK DESIGN PROBLEM

Patrice Marcotte
Centre de recherche sur les transports
Université de Montréal

ABSTRACT

In this paper we present a new formulation for a network design problem where the decision variables can take any real positive values, the objective is normative and flows are routed according to Wardrop's (descriptive) first principle of equal travel times. Based on intuitive ideas, four heuristic procedures are presented; in certain cases, upper bounds on the ratio of objective functions of the heuristic and (unknown) optimal solutions are derived, providing a yardstick to judge the theoretical performance of these heuristics.

INTRODUCTION

Network design problems (NDPs) may be divided into two main categories, depending on the situation being modeled:

a) Discrete NDP: the term 'discrete' refers to the nature of some of the control variables that are allowed to take only integer values. In the simplest form of the NDP, the discrete variables are binary variables associated with the arcs, and take the value one if the corresponding arc is part of the proposed network and zero otherwise. In general, these formulations lead to NP-hard problems, although some particular cases can be solved by polynomial-time algorithms (the shortest spanning tree problem is such an example). Since the flow variables are usually continuous, the discrete NDP leads to mixed integer programming models.

b) Continuous NDP: in the continuous models, all control

variables are continuous on a certain subset of n-dimensional euclidian space. This category includes the traffic signal setting problem, where the green split at an intersection may take any suitable value between zero and one; it also includes the problem of the choice of optimal capacities for a given network. The term 'continuous' is somewhat misleading since the objective function need not be a continuous function of the control variables.

For a fairly thorough review of the literature on NDPs, either discrete or continuous, the reader is referred to Wong [8].

In this paper we consider a particular version of the NDP (see Dantzig [1] and LeBlanc [2]) where the network infrastructure can be improved by choosing a set of arc capacities, which can take any real positive value. The basic assumptions of the model are the following:

a) The objective function consists of the weighted sum of the transportation costs (expressed in time units) and the investment costs (expressed in money units). The parameter that converts time value into money unit may be interpreted as the optimal Lagrange multiplier associated with a budget constraint. We assume that this parameter has been previously obtained from a calibration study.
b) Arc capacities can take any non-negative values and it is assumed that the initial capacities are zero.
c) Traffic assignment is carried out according to Wardrop's first principle of equal times (Wardrop [7]), which is the most adequate behavioral principle for highway or urban applications.
d) The trip-matrix of demands for origin/destination pairs of the network is known.
e) A single transportation mode is considered (private car).

Assumption (c) makes the problem extremely difficult, since the cost of an equilibrium assignment cannot be expressed in closed form in the general case, nor can its partial derivatives (according to flows) be obtained easily.

The remainder of the paper is concerned with a new

formulation of the continuous NDP that gives rise to various heuristic procedures whose analysis leads to interesting theoretical bounds that relate the value of the heuristic solution to the (unknown) optimal value of the NDP.

FORMULATION OF THE NDP

For a flow (of vehicles) on a network to be feasible, the following set of conservation equations and inequalities in node-arc formulation must be satisfied:

$$\forall \quad i \in N : \sum_{j \in N} v^k_{ij} - \sum_{j \in N} v^k_{ij} = \begin{cases} \sum_{\ell \in \Delta} g_{k\ell} & \text{if } i=k \\ -\, g_{ki} & \text{if } i \in \Delta \\ 0 & \text{otherwise} \end{cases} \qquad (1)$$

$$\forall \quad a \in A : v_a = \sum_{k \in \Omega} v^k_a \quad \text{and} \quad v^k_a \geq 0$$

where $a=(i,j)$ denotes an arc going from node i to node j

- N = set of nodes of the network
- A = set of arcs of the network
- Ω = set of origins, or flow-generating arcs ($\Delta \subseteq N$)
- Δ = set of destinations ($\Delta \subseteq N$)
- $g_{k\ell}$ = trip demand between origins $k \in \Omega$ and destinations $\ell \in \Delta$.

The set of constraints (1) represents a polyhedron, which we will denote by Φ. Thus a flow vector $v=(v_a)_{a \in A}$ will be feasible if and only if $v \in \Phi$. The NDP can then be written as:

$$\text{Min } F(v,z) = T(v,z) + G(z) \qquad (2)$$

$$\text{subject to (s.t.)} \quad v \in \Phi$$

and v is an equilibrium flow.

$T(v,z)$ represents the transportation cost and $G(z)$ the investment cost; $T(v,z)$ and $G(z)$ are separable functions:

$$T(v,z) = \sum_{a \in A} T_a(v_a,z_a) = \sum_{a \in A} v_a S_a(v_a,z_a) \text{ and } G(z) = \sum_{a \in A} G_a(z_a).$$

If, for each positive capacity z_a, $S_a(v_a,z_a)$ are non-decreasing functions of the flow vector v, the condition that v be an equilibrium flow is equivalent to saying that v is the optimal solution of the convex program:

$$\min_{v \in \Phi} \sum_{a \in A} \int_0^{v_a} S_a(x, z_a)\,dx.$$

It is easy to see (Smith [4]) that a flow v will solve that program if and only if v satisfies the following set of variational inequalities:

$$\sum_{a \in A} (v_a - v_a^{(i)}) S_a(v_a, z_a) \leq 0 \quad i=1,2,\ldots,n \tag{3}$$

for all extremal flows $v^{(i)}$ of Φ. Using (3), we can rewrite (2) as:

$$\min_{\substack{v \in \Phi \\ z \geq 0}} F(v,z) = T(v,z) + G(z) = v^T \cdot S(v,z) + G(z) \tag{4}$$

$$\text{s.t.} \quad (v - v^{(i)})^T S(v,z) \leq 0 \quad i=1,2,\ldots,n.$$

Since the constraints are non-convex this will in general be a non-convex mathematical program.

For the remainder of the paper, $T_a(v_a, z_a)$ will assume the particular functional form:

$$T_a(v_a, z_a) = v_a[\alpha_a + \beta_a (\frac{v_a}{z_a})^p]$$

as proposed by the United States Bureau of Public Roads, where $p \geq 1$. Also we will set $G_a(z_a) = \ell_a z_a^m$ with $m>0$. Depending on the value of the parameter m, the investment curve is either concave ($m \leq 1$), linear ($m=1$) or convex ($m \geq 1$). With these cost functions, we have:

USER-OPTIMAL NDP:

$$\min_{\substack{v \in \Phi \\ z \in 0}} F(v,z) = \sum_{a \in A} (v_a[\alpha_a + \beta_a (\frac{v_a}{z_a})^p] + \ell_a z_a^m \tag{5}$$

$$\text{s.t.} \quad \sum_{a \in A} (v_a - v_a^{(i)}) [\alpha_a + \beta_a (\frac{v_a}{z_a})^p] \leq 0$$

$$i=1,2,\ldots,n$$

An optimal solution to (5) will be written $[v(z^*), z^*]$.

HEURISTICS FOR THE USER-OPTIMAL NDP

Heuristic HO

For the sake of numerical tractability, some authors have used Wardrop's second principle of system-optimal (normative) assignment which leads to the formulation:

SYSTEM-OPTIMAL NDP:

$$\underset{\substack{v\in\Phi \\ z\geq 0}}{\text{Min}}\ F(v,z) = \sum_{a\in A} v_a[\alpha_a+\beta_a(\frac{v_a}{z_a})^p]+\ell_a z_a^m. \tag{6}$$

This corresponds to the statement that the users are not free to choose their own routes, but must use the routes that minimize the total, or average, travel cost.

This corresponds to dropping all the variational constraints of (5). When m is greater than one, this problem can be solved by standard convex assignment techniques; if m equals one, the problem reduces to an all-or-nothing assignment (Los [3]). The solution (v^{HO},z^{HO}) to (6) provides a lower bound on the optimal value of (5), whereas the feasible solution $(v(z^{HO}),z^{HO})$ obtained by assigning the flows on the network with capacity vector z^{HO} provides an upper bound: $F(v^{HO},z^{HO}) \leq F(v(z^*),z^*) \leq F(v(z^{HO}),z^{HO})$. The tightness of these bounds can be checked when solving a specific problem. The following proposition deals with an interesting special case where the system-optimal NDP and the user-optimal NDP are equivalent. In this case the congestion and investment functions are homogeneous throughout the network and proportional to the length of an arc; this is the rather unrealistic condition that all the arcs share the same physical characteristics.

Proposition 1: The solution to the system-optimal NDP with linear investment costs,

$$\underset{\substack{v\in\Phi \\ z\geq 0}}{\text{Min}}\ F(v,z) = \sum_{a\in A} d_a v_a[1+\beta(\frac{v_a}{z_a})^p]+\mu d_a z_a \tag{7}$$

(d_a represents the physical length of arc a),

leads to an equilibrium flow pattern, i.e. $v(z^{HO}) = v^{HO}$.

Proof: Since the z_a-variables are unconstrained, we must have at the optimum:

$$\frac{\partial F}{\partial z_a} = 0, \text{ i.e.: } -\beta d_a(\frac{v_a}{z_a})^{p+1} + \mu d_a = 0$$

$$\Rightarrow \frac{v_a}{z_a} = (\frac{\mu}{\beta})^{1/(p+1)} \Rightarrow z_a = v_a (\frac{\beta}{\mu})^{1/(p+1)}$$

which is always non-negative. If $v_a = z_a = 0$ we set, by continuity, $\frac{0}{0} = (\frac{\mu}{\beta})^{1/(p+1)}$. Putting this value for z_a back in (7) gives the master problem:

$$\underset{v \in \Phi}{\text{Min}} \sum_{a \in A} d_a v_a [1+\beta (\frac{\mu}{\beta})^{p/(p+1)} + (\frac{\beta}{\mu})^{1/(p+1)}]$$

which is equivalent to the minimization of the linear function $\sum_{a \in A} d_a v_a$, whose solution v^{HO} is obtained by computing shortest path trees rooted at the origin points of Ω. At the optimum, the transportation cost on arc a is $d_a [1+\beta (\frac{\mu}{\beta})^{p/(p+1)}]$ and thus proportional to its physical length; since the current solution v^{HO} also solves that problem, it is an equilibrium flow pattern.‖

Though there may be many solutions to the previous problem, the flow vector v^{HO} corresponding to a given optimal capacity vector z^{HO} will be unique, provided that p is strictly positive.

We define, for a given heuristic H, the quantities

$$R_m^p(H) = \underset{\alpha,\beta,\ell}{\text{Sup}} \frac{\text{total cost of the heuristic solution}}{\text{total cost of the optimal solution}}$$

$$= \underset{\alpha,\beta,\ell}{\text{Sup}} \frac{F(v(z^H), z^H)}{F(v(z^*), z^*)}$$

and $$r_m^p(H) = \underset{\alpha,\beta,\ell}{\text{Sup}} \frac{\text{transportation cost of the heuristic solution}}{\text{transportation cost of the optimal solution}}$$

$$= \underset{\alpha,\beta,\ell}{\text{Sup}} \frac{T(v(z^H), z^H)}{T(v(z^*), z^*)}$$

Proposition 2: $R_1^{\infty}(HO) \geq 2$ and $\lim_{p \to \infty} \frac{1}{p} \cdot r_m^{p^{\gamma}}(HO) \geq 1, \ \forall \ \gamma > 1$.

Proof: Let us consider the following 3-arcs network:

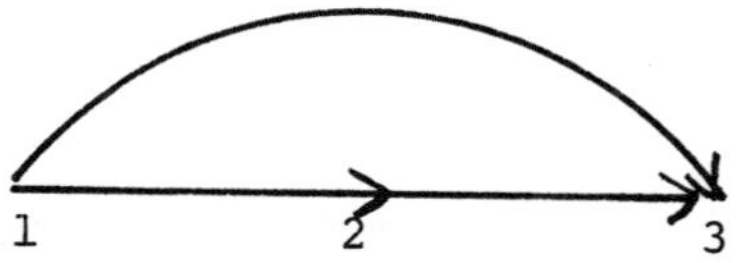

The non-zero elements of the trip-matrix are: $g_{12}=g_{23}=p$ and $g_{13}=1$. Transportation costs are given by $T_{12}(x)=T_{23}(x)=\frac{1}{2}(1+x^{p^\gamma})$, $T_{13}(x)=p$; and investment costs by $G_{12}(x)=G_{23}(x)=\frac{px}{2}$ and $G_{13}(x)=0$. The system-optimal problem is then expressed as:

$$\min_{\substack{v\in\Phi \\ z\geq 0}} pv_{13}+\frac{v_{12}}{2}\left[1+\left(\frac{v_{12}}{z_{12}}\right)^{p^\gamma}\right]+\frac{v_{23}}{2}\left[1+\left(\frac{v_{23}}{z_{23}}\right)^{p^\gamma}\right]+\frac{p}{2}(z_{12}+z_{23}).$$

As before, we differentiate with respect to the capacities z_a to obtain:

$$\frac{p^\gamma}{2}\left(\frac{v_{12}}{z_{12}}\right)^{p^\gamma+1}=\frac{p}{2}\Rightarrow\frac{v_{12}}{z_{12}}=\left(\frac{1}{p^{\gamma-1}}\right)^{1/(p^\gamma+1)}.$$

Similarly: $\frac{v_{23}}{z_{23}}=\left(\frac{1}{p^{\gamma-1}}\right)^{1/(p^\gamma+1)}$; the master problem is then:

$$\min_{v\in\Phi} pv_{13}+\frac{v_{12}+v_{13}}{2}\left[1+\left(\frac{1}{p^{\gamma-1}}\right)^{p^\gamma/(p^\gamma+1)}+p\cdot p^{(\gamma-1)/p^\gamma+1)}\right]$$

whose solution is given by: $v_{13}^{HO}=1$

$$v_{12}^{HO}=v_{23}^{HO}=p$$

The capacities then take the values:

$$z_{12}^{HO}=z_{23}^{HO}=p\cdot p^{(\gamma-1)/(p^\gamma+1)}.$$

This is clearly not an equilibrium solution; the equilibrium flow vector $v(z^{HO})$ is obtained when route (1,2)-(2,3) has its cost equal to p, i.e.:

$$1+\left(\frac{p+x}{p\cdot p^{(\gamma-1)/(p^\gamma+1)}}\right)^{p^\gamma}=p$$

where x represents the number of users shifting from route (1,3) to route (1,2)-(2,3). For large p, x is smaller than one, since:

$$\lim_{p\to\infty} \frac{1}{p} \cdot \left(\frac{p+1}{p \cdot p^{(\gamma-1)/(p^\gamma+1)}}\right)^{p^\gamma} = \lim_{p\to\infty} \frac{1}{p}\left(1+\frac{1}{p}\right)^{p^\gamma} \cdot \frac{1}{p^{(\gamma-1)[p^\gamma/(p^\gamma+1)]}}$$

$$\geq \lim_{p\to\infty} \frac{1}{p \cdot p^{\gamma-1}}\left(1+\frac{1}{p}\right)^{p^\gamma} = \lim_{p\to\infty} \frac{1}{p^\gamma} e^{p^{\gamma-1}} = \infty \quad \text{if} \quad \gamma>1.$$

The total cost of the user-optimal solution is, for $\gamma \approx 1$:

$$F[v(z^{HO}),z^{HO}] = p(1-x) + p(p+x) + p^2 \cdot p^{(\gamma-1)/(p^\gamma+1)} \approx 2p^2+0(p).$$

The optimal solution $[v(z^*),z^*]$ for the NDP is obtained by increasing capacities z_{12} and z_{23} to cater for trips going from origin 1 to destination 3.

Thus: $z^*_{12}=z^*_{23}=(p+1)p^{(\gamma-1)/(p^\gamma+1)}$. The cost is then:

$$F[v(z^*),z^*]=(p+1)\left[1+\left(\frac{1}{p^{\gamma-1}}\right)^{(p^\gamma)/(p^\gamma+1)}\right]+p(p+1)p^{(\gamma-1)/(p^\gamma+1)}$$

$$\approx 0(p)+p^2+0(p)=p^2+0(p)$$

Therefore we get the results: $R_1^\infty(HO) \geq \lim_{p\to\infty} \frac{F[v(x^{HO}),z^{HO}]}{F[v(z^*),z^*]} = 2$

and $\lim_{p\to\infty} \frac{1}{p} \cdot r_1^{p^\gamma}(HO) \geq \lim_{p\to\infty} \frac{1}{p} \cdot \frac{T[v(z^{H0}),z^{H0}]}{T[v(z^*),z^*]}$

$$= \lim_{p\to\infty} \frac{p^2+0(p)}{p(p+1)\left[1+\left(\frac{1}{p^{\gamma-1}}\right)^{(p^\gamma)/(p^\gamma+1)}\right]} = 1$$

which completes the proof.||

Heuristic H1: The idea behind this procedure is to solve iteratively modified versions of problem (5), by explicitly dealing with one active constraint, and dropping the others. At each iteration one checks whether the current solution is an equilibrium solution; if it is not, a new subproblem that satisfies the previously violated constraint is solved. The resulting algorithm is:

Algorithm H1:

Step 1. k←-1; LB=0; UB=∞; solve:

$$\underset{\substack{v\in\Phi\\ z\geq 0}}{\text{Min}}\ F(v,z) = \sum_{a\in A} v_a[\alpha_a+\beta_a(\frac{v_a}{z_a})^p]+\ell_a z_a^m$$

whose solution is $(\bar{v},\bar{z})$.

Step 2. k←k+1; solve the all-or-nothing problem:

$$\underset{v\in\Phi}{\text{Min}} \sum_{a\in A} v_a[\alpha_a+\beta_a(\frac{\bar{v}_a}{\bar{z}_a})^p] \text{ whose solution is } v^{(1)}.$$

Solve: $$\underset{v\in\Phi}{\text{Min}} \sum_{a\in A} \int_0^{v_a} [\alpha_a+\beta_a(\frac{t}{\bar{z}_a})^p]dt$$

whose solution is $v(\bar{z})$.

Set: $UB^{(k)}=F[v(\bar{z}),\bar{z}]$, $LB^{(k)}=F(\bar{v},\bar{z})$,

$UB=MIN\ [UB,UB^{(k)}]$, $LB=Max\ [LB,BL^{(k)}]$.

If UB-LB is small, STOP; the best solution corresponds to vectors $v(\bar{z})$ and $\bar{z}$ obtained at the iteration k for which $UB=UB^{(k)}$.

Step 3. Solve: $$\underset{v\in\Phi}{\text{Min}} \sum_{a\in A} v_a[\alpha_a+\beta_a(\frac{v_a}{z_a})^p]+\ell_a z_a^m$$

$$\text{s.t.} \sum_{a\in A} [v_a-v_a^{(1)}][\alpha_a+\beta_a(\frac{v_a}{z_a})^p] \leq 0 \qquad (8)$$

whose solution is $(\bar{v},\bar{z})$; go to step 2.

At step 1, the procedure is initialized with flows and capacities $(v,z) = (v^{HO},z^{HO})$. If v^{HO} is an equilibrium flow, then v^{HO} is a solution of the all-or-nothing problem and of the assignment problem at step 2; the algorithm then stops with UB-LB=0. If the algorithm converges, i.e., if UB-LB→0 as k→∞, it converges towards an optimal solution $[v(z^*),z^*]$ to the NDP; however, the convergence of this algorithm remains an open problem. Since the constraint of problem (8) of step 3 is active, the optimum must be reached at a point where $\sum_{a\in A} [v_a-v_a^{(1)}][\alpha_a+\beta_a(\frac{v_a}{z_a})^p]=0$. By dualizing the variational constraint we obtain:

$$\underset{\substack{\lambda\in R \\ z\geq 0}}{\mathrm{Max}}\ \underset{v\in\Phi}{\mathrm{Min}}\ F(v,z,\lambda) = \sum_{a\in A} [\alpha_a+\beta_a(\frac{v_a}{z_a})^p][v_a+\lambda(v_a^{(1)}-v_a)]+\ell_a z_a^m.$$

This last function has interesting properties that would be too long to establish here, and which will be exposed in a forthcoming paper.

<u>Heuristic H2</u>. This heuristic has been proposed by many authors, including Steenbrink [5] and Tan [6]; the latter names it "Iterative Optimization Assignment"; it consists in iteratively solving an optimization problem and a user-optimal assignment, as follows:

<u>Algorithm H2</u>. Let $\bar{v}$ be any feasible flow vector.

Step 1. Optimization step:

$$\underset{z\geq 0}{\mathrm{Min}}\ F(\bar{v},z)=T(\bar{v},z)+G(z) \quad \text{whose solution is } \bar{z}.$$

Step 2. Assignment step:

$$\underset{v\in\Phi}{\mathrm{Min}}\ \sum_{a\in A} \int_0^{v_a} T(x,\bar{z}_a)dx \quad \text{whose solution is } v(\bar{z}).$$

If $\|\bar{v}-v(\bar{z})\|<\varepsilon$ STOP; the ε-optimal solution is $[v(\bar{z}),\bar{z}]$; otherwise, set $\bar{v}\leftarrow v(\bar{z})$ and go to step 1.

<u>Proposition 3</u>. If the congestion functions $T_a(v_a,z_a)$ are of BPR type and $G_a(z_a)$ are convex, algorithm H2 converges towards the solution of the optimization problem:

$$\underset{\substack{v\in\Phi \\ z\geq 0}}{\mathrm{Min}}\ H(v,z) = \sum_{a\in A} v_a[\alpha_a + \frac{\beta_a}{p+1}(\frac{v_a}{z_a})^p] + \frac{G_a(z_a)}{p+1}$$

<u>Proof</u>. For $v=\bar{v}$ fixed, $\underset{z\geq 0}{\mathrm{Min}}\ H(\bar{v},z)$ is equivalent to

$$\underset{z\geq 0}{\mathrm{Min}}\ \sum_{a\in A} \bar{v}_a[\alpha_a+\beta_a(\frac{\bar{v}_a}{z_a})^p]+G_a(z_a) \quad \text{which is exactly the}$$

minimization performed at step 1 of algorithm H2.

For $z=\bar{z}$ fixed, $\underset{v\in\Phi}{\mathrm{Min}}\ H(v,\bar{z})$ is equivalent to

$$\underset{v\in\Phi}{\mathrm{Min}}\ \sum_{a\in A} v_a[\alpha_a+\beta_a(\frac{v_a}{z_a})^p] \quad \text{which is the assignment}$$

performed at step 2. Algorithm H2 thus minimizes $H(v,z)$ by successively fixing the values of vectors v and z; convexity of $H(v,z)$ ensures convergence of the procedure.‖

However appealing it may look at first glance, heuristic H2 may lead to quite bad solutions for the NDP, as is shown by the next proposition.

Proposition 4. $R_1^p(H2) = p+1$.

At step 1 of algorithm H2, the solution of the minimization is given by $\bar{z}_a = (\frac{p\beta_a}{\ell_a})^{1/(p+1)} v_a$ if $G(z_a) = \ell_a z_a$. Heuristic H2 can then be expressed as:

$$\underset{v\in\Phi}{\text{Min}} \sum_{a\in A} v_a[\alpha_a + \frac{\beta_a}{p+1}(\frac{v_a}{\bar{z}_a})^p] + \ell_a \bar{z}_a$$

$$= \sum_{a\in A} v_a[\alpha_a + \beta_a(\frac{\ell_a}{p\beta a})^{p/(p+1)}] \tag{9}$$

whose solution is v^{H2}, giving $[v^{H2}, z^{H2}] = [v(z^{H2}), z^{H2}]$ as the solution of heuristic H2.

Let's consider the corresponding system-optimal problem:

$$\underset{\substack{v\in\Phi \\ z\geq 0}}{\text{Min}} \sum_{a\in A} v_a[\alpha_a + \beta_a(\frac{v_a}{z_a})^p] + \ell_a z_a,$$

whose solution is given by $(v^{(s)}, z^{(s)})$ where

$z_a^{(s)} = (\frac{p\beta_a}{\ell_a})^{1/(p+1)} v_a^{(s)}$ and $v^{(s)}$ minimizes

$$\sum_{a\in A} v_a[\alpha_a + (p+1)\beta_a(\frac{\ell_a}{p\beta_a})^{p/(p+1)}].$$

We can write:

$$F(v(z^{H2}), z^{H2}) = F(v^{H2}, z^{H2}) \leq (p+1) \sum_{a\in A} v_a^{H2}[\alpha_a + \beta_a(\frac{\ell_a}{p\beta_a})^{p/(p+1)}]$$

$$\leq (p+1) \sum_{a\in A} v_a^{(s)}[\alpha_a + \beta_a(\frac{\ell_a}{p\beta_a})^{p/(p+1)}] = (p+1)F(v^{(s)}, z^{(s)})$$

$\leq (p+1)F(v(z^*),z^*)$ where z^* is the optimal solution to the NDP

$\Rightarrow R_1^p(H2) \leq p+1$.

To prove that $R_1^p(H2) \geq p+1$ we will use an example constituted of one O-D pair linked by two distinct paths.

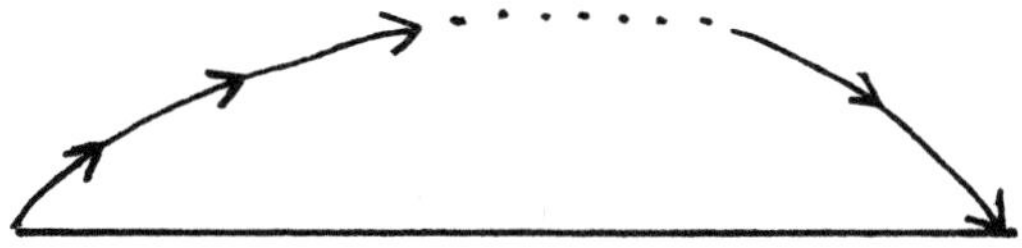

The trip demand is one flow unit between the origin and the destination. Path (1) is made of n arcs for which $\alpha_a = \beta_a = 1$ and $\ell_a = p(\frac{1}{n})^{(p+1)/p}$. Path (2) has a single arc with parameters $\alpha_a = \beta_a = 1$ and $\ell_a = p(n-1)^{(p+1)/p}$. Function (9) is minimized by $v^{H2} = (v^{H2}_{(1)}, v^{H2}_{(2)}) = (0,1)$, since $n(1 + \frac{1}{n}) > 1+n-1$. The actual cost of this solution is

$$F(v^{H2}, z^{H2}) = v^{H2}_{(2)} \quad [\alpha_a + (p+1)\beta_a (\frac{\ell_a}{p\beta_a})^{p/(p+1)}]$$

$$= 1[1+(p+1)(n-1)].$$

However the optimal solution to the NDP coincides with the system-optimal solution and is obtained when $v^* = (v^*_{(1)}, v^*_{(2)}) = (1,0)$; the corresponding total cost is $F(v(z^*),z^*) = F(v^*,z^*) = n[1+(p+1)\cdot\frac{1}{n}] = n+p+1$. Therefore:

$$\frac{F(v(z^{H2}), z^{H2})}{F(v(z^*), z^*)} = \frac{(n-1)(p+1)+1}{n+p+1} \underset{n\to\infty}{\to} p+1 .$$

This completes the proof.||

<u>Heuristic H3</u>. The idea behind heuristic H3 is to obtain a set of equilibrium flows v^{H3} identical to the system-optimal flows $v^{(s)}$ by adjusting the capacity vector z. Though the resulting solution will almost always be sub-optimal, it will yield interesting theoretical bounds that remain valid even in the case of concave investment cost functions.

Proposition 5. $R_m^p \le \frac{m(p+1)}{m+p} + \frac{p}{(m+p)(p+1)^{m/p}}$

for all m and p positive.

Proof. The system-optimal problem is:

$$\min_{\substack{v\in\Phi\\ z\ge 0}} \sum_{a\in A} v_a[\alpha_a+\beta_a(\frac{v_a}{z_a})^p]+\ell_a z_a^m$$

which, after setting z-partial derivatives equal to zero, reduces to

$$\min_{v\in\Phi} \sum_{a\in A} v_a[\alpha_a+\beta_a(\frac{m\ell_a}{p\beta_a})^{p/(m+p)} v_a^{[p(m-1)]/(m+p)}]$$

$$+ \ell_a(\frac{p\beta_a}{m\ell_a})^{m/(m+p)} v_a^{[m(p+1)]/(m+p)}$$

which, after some algebra, becomes:

$$\min_{v\in\Phi} \sum_{a\in A} v_a[\alpha_a+d_a v_a^{[p(m-1)]/(m+p)}] \tag{10}$$

with $d_a=(p+m)(\frac{\beta_a}{m})^{m/(m+p)}(\frac{\ell_a}{p})^{p/(m+p)}$. Let $v^{(s)}$ be the solution of (10). In order for the flow v^{H3} to be in equilibrium and identical to $v^{(s)}$, the capacity vector z^{H3} must be chosen in such a way that the following variational inequality is satisfied:

$$\sum_{a\in A}[v_a^{(s)}-v_a^{(i)}][\alpha_a+\beta_a(\frac{v_a^{(s)}}{z_a(v^{(s)})})^p] \le 0 \quad i=1,2,.,n \tag{11}$$

where, as before, the $v^{(i)}$ are the extreme points of the polyhedron Φ. If $z(v^{(s)})$ is a diagonal function, (11) is equivalent to saying that $v^{(s)}$ is the optimal solution of problem (10) and of:

$$\min_{v\in\Phi} \sum_{a\in A} \int_0^{v_a}\{\alpha_a+\beta_a[\frac{t}{z_a(t)}]^p\}dt. \tag{12}$$

A straightforward manner to guarantee that (10) and (12) will have the same solution vector $v^{(s)}$ is to choose the functions $z_a(t)$ such that:

$$\int_0^{v_a} [\alpha_a + \beta_a (\frac{t}{z_a(t)})^p] dt \equiv v_a [\alpha_a + d_a v_a^{[p(m-1)]/(m+p)}] \quad \forall\ v_a \geq 0.$$

If we restrict ourselves to functions of the form $z_a(t) = b_a t^n$ we get:

$$\beta_a \int_0^{v_a} [\frac{t}{z_a(t)}]^p dt = \frac{\beta_a}{b_a^p} \int_0^{v_a} t^{p(1-n)} dt = \frac{\beta_a}{[1+p(1-n)]b_a^p} v_a^{1+p(1-n)}$$

$$=> 1+p(1-n) = \frac{m(p+1)}{p+m} => n = \frac{p+1}{p+m}$$

Also: $$\frac{\beta_a}{[1+p(1-n)b_a^p} = d_a => b_a = [\frac{\beta_a}{d_a} \cdot \frac{m+p}{m(p+1)}]^{1/p}$$

$$=> z_a^{H3} = [\frac{\beta_a}{d_a} \cdot \frac{m+p}{m(p+1)}]^{1/p} . v_a^{(s)\ (p+1)/(p+m)}$$

The cost of this solution is:

$$F(v^{(s)}, z^{H3}) = F(v^{H3}, z^{H3})$$

$$= \sum_{a \in A} v_a^{(s)} \left[\alpha_a + \beta_a \frac{d_a}{\beta_a} \frac{m(p+1)}{m+p} v_a^{(s)\ p(1-[p+1]/[p+m])} \right]$$

$$+ \ell_a [\frac{\beta_a}{d_a} \cdot \frac{m+p}{m(p+1)}]^{m/p} . v_a^{(s)\ [m(p+1)]/[p+m]}$$

$$= \sum_{a \in A} v_a^{(s)} \left[\alpha_a + \left\{ \frac{m(p+1)}{p+m} d_a + \ell_a (\frac{\beta_a}{d_a} \cdot \frac{m+p}{m(p+1)})^{m/p} \right\} \cdot v_a^{(s)\ [p(m-1)]/[m+p]} \right]$$

For $\alpha=0$ one has:

$$\frac{F(v^{H3}, s^{H3})}{F(v^{(s)}, z^{(s)})} = \frac{1}{d_a} [\frac{m(p+1)}{m+p} d_a + \ell_a (\frac{\beta_a}{d_a} \cdot \frac{m+p}{m(p+1)})^{m/p}]$$

which, after substituting for d_a, yields:

$$R_m^p \le \frac{F(v^{H3},z^{H3})}{F(v^{(s)},s^{(s)})} = \frac{m(p+m)}{m+p} + \frac{p}{(m+p)(p+1)^{m/p}} \quad \|$$

It is interesting to notice that the proof makes no assumption whatsoever about the values of m and p, which need only to be positive. For particular values of m and p of special interest we have:

$$R_{1/2}^1 = 1.138 \qquad R_{1/2}^4 = 1.282$$

$$R_1^1 = 1.250 \qquad R_1^4 = 1.535$$

Corollary 6.

$$\lim_{m\to 0} R_m^p = 1 \quad ,$$

This result is intuitive: when m is small, all the arcs that are built have large capacities, since the marginal cost of adding capacity to an arc is rapidly decreasing; hence congestion is low and system-optimum is close to equilibrium.

Corollary 7.

$$\lim_{p\to\infty} R_1^p = 2 \ .$$

CONCLUSION

The continuous NDP with user-optimal behavior of the users of the network represents a formidable theoretical problem. In this article we have presented several heuristics to tackle it; the first one, H0, has been tested on a practical problem by the author in Los [3] where it is reported that the flow v^{H0} is almost an equilibrium flow. This heuristic would be clearly superior to H2 and H3 if it could be shown that $R_m^p(H0)$ is always smaller than $R_m^p(H3)$; this remains an open question. Corollary 6 suggests that system-optimality is a good approximation to user-optimality when the investment cost function is concave; however, in this case, the system-optimal master problem is concave and its global optimum is difficult to obtain. Algorithm H2, which has often been proposed in practice, has by far the worst theoretical bound; in fact, insisting that there exist system-optimal relations between the flows and the capacities may have adverse effects on total congestion. The bounds for heuristic H3 can possibly be sharpened by choosing different capacity functions $z^{H3}(v^{H3})$ for which problems (10) and (12) are no longer identical, but still admit the same solution vector $v^{(s)}$. Finally, heuristic H1 needs to be analyzed in more detail and tested on actual networks; this will be the subject of another report.

REFERENCES

[1] Dantzig, G.B. et al., "Formulating and solving the network design problem by decomposition", Transportation Research-B, 13B, pp.5-17 (1979).

[2] LeBlanc, L.J. and Abdulaal, M., "Continuous equilibrium network design models", Transportation Research-B, 13B, pp.19-32 (1979).

[3] Los, M., "A discrete-convex programming approach to the simultaneous optimization of land-use and transportation", Transportation Research-B, 13B, pp. 33-48 (1979).

[4] Smith, M.J., "The existence, uniqueness and stability of traffic equilibria", Transportation Research-B, 13B, pp. 295-304 (1979).

[5] Steenbrink, P.A., "Optimization of Transportation Networks", Wiley, New York (1979).

[6] Tan, H.N., Gershwin, S.B., Athans, M., "Hybrid optimization in urban urban traffic networks", MIT report no. DOT-TSC-RSPA-79-7 (1979).

[7] Wardrop, I.G., "Some theoretical aspects of road traffic research", Proc. Inst. Civil Engineers, Part II, 1, pp. 325-278 (1952).

[8] Wong, R.T., "A survey of network design problems", MIT working paper OR-080-78 (1978).

AFTERWORDS

Submitted by G. F. Newell

The author explores some interesting questions regarding the comparative properties of the system and user optimal network design problems. Within the class of cost functions considered, however, only the concave systems are of much practical interest.

In the case of a linear investment cost, proposition 1 can be generalized. If

$$S_a(x, z_a) = d_a S(v_a/z_a) \tag{1}$$

with S <u>any</u> function of v_a/z_a (independent of a), and

$$G_a(z_a) = \mu d_a z_a , \tag{2}$$

then

$$F(v, z) = \sum_a d_a v_a [S(x_a) + \mu/x_a] , \tag{3}$$

with

$$x_a = v_a/z_a ,$$

and

$$\min_{\substack{v\in\Phi \\ z\geqslant 0}} F(v, z) = \left\{\min_{v\in\Phi} \sum_a d_a v_a\right\}\left\{\min_{x>0}[S(x) + \mu/x]\right\} . \tag{4}$$

In essence, this means that each facility of "size" z_a costs z_a times a unit facility and can carry z_a times as much flow at the same cost per km. The system acts as if each facility of size z_a is like z_a facilities in parallel, for any z_a (not necessarily integer).

The form (1), (2) is quite artificial. Aside from the hypothesis of homogeneity, that $G(\cdot)$ and μ are independent of a, the linear form (2) is unrealist If one were to include in the set of arcs anything that one might build, the formal solution for the optimal network would give a $z_a > 0$ for nearly everything one proposes. The value of a z_a, however, will likely correspond to an artificial facility with a "width" less than one lane.

To overcome the latter difficulty, one could generalize (2) so that

$$G_a(z_a) = \begin{cases} G_a + \mu d_a z_a & z_a > 0 \\ 0 & z_a = 0 , \end{cases} \tag{5}$$

for arbitrary initial costs G_a. The evaluation of the optimal network, i.e., the set of a with $z_a > 0$, is a potentially difficult integer programming problem, but proposition 1 is still true, i.e., the system and user optimal assignments are equal.

That the system and user optimal network designs are the same is certainly true for a much wider class of cost functions than the above but it remains unclear whether it is (nearly) true for (most) realistic cost functions.

BLOCKING EFFECTS FOR SYNCHRONIZED SIGNALS

G. F. Newell
Institute of Transportation Studies
University of California
Berkeley, California

We consider various common schemes for synchronization of traffic signals along an arterial highway carrying two-way traffic with negligible turning movements. For flows close to capacity it is usually not possible for traffic to flow in both directions with no interruption. If queues form at a signal, there is a danger that the queue will back-up and block the upstream signal, possibly causing a loss in capacity. It is shown, in particular, that the simultaneous synchronization does not cause blocking, but, if signals are set for progression in one direction, blocking could occur in the opposite direction. Blocking may also occur for the double or triple alternate synchronization schemes.

1. INTRODUCTION

In choosing off-sets and cycle times for traffic signals along an arterial highway carrying two-way traffic, one would ideally like to choose the off-sets between adjacent signals equal to the transit time between these signals (perfect progression), but for both directions simultaneously. Unfortunately this is usually not feasible. If the transit time between a pair of signals were equal to some number τ in one direction and τ' in the opposite direction, then it would be possible to have two-way progression if the cycle time were equal to $\tau + \tau'$ (or $\tau + \tau'$ is any integer multiple of the cycle time). In typical urban areas, however, where there are signals at every intersection, $\tau + \tau'$ is likely to be less than 30 seconds, and a cycle time of 30 sec. (or less) is usually too short to provide the necessary capacity for the highway traffic or to provide sufficient time for pedestrians to cross.

Even if the transit times τ_j and τ'_j between the jth pair of adjacent signals were large enough so one could operate with a cycle time of $\tau_j + \tau'_j$, a common cycle time for all signals cannot be equal to $\tau_j + \tau'_j$ for all j unless the $\tau_j + \tau'_j$ are all equal.

In the standard traffic engineering references [1-3] the most commonly discussed schemes of synchronization are:

1) simultaneous
2) alternating
3) double (or triple) alternate
4) one-way progression
5) maximize bandwidth
6) minimize total travel time (or "cost").

The usual space-time representations of the first four schemes are shown in Figure 1 for a sequence of equally spaced intersections having the same red-green splits (but not necessarily with the red times equal to the green times).

There are various computer programs designed to determine the optimal settings according to scheme (6), for those who have access to a computer and are willing to measure all the parameters necessary to run the program. The other schemes should, under appropriate circumstances, approximate the "optimal" and have the advantage of easy implementation. Most traffic engineers would obviously prefer a simple nearly optimal scheme to some complex recipe that requires frequent adjustment.

Except in the rare situation in which the cycle time (or a multiple of it) can be chosen as the transit time $\tau = \tau'$ so as to achieve a two-way progression (which typically would apply only for signals at least a half mile apart), the simultaneous synchronization may be advantageous when the demand is close to capacity in both directions and the cycle time is considerably larger than the transit time between intersections, a common rush hour

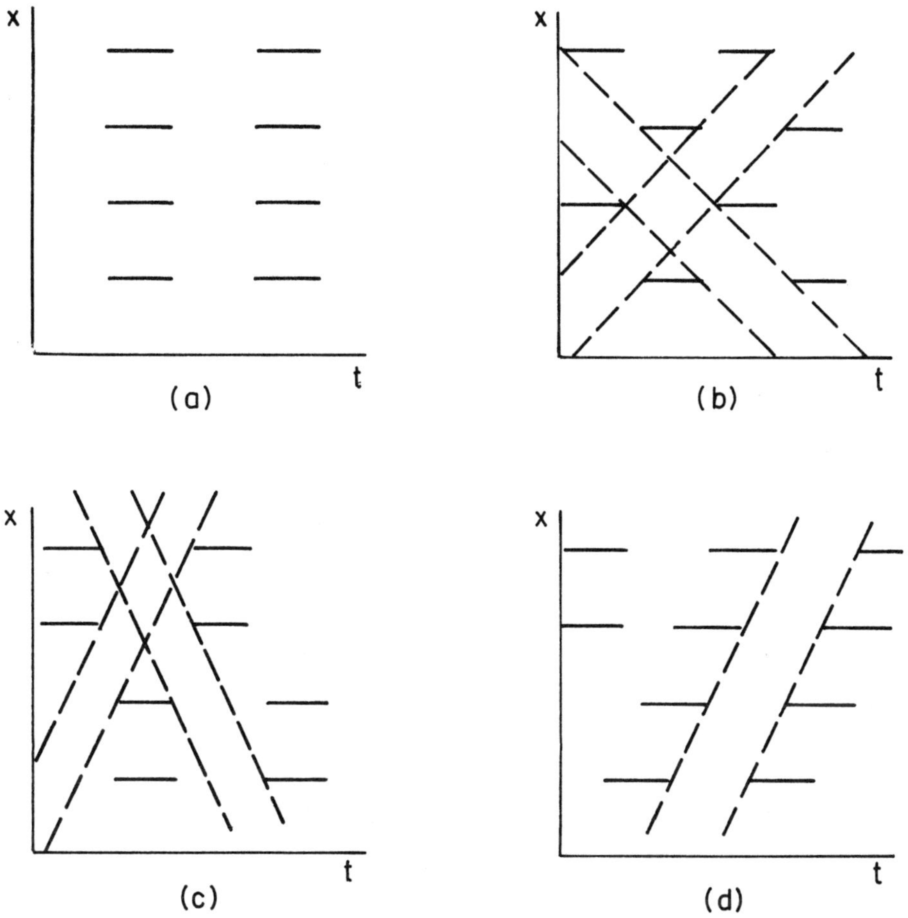

Figure 1 - Signal synchronization schemes (a) simultaneous (b) alternating (c) double alternate (d) one-way progression

situation in city centers. Although a driver can travel only a few blocks before being stopped, the total delay under such circumstances is quite insensitive to the offsets[4]. The advantages of this scheme are that both travel directions are treated equally, the settings are independent of travel time or block lengths, and there is usually little difficulty in coordinating the signals on one highway with other signals in the network of parallel or cross streets. Opponents of the scheme claim that this encourages excessive speeds; a driver will try to go as far as possible during the green. On the other hand, a lead car which accelerates quickly will minimize the start-up loss time and help to increase the flow during the green time, thus increasing the capacity.

The alternating scheme (b) is the ideal setting mentioned above when the cycle time can be chosen to be $\tau + \tau'$. The two-way progression is indicated by the through bands (broken lines) of width equal to the green time. If such a cycle time is unacceptably short, this scheme is not desirable. The average time required to travel one block (including stops) will become half the cycle time, independent of the block length.

In the double (triple) alternate schemes the cycle is chosen as the sum of the times to traverse two (three) blocks in both directions. Such a cycle time, two (three) times that of the (single) alternating scheme, is often in an acceptable range. As shown in Figure 1c, this double alternate scheme creates equal through bands in both directions simultaneously. If R and G are the effective red and green times, respectively, the through bands have a width:

$$(3G - R)/4, \quad \text{for} \quad 3R > G\ ;$$

for $R = G$ the width is $G/2$. In the unlikely case $3R < G$, there are two through bands per cycle.

The double or triple alternate schemes are advantageous

if the block lengths are nearly equal and the demand is low. It is possible to increase the bandwidth in one direction at the expense of an equal decrease in the opposite direction[5] but, for the traffic to move freely, the flow in the two directions must fit in the combined band of width $(3G - R)/2$ for the double alternate scheme. Since the capacity flow in the two directions is that which can pass during a time G, the traffic can flow in the bands only if the two direction demand/capacity is less than $(3G - R)/4G$, $1/2$ for $R = G$. For the triple alternate scheme the bandwidth is even smaller, only $1/3\ G$ for $R = G$.

If the demand exceeds that which can flow in the band the usual motivation for these schemes is lost. They are not likely any longer to minimize delays or stops. Indeed, as we shall see later, they may even cause blocking which will reduce the capacity.

The double or triple alternate schemes are special cases of the more general schemes designed to maximize the combined bandwidths in two directions, particularly for situations in which the block lengths and red-green splits may differ from one intersection to the next. In typical applications the bandwidths are only a fraction of the green times. As with the double or triple alternate schemes, these more general schemes are appropriate only if the demand can be accommodated in the bands.

In the one-way progression the off-sets are chosen equal to the transit times in one direction. This can, of course, be done for any block lengths and is independent of the cycle time. Usually the progression is given to the direction of heavier flow. Whether or not this is acceptable depends upon how badly the opposing traffic is treated. Typically there is no through band for the opposing flow. In addition to the fact that this traffic suffers considerable delay when the demand is below capacity, there is also a danger that the queues may cause blocking if the cycle time is too long.

The main objective here is to investigate the possibility that a queue behind one signal will back up and block the upstream intersection, thereby reducing the capacity of the system. To calculate the physical length of the queue as a function of all the signal settings and the dynamical behavior of the traffic would not be a very profitable exercise. First of all, existing fluid or car-following theories are of questionable accuracy. But even if they were reliable, to evaluate queue lengths and effects of blocking for all possible off-sets, cycle times, etc. would be hopelessly tedious. Instead, we will consider some qualitative aspects, particularly as they relate to the common signal strategies described above.

We will be concerned mostly with the flow along a single arterial highway carrying two-way traffic with negligible turning movements. Blocking effects will be most troublesome if the demand is near or exceeds the capacity. Under such conditions the first objective is to make the capacity (through-put) as large as possible, subject to certain restrictions.

If all signals operate on a common cycle time C, the jth signal has a saturation flow s_j in one direction and s'_j in the opposite direction, and an effective green time G_j, then the maximum flows that can pass the jth signal are

$$\frac{s_j G_j}{C} \quad \text{and} \quad \frac{s'_j G_j}{C}$$

for the two directions. If there is no blocking, the maximum flows that can pass through the sequence of signals is:

$$\min_j \frac{s_j G_j}{C} \quad \text{and} \quad \min_j \frac{s'_j G_j}{C} .$$

This capacity can be increased if one can increase the smallest of the $s_j G_j/C$ (or $s'_j G_j/C$). Obviously an increase of the appropriate s_j will achieve this, but we will assume that the s_j has already been made as large as possible (by widening the roadway, ban of parking, etc.). An increase of G_j for fixed C can be achieved only by taking green time from the cross street. This can be done up to the point at which the capacity for the cross street traffic has been reduced to the existing flow.

Having done these things, the only remaining freedom is to increase C (for a fixed red-green split). This will reduce the fraction of time lost in switching (yellow time and/or start-up time). As a practical matter, however, one usually sets some bound on C (about two minutes) because the gain in capacity from further increase in C is quite small and drivers do not like long cycle times (if, for example, the time lost per cycle with a two-minute cycle time is 6 percent, one only gains 3 percent by going to a four-minute cycle time).

In the absence of any blocking, the capacity described above is independent of the off-sets. One could, potentially, gain some capacity at an intersection if the signal turns green just as a platoon arrives, provided there is no residual queue from the previous cycle or from turning traffic. This gain comes from eliminating some of the start-up time, but any flow pattern which tries to exploit this extra capacity will be unstable. A stray car which is stopped at the intersection will destroy the gain, cause an overflow into the next cycle and in all subsequent cycles.

If, having made the capacity as large as possible, we still find that demand exceeds capacity, queues will grow somewhere. The time at which any car passes the bottleneck is essentially independent of how it got there, i.e., of the signal off-sets upstream (provided only that a car is at the intersection when it is its turn to be served). If there is some place upstream of the bottleneck where one can more conveniently store vehicles, then one might use the signals upstream to regulate the flow; in effect, transfer the bottleneck to where the queue can be stored. The off-sets can then be chosen at least to provide as smooth a flow as possible downstream of the main queue. If, on the other hand, one can obtain sufficient capacity to accommodate the flow (barely), the off-sets can be chosen to minimize the travel time or some other objective.

Blocking could affect the capacity of any intersection and create a bottleneck at some intersection other than as described above, but there will certainly be a loss in system capacity if the signal of the smallest $s_j G_j/C$ (or $s_j' G_j/C$) cannot maintain a flow s_j during the green. This may happen either because a queue downstream backs up and blocks the output or a queue upstream blocks an intersection which subsequently cannot supply enough cars to keep the bottle-neck signal busy.

2. SIMULTANEOUS SYNCHRONIZATION

We consider first the case of simultaneous synchronization partly for its own sake and partly as a point of reference for other schemes.

Figure 2 shows some representative vehicle trajectories passing through a sequence of simultaneous signals with unequal spacing but $\tau_j < G$ for all j. For $\tau_j > G$, the figure would be somewhat different, but this is an unlikely situation and will not be considered. The figure shows some trajectories that would exist for saturated flow. If the flow is less than capacity, one need only erase a few trajectories at the end of the platoons. Such a simultaneous synchronization for one direction is also simultaneous for the opposite direction; anything that is true for one direction is true for the other. The trajectories for opposing direction correspond to turning Figure 2 upside down (and reversing the order of the signal spacings).

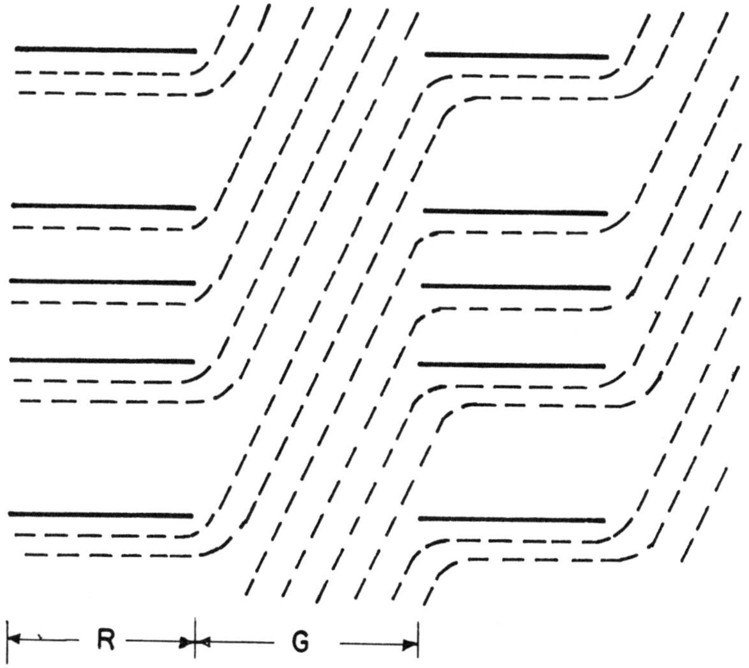

Figure 2 - Typical vehicle trajectories for simultaneous synchronization.

From Figure 2 or variations of it we can readily observe the following features:

a) A vehicle leaving any signal after being stopped will travel approximately a time G before being stopped again (the average for all vehicles is almost exactly G). The stopped time of a car is approximately R regardless of its position in the queue. Thus each car spends a fraction of time G/C moving and R/C stopped. The signals cause the average trip time to increase by a factor C/G (the same factor that determines the capacity) independent of the spacing between signals, the flow, or the cycle time itself.

b) For any flow less than capacity, the trajectory plot shown in Figure 2 is stable. If one were to inject a few extra cars into the system (or remove a few), the resulting disturbance would propagate with the traffic stream and new cars fed into the system from upstream would regenerate the same types of trajectories.

c) There is no blocking, regardless of the cycle time or spacings between intersections. As one can see from Figure 2, if the flow is barely at capacity, the number of cars stopped by any jth signal is the number which were in the space between the j - 1th and jth signals and moving when the signals turn red. Since a platoon compresses as it is stopped, there is obviously sufficient space to store the queue.

If k_j is the jam density, i.e., the density in the queue, and k_m is the average density of moving cars when the signal turns red (which should be close to the density corresponding to maximum flow), the space consumed by the stationary queue is only a fraction k_m/k_j of the available space (independent of the signal spacing, cycle time, etc.).

d) If the flow increases, approaches, and tries to exceed the capacity, the flow pattern may become unstable (regardless of the off-sets). Since, for a main street flow barely at capacity, there will be empty space on the highway after the queue compresses, there is room for cars to turn onto the highway from the cross streets. Even though the turning rates may be very (arbitrarily) low, even a single car injected into the system during one signal cycle will cause a residual queue of one car for the next cycle and all subsequent cycles. If the extra car is inserted behind the jth signal, the residual queue may stay at the jth intersection (over and over again) or it could propagate back to the j - 1th intersection and possibly continue back to the input.

There are many possible equilibrium flow patterns that could develop if the demand on the system is at or slightly above capacity. Figure 3 illustrates a type of stable flow pattern that might evolve from Figure 2. Unfortunately the possible patterns that could evolve depend upon the spacings between signals, the cycle times, and the dynamics of queue propagation.

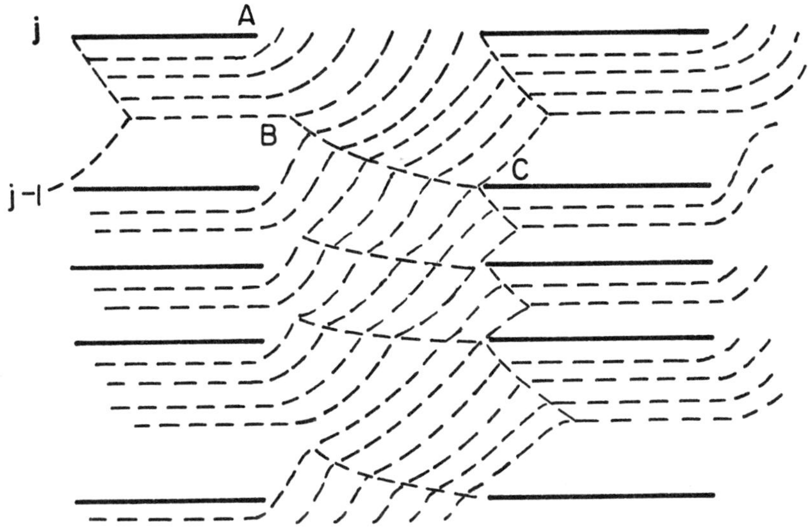

Figure 3 - A stable flow pattern at saturation for simultaneous synchronization.

In Figure 3 there are four cars in the queue at the jth signal at the start of the green instead of only two cars as in Figure 2. These extra cars may have drifted in from the cross streets or parking lots. The green signal at A first serves the cars in queue. Meanwhile, the platoon from upstream joins the back of the queue at B and a deceleration wave (shock) propagates upstream. If this wave reaches the j - 1th signal while that signal is still green, it could disrupt the output from the j - 1th signal causing one (or more) fewer cars than normal to leave that signal during the green. The extra cars from the jth signal would thus generate an overflow at the j - 1th signal. This, in turn, could continue to generate an overflow at the j - 2th signal, etc., until the disturbance propagates back to the main street source where there is presumably already a queue of waiting cars.

If, on the other hand, the stopping wave from B does not reach the j - 1th intersection before that signal turns red and cuts off the flow, the residual queue stays at the jth signal. If the output from the j - 1th signal matches that of the jth signal, we will continue to have the same residual queue every cycle, but eventually other cars may sneak in from the cross streets.

Figure 3 shows a stable pattern in which the deceleration wave reaches the j - 1th signal just as it turns red, point C . If any new car now tries to sneak in, it will create a disturbance that will propagate upstream, eventually to the entrance queue. If any car turns off the road creating empty space, another car will eventually sneak in to take its place.

Figure 3 shows residual queues at all intersections that repeat every cycle. If the cycle time is long enough and the spacing between signals is short, any disturbance from the jth signal will propagate upstream; there will be no residual queue except from the main street flow as illustrated in Figure 2. If, however, the spacing between intersections is sufficiently long and the cycle time is short, the residual queue could be so long that it would take more than one green time to serve the residual queue. There will be several deceleration waves between the j - 1th and jth signals , and any main street car must wait several cycles to travel between the two signals.

If the demand on the system exceeds the capacity, a queue will grow somewhere; and if cars have access to an empty space, they will use it. The residual queues shown in Figure 3 increase the travel time in the section of highway shown on that figure as compared with Figure 2; but if the residual queue were not in this section of highway, it would be upstream somewhere.

Similar figures can be drawn for other signal off-sets. We will not try to analyze here any possible advantages of various synchronization schemes when demand exceeds capacity. It appears, however, that the scheme of simultaneous synchronization is the one that minimizes the size of the stationary residual queues between signals and, therefore, also the travel time over the section of highway under consideration. This may be either the most desirable or the least desirable depending upon where one wishes to store excess vehicles. The time at which any vehicle leaves the system is essentially independent of where the queue is stored.

This is not an illustration of what we will later describe as "blocking." Every intersection is kept busy (most of the time).

3. PROGRESSION

As the off-sets of Figure 2 are retarded, the queues decrease for the traffic direction shown. There will (ideally) be no delay if the off-sets are equal to the transit times. But suppose we were to choose the off-sets to favor the opposite direction of the travel, i.e., the off-sets are chosen to be $-\tau_j$ as in Figure 4.

Figures 2 and 3 were drawn with $k_m/k_j = 1/4$ to exaggerate the compression of the platoon and the possibility of stable residual queues when the system is oversaturated. Figure 4, however, is drawn with $k_m/k_j = 1/2$. If the speed decreases linearly with the density as in the Greenshield relation, then the maximum flow does occur at a density $k_m = 1/2\ k_j$. Although there is some empirical basis for this choice, one cannot take it too seriously.

If the flow is (barely) at saturation, $k_m/k_j = 1/2$, and $G > 2\tau_j$ for all j, then there will be just enough space to store the queue at every signal as shown in Figure 4 (the number of cars in each queue is twice that of Figure 2). More generally, the number of cars in a queue

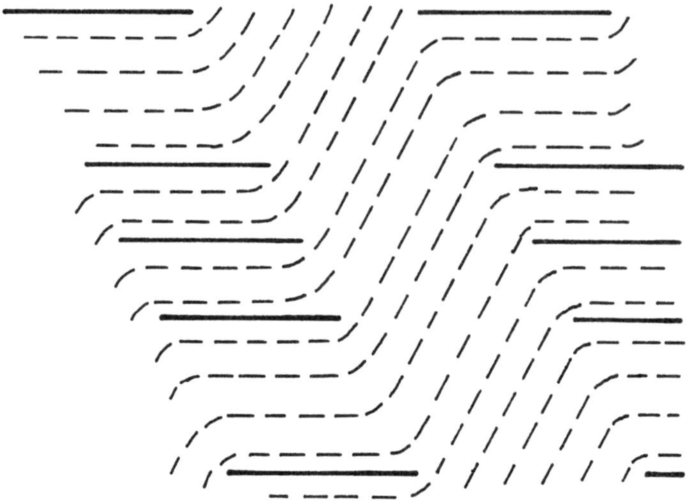

Figure 4 - Flow pattern corresponding to progression in the opposite direction.

j

j-1

j-2

j-3

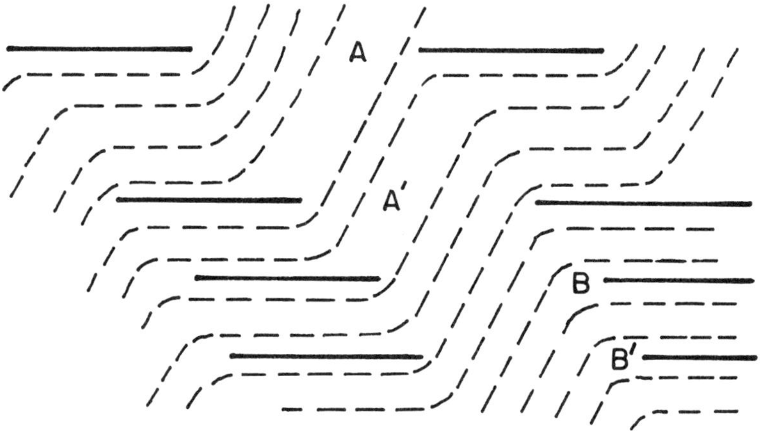

Figure 5 - A typical blocking pattern.

behind the jth signal when it turns green will be the smaller of $2s\tau_j$ or sG at saturation.

If $k_m/k_j > 1/2$ with the off-sets of Figure 4, or the off-sets in Figure 4 are advanced slightly, or there should be fluctuations in the output from the signals of Figure 4, then blocking may occur. A necessary condition for blocking to occur is that the space between some pair of adjacent signals cannot store sG cars, a condition which is likely to be true during the rush hour when the green time is made as large as possible in an attempt to achieve a high capacity. It is also necessary that the off-sets be in a certain range.

A typical blocking pattern is shown in Figure 5. Some jth signal can store less than sG cars. The highway downstream is empty, so this signal can clear the cars in the queue, but the off-sets are such that the queue empties at A before the platoon from the j - 1th signal arrives. Some green time is therefore wasted because cars cannot get to it.

In equilibrium, the flow past all signals must be the same. If all signals have the same green time but some time is wasted at the jth signal, the same time must be wasted at every other signal. In Figure 5, the off-set between j - 2 and j - 1 has been chosen so that the queue vanishes also at the j - 1th signal at A'. The output from j - 1 matches that of j. The off-set between j - 3 and j - 2, however, has been chosen so that the platoon from j - 3 arrives when (or before) the queue has been served.

The output from j - 2 must match that at j - 1 but now we see at point B that the storage becomes full. The output from the j - 2th signal is blocked; a car that would normally have passed the signal has no place to go and must wait. The figure also shows a similar blocking of the $j - 3$th signal at B' because more cars arrive from upstream than can pass through the system.

In a long sequence of signals one can have a wide variety of trajectory patterns with some signals being underutilized as at A or A' and others blocked as at B or B' or even both effects at the same signal in the same cycle. If these effects do occur and the system is oversaturated because of it, the penalty is very high; queues will grow somewhere upstream. This can always be avoided, however, possibly at the expense of some increase in travel time for the opposite travel direction but without loss of capacity for the opposite direction. Blocking does not occur for the simultaneous synchronization or for any off-sets intermediate between those of Figures 2 and 4.

To a driver or some naive observer it may appear that the trouble spot is at points such as B or B' but actually the cause of the blocking is at A or A'. There is not much point in analyzing the delays and capacities for such a system. It suffices to recognize the pattern when it exists and how to avoid it. Unfortunately, standard traffic engineering manuals do not give any warning of this.

4. DOUBLE (TRIPLE) ALTERNATE

Although the double (or triple) alternate synchronization scheme loses its advantage once the flow exceeds that which can be accommodated in the through bands, manuals again fail to warn engineers of its limitations. Apparently many signal systems set with double alternate synchronization to accommodate off-peak traffic are left with the same settings even during the rush hour.

Figure 6 shows a double alternate signal setting with equal red and green times, equal spacing between signals and a cycle time equal to 4τ, the standard "textbook" illustration that gives through bands of width $G/2$ in both directions. If, however, this were used to carry a flow of barely sG (if possible) and $k_m/k_j = 1/2$, the flow pattern shown in Figure 6 would result.

The pattern between the simultaneous signal pairs is similar to Figure 2, that between the alternating pairs is similar to Figure 4. There is barely enough space to store the queue at point B . Statistical fluctuations in output from various signals could cause occasional blocking. There would also be blocking for $k_m/k_j > 1/2$.

It is not necessary that the red and green times be equal for a double alternate synchronization. For an arterial highway carrying heavy traffic one might choose the green time larger than the red, as in Figure 7, in order to have wider through bands and possibly also higher capacity.

If, for a fixed cycle time $C = 4\tau$, the green time is increased (at the expense of the cross street) and the flow is kept at its maximum value, the queue between the alternating signal pair increases until blocking occurs. If blocking occurs, the number of vehicles that can pass this signal pair will be determined by the storage rather than the green time. Any further increase in green time contributes nothing to the capacity.

Figure 7 has the same intersection spacing and cycle time as Figure 6 but a longer green (both have $k_m/k_j = 1/2$). The flow in Figure 6 was already on the verge of blocking. In Figure 7 signal $j - 1$ interrupts the flow so that no car can reach signal j at point A . Signal j can only serve the cars stored in the queue. Of course, the flow must be the same past every signal. The lack of cars at A creates a void also at signal $j + 1$ at point C . The maximum main street flow in Figure 7 is the same as in Figure 6, but Figure 7 provides less time for the cross street traffic.

Certainly the double alternate synchronization is not a satisfactory scheme for flows near capacity. The triple

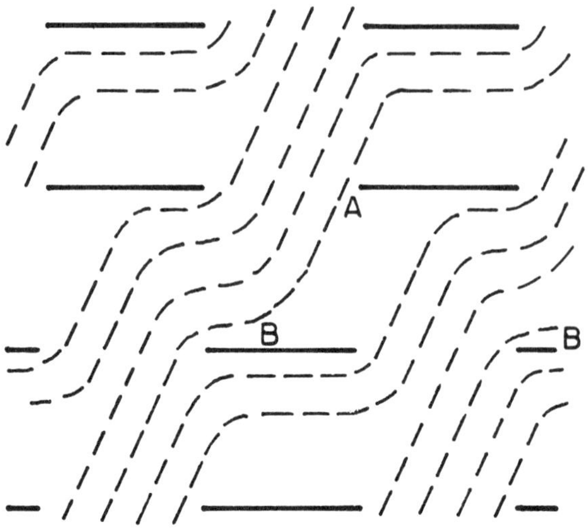

Figure 6 - Double alternate with equal red and green.

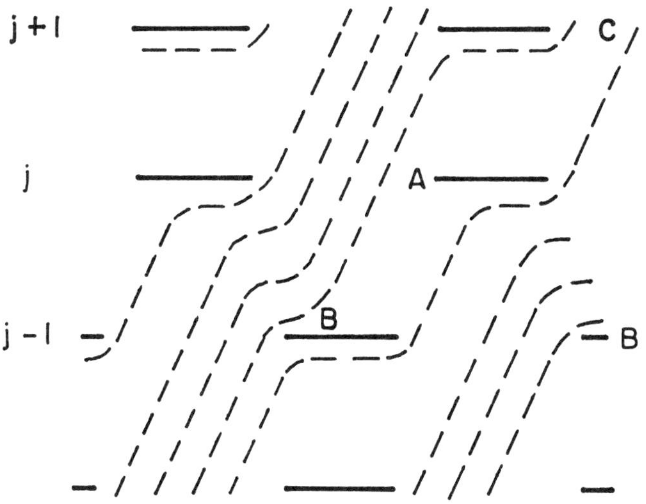

Figure 7 - Double alternate with G > R .

alternate scheme is even worse. For $R = G$, the capacity of the triple alternate scheme is limited by blocking even for $k_m/k_j \geq 1/3$ (a condition which may well be true for all traffic conditions). That the through bands have a width of only $G/3$ also makes this scheme questionable even for light flows.

5. CONCLUSION

Most of the things described here should be "well-known" to anyone who has actually worked with signal settings but, as is often true of the transportation literature, many things that a practitioner should know are not published in the standard references. Apparently even the computer programs for signal settings, presumably designed to minimize travel time or cost, treat cars as if they had zero length and disregard the physical length of the queue. If blocking occurs, the programs would not recognize it and could, therefore, completely miscalculate the delay (even the capacity).

To estimate the physical length of a queue under all conditions is quite difficult since existing theories are of questionable accuracy. Even if computer programs were designed to include blocking effects, they should be viewed with some doubts. Having set the signals according to any recipe, however, one should observe the system and see how it performs.

When blocking occurs the system is likely to be nearly saturated and have excess cars turning onto the main street. When this happens one should not focus attention on the chaos that may exist where the blocking occurs. One should test to see if any signals downstream of the blocking have green time that is not used because the queue vanishes. Off-sets should certainly be chosen so as to guarantee that any key intersection has cars to serve at all times during the green.

REFERENCES

1. J. E. BAERWALD, ed., Traffic Engineering Handbook, Institute of Traffic Engineers, 1965.

2. J. E. BAERWALD, ed., Transportation and Traffic Engineering Handbook. Institute of Traffic Engineers, Prentice-Hall, Englewood Cliffs, N. J., 1976.

3. F. V. WEBSTER and B. M. COBBE, Traffic Signals. Her Majesty's Stationery Office, London, 1966.

4. G. F. NEWELL, "Traffic Signal Synchronization for High Flows on a Two-way Street," Beiträge zur Theorie des Verkehrsflusses (Proc., Fourth Int. Symp. on the Theory of Traffic Flow), Karlsruhe, Germany, pp. 87-92, 1968.

5. J. D. C. LITTLE, "The Synchronization of Traffic Signals by Mixed-Integer Linear Programming," Operations Research, 14, pp. 568-594, 1966.

AFTERWORDS

Comment by R.H. Hewitt, University of Glasgow.

I think that Figure 5 of Professor Newell's paper is rather oversimplified as it assumes that vehicles always travel at the optimum speed. I have redrawn this figure assuming Greenberg's relationship between speed and concentration, which I have interpreted as meaning that the speed of a vehicle will be proportioal to the reciprocal of the distance headway between it and the preceding vehicle. In the revised diagram, both vehicles between points A and A' accelerate to a higher speed because of the gap in the traffic at A, resulting in both passing during the green at the jth signal. There is a chain-effect of increased speeds down the line. However, only 5 vehicles manage to pass the j -1th signal creating a gap at C. During the next cycle, which is not shown, the gap at A will be larger by one vehicle space, and the speeds of the following vehicle correspondingly greater, which may result in the queue between the j and j - 1th signals reverting from three to four hence creating an oscillation. Obviously there will still be a blocking effect, but it may not cause quite as much delay as would appear from the diagram in the paper.

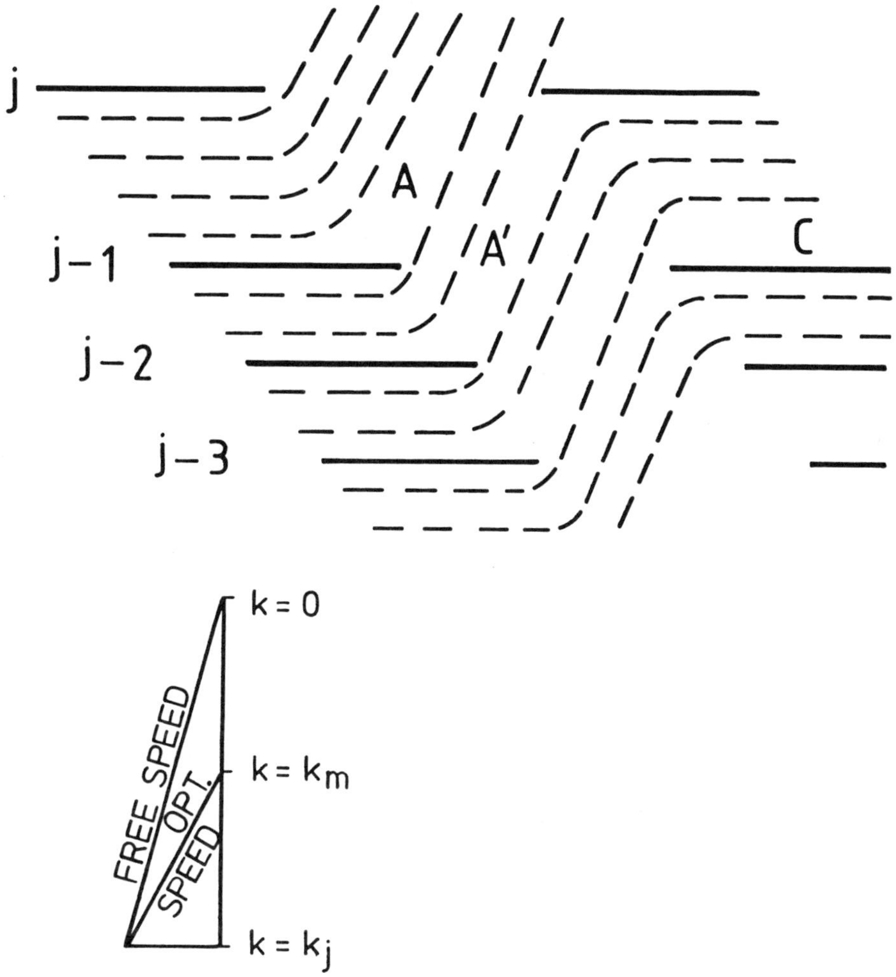
j
j-1
j-2
j-3
A
A'
C
k = 0
k = k_m
k = k_j
FREE SPEED
OPT. SPEED

A THEORETICAL STUDY OF TRAFFIC ASSIGNMENT AND TRAFFIC CONTROL

M. J. SMITH

Department of Mathematics
University of York
Heslington, York, England

The paper considers the steady state assignment problem for a signal-controlled road network. Examples are provided to show that current control policies may cause costly instabilities and do not ensure the existence of a consistent equilibrium. The paper gives a large collection of new control policies which, under natural conditions, ensure stable behaviour and the existence of a consistent equilibrium. It is shown that for certain simple networks total travel cost is sometimes sharply reduced by using one of the new policies.

1. INTRODUCTION

This paper presents a theory which combines traffic control and route-choice. The need for such a theory has been pointed out by Allsop [1] and Gartner [10]. A collection of new responsive traffic control policies arises naturally from our theory.

1.1. Motivation

Choosing control parameters which minimise, or approximately minimise, delay for the observed traffic pattern may increase delay once a new traffic equilibrium has been established. This is a well-known feature of the interaction between traffic control and route-choice; see, for example, Braess [3], Murchland [19], Smith [27] or Dickson [8]. This feature represents a substantial difficulty for the traffic engineer.

Our first objective is to show that, within a fixed demand mathematical model, far worse behaviour is possible; ordinary, currently used, traffic control policies can make quite straightforward networks unstable in a most undesirable, costly, manner. The instability is such that, as time passes, equilibrium is not approached and delays tend to infinity. Congestion continually increases as a result of the continual application of the control policy.

If such behaviour occurs, or begins to occur, in an actual road network as opposed to a mathematical model then the traffic engineer is presented with a second substantial difficulty.

Both difficulties present opportunities. Because the second difficulty may cause large unnecessary delays and may also be easier to circumvent, it seems natural to seek traffic control policies which certainly do not cause costly instabilities.

Our second objective is to actually provide a collection of alternative policies which do not provoke costly instabilities of the type considered here. The policies would perhaps be worth trying (after appropriate computer tests) in any case, but particularly when it appears that an existing policy may be increasing, rather than reducing, congestion.

Our third objective arises naturally from the first two; it is to show that, under natural conditions, there is always an equilibrium consistent with each of our new policies.

This result may appear to be of slight consequence, because the existence of equilibria is often taken for granted. Furthermore, many existence results appear in the literature, one of the first being proved by Beckmann [2] and one of the latest by Dafermos [7]. However, as soon as we have responsive signals, Beckmann's theory ceases to apply and as soon as we insist on explicit capacity constraints at junctions, Dafermos' theory ceases to apply. Thus, at least from a theoretical viewpoint, even slowly responsive traffic signals combined with capacity constraints at junctions force a re-examination of the existence problem. Because these two features, the signals and the constraints, are everyday features of any real urban road network, we may reasonably expect our theoretical re-examination to provide some practical insights.

Two alternative approaches to the traffic control problem

There seem to be two rather distinct approaches to the problem of urban traffic control when the origin-destination demand is considered to be fixed and a single criterion function is given. In the first approach, one determines the equilibrium traffic pattern for fixed signal-settings, examines the value of the criterion at this and at nearby equilibria, and then alters the signal-settings so as to reduce the criterion, taking proper account of changes in the equilibrium consequent upon changes in the signal-settings.

This first approach to the traffic control problem has been suggested by Allsop [1] and Gartner [10]. Gershwin and Tan [11] have found conditions on the signal-settings which are necessarily satisfied when the signal-settings minimise a given criterion at equilibrium; their conditions yield a direction in which the signal-settings may profitably be moved when the settings are not locally optimal. Dickson [9] independently found the direction which reduces the criterion (at equilibrium) at a maximum rate. In order to calculate such directions the origin-destination distribution must usually be known.

In the second approach, one seeks simpler traffic control policies, such as those in this paper, which achieve a more modest objective, hoping that as a byproduct, the initially given criterion will be significantly reduced.

In order to carry through the first approach, it is necessary to have estimates of the origin-destination distribution. As we shall see in this paper, such estimates are not necessary for the second approach. Furthermore, the first approach may yield only a local minimum of the criterion. (This is shown by the counterexample in Smith [27].) On the other hand, there is no real hope of precisely minimising the criterion if the second approach is adopted.

In this paper we pursue the second of these two approaches, hoping for a robust, practical and reasonably satisfactory solution to a difficult problem.

1.2. Theoretical Context

The paper lies in the interface between traffic assignment and traffic control. The traffic control policies which arise in the paper encourage drivers to choose routes which are economical in their consumption of junction space.

The paper originated from a study of road pricing, which also lies on our interface. (See Smith [25].) We have sought to apply the basic ideas behind road pricing without recourse to financial charging mechanisms; traffic signals and time replace toll booths and money. Road pricing itself has been studied in detail by many authors, including Pigou [23], Knight [16], Smeed [24], Tanner [30], Marchand [17], Netter [21,22] and Dafermos [6].

We develop ideas which appear in Dafermos [5], Netter [21] and Smith [26].

1.3. Practical Context

There has been much practical research into area traffic control. See Holroyd and Robertson [15], Holroyd [14], Holroyd, Joyce and Hillier [13], Tarnoff [31] and Henry, Ferlis and Kay [12] for a partial summary of

British and U.S. experience. Overall, responsive systems have so far proved somewhat disappointing. Recently, however, SCOOT has been more encouraging.

There has been less research into the control of a single intersection. A study, by De la Breteque, Duraz and Jezequel [4] shows that Miller's algorithm (see Miller [18]) compares favourably with a number of other algorithms.

Practical implications

The control policies which arise from this theoretical study are new. Furthermore, these policies do, by virtue of the theoretical work presented in this paper, have a suggestive theoretical basis. Thus we have, in building this theory, provided the traffic engineer with some specific ideas, firmly rooted in theory, which strike out in new directions.

Tests are required in order to determine whether our control policies would provide significant practical benefits in actual towns. These tests could be performed on a real road network or they could be performed on a computer simulation which embodies route-choice.

1.4. Assumptions

In both the examples and the theory which follow we make a number of assumptions. The two main ones are that lost-times are zero and that queue-storage capacity is not limited.

The first of these assumptions is not vital to our theory, but the second one is. It would be interesting to seek extensions of the theory presented here which allow for finite queue-storage capacities.

2. FOUR EXAMPLES

We shall discuss four simple examples. These examples are designed to illustrate the basic principles on which this paper is based; more complicated or realistic examples would not as readily achieve this purpose. It would be advisable to consider more complicated and realistic models before attempting to apply the theory of this paper in practice.

Two of the examples illustrate the costly instabilities which can arise when ordinary traffic control policies are used.

The other two examples illustrate the beneficial, stabilising, effect of our new policies.

A graph is provided which roughly indicates the cost benefits deriving from the new policies, for the networks considered in the examples.

2.1. The First Example

We give here a network, some feasible demands and an ordinary control policy. In the example, a costly instability is present and there is no consistent equilibrium for certain feasible demands.

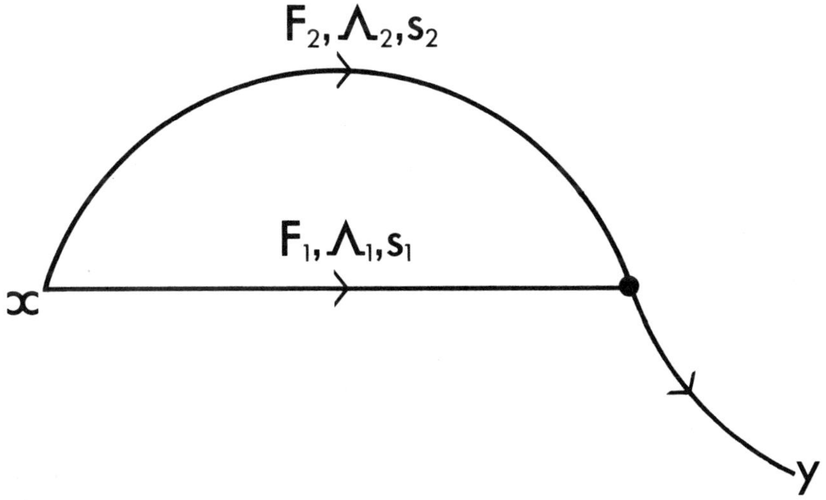

Figure 1. Network geometry for the first example.

Supply

Figure 1 shows the network geometry. The junction is signal-controlled, and two routes join the origin x to the destination y. The first route, R_1, has a saturation flow $s_1 = 1$ at the junction and the second (wider but longer) route , R_2, has a saturation flow $s_2 = 2$ at the junction. The link with no saturation flow marked has a very large saturation flow.

Figure 1 shows the average flows (vehicles per second) along the links in terms of the route-flows F_1 along R_1 and F_2 along R_2, and the proportions, Λ_1 and Λ_2, of green-time awarded to those links terminating at the junction.

We suppose that if a link (terminating at the junction) has flow f, green-time proportion λ and saturation flow s then the average delay d (seconds) felt by a single vehicle at its end is given by:

$$d = d(f,\lambda,s) = \frac{9}{20}\left\{ \frac{c(1-\lambda)^2}{1-f/s} + \frac{f}{\lambda s(\lambda s-f)} \right\}, \tag{1}$$

where c seconds is the cycle time of the signal. This is Webster's two-term delay formula. There is to be no delay at the end of the third link.

The undelayed travel time for route R_1 is to be r_1 (seconds) and for R_2 is to be r_2 seconds. We suppose that

$$r_2 = r_1 + 100.$$

Demand

Let the demand for travel from x to y be k (average number of vehicles per second). Then the flows F_1, F_2 satisfy:

$$F_1 + F_2 = k, F_1 \geqslant 0 \text{ and } F_2 \geqslant 0. \tag{2}$$

Let the signal allocate a total proportion of 1 (this is where we suppose lost-times to be zero) to the two conflicting approach links, and let there be a minimum green time proportion of 0·1. Then the green-time proportions Λ_1 and Λ_2 satisfy:

$$\Lambda_1 + \Lambda_2 = 1, \quad \Lambda_1 \geqslant 0.1 \quad \text{and} \quad \Lambda_2 \geqslant 0{\cdot}1. \tag{3}$$

(The similarity between (2) and (3) suggests that we place the green-time constraint under the "demand" heading.)

We shall denote by $\mathscr{D}_1$ the set of those route-flow vectors $\underline{F}$ satisfying (2) and by $\mathscr{D}_2$ the set of those green-time vectors $\underline{\Lambda}$ satisfying (3).

Feasibility

The vector $\underline{F}$ will be called demand-feasible if and only if $\underline{F} \in \mathscr{D}_1$

The vector $\underline{F}$ will be called supply-feasible if and only if there is a vector $\underline{\Lambda} \in \mathscr{D}_2$ such that

$$0 \leqslant F_1 < \Lambda_1 s_1 \quad \text{and} \quad 0 \leqslant F_2 < \Lambda_2 s_2. \tag{4}$$

The set of all supply-feasible route-flows will be denoted by $\mathscr{S}$.

A flow $\underline{F}$ is feasible if and only if $\underline{F}$ is both demand-feasible and supply-feasible.

Control policy

In this example we shall suppose that Webster's Method is used to determine $\underline{\Lambda}$ if an $\underline{F}$ in $\mathscr{S}$ is given, subject to $\underline{\Lambda} \in \mathscr{D}_2$ (See appendix 1 for the precise statement of this policy.) This $\underline{\Lambda}$ shall be called $\underline{\Lambda}^W(\underline{F})$.

Route-choice

We suppose that drivers tend to change their route for a quicker one, if a quicker one is available. If no one has an alternative route with a lower journey time, then, and only then, we have an equilibrium.

We also suppose that the journey time C^W_i along route R_i is the sum of the junction delay d^W_i and the undelayed travel time r_i, where d^W_i is determined from (1) with

$$c = 120, \ \lambda = \Lambda^W_i(\underline{F}), \ f = F_i \quad \text{and} \quad s = s_i.$$

(Here we substitute the policy functions $\Lambda^W_i(\cdot)$ in the delay formula.)

Dynamics and consistent equilibrium when $k < 1{\cdot}1$

If $k < 1{\cdot}1$ there is exactly one equilibrium consistent with the control policy. If this is (F_1^*, F_2^*), then, in fact,

$$\Lambda^W_1(\underline{F}^*) = 0{\cdot}9, \ \Lambda^W_2(\underline{F}^*) = 0{\cdot}1 \quad \text{and} \quad F_1^* > 10F_2^*.$$

The shorter, narrower route is by far the more popular.

Any feasible (non-equilibrium) flow $\underline{F} \neq \underline{F}^*$ will move towards $\underline{F}^*$ as drivers swap to the quicker route if the Webster setting $\underline{\Lambda}^W(\underline{F})$ is employed.

Here there is a consistent equilibrium and no costly instabilities arise.

Dynamics and consistent equilibrium when $k = 7/6$

Let $k = 7/6$ and $\underline{F}$ be any feasible flow. Then any $\underline{\Lambda}$ in $\mathcal{D}_2$ satisfies $\Lambda_2 \geqslant 0{\cdot}1$ and, hence, $\Lambda_1 \leqslant 0{\cdot}9$.

Since $\underline{F}$ is demand-feasible,

$$F_1 + F_2 = 7/6$$

and since $\underline{F}$ is supply-feasible,

$$F_1 < \Lambda_1(\underline{F})s_1 \leqslant 0{\cdot}9. \qquad (5)$$

Therefore $F_2 > 0$ for any feasible flow $\underline{F}$.

Now, for any feasible $\underline{F}$,

$$d^W_1 < d^W_2 + 100.$$

(See appendix 2 for a proof of this result.) Also

$$r_1 = r_2 - 100,$$

and so we must have

$$c^W_1 = r_1 + d^W_1 < r_2 + d^W_2 = c^W_2. \qquad (6)$$

It follows from (5) and (6) that any feasible $\underline{F}$ satisfies

$$F_2 > 0 \quad \text{and} \quad c^W_2 > c^W_1.$$

Therefore any feasible $\underline{F}$ is not an equilibrium as there are drivers, those on R_2, who have a quicker alternative route.

Since drivers tend to swap to a quicker route, F_1 will tend to increase and F_2 will tend to decrease. The route-flow will converge to the route-flow (5/6,2/6), which is not supply-feasible. As the route-flow converges to (5/6,2/6), delays tend to infinity.

We have a costly instability and no consistent equilibrium.

Equilibrium and dynamics when $1{\cdot}1 \leqslant k \leqslant 7/6$

The arguments just provided extend to this case; there is no consistent equilibrium and there is a costly instability.

Following Smith [27] it would be natural to define the capacity of our network, with our version of Webster's Method, to be 1·1. If $k < 1{\cdot}1$ then the network provides in a stable way for the demand k, while if $k = 1{\cdot}1$ the network does not provide in a stable way for the demand k.

2.2. The Second Example

We let k = 7/6. The only difference between this example and the last is that here we use policy P_o, which is fully specified later in the paper and here gives rise to a signal-setting $\underline{\Lambda}^o(\underline{F})$ (for any relevant supply-feasible $\underline{F}$) satisfying

$$s_1 d(F_1, \Lambda^o_1(\underline{F}), s_1) = s_2 d(F_2, \Lambda^o_2(\underline{F}), s_2).$$

We define d^o_i for i = 1,2 by:

$$d^o_i = d^o_i(\underline{F}) = d(F_i, \Lambda^o_i(\underline{F}), s_i).$$

It is clear that, since $s_1 = 1$ and $s_2 = 2$,

$$d^o_1 = 2d^o_2.$$

Therefore

$$d^o_1 - d^o_2 = d^o_2 = \frac{1}{2} d^o_1$$

is very large and positive if F_1 is close to 5/6 (and F_2 is close to 2/6). Thus there is a strong encouragement for drivers on route R_1 to transfer to route R_2 if F_1 is close to 5/6, and a flow (F_1, F_2) close to (5/6,2/6) is repelled from this unfeasible flow.

Our policy P_o stabilises this network for k = 7/6 - the previous continuing attraction of the narrower route R_1 is absent and a stabilising attraction for the wider route R_2 is present.

The policy P_o also ensures the stability of the network when $1{\cdot}1 \leq k < 1{\cdot}9$. If $k \geq 1{\cdot}9$ there is no feasible flow: $\mathcal{S} \cap \mathcal{D}_1$ is empty.

It seems natural to say that, with P_o, the capacity of the network is now 1·9 and, since 1·9 cannot be improved by any other policy, that P_o maximises the capacity of the network.

2.3. Comparing Delays

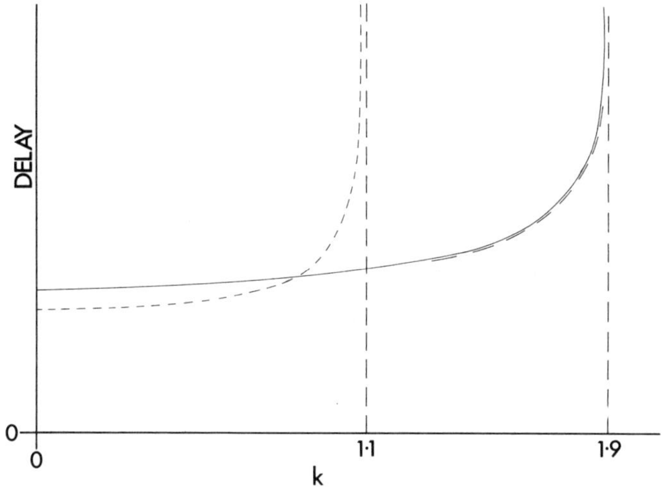

Figure 2. Comparing the performance of Webster's Method (broken curve) and P_o (unbroken curve).

Figure 2 shows roughly how the delay incurred per vehicle varies as k increases, in each of the two previous examples. The delay is evaluated at the equilibrium consistent with Webster's Method in the first example (dotted graph) and at the equilibrium consistent with P_o in the second example (solid line). (It so happens that, with this network, there is at most one consistent equilibrium in each case.)

There is no equilibrium consistent with Webster's Method if $1{\cdot}1 \leqslant k \leqslant 7/6$. For some values of $k > 7/6$ there will again be a consistent equilibrium (hence the disconnected part of the dotted graph) but this disconnected part could never be reached in practice if Webster's Method is consistently used and k steadily increases from zero. (Here the network capacity is reasonably defined to be 1·1.)

The dotted curve is roughly 14% below the solid curve for most values of k less than 1.1. For some values of k the dotted curve is far above the solid one, while for other values of k there is no Webster-consistent equilibrium at all. Webster's Method is, for this network, often better than P_o; but it is often far worse.

"Delay-minimising" settings

Similar results would obtain if the signal-settings are chosen to minimise delay for the current traffic pattern, instead of according to Webster's Method. The policy P_o would still often confer large benefits.

2.4. The Third Example

As in the first example, we give a network, some feasible demands and an ordinary control policy. In this example, for certain demands, a costly instability is present and there is no consistent equilibrium. This example is somewhat more realistic than those given earlier.

Supply

Figure 3 shows the network geometry, in which we shall only consider the six most direct routes; there is no real loss of generality here.

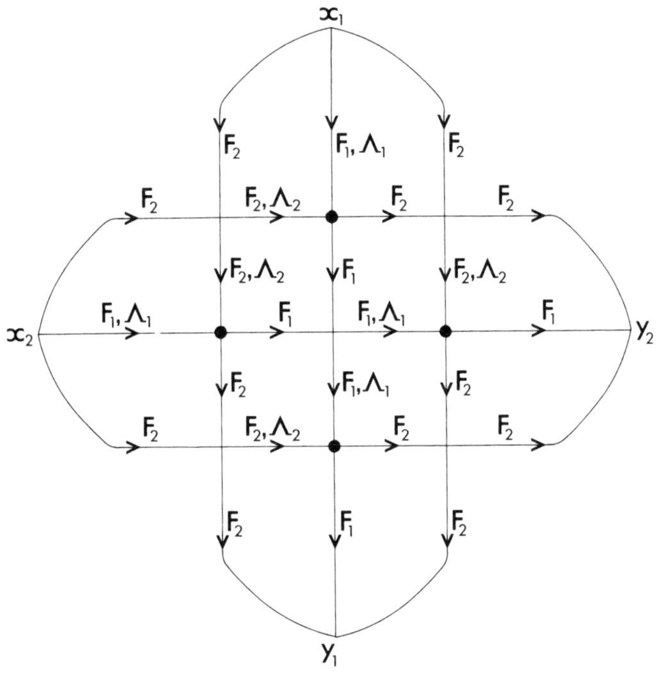

Figure 3. Network geometry for the third example.

All junctions are signal controlled. The emphasised junctions have approaches with a saturation flow $s_1 = 1 = s_2$. The unemphasised junctions have very generous approaches and impose a delay of no more than 60 seconds on each vehicle.

Figure 3 shows the average flows along the links in terms of the route-flows F_1, F_2, F_2, F_1, F_2 and F_2. (We shall restrict our attention to only such rather symmetrical route-flows.) Figure 3 shows the green-time proportions awarded to those links with a small saturation flow. (We shall restrict our attention to only such rather symmetrical proportions.) For these critical links, we suppose the average delay d is determined in terms of the link flow f, the green-time proportion λ and the saturation flow s by the formula (1), with c = 120.

The undelayed travel times for the four outer routes and for the two inner routes are to be r_2, r_1 seconds respectively where

$$r_2 = r_1 + 160.$$

<u>Demand</u>

Let the demand for travel from x_1 to y_1 and from x_2 to y_2 be k (vehicles per second) in each case. Then the route-flow

$$\underline{F} = (F_1, F_2, F_2, F_1, F_2, F_2) \tag{6}$$

shown satisfies

$$F_1 + 2F_2 = k, \; F_1 \geqslant 0 \quad \text{and} \quad F_2 \geqslant 0. \tag{7}$$

Let the signal allocate a total proportion of 1 to each conflicting pair of links, and let there be a minimum green-time proportion of 0·1. Then

$$\begin{aligned} &\underline{\Lambda} = (\Lambda_1, \Lambda_2, \Lambda_2, \Lambda_1, \Lambda_2, \Lambda_2), \\ &\Lambda_1 + \Lambda_2 = 1, \Lambda_1 \geqslant 0{\cdot}1 \text{ and } \Lambda_2 \geqslant 0{\cdot}1. \end{aligned} \tag{8}$$

Let $\mathcal{D}_1$ be the set of those vectors $\underline{F}$ which satisfy (6) and (7). Let $\mathcal{D}_2$ be the set of those green-time proportions $\underline{\Lambda}$ which satisfy (8).

Feasibility

This is defined precisely as before, in section 2.1.

Control policy

In this example we shall suppose that Webster's Method, subject to a 10% minimum proportion, is used to determine $\underline{\Lambda}$ for any supply-feasible $\underline{F}$. See appendix 1 for a specification of the policy.

Route-choice

Again this is exactly as in section 2.1.

Dynamics and consistent equilibrium when $k < 1{\cdot}1$

If $k < 1{\cdot}1$ there is exactly one equilibrium consistent with the control policy. If this is

$$\underline{F}^* = (F_1^*, F_2^*, F_2^*, F_1^*, F_2^*, F_2^*)$$

then

$$\Lambda^W{}_1(\underline{F}^*) = 0{\cdot}9, \ \Lambda^W{}_2(\underline{F}^*) = 0{\cdot}1 \text{ and } F^*{}_1 > 10F^*{}_2.$$

The shorter, narrower routes are by far the most popular.

Any feasible flow $\underline{F} \neq \underline{F}^*$ will move towards $\underline{F}^*$ as drivers swap to the quicker route if the Webster setting $\underline{\Lambda}^W(\underline{F})$ is employed.

Here there are no costly instabilities and there is a consistent equilibrium.

Dynamics and consistent equilibrium when $k = 7/6$

Let $k = 7/6$ and $\underline{F}$ be any feasible flow. Then, as before, $F_2 > 0$. Also, now,

$$2d^W_1 < d^W_2 + 100$$

and, since

$$r_1 = r_2 - 160,$$

we must have

$$c^W_1 \leqslant r_1 + 2d^W_1 + 60 < r_2 + d^W_2 \leqslant c^W_2.$$

It follows from this, together with $F_2 > 0$, that any feasible $\underline{F}$ is not an equilibrium as there are drivers on the outer routes who have a quicker alternative.

As before, there is no consistent equilibrium. The flow pattern converges to

$$(5/6,\ 1/6,\ 1/6,\ 5/6,\ 1/6,\ 1/6)$$

as time passes. This limit flow is not supply-feasible (it is on the unfeasible boundary of the supply-feasible set) and delays tend to infinity as it is approached; we have a costly instability.

As before, these arguments extend to the case $1{\cdot}1 \leqslant k \leqslant 7/6$.

With Webster's Method the capacity of this network is 1·1.

2.5. The Fourth Example

We let k = 7/6. The only difference between this example and the last is that here we use policy P_o, which for any feasible $\underline{F}$ gives rise to a signal-setting $\underline{\Lambda}^o(\underline{F})$ satisfying (for relevant feasible $\underline{F}$)

$$s_1 d(F_1, \Lambda^o_1(\underline{F}), s_1) = s_2 d(F_2, \Lambda^o_2(\underline{F}), s_2).$$

We define d^o_i by:

$$d^o_i = d^o_i(\underline{F}) = d(F_i, \Lambda^o_i(\underline{F}), s_i).$$

Because $s_1 = s_2 = 1$,

$$d^o_1 = d^o_2.$$

Therefore

$$2d^o_1 - d^o_2 = d^o_1$$

is very large and positive if F_1 is close to 5/6 (and F_2 is close to 1/6). Thus there is a strong encouragement for drivers on the inner routes to swap to the outer ones if F_1 is close to 5/6. The flow $\underline{F}$ is repelled from (rather than attracted to) the unfeasible flow

$$(5/6,\ 1/6,\ 1/6,\ 5/6,\ 1/6,\ 1/6).$$

The above argument also works if $1{\cdot}1 \leqslant k < 1{\cdot}9$, so the network can now certainly provide in a stable way for any $k < 1{\cdot}9$ and P_o maximises the capacity of this network.

2.6. Comparing Delays

Figure 2 represents the comparison between Webster's Method and P_o for this network too.

3. FORMULATION OF THE NETWORK

Let [N,L] be a mathematical network which represents a road network in the following way. Each single carriageway is represented by a single link in L. Carriageways which meet at a junction are represented by links meeting at a node in N. Junctions, any other bottlenecks, origin zones and destination zones are represented by nodes in N.

Each link A_i will have the following associated parameters:

f_i flow (vehicles per second),

λ_i green-time proportion,

s_i saturation flow (vehicles per second),

c_i total travel cost, per vehicle, on A_i (seconds).

The cost c_i of travel along A_i is to be a given function of f_i and λ_i. So we let c_i (.,.) be defined for all (f_i, λ_i) such that

$$0 \leq f_i < \lambda_i s_i.$$

3.1. Two Assumptions

In order to make our theory as simple as possible initially, we assume that lost-times are zero and that the cycle-time does not enter into the cost formulae. Both of these assumptions may be dispensed with within a significantly extended version of our theory, but this extension is significantly more complicated and the fundamental ideas of our theory would be significantly obscured by the increased complexity.

The simplicity of the theory here depends to some degree on the fact that λ_i is the unique control parameter corresponding to A_i, and to some degree on the convexity of the space of feasible control vectors $\underline{\lambda}$. The uniqueness of λ_i would be lost if cycle-times were considered and the convexity would be lost if there were non-zero lost-times.

The extended, more complicated, theory contains no new ideas; it is just technically more difficult.

4. ASSIGNMENT AND CONTROL

Our objective is to give conditions on the link cost functions $c_i(.,.)$ and a policy for choosing the control vector $\underline{\lambda}$ which ensure that any feasible assignment problem has a solution. These conditions, and the policy, will also ensure that the instabilities of our examples cannot occur. We shall prove the existence result, and rely on the previous examples to demonstrate the stabilising effect of our policies.

4.1. Salient Features of the Model

Feasibility.

The set S_i of those ordered pairs (f_i,λ_i) which are feasible for link A_i is defined as follows:

$$S_i = \{(f_i,\lambda_i);\quad 0 \lesssim f_i < \lambda_i s_i\}.$$

The unfeasible boundary of S_i is $B_i \cup \{(0,0)\}$ where

$$B_i = \{(f_i,\lambda_i);\quad 0 < f_i = \lambda_i s_i\}.$$

Properties of the link cost function.

We suppose here that each feasible pair (f_i,λ_i) determines a unique cost, or travel time, $c_i(f_i,\lambda_i)$.

We also suppose that:

$$c_i : S_i \to \mathbb{R} \text{ is continuous.} \tag{9}$$

$$c_i(f_i,\lambda_i) \to +\infty \text{ as } f_i \to \lambda_i s_i- \tag{10}$$

for each fixed $\lambda_i > 0$.

Given the cost function

$$c_i : S_i \to \mathbb{R}_+,$$

we define the signal-setter's cost function

$$c^\dagger_i : S_i \to \mathbb{R}_+$$

by putting, for each $(f_i,\lambda_i) \in S_i$,

$$c^{\dagger}_i(f_i,\lambda_i) = s_i c_i(f_i,\lambda_i).$$

It turns out that $c^{\dagger}_i$ arises naturally, and is particularly relevant for the signal-setter. (See the following lemma.)

Further "cost" functions

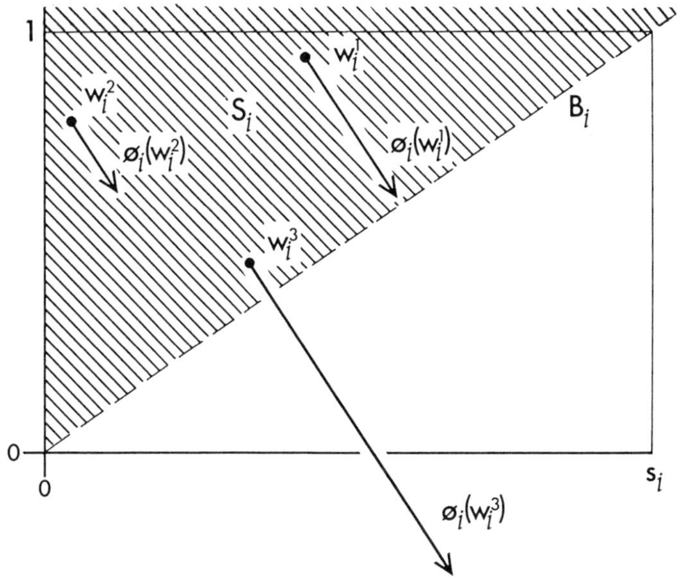

Figure 4. The set S_i is shaded and some vectors $\phi_i(w_i)$ are indicated. These all make a right angle with the unfeasible boundary.

We define the extended "cost" functions ϕ_i and ϕ as follows: for each $w_i = (f_i,\lambda_i) \in S_i$, we let (see figure 4)

$$\phi_i(w_i) = (c_i(f_i,\lambda_i), - s_i c_i(f_i,\lambda_i)). \tag{11}$$

Also we write, if m is the number of links in the network,

$$\underline{w} = (w_1,w_2,\ldots,w_m)$$

and, for each $\underline{w} \in S_1 \times S_2 \times \ldots \times S_m$,

$$\phi(\underline{w}) = (\phi_1(w_1), \phi_2(w_2), \ldots, \phi_m(w_m)). \tag{12}$$

Then, putting

$$S = S_1 \times S_2 \times \ldots \times S_n,$$

we have new functions

$$\phi_i : S_i \to \mathbb{R}^2$$

and a new function

$$\phi : S \to \mathbb{R}^{2m}$$

4.2. A Preliminary Result

In the following lemma we shall, on occasion, think of

$$\begin{aligned} w &= (w_1, w_2, \ldots, w_m) \\ &= ((f_1,\lambda_1), (f_2,\lambda_2), \ldots, (f_m,\lambda_m)) \end{aligned}$$

as a typical point of $\mathbb{R}_+^{2m}$. We shall also write $\bar{\phi}_i$ for the vector $\phi_i/||\phi_i||$.

Lemma 1. Let D be a compact convex subset of $\mathbb{R}_+^{2m}$ such that $\lambda_i \geqslant \varepsilon > 0$ (for $1 \leqslant i \leqslant m$) whenever $\underline{w} \in D$. Let each c_i: $S_i \to \mathbb{R}$ $(i = 1,2,\ldots,m)$ satisfy (9) and (10), and let ϕ: $S \to \mathbb{R}^{2m}$ be defined by (11) and (12). Finally, let $S \cap D$ be non-empty. Then there is $\underline{v}$ $S \cap D$ such that

$-\phi(\underline{v})$ is normal, at $\underline{v}$, to D

Proof. We shall prove that ϕ_i satisfies the directional constraint (7) and the magnitude constraint (8) in Smith [28]; this is the main step in the proof of the lemma.

There is always one unit normal, at any $w_i \in B_i$, to $\bar{S}_i$: namely,

$$\frac{(1,-s_i)}{\sqrt{1+s_i^2}} = n_i \text{ (say)}.$$

Also, the direction

$$\tilde{\phi}_i(w_i) = \frac{(c_i(w_i), -s_i c_i(w_i))}{\sqrt{c_i^2(w_i) + s_i^2 c_i^2(w_i)}}$$

$$= n_i.$$

Hence the directional condition :

For each $z_i \in B_i$,

$$\lim_{\substack{w_i \to z_i \\ w_i \in S_i}} \tilde{\phi}_i(w_i) = n_i$$

is trivially satisfied. This condition implies that

For each $z_i \in B_i$,

$$\lim_{\substack{w_i \to z_i \\ w_i \in S_i}} \tilde{\phi}_i(w_i) \subset N(z_i) \tag{13}$$

where $N(z_i)$ is the set of unit normals to $\bar{S}_i$ at z_i (here $N(z_i) = \{n_i\}$), and this is the directional condition (7) in Smith.

Furthermore the magnitude condition:

For each $z_i \in B_i$,

$$\lim_{w_i \to z_i} ||\phi_i(w_i)|| = +\infty \tag{14}$$

is also satisfied; because $||c_i(w_i)||$ tends to $+\infty$ as $w_i \to z_i \in B_i$.

Thus each ϕ_i satisfies the directional constraint and the magnitude constraint, (7) and (8), in Smith [28]. Similar arguments to those employed in the proof of theorem 2 in Smith [26] and in the proof of the theorem in Smith [28] may now be used to complete the proof of this lemma. The former theorem uses Brouwer's fixed point theorem.

4.3. Some Further Features of the Model

The supply-feasible set

The set

$$S = S_1 \times S_2 \times \ldots \times S_m$$

and the set

$$S^1 = \{(\underline{f},\underline{\lambda});\ (f_i,\lambda_i) \in S_i \text{ for } 1 \leqslant i \leqslant m\}$$

are different forms for the supply-feasible set. A typical member of S is

$$(f_1,\lambda_1,f_2,\lambda_2,\ldots,\ f_m,\lambda_m)$$

and a typical member of S^1 is

$$(f_1,f_2,\ldots,f_m,\lambda_1,\lambda_2,\ldots,\lambda_m).$$

We shall frame our definitions in terms of S^1 and then exploit the obvious relationship between S and S^1.

Two network cost-flow functions

We define

$$\underline{c} : S^1 \to \mathbb{R}^m_+ \ ,\ \underline{c}^\dagger : S^1 \to \mathbb{R}^m_+$$

by putting, for each $(\underline{f},\underline{\lambda}) \in S^1$,

$$c(\underline{f},\underline{\lambda}) = (c_1(f_1,\lambda_1),\ c_2(f_2,\lambda_2),\ldots,c_m(f_m,\lambda_m))$$

and

$$c^\dagger(\underline{f},\underline{\lambda}) = c_1^\dagger(f_1,\lambda_1),c_2^\dagger(f_2,\ _2),\ldots,c_m^\dagger(f_m,\lambda_m)).$$

These functions correspond to the usual network cost-flow function, only now we have an additional control vector $\underline{\lambda}$ and the additional "cost-flow" function $\underline{c}^\dagger(\cdot,\cdot)$.

Two demand sets

We shall suppose that we are given two closed convex subsets D_1, D_2 of $\mathbb{R}^m_+$. The flow $\underline{f}$ is to be confined to D_1 and the control vector $\underline{\lambda}$ is to be confined to D_2. These two sets together play the same role as D in Smith [26].

Feasibility

We shall say that the triple (S^1, D_1, D_2) is feasible if and only if there is an $(\underline{f}, \underline{\lambda}) \in S^1$ such that

$$\underline{f} \in D_1 \quad \text{and} \quad \underline{\lambda} \in D_2.$$

This is the case whenever

$$S^1 \cap (D_1 \times D_2) \text{ is non-empty.}$$

Under these circumstances the "demand" constraints

$$\underline{f} \in D_1,\ \underline{\lambda} \in D_2$$

and the m supply constraints

$$0 \leqslant f_i < \lambda_i s_i$$

can all be simultaneously satisfied.

4.4. The Central Result

Theorem 1. Let D_1, D_2 be compact convex subsets of $\mathbb{R}^m_+$ such that $\lambda_i \geqslant \varepsilon > 0$ (for $1 \leqslant i \leqslant m$) whenever $\underline{\lambda} \in D_2$. Let each

$$c_i : S_i \to \mathbb{R}$$

$(i = 1,2,\ldots,m)$ satisfy (9) and (10). Further, let the triple (S^1, D_1, D_2) be feasible. Then there is an $\underline{f} \in D_1$ and a $\underline{\lambda} \in D_2$ such that $(\underline{f}, \underline{\lambda}) \in S^1$,

$$-\underline{c}(\underline{f}, \underline{\lambda}) \text{ is normal, at } \underline{f}, \text{ to } D_1$$

and

$\underline{c}^{\dagger}(\underline{f},\underline{\lambda})$ is normal, at $\underline{\lambda}$, to D_2.

Proof. We just need to apply lemma 1 with

$$D = \{\underline{w};\ w_i = (f_i,\lambda_i) \text{ where } \underline{f} \in D_1 \text{ and } \underline{\lambda} \in D_2\}.$$

We have that $S \cap D \neq \phi$ since $S^1 \cap (D_1 \times D_2) \neq \phi$. Hence there is $\underline{w} \in S \cap D$ such that

$-\phi(\underline{w})$ is normal, at $\underline{w}$, to D.

Writing this condition in terms of $\underline{f}$, $\underline{\lambda}$, $\underline{c}$ and $\underline{c}^{\dagger}$, we obtain:

$-\underline{c}(\underline{f},\underline{\lambda})$ is normal, at $\underline{f}$, to D_1

and

$\underline{c}^{\dagger}(\underline{f},\underline{\lambda})$ is normal, at $\underline{\lambda}$, to D_2.

Furthermore, the condition $\underline{w} \in S \cap D$ becomes, since $w_i = (f_i,\lambda_i)$,

$(\underline{f},\underline{\lambda}) \in S^1$, $\underline{f} \in D_1$ and $\underline{\lambda} \in D_2$.

Corollary. Consider the following policy:

For any flow $\underline{f}$ in D_1 choose

$\underline{\lambda} \in D_2$ so that $(\underline{f},\underline{\lambda}) \in S^1$ (15)

and $\underline{c}^{\dagger}(\underline{f},\underline{\lambda})$ is normal, at $\underline{\lambda}$, to D_2.

We shall call this policy "P". Then, by theorem 1, there is (under natural conditions) a pair $(\underline{f},\underline{\lambda})$ such that

(i) $\underline{f}$ is a feasible equilibrium when the control vector is $\underline{\lambda}$ and

(ii) $\underline{\lambda}$ is a feasible control vector which could be chosen by a signal-setter following policy P.

If such a pair arose in an actual road network then all drivers would be using least cost routes, and so would have no incentive to change route; and the signal-setting would be consistent with the policy P, and so a signal-setter obeying P would have no incentive to change the signal-settings.

A generalisation of the policy P

Having found one policy for which there is an equilibrium consonant with the policy, it seems natural to seek others with this property. Then the signal-setter may be offered a variety of possible policies, some of which may suit special circumstances better than others; all, however, still guaranteeing the existence of a consistent equilibrium.

In order to do this let

$$b : S \to \mathbb{R}^m_+$$

be any bounded continuous function. For each such function let $b = (b_1, b_2, \ldots, b_m)$ and

$$\phi^b_i(\underline{w}) = (c_i(w_i),\ s_i(b_i(\underline{w}) - c_i(w_i))).$$

Further let

$$\phi^b(\underline{w}) = (\phi^b{}_1(\underline{w}),\ \phi^b{}_2(\underline{w}), \ldots, \phi^b{}_m(\underline{w})).$$

Then

$$\phi^b : S \to \mathbb{R}^{2m}$$

satisfies both the directional constraint (13) and the magnitude constraint (14), because ϕ does and the bounded perturbation b cannot upset either.

Let

$$c_i^b(\underline{f},\underline{\lambda}) = s_i\ (c_i(w_i) - b_i(\underline{w}))$$

and

$$c^b(\underline{f},\underline{\lambda}) = (c_1^b(\underline{f},\underline{\lambda}),\ c_2^b(\underline{f},\underline{\lambda}),\ldots,c_m^b(\underline{f},\underline{\lambda})).$$

Then for each bounded continuous function $b : S \rightarrow \mathbb{R}^m_+$ we obtain the following policy P(b):

For any flow $\underline{f}$ in D_1 choose
$\underline{\lambda} \in D_2$ so that $(\underline{f},\underline{\lambda}) \in S^1$ and (16)
$\underline{c}^b(\underline{f},\underline{\lambda})$ is normal, at $\underline{\lambda}$, to D_2.

Because ϕ^b satisfies the directional constraint (13) and the magnitude constraint (14), any feasible assignment problem has a solution consistent with P(b), as with P.

Let

$$\mathscr{P} = \{P(b);\ b : S \rightarrow \mathbb{R}^m_+ \text{ is bounded and continuous.}\}$$

Then $\mathscr{P}$ is the collection of policies we seek; it is a pretty large collection although the usual policies like Webster's Method do not belong to it.

4.5. The Policy P_o

Here we split the cost c_i of travel along A_i into two parts, the running cost r_i and the junction delay d_i at the end of A_i. We suppose that the given delay functions $d_i : S_i \rightarrow \mathbb{R}$ satisfy conditions (9) and (10), and that each running cost r_i depends in a bounded continuous manner on the overall flow pattern $\underline{f}$.

For any

$$\underline{w} = (f_1,\lambda_1,\ f_2,\lambda_2,\ldots,f_m,\lambda_m) \in S$$

we may then put

$$b(\underline{w}) = (r_1(\underline{f}\ ),\ r_2(\underline{f}),\ldots,r_m(\underline{f})).$$

Our condition that the r_i are bounded continuous now implies that

$$b : S \rightarrow \mathbb{R}^m_+$$

is bounded continuous, and we may apply the ideas of section 4.4. (Our fundamental theorem 1 is still true in the context of this section, even though $c_i = r_i + d_i$ depends here on w_j with $j \neq i$. The proof is essentially the same as before.)

Letting $d^\dagger_i = s_i d_i$, we obtain the following policy P(b):

$$\begin{aligned}&\text{For any flow } \underline{f} \text{ in } D_1 \text{ choose}\\ &\underline{\lambda} \in D_2 \text{ so that } (\underline{f},\underline{\lambda}) \in S^1 \text{ and}\\ &\underline{d}^\dagger(\underline{f},\underline{\lambda}) \text{ is normal, at } \underline{\lambda}, \text{ to } D_2.\end{aligned} \tag{17}$$

Here we have used the fact that $\underline{c} - b = \underline{c} - \underline{r} = \underline{d}$. This policy we shall call P_o.

In order to operate P_o, the signal-setter does not require knowledge of the costs being imposed along a (possibly) long road link. Only very local information is required at each separate junction; namely the flows, delays and signal-settings on the approaches to that junction. The policy P_o would be suited to automatic implementation.

We shall concentrate our attention, for the moment, on P_o; similar results are obtainable for any other policy in $\mathcal{P}$.

An alternative form for P_o

It is well known that there are equivalent route and link formulations of the traffic assignment problem; this suggests that we seek a formulation of P_o corresponding to the route formulation of the traffic assignment problem.

In order to do this, we shall suppose that we are given all the sets of links which may be shown green simultaneously, and that these sets are fixed (independent of traffic flow). Let

$$G_1, G_2, \ldots, G_\ell$$

be these sets. (We suppose here that each approach to each junction is represented by a link in L, that the movements possible from a given approach are known and fixed, and that the usual practical considerations have been taken into account in determining the G_q. We also suppose that flows at one junction do not affect costs at others. Finally, we suppose that lost times, or clearance times, are zero. Our earlier comments on our assumptions apply particularly here; the one really essential assumption is that flows at one junction do not affect costs at others. The other assumptions may be relaxed, at least to some degree, within a suitably modified theory; but it is not clear that our theory can be modified so as to allow for queues at one junction blocking a previous junction.) We shall refer to each G_q as a clique. (See Stoffers [29] and Muchland and Tully [20].)

Let Λ_q be the proportion of green-time awarded to clique G_q. Then, because we are supposing that lost-times are zero, we may impose the natural constraint

$$\sum_{q=1}^{\ell} \Lambda_q = 1$$

on the non-negative clique green time proportions. Let the minimum green time proportion be $\varepsilon > 0$ for each link A_i. Also let

$$\alpha_{iq} = \begin{cases} 1 \text{ if link } A_i \text{ belongs to clique } G_q \\ 0 \text{ otherwise} \end{cases}$$

and let $\underline{A}_2$ denote the matrix (α_{iq}).

Then

$$\lambda_i = \sum_q \alpha_{iq} \Lambda_q,$$

$$\underline{\lambda} = \underline{A}_2 \underline{\Lambda}$$

and the set of feasible clique green time proportions,

$$\mathcal{D}_2 = \{\underline{\Lambda};\ \sum_{q=1}^{\ell} \Lambda_q = 1,\ \Lambda_q \geqslant 0 (q = 1,2,\ldots,\ell) \text{ and } \lambda_i \geqslant \varepsilon\ (i=1,2,\ldots,m)\}.$$

Furthermore the set of feasible link green-time proportions,

$$D_2 = \{\underline{A}_2\underline{\Lambda};\ \underline{\Lambda} \in \mathcal{D}_2\}.$$

With the agreement that

$$\underline{\lambda} = \underline{A}_2\underline{\Lambda} \quad \text{and } \underline{c}^{\dagger}(\underline{f},\underline{\Lambda}) = \underline{d}^{\dagger}(\underline{f},\underline{\lambda})\underline{A}_2,$$

it is clear that

$$\underline{d}^{\dagger}(\underline{f},\underline{\lambda}) \text{ is normal, at } \underline{\lambda}, \text{ to } D_2 \tag{18}$$

if and only if

$$\underline{c}^{\dagger}(\underline{f},\underline{\Lambda}) \text{ is normal, at } \underline{\Lambda}, \text{ to } \mathcal{D}_2. \tag{19}$$

(The argument required here is given explicitly in Smith [26], on pages 299-300.)

The equation

$$\underline{c}^{\dagger} = \underline{d}^{\dagger}\underline{A}_2$$

may be thought of as the definition of the vector

$$\underline{c}^{\dagger} = (c^{\dagger}_1, c^{\dagger}_2,\ldots,c^{\dagger}_{\ell})$$

of clique costs.

The conditions (18) and (19) give rise to equivalent link and clique formulations of P_o.

<u>The policy P_o when $\varepsilon = 0$.</u>

If $\varepsilon = 0$,

$$\mathscr{D}_2 = \{\underline{\Lambda};\ \sum_{q=1}^{\ell} \Lambda_q = 1 \text{ and } \Lambda_q \geqslant 0 (q=1,2,\ldots,\ell)\}.$$

The condition (19) may now be written:

$$C^{\dagger}_p(\underline{f},\underline{\Lambda}) > C^{\dagger}_q(\underline{f},\underline{\Lambda}) \text{ implies } \Lambda_q = 0$$

or

<u>less costly cliques receive no green time.</u>

The parallel with Wardrop's route-choice principle:

<u>more costly routes are not used,</u>

is striking.

There is a theoretical problem with this formulation because our proof of theorem 1 depends on ε being positive. If $\varepsilon = 0$ both the definition of equilibrium and the proof of theorem 1 must be altered. I feel that such alterations are entirely possible but this paper seems technical enough as it is, without further technical additions. In any case, I cannot yet provide the required alterations.

4.6. A Queueing Approximation to P_o When $\varepsilon = 0$

In section 4.4 we indicated that a bounded continuous "error" in the estimate of the actual link-cost function used in the definition of P did not affect the basic property possessed by P. (This allowed us to construct a large family $\mathscr{P}$ of policies with satisfactory properties.)

It is possible to push these ideas further, as follows.

Let $q_i \geqslant 0$ be an estimate of the number of vehicles queueing at the end of link A_i. Then a natural estimate for d_i is $q_i/\lambda_i s_i$. In fact, under natural conditions,

$$d_i - q_i/\lambda_i s_i$$

is a bounded continuous function of the triple (f_i, q_i, λ_i). Imitating the work of section 4.5, and supposing that $\varepsilon = 0$, we are led to the following queueing approximation to P_o, which we shall call $\tilde{P}_o$, as follows.

Here we define (e stands for estimate)

$$d^e{}_i = q_i/\lambda_i s_i$$

and

$$d^\dagger{}_i = s_i d^e{}_i = q_i/\lambda_i.$$

Now we put, where $\underline{\lambda} = \underline{A}_2\underline{\Lambda}$,

$$\underline{c}^\dagger(\underline{q},\underline{\Lambda}) = \underline{d}^\dagger \underline{A}_2.$$

Then $\tilde{P}_o$ (with this estimate for d_i and $\varepsilon = 0$) is as follows:

$$\begin{array}{l} \text{For any flow } \underline{f} \text{ in } D_1 \text{ choose} \\ \underline{\Lambda} \in \mathscr{D}_2 \text{ so that} \\ c^\dagger{}_r(\underline{q},\underline{\Lambda}) > c^\dagger{}_s(\underline{q},\underline{\Lambda}) \text{ implies } \Lambda_s = 0. \end{array} \qquad (20)$$

This special queueing approximation to P_o is of particular interest because the $d^\dagger{}_i$, and hence the $c^\dagger{}_r$, do not depend on the s_i.

CONCLUSION

We have given examples of simple networks for which current signal-setting methods produce results which are undesirable for the motorist, because high costs are imposed, and undesirable for the traffic authority, because traffic capacity is inefficiently used.

We have also introduced a collection of policies $\mathscr{P}$ for which these undesirable results (by our theorem) can never occur; our simple example networks benefit from such policies.

It may be that large realistic networks would not behave like our simple models, and hence that the benefits which accrue in our examples, when policies in $\mathcal{P}$ are used, would not be found in large realistic networks when these policies are used. It is, though, not clear why large networks should usually avoid the traps demonstrated in our simple examples.

The only real way to discover whether the new policies in $\mathcal{P}$ would confer benefits, compared to current signal-setting policies, is to compare total travel times and other significant criteria when a network is controlled according to, say, P_o and Webster's Method. Such a comparison could be done in a computer simulation or on an actual road network.

The policies which have arisen in this paper require no origin-destination information; furthermore, the approximation $\tilde{P}_o$ to P_o does not require knowledge of the saturation flows of the carriageways. These features may prove useful in practice, because origin-destination distributions are rarely constant over time and saturation flows may well change with weather conditions, or traffic incidents.

Further theoretical work is required to take account of gueue-storage constraints.

REFERENCES

1. R E Allsop, "Some possibilities for using traffic control to influence trip distribution and route-choice". Paper presented at the Sixth International symposium on Transportation and Traffic Theory, Sydney, Australia (August) and published in the proceedings: Transportation and Traffic Theory, 345-374, Elsevier, Amsterdam and New York, 1974.

2. M J Beckmann, C B McGuire and C B Winsten, Studies in the Economics of Transportation. Yale University Press, New Haven, Conn., 1956.

3. D Braess, "Uber ein Paradoxen ans der Verkehrsplanung". Unternehmensforschung 12, 258-268 (1968).

4. L De La Breteque, M Duraz and R Jezequel, Commande adaptive d'un carrefour isolé, étude comparative de quelques algorithmes. Institut de Recherche des Transports Report No. 32, 1978.

5. S C Dafermos, "An extended traffic assignment model with applications to two-way traffic". Trans. Sci. 5, 376-389 (1971).

6. S C Dafermos, "Toll patterns for multiclass-user transportation networks". Trans. Sci. 7, 211-223 (1973).

7. S C Dafermos, "Traffic equilibrium and variational inequalities". Trans. Sci. 14, 42-54 (1980).

8. T J Dickson, "A note on traffic assignment and signal timings in a signal-controlled road network". Trans. Res. (To appear.)

9. T J Dickson, "Assignment and signal controls". Unpublished note, 1977.

10. N Gartner, "Area traffic control and network equilibrium". Paper presented at the conference on traffic equilibrium methods, Montreal, November 1976 and published in the proceedings: Traffic Equilibrium Methods, 119-131, Springer Verlag, New York, 1976.

11. S B Gershwin and H N Tan, Hybrid optimization: optimal traffic control constrained by drivers' route-choice behaviour. M.I.T. Laboratory for Information and Decision Systems Report LIDS-P-870, 1979.

12. R D Henry, R A Ferlis and J L Kay, Evaluation of UTCS control strategies - executive summary. Report FHWA-RD-67-149, Federal Highway Administration Offices of Research and Development, Washington D.C.

13. J Holroyd, Joyce and J A Hillier, The Glasgow Experiment: PLIDENT and after. RRL Report LR384, Transport and Road Research Laboratory, Crowthorne, 1971.

14. J Holroyd, Comparison between a dynamic signal plan generation system of area traffic control and a fixed time system. TRRL Report LR487, Transport and Road Research Laboratory, Crowthorne, 1972.

15. J Holroyd and D I Robertson, Strategies for area traffic control systems: present and future. TRRL Report LR569, Transport and Road Research Laboratory, Crowthorn, 1973.

16. F H Knight, "Some fallacies in the interpretation of social cost". Quarterly Journal of Economics 38, 582-606 (1924).

17. M Marchand, "A note on optimal tolls in an imperfect environment". Econometrica 36, 575-581 (1968).

18. A J Miller, "A computer controlled system for traffic networks". Paper presented at the 2nd International Symposium on the Theory of Road Traffic Flows and published in the proceedings, London, 1963.

19. J D Murchland, "Braess's paradox of traffic flow". Trans. Res. 4, 391-394 (1970).

20. J D Murchland and Z I Tully, "Calculation and use of the critical cycle time for a single traffic controller". Paper presented at the PTRC Summer Annual Meeting, University of Warwick, and published in the proceedings.

21. M Netter, "Equilibrium and marginal cost pricing on a road network with several traffic flow types Paper presented at the 5th International Symposium on Traffic Flow and Transportation, Berkeley, and published in the proceedings, <u>Traffic Flow and Transportation</u>, Elsevier, Amsterdam and New York, 1971.

22. M Netter, "Affectations de trafic et tarification an côut marginal social: critique de quelques idées admises". <u>Trans. Res. 6</u>, 411-429 (1972).

23. A C Pigou, <u>The Economics of Welfare</u>. Mac illan, New York, 1920.

24. R J Smeed, <u>Road Pricing: The Economic and Technical Possibilities</u>, HMSO, London, 1964.

25. M J Smith, "The marginal cost taxation of a transportation network". <u>Trans. Res. 13B</u>, 237-242 (1979).

26. M J Smith, "The existence, uniqueness and stability of traffic equilibria". <u>Trans. Res. 13B</u>, 295-304 (1979).

27. M J Smith, "A local traffic control policy which automatically maximises the overall travel capacity of an urban road network". Proceedings of the International Symposium on Traffic Control Systems, Berkeley, 1979 (UCB-ITS-P-79-4) and <u>Traffic Engineering and Control, 21</u>, 298-302 (1980).

28. M J Smith, "The existence of an equilibrium solution to the traffic assignment problem when there are junction interactions". <u>Trans. Res. 15B</u> (1981).

29. K E Stoffers, "Scheduling of traffic lights - a new approach". <u>Trans. Res. 2</u>, 199-234 (1968).

30. J C Tanner, "Pricing the use of roads - a mathematical and numerical study". Proceedings of the 2nd International Symposium on the Theory of Traffic Flow and Transportation, 317-337, OECD, London, 1965.

31. P J Tarnoff, "Concepts and Strategies - Urban Street Systems". Proceedings of the International Symposium on Traffic Control Systems, Berkeley, 1979 (UCB-ITS-P-79-4).

APPENDIX 1. A VERSION OF WEBSTER'S METHOD

For any feasible $\underline{F}$, evaluate

$$y_1 = F_1/s_1 \qquad \text{and} \qquad y_2 = F_2/s_2.$$

Then let

$$\Lambda_1 = \begin{cases} 0{\cdot}1 & \text{if } y_1/(y_1+y_2) \leqslant 0{\cdot}1 \\ 0{\cdot}9 & \text{if } y_1/(y_1+y_2) \geqslant 0{\cdot}9 \\ y_1/(y_1+y_2) & \text{otherwise,} \end{cases}$$

and $\Lambda_2 = 1 - \Lambda_1$. This $\underline{\Lambda} = (\Lambda_1,\Lambda_2)$ we denote by $\underline{\Lambda}^W(\underline{F})$ in example 1. In example 3 we denote

$$\underline{\Lambda} = (\Lambda_1,\Lambda_2,\Lambda_2,\Lambda_1,\Lambda_2,\Lambda_2) \text{ by } \underline{\Lambda}^W(\underline{F}).$$

APPENDIX 2. PROOF THAT $d^W{}_1(\underline{F}) - d^W{}_2(\underline{F}) < 100$

Let $\underline{F}$ satisfy

$$F_1 + F_2 = 7/6,\ F_1 \geqslant 0,\ F_2 \geqslant 0$$

and

$$F_1/1 + F_2/2 < 1,$$

so that $\underline{F}$ is feasible.

As in section 2.1, let

$$d^W{}_1(\underline{F}) = d\ (F_1,\Lambda^W{}_1(\underline{F}),\ 1)$$

and

$$d^W{}_2(\underline{F}) = d\ (F_2,\Lambda^W{}_2(\underline{F}),\ 2).$$

It is easy to check, by considering separately the cases

$$F_1 \geqslant 7/12,\ F_1 \leqslant 7/102$$

and

$$7/102 \leqslant F_1 \leqslant 7/12,$$

that (for any feasible $\underline{F}$)

$$d^W_1(\underline{F}) - d^W_2(F) \leqslant \max_{\frac{7}{102} \leqslant F_1 \leqslant \frac{7}{12}} d^W_1(\underline{F})$$

$$\leqslant \frac{9}{20}\,[c + \frac{1}{8/7 \cdot 1/7 \cdot 7/102}]$$

$$< 100$$

since c = 120. This is the required result.

THE GARAGE CONSTRAINED - BALANCED VEHICLE SCHEDULE MINIMUM FLEET SIZE PROBLEM

HELMAN I. STERN*

Department of Industrial Engineering and Management
Ben Gurion University of the Negev
Beersheva, Israel

AVISHAI CEDER

Transportation Research Institute
Technion-Israel Institute of Technology
Technion City, Haifa, Israel

ABSTRACT

The problem of finding the least number of vehicles required to service a given schedule of transportation trips is the minimum fleet size problem. The first part of this paper reviews two common approaches for solving this problem (with and without deadheading): (i) the deficit function approach, and (ii) the well known Ford and Fulkerson array labeling technique. In the second part of this paper the problem is extended to include garage limitations and balanced schedule constraints. This development is pursued first as a problem in deficit function theory, and secondly as a mathematical programming problem. A simple solution procedure is proffered for a special case.

*Visiting during 1980-81, School of Business, SUNYA, Albany, N.Y.

1. INTRODUCTION

Consider a schedule of required transportation trips. Each trip is defined by a starting terminal, starting time, ending terminal, and ending time. The problem is to find the minimum number of vehicles (the fleet size) that can carry out all the trips in the schedule. As this is most efficiently done if single vehicles are allowed to service several trips in succession, a second part to the problem is to find the set of trips assigned to each vehicle in the fleet. A vehicle schedule is an ordered set of trips called a chain in which each set of adjacent trips can be feasibly serviced by the same vehicle. For example, if trip j immediately follows i, then: (a) the starting time of trip j must be later than or the same as the ending time of trip i, and (b) the starting and ending terminal of j and i must be identical. If the second condition is not met some systems allow the vehicle to run empty from i to j. If the travel time for this empty trip can be completed before the starting time of trip j then both trips may appear in the same chain and be assigned the same vehicle. Bus operations are an example of transportation systems which often carry out empty trips under the name of deadheading (DH) trips. On the other hand airline companies rarely, if ever, "deadhead" their aircraft due to the high cost of running empty. Consequently, one may refer to the minimum fleet size problem with and without DH trips. When DH is allowed one can reach the counterintuitive result of decreasing the required resources (fleet size) by introducing more work into the system (adding deadheading trips). This is under the assumption that the capital cost of saving a vehicle far outweighs the cost of any increased operational cost (driver and vehicle travel cost) imposed by the introduction of deadheading trips.

The solution procedures for solving the minimum fleet size problem have essentially followed two approaches. The first has its roots in the classic work of Ford and Fulkerson [8], namely, placing the maximum number of 1's in an array of admissible and inadmissible cells. See, for example, Dantzig and Fulkerson [7], Saha [20] and Hipp [14]. Without DH the problem is exceedingly simple and solved through application of the staircase rule of Ford and Fulkerson [8] or the first in/first out rule. When deadheading is included, an array labeling technique is used. When costs are associated with the joining of two trips, the problem can be formulated as a minimum cost flow problem of various forms and solved by the out-of-kilter assignment or transportation algorithms. This approach underlies the work of Allan-Anderson [1], Gavish [10], Kirkman [16], Lampkin [17], and Orloff [19]. Computer packages described in Bennington and Rebibo [3] and Hoffstadt [15] are based on network flow techniques. Heuristic procedures are described by Bodin and Dial [4], and Smith and Wren [23].

Meanwhile a second stream of work approached from another direction. One tributary starting from Bartlett [2] and Salzborn [22] joined with that of Linus and Maksim [18], Gertsbach and Stern [12] and Gertsbach and Gurevich [11] to form a deficit function approach. These authors dealt with the minimum fleet size problem (without DH trips) through the formal construction of a step function representation of a trip schedule, and its decomposition into a set of vehicle chains. Deficit function theory was extended to include the introduction of DH trips by Stern and Ceder [25] and Ceder and Stern [5]. For the most part, researchers in either school have proceeded to develop their ideas independently without recognizing the correspondences between the two basic methodologies (although Gertsbach and Stern [12] and Salzborn [21] have alluded to the connection).

In the next section, the main results for both techniques are presented along with a simple illustrative example. In section 3, the deficit function formulation of the basic minimum fleet size problem with DH trips is extended to include (a) garage limitations on the number of vehicles that can operate from each terminal, and (b) a requirement that the schedule be balanced, i.e., the number of vehicles that start and end at the same terminal must be equal. The problem of minimizing the fleet size under these constraints is called the garage constrained - balanced schedule (GC-BS) problem. In section 4 this problem is formulated as a mathematical programming problem and shown to be equivalent to the deficit function formulation. A simple solution procedure is proffered for a special case. Section 5 contains concluding remarks.

2. TWO MAJOR TECHNIQUES FOR SOLVING THE MINIMUM FLEET SIZE PROBLEM

Before embarking on a discussion of the two major solution techinques (array labeling and deficit function) it is helpful to introduce a number of definitions and corresponding notation.

Preliminary Definitions and Notation

Let $N = \{i:i=1,\ldots,n\}$ denote a set of required trips. Trips are conducted between a set of terminals $K = \{k:k=1, \ldots,q\}$. Each trip is to be serviced by a single vehicle, and each vehicle can service any trip. Each trip i can be represented as a 4-triple, (p^i, t_s^i, q^i, t_e^i), in which the ordered elements denote departure terminal, departure (start) time, arrival terminal, and arrival (end) time. It is assumed that each trip i lies within a schedule horizon $T = [T1, T2]$, i.e., $T1 \leq t_s^i < t_e^i \leq T2$. The set of all trips $S = \{(p^i, t_s^i, q^i, t_e^i): p^i, q^i \varepsilon K, i \varepsilon N\}$ is said to be a multiterminal fixed trip schedule or timetable. Two trips, i and j, may be serviced sequentially (feasibly joined) by the same vehicle if and only if (a) $t_e^i \leq t_s^j$ and (b) $q^i = p^j$. If i is feasibly joined to j, then it is said that i is the predecessor of j and j is the successor of i. An ordered sequence of trips $i_1, i_2, \ldots, i_w$ such that each adjacent pair of trips satisfies (a) and (b) is called a string or chain. It follows that a chain is a set of trips which can be serviced by a single vehicle. It is often helpful to describe a chain in expanded form as an alternating sequence of terminal and trip indices; i.e., $(k_1, i_1, k_2, i_2, \ldots, k_v, i_v, k_{v+1}, \ldots, i_w, k_{w+1})$. It is understood that for a given trip i_v, $k_v = p^{i_v}$ and $k_{v+1} = q^{i_v}$. A set of chains which includes each trip i in N exactly once is said to be a vehicle schedule. The problem of finding the minimum number of chains for a fixed schedule S is defined as the minimum fleet size problem (MFS).

Define a deadheading (DH) trip as an empty (nonrequired) trip from some terminal p to some terminal q in time $\tau(p,q)$. If it is permissible to introduce DH trips into the schedule, then conditions (a) and (b) for the feasible joining of two trips i and j may be replaced by:

$$t_e^i + \tau(q^i, p^i) \leqq t_s^j \ . \qquad (1)$$

Denote the minimum fleet size problem with DH as MFS-DH.

The Array Labeling Technique

A trip joining array for S may be constructed by associating the ith row with the arrival epoch of the ith trip, and the jth column with the departure epoch of the jth trip. Cell (i,j) will be admissible if i and j may be feasibly joined. Otherwise, (i,j) will be an inadmissible cell. Let x_{ij} be a 0-1 variable associated with cell (i,j) and consider the following problem:

Problem P1

$$\text{Max } Z1 = \sum_{i \varepsilon N} \sum_{j \varepsilon N} x_{ij} \qquad (2)$$

subject to:

$$\sum_{j \varepsilon N} x_{ij} \leqq 1, \qquad i \varepsilon N \qquad (3)$$

$$\sum_{i \varepsilon N} x_{ij} \leqq 1, \qquad j \varepsilon N \qquad (4)$$

$$\left.\begin{array}{ll} x_{ij} \ \varepsilon \ \{0,1\}, & \text{all } (i,j) \text{ admissible} \\ x_{ij} = 0, & \text{all } (i,j) \text{ inadmissible} \end{array}\right\} \qquad (5)$$

A solution with $x_{ij} = 1$ indicates that trips i and j are joined. The objective function maximizes the number of such joinings. Constraint (3) ensures that each trip may be joined with, at most, one successor trip. Similarly, constraint (4) indicates that each trip may be joined with, at most, one predecessor trip. The following theorem states that maximizing (2) in P1 is tantamount to minimizing the number of chains for a trip schedule of size n.

Theorem 1. Let C and n denote the number of chains and trips respectively. Then Min C = n - Max Z1.

Proofs of theorem 1 can be found in [8] and [20]. Note that P1 is equivalent to the problem of placing the maximum number of 1's in a trip joining array such that no more than one number 1 appears in each row and column. For MFS the maximal placement of 1's can be carried out in a very simple fashion. By permuting the rows and columns of the array, a block diagonal structure can be arranged such that each block contains its admissible cells in a staircase structure. The staircase rule of Ford and Fulkerson (8) can now be applied to each block to obtain a maximal placement.

Consider the three terminal MFS and MFS-DH problems defined by the data in Tables I and II.

TABLE I

Trip Schedule S for the Example MFS Problem

Trip Number i	Departure Terminal p^i	Departure Time t_s^i	Arrival Terminal q^i	Arrival Time t_e^i
1	k	1	k	2
2	a	4	m	12
3	m	6	a	8
4	k	11	a	17
5	a	13	k	15

TABLE II

Deadheading Travel Time Matrix Used to Extend the Problem to the MFS-DH Problem

Departure Terminal \ Arrival Terminal	a	m	k
a	0	4	3
m	1	0	4
k	2	4	0

Figure 1(a) contains the trip joining array for the sample MFS problem. Note that the minimum number of chains is three as only two 1's can be placed in the array.

For the MFS-DH problem,the emergence of new admissible cells destroys the special structure and requires a more complex solution technique, such as the array labeling procedure of Ford and Fulkerson (8). For an equivalent set notation version of the array labeling technique see Friedman (9) and Stern (24). Note that additional admissible cells may allow an increase in the number of 1's in the array and concomitant reductions in the number of chains. The array labeling operations are applied to the sample problem when DH is allowed in Figure 2. Note that the number of chains has been reduced to 2 after the introduction of three DH trips. The following corollary should now be evident.

Corollary 1. Min C for the MFS problem is an upper bound for the MFS-DH problem.

Up to this point the procedure described determines the minimum number of chains and a set of internal trip joinings. To actually construct the vehicle schedule a second phase is

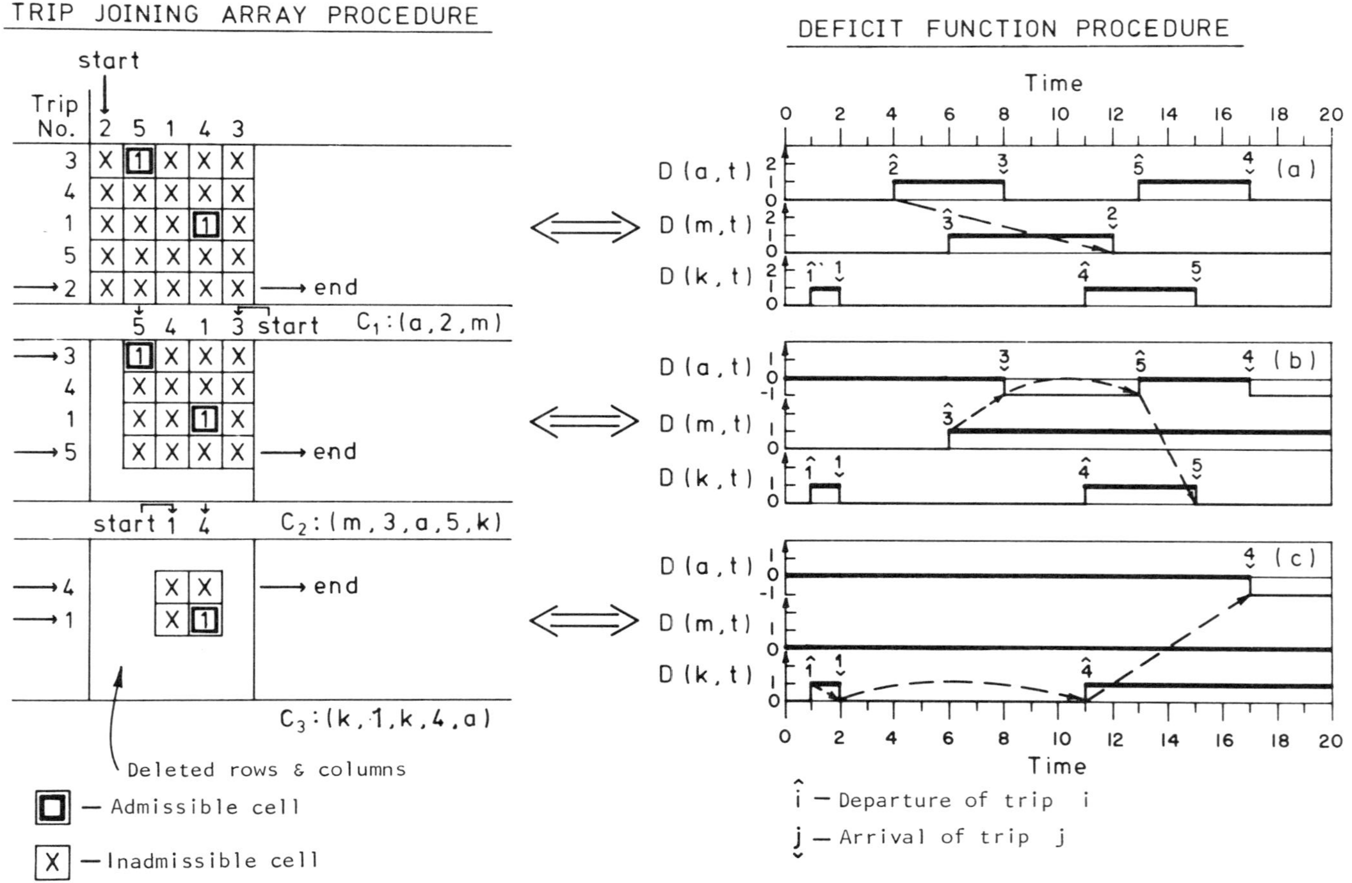

Figure 1. Multiterminal chain construction on trip joining array and deficit function for the MFS problem without DH:
(a) Selection of first chain
(b) Selection of second chain after extracting first chain
(c) Selection of third chain after extracting second chain.

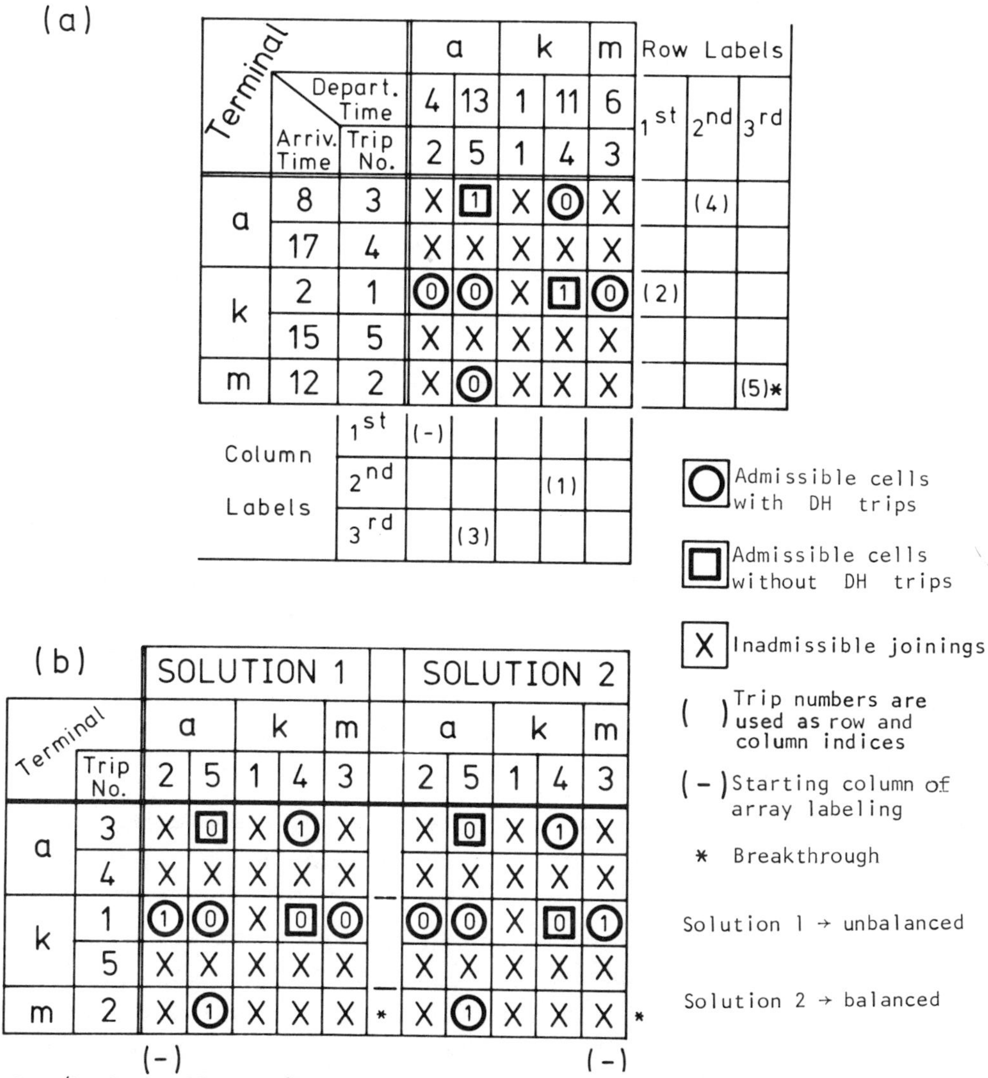

C_1: (k, 1, k, DH_1, a, 2, m, DH_2, a, 5, k)

C_2: (m, 3, a, DH_3, k, 4, a)

C_1: (k, 1, k, DH_4, m, 3, a, DH_3, k, 4, a)

C_2: (a, 2, m, DH_2, a, 5, k)

Figure 2. Array Labeling for the MFS-DH Problem and Optimal Solutions:
(a) Labeling to Increase the Placement of 1's (given the non DH Solution) in the Trip Joining Array with Admissible Deadheading Cells
(b) Optimal Solution Obtained when Deadheading Trips are Added to the Multiterminal Trip Joining Array.

required. A detailed description of the array chain construction procedure is found in Saha (20). Briefly, chains are constructed by selecting an empty column (a column without a 1 in it) and tracing out the trip joinings until an empty row is found. After each chain is identified the array is reduced by deleting the columns and rows corresponding to the trips in the chain. The procedure is repeated until the array contains no active rows or columns. This procedure is illustrated in Figure 1 (left side) for MFS solution to the sample problem. Optimal chain constructions for the two MFS-DH solutions are recorded in Figure 2(b).

The Deficit Function Approach

A deficit function is a step function defined over the schedule horizon which increases by one at the time of each trip departure and decreases by one at the time of each trip arrival. Let D(k,t) denote the cumulative number of departures less the cumulative number of arrivals at terminal k over the interval [T1,t]. It is possible to partition the schedule horizon of D(k,t) into a sequence of alternating hollow and maximal intervals (H_o^k, M_1^k, H_1^k,....,H_i^k, M_{i+1}^k,.... ....,$M_{n(k)}^k$, $H_{n(k)}^k$). Maximal intervals, M_i^k, define the intervals of time over which D(k,t) takes on its maximum value. Denote this maximum value as $\bar{D}(k) = \text{Max}\ \{D(k,t): t \varepsilon T\}$. The value $\bar{D}(k)$ is referred to as the deficit at k. Hollows, H_i^k, are intervals where $D(k,t) < \bar{D}(k)$. The sum of all deficit functions over k is defined as the overall deficit function $g(t) = \sum_{k \varepsilon K} D(k,t)$. Deficit functions for the three terminal example are illustrated in Figure 3. Determining the minimum fleet size from the set of deficit functions is quite simple - one merely adds the deficits over all the terminals. This result was apparently independently derived by Bartlett [2], Linus and Maksim [18], Gertsbach and Gurevich [11] and Salzborn [22]. It is formally stated below as Theorem 2.

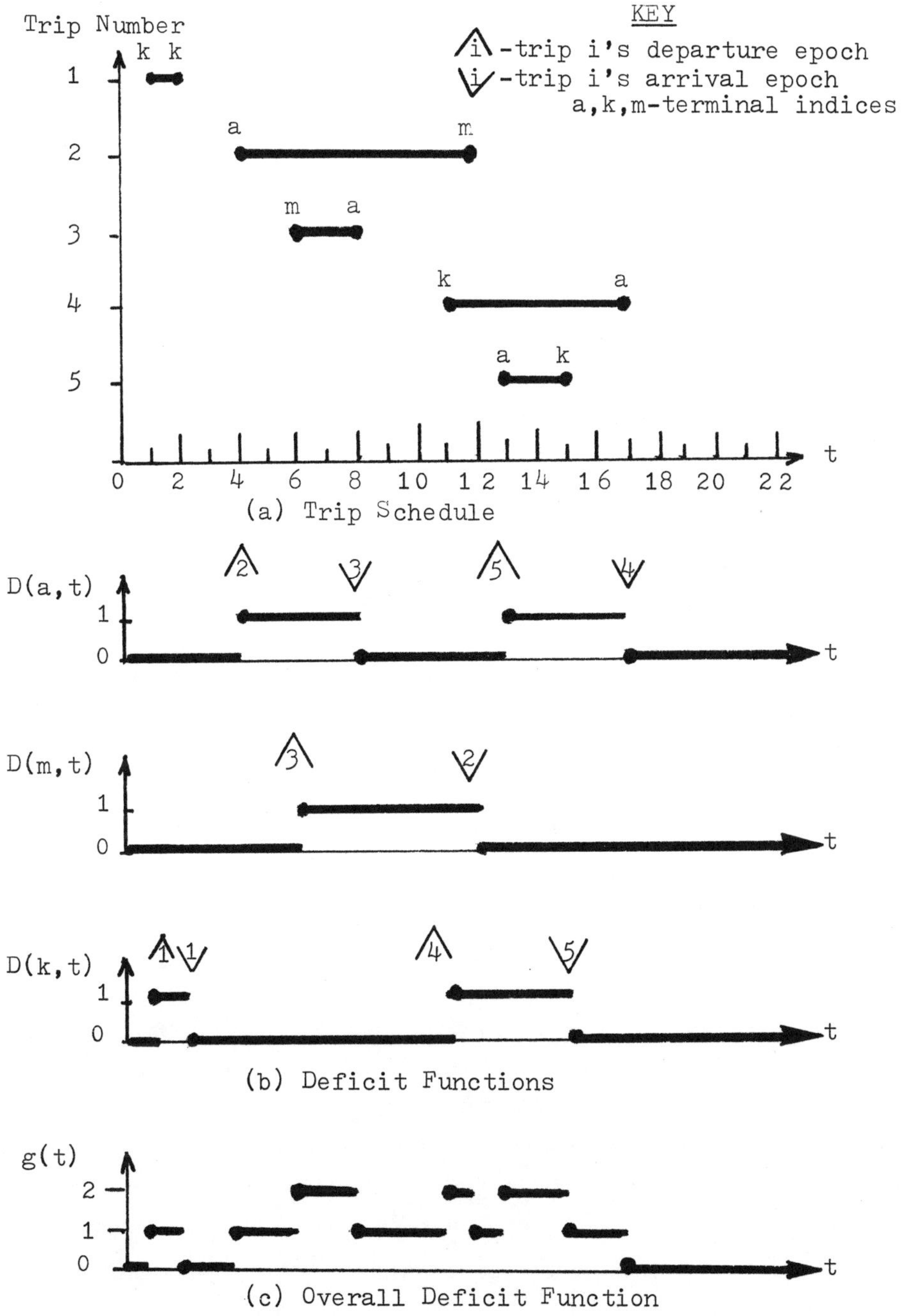

Fig.3. Three Terminal Example of Trip Schedule and Associated Deficit Functions.

Theorem 2. Fleet Size Formula

If, for a set of terminals K and a fixed set of required trips N, all trips start and end within the schedule horizon T = [T1,T2] and no DH insertions are allowed, then the minimum number of vehicles required to service all trips in N is equal to the sum of all the deficits.

$$\bar{C} = \text{Min } C = \sum_{k \varepsilon K} \bar{D}(k) = \sum_{k \varepsilon K} \max_{t \varepsilon T} D(k,t) \,. \qquad (6)$$

Like the array labeling technique, the deficit function approach requires two phases. The first phase consists of constructing the set of deficit functions. In the second phase, the vehicle schedules are obtained through a chain extraction procedure. This procedure is carried out on the deficit functions themselves and is described in Gertsbach and Gurevich (11). The procedure is illustrated in Figure 1 (right side) for the example MFS problem.

Stern and Ceder [5, 25] have treated the MFS-DH problem in the context of deficit function theory. Through the insertion of DH trips, the fleet size may be reduced below the level described in Theorem 2, which provides an upper bound on the MFS-DH problem. The procedure is based on the search for a unit reduction deadheading chain (URDHC) which, when inserted into the schedule, reduces the fleet size by a unit amount. A URDHC is a sequence of DH trips of the form $(k_1, DH_1, k_2, DH_2, \ldots k_i, DH_i, k_{i+1}, \ldots, DH_w, k_{w+1})$, where DH_i is a DH trip from terminal k_{i+1} to k_i. The DH trips are inserted into the deficit functions from hollow to hollow with DH_1 arriving at the first hollow H_o of $D(k_1,t)$ and DH_w departing from the last hollow of $D(k_{w+1},t)$. The procedure inserts URDHC's until no more can be found or a lower bound on the minimum fleet is reached. The lower bound is stated below.

Theorem 3. DH Trip Lower Bound Theorem

(a) For a set of terminals K, and a trip schedule S, the maximum number of trips in simultaneous operation provides a lower bound on the minimum fleet size.

$$\bar{G} = \max_{t \in T} g(t) \leqq \bar{C} . \qquad (7)$$

(b) If $\bar{G} = \bar{C}$, it is impossible to reduce the minimum fleet size $\bar{C}$ through the introduction of DH trips.

Proof.

(i) In general, it is known that:

$$\max_{t \in T} \sum_{k \in K} D(k,t) \leqq \sum_{k \in K} \max_{t \in T} D(k,t).$$

From the definition of g(t) and (6),

$$\max_{t \in T} g(t) \leqq \bar{C}.$$

(ii) Let S represent the original schedule of given trips. Add some trip i starting at t(s) and ending at time t(e) to the schedule (i can be a deadheading trip). Call the new schedule $S' = S \cup \{i\}$. The symbol S or S' will be added to the argument of the deficit and overall deficit functions to indicate the set of trips over which they are constructed.

Since

$$g(t,S') = \begin{cases} g(t,S), & \text{for } t \notin [t(s),t(e)] \\ g(t,S)+1, & \text{for } t \in [t(s),t(e)], \end{cases}$$

$$\max_t g(t,S') \geqq \max_t g(t,S),$$

$$\text{and } \bar{G}(S') \geqq \bar{G}(S),$$

from (i), $\bar{C}(S') \geqq \bar{G}(S')$ and, by the assumption in (b),
$\bar{C}(S) = \bar{G}(S)$.

Therefore,
$$\bar{C}(S') \geqq \bar{C}(S).$$

Q.E.D.

From Figure 3, it can be seen that the lower bound on the MFS-DH problem is 2. The reduction to 2 is achieved through the construction and insertion of the URDHC (a, DH_1, k, DH_3, a, DH_2, m) as illustrated in Figure 4. The procedure

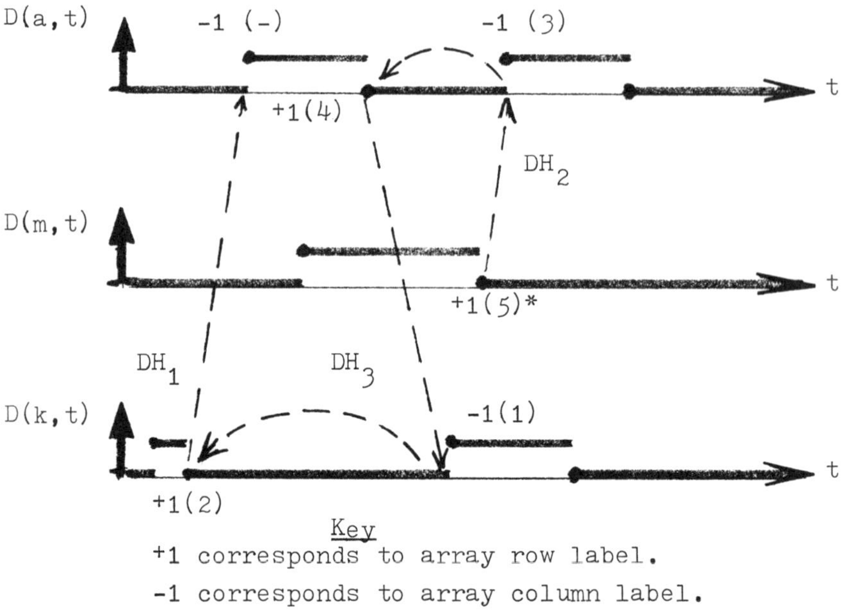

(a) Construction of URDHC and Array Labeling Carried Out on Deficit Functions.

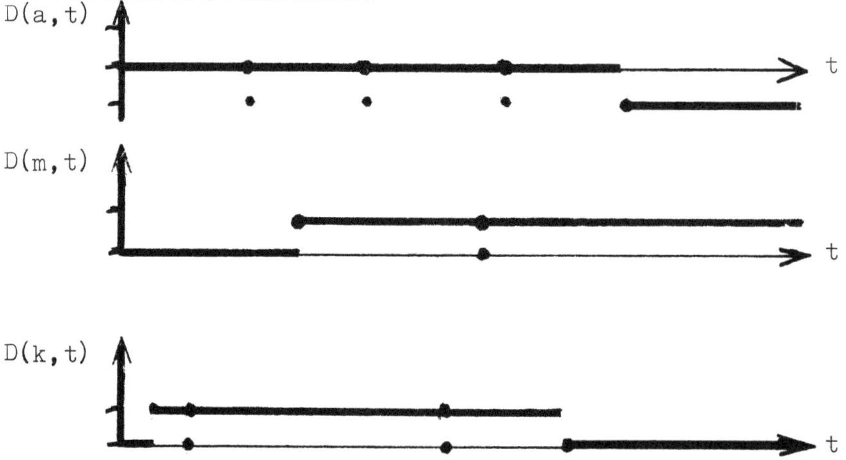

(b) Deficit Functions After Insertion of URDC (Corresponds to Array Flow Change).

Fig.4. Deficit Function Equivalence of Labeling and Flow Change Routines for Solution 1.

is fully described in Ceder and Stern (5).

Comparison of the Array Labeling and Deficit Function Procedures

The following are some correspondences between the array labeling and deficit function MFS solution procedures.

Correspondence 1. A column without a 1 identifies the initial trip departure of a chain. This is in one to one correspondence to a departure point in H_o of some deficit function.

Correspondence 2. Each placement of a 1 in the array corresponds to an internal joining on a chain and the pairing of an arrival with a departure from the same hollow of a deficit function.

Correspondence 3. A row without a 1 identifies the final trip arrival of a chain. This is in one to one correspondence with an arrival point in the final hollow of some deficit function.

Correspondence 4. A chain construction on a trip joining array corresponds to a chain extraction on the set of deficit functions. The steps are illustrated side by side in Figure 1.

The following are some correspondences between the array and deficit function procedures for inserting DH trips during the solution of the MFS-DH problem.

Correspondence 5. Finding and adding a 1 to the array when DH cells are made admissible implies a saving of one vehicle. This is in correspondence to finding a URDHC and inserting it into the set of deficit functions. The steps of each of these procedures can be seen to be in one to one correspondence by observing Figures 2 and 4. Here, it can be observed that array labeling and breakthrough correspond to the steps of constructing a URDHC. An array flow change corresponds to a URDHC insertion.

When the procedures are implemented, there is one slight difference. In the array technique, the minimum number of chains and the selection of all internal joinings are determined concomitantly before the actual chain constructtion phase. Construction of the deficit functions establishes the minimum fleet size, but the actual trip joinings are deferred until the chain construction phase.

3. GARAGE CONSTRAINED - BALANCED SCHEDULES

A vehicle schedule (set of chains) is said to be balanced if for every terminal k, the number of vehicles starting their schedules from k equals the number of vehicles ending their schedules at k (not necessarily the same vehicles). Otherwise, the vehicle schedule is unbalanced. An example of a balanced schedule appears in Figure 1 for the MFS problem. The solution indicates that exactly one vehicle chain starts and ends at each of the three terminals (albeit not the same chain for each terminal). However, solution 1 to the MFS-DH problem, which saved a vehicle at terminal a, resulted in an unbalanced schedule (see Figure 2). The second solution to MFS-DH, which saved a vehicle at terminal m, remains balanced.

To determine if a schedule is balanced, it is not necessary to carry out the actual chain constructions. The assessment may be made through an examination of the schedules' deficit functions. Define vs(k) and ve(k) as the number of vehicles required at the start, and remaining at the end of the schedule horizon, for terminal k, respectively. Then the schedule is balanced if vs(k) = ve(k), for all k.

<u>Theorem 4.</u> Balanced Schedule Condition

A vehicle schedule is balanced if and only if the value of each deficit function at the end of the schedule horizon is zero, i.e.:

$$D(k,T2) = 0, \; k \in K.$$

Proof. (a) Assume $D(k,T2) = 0$, then the net number of arrivals from the end of the last maximal interval is $\bar{D}(k)$. As these arrivals are not followed by departures in any chain construction, they represent the number of chain ends at k; i.e., $ve(k) = \bar{D}(k)$. Also, it follows from Theorem 2 that $vs(k) = \bar{D}(k)$. (b) Assume $vs(k) = ve(k)$. The following two identities,

$$vs(k) = \bar{D}(k) \geqq 0, \quad k \varepsilon K,$$
$$ve(k) = \bar{D}(k) - D(k,T2) \geqq 0, \quad k \varepsilon K,$$

imply:

$$D(k,T2) = vs(k) - ve(k) = 0, \quad k \varepsilon K.$$

Q.E.D.

Corollary 2. Unbalanced Schedules

1. If $D(k,T2) < 0$, then a surplus of $D(k,T2)$ vehicles is present at k at the end of the day.

2. If $D(k,T2) > 0$, then a shortage of $D(k,T2)$ vehicles occurs at k at the end of the day.

3. Under a system that conserves vehicles, the following identity is true:

$$\sum_{k \varepsilon K1} D(k,T2) = - \sum_{k \varepsilon K2} D(k,T2), \tag{8}$$

where K1 and K2 represent the sets of shortage and surplus terminals, respectively.

4. The degree of imbalance may be measured as:

$$\text{Total Imbalance} = (1/2) \sum_{k \varepsilon K} |D(k,T2)|. \tag{9}$$

Observation of the set of deficit functions for the MFS problem in Figure 3 indicates that the schedule will be in balance, as all $D(k,T2) = 0$. However, examination of the deficit functions corresponding to solution 1 of the MFS-DH problem (see Figure 4(b)) indicates a surplus of 1 vehicle at terminal a, and a shortage of 1 vehicle at terminal m at the end of the schedule. This results in a total imbalance of 1. In order to bring unbalanced schedules into balance it is necessary to either add new DH trips to the schedule or rearrange the existing number of deadheading trips into the proper position. For example, a return deadheading trip from a to m could have been added to the end of chain C_2 (see Figure 2(b)) to bring solution 1 into balance. Alternatively, the replacement of DH_1 by DH_4 and some rearrangement converts solution 1 (unbalanced) to solution 2 (balanced).

Balancing the schedule, however, may not be enough to satisfy practical situations in which limitations exist on the extent of overnight garage or parking facilities. Let Q(k) represent such a limitation at terminal k. Then a feasible schedule must also satisfy the garage constraints:

$$vs(k) \leqq Q(k), \; ve(k) \leqq Q(k) \text{ for all } k.$$

If the schedule is balanced, this reduces to $\bar{D}(k) \leqq Q(k)$ and the garage constrained balance schedule (GC-BS) problem may be stated in deficit function form as:

<u>Problem P2</u> The GC-BS Minimum Fleet Size Problem

Insert as many deadheading trips as necessary into the schedule of required trips in order that

$$\text{Min } Z2 = \text{Min} \sum_{k \varepsilon K} \bar{D}(k), \tag{10}$$

subject to:

$$D(k,T2) = 0, \qquad k\varepsilon K, \tag{11}$$

$$\bar{D}(k) \leqq Q(k), \; k\varepsilon K. \tag{12}$$

Alternatively, using the identities of the proof (part b) of Theorem 4, the problem may be restated as:

$$\text{Min } Z2 = \text{Min} \sum_{k \varepsilon K} vs(k), \tag{13}$$

subject to:

$$vs(k) - ve(k) = 0, \quad k\varepsilon K, \tag{14}$$

$$0 \leqq vs(k) \leqq Q(k), \; k\varepsilon K, \tag{15}$$

$$0 \leqq ve(k) \leqq Q(k), \; k\varepsilon K. \tag{16}$$

In this case either (15) or (16) is redundant but both are included in order to facilitate the development of the next section.

4. MATHEMATICAL PROGRAMMING FORMULATION OF THE GC-BS PROBLEM

In order to extend problem P1 to include garage and vehicle balancing constraints, define a number of dummy overnight trips that travel backwards in time. For each terminal k, let n+k represent an undisclosed number of trips of the form $(n+k, t_s^{n+k}, n+k, t_e^{n+k})$, such that $t_e^{n+k} \leqq T1 < T2 \leqq t_s^{n+k}$. The times t_s^{n+k} and t_e^{n+k} represent the latest time at which a vehicle can arrive at garage k after servicing its last required trip and the earliest time a vehicle can leave garage k to service its first required trip, respectively. Such times may be stipulated in a driver union contract. The following additional variables are now introduced:

$Y_{n+k,j}$ = 1 if a vehicle departs for garage k (after completing dummy trip n+k) to service j, its first required trip, and 0 otherwise.

$Y_{i,n+k}$ = 1 if a vehicle services its last required trip as trip i before arriving at garage k to park for the night (start dummy trip n+k) and 0 otherwise.

z_{kk} = an integer variable whose value represents the number of unused spaces at garage k ($0 \leqq z_{kk} \leqq Q(k)$, $k \varepsilon K$).

Define the travel time of the morning and evening DH trips from and to garage k as: $\tau(n+k, p^j)$ and $\tau(q^i, n+k)$, respectively. The joining (n+k,j) will be considered admissible if $t_e^{n+k} + \tau(n+k, p^j) \leqq t_s^j$ for $j \varepsilon N$, $k \varepsilon K$. Otherwise, (n+k,j) will be inadmissible and $Y_{n+k,j} = 0$. The joining (i,n+k) is admissible if $t_e^i + \tau(q^i, n+k) \leqq t_s^{n+k}$ for $i \varepsilon N$, $k \varepsilon K$. Otherwise, (i,n+k) is inadmissible and $Y_{i,n+k} = 0$. Note that (n+k,j) and (i,n+k) are always admissible if $p^j = k$, and $q^i = k$, respectively. Let A represent the set of admissible joinings of the above type in addition to those between required trips (i,j) as described earlier. The mathematical programming version of the GC-BS problem with deadheading trips may now be stated.

Problem P3

$$\text{Max } Z3 = \sum_{i \in N} \sum_{i \in N} x_{ij}, \tag{17}$$

subject to:

$$\sum_{j \in N} x_{ij} + \sum_{k \in K} y_{i,n+k} = 1 \qquad i \in N, \tag{18}$$

$$\sum_{i \in N} x_{ij} + \sum_{k \in K} y_{n+k,j} = 1 \qquad j \in N, \tag{19}$$

$$\sum_{j \in N} y_{n+k,j} + z_{kk} = Q(k) \qquad k \in K, \tag{20}$$

$$\sum_{i \in N} y_{i,n+k} + z_{kk} = Q(k) \qquad k \in K, \tag{21}$$

$$z_{kk} \in \{0, 1, \ldots, Q(k)\} \qquad k \in K, \tag{22}$$

$$\left.\begin{array}{lll} x_{ij} \in \{0,1\} & & (i,j) \in A, \\ x_{ij} = 0 & & (i,j,) \notin A, \end{array}\right\} \tag{23}$$

$$\left.\begin{array}{lll} y_{i,n+k} \in \{0,1\} & & (i,n+k) \in A, \\ y_{i,n+k} = 0 & & (i,n+k) \notin A, \\ y_{n+k,j} \in \{0,1\} & & (n+k,j) \in A, \\ y_{n+k,j} = 0 & & (n+k,j) \notin A. \end{array}\right\} \tag{24}$$

The set A represents the set of admissible trip joinings. The variables x_{ij} are the same as defined previously in problem P1. Constraint (18) ensures that each required trip i is joined to exactly one successor trip. The successor trip may be either another required trip j or a dummy trip n+k. The latter implies that trip i is the last trip serviced by a vehicle before returning to garage k. Constraint (19) ensures that each required trip j must be joined to exactly one predecessor trip. The predecessor trip may be either a required trip i or a dummy trip n+k. If joined to the latter trip, j is the first trip serviced by a vehicle departing from garage k. The objective function (17) is identical to that of P1. From the definitions of $y_{n+k,j}$ and $y_{i,n+k}$, the first terms of (20) and (21) represent:

$\sum_{j \in N} y_{n+k,j}$ = the number of vehicles that start their daily schedules from garage k, and

$\sum_{i \in N} y_{i,n+k}$ = the number of vehicles that end their daily schedules at garage k.

Both terms are also equal to the number of occupied overnight parking spaces (used capacity) at garage k.

Theorem 5. Problem P3 solves the GC-BS Problem

Problem P2 and P3 are equivalent. The solution P3 minimizes the fleet size subject to GC-BS constraints.

Proof. Make the following substitutions in P3:

$$\sum_{j \in N} y_{n+k,j} = vs(k) \qquad k \varepsilon K,$$

$$\sum_{i \in N} y_{i,n+k} = ve(k) \qquad k \varepsilon K.$$

Any solution satisfying (20) and (21) satisfies the balance constraint, since subtracting (21) from (20) for each k yields:

$$vs(k) - ve(k) = 0 \quad , \quad k \varepsilon K. \tag{14}$$

Constraints (20) and (21) are also equivalent to the garage constraints (15) and (16) when the nonnegative slack variables z_{kk} are dropped. Summing (19) over j shows that the objective functions for both problems are equivalent; i.e.,

$$\text{Max} \sum_{j \varepsilon N} \sum_{i \varepsilon N} x_{ij} = n + \text{Max} \left(- \sum_{j \varepsilon N} \sum_{k \varepsilon K} y_{n+k,j}\right) = n - \text{Min} \sum_{k \varepsilon K} vs(k).$$

Q.E.D.

Note that P3 is a capacitated transportation problem. As such, it has a bipartite graph representation as shown in Figure 5. The set of supply nodes is in one to one correspondence with the arrival epochs of trips i = 1,2,...,n and n+1, n+2,...,n+k,...,n+q. Similarly, the set of demand nodes is in one to one correspondence with the departure epochs of each of the required and dummy trips. Since the

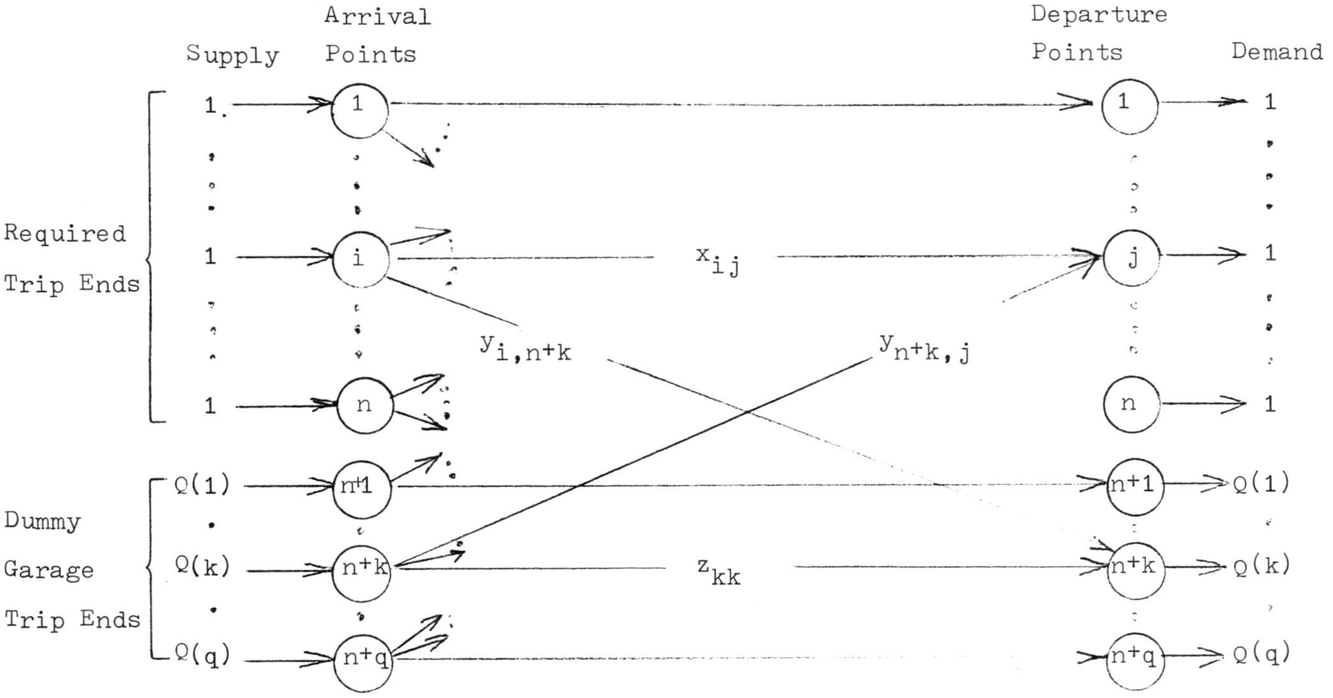

Notes
1. Only admissible arcs appear.
2. If $Q(k)=0$, nodes n+k and their incident arcs are dropped.

Fig. 5 Underlying Bipartite Network for the GC-BS Problem.

supplies and demands are integer, the solution will be integer upon relaxing the integer requirements on the variables. Any of the standard transportation algorithms may be used to solve P3. However, it is possible that P3 is infeasible since not all possible arcs from the supply node set to the demand node set are admissible. Feasibility conditions are stated in the multiterminal supply-demand theorem found in Ford and Fulkerson (8). A special case in which a feasible and optimal solution to P3 can be found rather easily will be presented. The solution procedure will make use of the following two corollaries concerned with an upper bound on P3.

Corollary 3. Upper Bound on P3 (lower bound on fleet size)

The optimal solution $Z1^o$ for problem P1 provides an upper bound on problem P3; i.e., $Z1^o \geqq Z3^o$ (or $C^o \leqq \min_k \Sigma vs(k)$).

Proof. It is sufficient to observe that upon removal of the sets of variables $\{y_{n+k,j}\}$, $\{y_{i,n+k}\}$, $\{z_{kk}\}$ P3 is reduced to (17), subject to the inequality equivalents of (18) and (19). This is exactly problem P1.

Q.E.D.

Problem P3(X^o) Find a feasible solution $\{y_{n+k,j}\}$, $\{y_{i,n+k}\}$, $\{z_{kk}\}$ to (22), (24) and

$$\sum_{k\varepsilon K} y_{i,n+k} = R_i \qquad i\varepsilon N, \qquad (18a)$$

$$\sum_{j\varepsilon N} y_{n+k,j} + z_{kk} = Q(k) \qquad k\varepsilon K, \qquad (20)$$

$$\sum_{k\varepsilon K} y_{n+k,j} = R_j \qquad j\varepsilon N, \qquad (19a)$$

$$\sum_{i\varepsilon N} y_{i,n+k} + z_{kk} = Q(k) \qquad k\varepsilon K, \qquad (21)$$

where $R_i = 1 - \sum_{j\varepsilon N} x^o_{ij}$, $i\varepsilon N$, and $R_j = 1 - \sum_{i\varepsilon N} x^o_{ij}$, $j\varepsilon N$.

Corollary 4. Attainment of the Upper Bound on P3

Given an optimal solution $\{x^o_{ij}\}$, $Z1^o$ to P1, if there exists a feasible solution $\{\bar{y}_{n+k,j}\}$, $\{\bar{y}_{i,n+k}\}$, $\{\bar{z}_{kk}\}$ to $P3(X^o)$, then $Z3^o$, $\{x^o_{ij}\}$, $\{\bar{y}_{n+k,j}\}$, $\{\bar{y}_{i,n+k}\}$; $\{\bar{z}_{kk}\}$ is optimal for P3 and P3 attains its upper bound; i.e., $Z3^o = Z1^o$.

Note that $P3(X^o)$ contains the constraints of a transportation problem. The supply and demand constraints correspond to (18a), (20) and (19a), (21), respectively. Even though the total supply equals the total demand, $\sum_i R_i + \sum_k Q(k) = \sum_j R_j + \sum_k Q(k)$, a feasible solution may not exist as some of the possible arcs from supply nodes to demand nodes may be inadmissible.

A Procedure for Solving a Special Case of P3

Let $\{x^o_{ij}\}$, $Z1^o$ be optimal for P1. Then, from Theorem 1, there exist $C^o = n - Z1^o$ chains. Construct a set of chains using $\{x^o_{ij}\}$ according to the array reduction procedure described earlier. For each chain $u = 1,\ldots,C^o$, associate an ordered pair, (j_u, i_u), such that j_u and i_u represent the first and last trips in chain u, respectively.

Theorem 6. Let C^o be the minimum number of chains corresponding to the solution to P1. If conditions (a) and (b) shown below are true, then there exists a set of C^o chains satisfying the GC-BS constraints which provide an optimal solution to P3.

(a) All joinings of the form (i,n+k) and (n+k,j) are admissible.

(b) The total garage capacity is at least as large as C^o.

Proof. The proof will be by construction.

Since

$$R_i = \begin{cases} 1, \forall\, i = i_u, u=1,\ldots,C^o \\ 0, \forall\, i \neq i_u \end{cases} \quad \text{and } R_j = \begin{cases} 1, \forall\, j = j_u, u=1,\ldots,C^o \\ 0, \forall\, j \neq j_u, \end{cases}$$

it is possible to define the index sets:

$$I = \{i: R_i = 1\} \qquad\qquad J = \{j: R_j = 1\}$$

and

$$\bar{I} = \{i: R_i = 0\} \qquad\qquad \bar{J} = \{j: R_j = 0\}.$$

Hence, $y_{i,n+k} = 0$, $i\varepsilon\bar{I}$ and $\bar{y}_{n+k,j} = 0$, $j\varepsilon\bar{J}$,

and $P3(X^o)$ reduces to:

$$\sum_{k\varepsilon K} y_{i,n+k} = 1 \qquad i\varepsilon I, \tag{18b}$$

$$\sum_{j\varepsilon J} y_{n+k,j} + z_{kk} = Q(k) \qquad k\varepsilon K, \tag{20}$$

$$\sum_{k\varepsilon K} y_{n+k,j} = 1 \qquad j\varepsilon J, \tag{19b}$$

$$\sum_{i\varepsilon I} y_{i,n+k} + z_{kk} = Q(k) \qquad k\varepsilon K, \tag{21}$$

$$z_{kk} \;\varepsilon\; \{0, 1, \ldots Q(k)\} \qquad k\varepsilon K, \tag{22}$$

$$\left.\begin{array}{ll} y_{i,n+k} \;\varepsilon\; \{0,1\} & i\varepsilon I,\ k\varepsilon K, \\ y_{n+k,j} \;\varepsilon\; \{0,1\} & j\varepsilon J,\ k\varepsilon K. \end{array}\right\} \tag{24a}$$

A number of feasible solutions exist for the above constraints. One of them may be constructed by the following procedure:

Step 1. Let

$$\bar{z}_{kk} = \begin{cases} 0 \quad \text{if } k = 1,2,\ldots,s-1 \\ \sum_{k=1}^{s} Q(k) - C^o \text{ if } k = s \\ Q(k) \text{ if } k = s+1,\ldots,q, \end{cases}$$

where s is the smallest integer such that $\sum_{k=1}^{s} C(k) \geq C^{o}$.

Step 2. To satisfy (18b) and (21) let:

$$\bar{y}_{i,n+1} = 1, \qquad i = i_1, \ldots, i_{C(1)},$$
$$\bar{y}_{i,n+2} = 1, \qquad i = i_{C(1)+1}, \ldots, i_{C(2)},$$
$$\vdots$$
$$\bar{y}_{i,n+s} = 1, \qquad i = i_{C(s-1)+1}, \ldots, i_{C^{o}} .$$

Step 3. To satisfy (20) and (19b) let:

$$\bar{y}_{n+1,j} = 1, \qquad j = j_1, \ldots, j_{C(1)},$$
$$\bar{y}_{n+2,j} = 1, \qquad j = j_{C(1)+1}, \ldots, j_{C(2)},$$
$$\vdots$$
$$\bar{y}_{n+s,j} = 1, \qquad j = j_{C(s-1)+1}, \ldots, j_{C^{o}} .$$

Set all other y variables in (18b), (21), (20) and (19b) to 0. This completes the proof of Theorem 6.

This construction uses s garages. Chains u=1 through C(1) are serviced by vehicles operating out of garage 1, chains C(1)+1 through C(2) operate out of garage 2, etc. Garages 1 through s-1 are used to capacity. Garage s may be partially used, and garages s+1 through q are not used. Not only is the solution balanced, but it has the special feature that each vehicle servicing a chain of required trips starts and ends its schedule from the same garage.

Alternative Optimal Solutions

The above construction assigned vehicle chains to garages in sequence until each garage was filled. This sequence was arranged according to the initial arbitrary garage index k=1,...,q. Alternatively, if one reorders the garage indices according to nonincreasing capacity, a solution using the minimum number of garages may be obtained. Also, by evenly distributing the C^{o} vehicle chains among the available garages, a balanced garage utilization type of solution is achieved. Vehicle-garage assignments could also be made on the basis of minimizing the total garage pull-out and pull-in DH time (or some form of operational cost such as

DH plus idle time cost). This however, could lead to a sub-optimization as the total garage pull-out and put-in costs should be added to the operational cost of trip joining within the original vehicle chains determined from P1. A search over all minimum fleet solutions to P1 would be necessary before the secondary objective of minimum total operational cost could be assessed. The problem could be solved optimally in one step by defining a minimum cost objective function for P3. One would assign a very large negative cost to each of the {z} variables and the appropriate operational cost to the {x} and {y} joining variables. The resulting transportation problem could also be solved when condition (a) in Theorem 6 is relaxed to reimpose the garage morning departure and evening arrival time constraints. For this case, however, the fleet size may have to be increased above the lower bound (as described in Corollary 3) in order to retain feasibility.

5. CONCLUDING REMARKS

A review of the array labeling and deficit function approaches for solving the minimum fleet size problem (with and without deadheading) has been given. A simple example was used as an expository device to illustrate the correspondences between the two procedures. The problem was extended to include garage capacity and balanced schedule constraints. Deficit function and mathematical programming (transportation problem) formulations of this problem were derived. The structure of the second formulation was exploited to provide a simple procedure for a special case.

The deficit function approach for solving the MFS-DH problem is the basis of a heuristic algorithm which is being developed for use in an interactive man-machine bus scheduling operation (5), (25). This algorithm is being extended to include garage constraints and the requirement for a balanced schedule in the context of problems with large numbers of terminals (6). This algorithm will also include the capability to handle vehicles of different types. The addition of nonidentical vehicle types, each capable of servicing only a subset of the required trips, may also be included in the mathematical programming formulation presented here. Such an extension requires the network problem to be stated as a form of multi-commodity flow problem along the lines described in Golden, et. al. (13) for the delivery truck problem.

REFERENCES

1. P. Allan-Anderson, O. Holst and P.V. Nexielsen, "Scheduling in Transportation: Two Case Studies," Research Report No. 10/1978, The Institute of Mathematical Statistics and Operations Research, The Technical University of Denmark, Lyngby, Denmark, 1978.
2. T.E. Bartlett, "An Algorithm for the Minimum Number of Transport Units to Maintain a Fixed Schedule," Naval Res. Logist. Quart. 4,139-149 (1957).
3. G.E. Bennington and K.K. Rebibo, "Overview of the RUCUS Vehicle Scheduling Program (Blocks)," The Mitre Corp., McLean, Va. 1975 (Paper presented at the Workshop on Automatic Techniques for Scheduling of Vehicle Operators for Urban Public Transportation Services, Chicago, 1970).
4. L.D. Bodin and R.B. Dial, "Hierarchical Procedures for Determining Vehicle and Crew Requirements for Mass. Transit Systems," Trans. Res. Rec. 746, 58-64 (1980).
5. A. Ceder and H.I. Stern, "Deficit Function Bus Scheduling with Dead-Heading Trip Insertions for Fleet Size Reduction," to appear in Trans. Sci. 16, 1 (1982).
6. A. Ceder and H.I. Stern, "Heuristic Reductions for Bus Fleet Scheduling with Large Numbers of Terminals," Paper presented at CORS/TIMS/ORSA Meeting, May 3-6, 1981, Toronto.
7. B. Dantzig and D.R. Fulkerson, "Minimizing the Number of Tankers to Meet a Fixed Schedule," Naval Res. Logist. Quart. 1,217-222 (1954).

8. L.R. Ford Jr. and D.R. Fulkerson, Flows in Networks, Princeton University Press, Princeton, N.J., 1962.

9. M. Friedman, "Combinatorial Problems in an Adaptive Public Transportation Network," Trans. Res. 12,305-308 (1978).

10. B. Gavish, P. Schweitzer, and E. Shlifer, "Assigning Buses to Schedules in a Metropolitan Area," Comp. and Opns. Res. 5,129-138 (1978).

11. I. Gertsbach and Y. Gurevich, "Constructing an Optimal Fleet for a Transportation Schedule," Trans. Sci. 11, 20-36 (1977).

12. I. Gertsbach and H.I. Stern, "Minimal Researches for Fixed and Variable Job Schedules," Opns. Res. 26,68-85 (1978).

13. B.L. Golden, T.L. Magnanti, and H.Q. Nguyen, Implementing Vehicle Scheduling Algorithms," Networks, 7, 113-148 (1977).

14. W.S. Hipp, "An Algorithm for Rapid Reassignment of Motor Freight Carriers," Software Age, 8 -14 (1968).

15. J. Hoffstadt, "Computerized Vehicle and Driver Scheduling for the Hamburger Hochbahn Aktiengesellschaft," in Computer Scheduling of Public Transport: Urban Passenger Vehicle and Crew Scheduling, A. Wren (Ed.), North-Holland, Amsterdam, 35-52, 1981.

16. F. Kirkman, "Problems of Innovation in the Transport Industry: A Bus Scheduling Program," Proceedings of PTRC Public Transport Analysis Seminar, 12, pg. 23, London, 1968.

17. W. Lampkin and P.D. Saalmans, "The Design of Routes, Service Frequencies, and Schedules for a Municipal Bus Undertaking: Case Study," Opnl. Res. Quart. 18, 375-397 (1967).

18. V.K. Linus and M.S. Maksim, "On the Problems of Constructing Routes," Proceedings of the Institute of Civil Aviation Engineering, 102, 36-45, 1967 (in Russian)

19. C.S. Orloff, "Route Constrained Fleet Scheduling," Trans. Sci. 10, 149-168 (1976).

20. J.L. Saha, "An Algorithm for Bus Scheduling Problems," Opnl. Res. Quart. 21, 463-474 (1970).
21. F.J.M. Salzborn, "A Note on Fleet Routing Models for Transportation Systems," Trans. Sci. 6, 3, 335-337 (1972).
22. F.J.M. Salzborn, "Minimum Fleet Size Models for Transportation Systems," in Transportation and Traffic Theory, D,J, Buckley (Ed.), 607-624, Reed, Sydney, 1974.
23. B.M. Smith and A. Wren, "Vampires and Tasc: Two Successfully Applied Bus Scheduling Programs," in Computer Scheduling of Public Transport: Urban Passenger Vehicle and Crew Scheduling, A. Wren (Ed.), North-Holland, Amsterdam, 97-124, 1981.
24. H.I. Stern, "Bus and Crew Scheduling " (note), Trans. Res. 14A, 154 (1980).
25. H.I. Stern and A. Ceder, "A Deficit Function Approach for Bus Scheduling", in Computer Scheduling of Public Transport: Urban Passenger Vehicle and Crew Scheduling, A. Wren (Ed.), North-Holland, Amsterdam, 85-96, 1981.

METHODS OF FORECASTING KILOMETRES PER CAR

J.C. Tanner

Transport and Road Research Laboratory, Crowthorne, U.K.

The object of the paper is to reconsider the methods used to forecast kilometres per car per year at an average national level. Such forecasts are used as an input to the national traffic forecasts. The first part summarizes various sets of time-series and cross-section data and draws certain conclusions about the effects of incomes, petrol prices and car speeds.

A simple constant-elasticity model is then developed; it is used to illustrate the effects of a range of assumptions about the future. An alternative model of car ownership and use is then proposed. This has a disaggregate causal structure and uses ideas of utility maximisation and constant travel budgets. The forecasts from both models cover a wide range; at the centre of the range of the assumptions, they indicate little change in kilometres per car up to the year 2010.

It is concluded that at its present stage of development the causal model is less satisfactory than the simple method.

1. INTRODUCTION

National forecasts of levels of car traffic in Great Britain have for some years been based on a two-stage model. The first stage calculates the future number of cars, and the second stage the average number of kilometres per car per year. The object of the present paper is to reconsider the methods of forecasting kilometres per car, and in particular to compare the existing type of model based on constant elasticities with a more complex causal model.

The existing methods for kilometres per car were developed and extended in references 1 to 4. The latest official forecasts(4) assume constant elasticities of + 0.1 with respect to income and alternatives of - 0.1 and - 0.2 with respect to petrol price.

Use of a simple constant-elasticity model presents problems because the available data do not enable the elasticities to be estimated with much confidence, nor is there any reason to expect them to remain constant in the future. In addition, there is no evidence at all on the effects of some variables that might be expected to influence the forecasts. These problems prompted an attempt to develop a causal model that represents the decision processes of individual households, and it is this model that forms the main substance of this paper.

Section 2 briefly summarises the available data on kilometres per car and how it is influenced by various factors. Section 3 puts these results together into the form of a simple model that can be used for forecasting. Section 4 develops the causal model. Section 5 summarises the forecasts obtained from both models for a range of assumptions about the future of the economy. A fuller account especially of the empirical data is being published elsewhere[5].

2. SOME EMPIRICAL EVIDENCE

Average kilometres per car in Great Britain have risen from 12,100 in 1953 to 14,300 in 1979. To forecast how the figure will change in the future it is necessary first to understand the factors that have caused this increase and then to take a view on how these and perhaps other factors will change in the future. During this period many factors have changed, and among those likely to be relevant to car use are real income per person, which has risen by 73 per cent, petrol prices, which have fluctuated without a systematic trend, car purchase costs, which have fallen by about one third, and average car travel speeds, which have increased by about a quarter.

Recent fluctuations in petrol prices have been accompanied by measurable fluctuations in kilometres per car, giving a short-term price elasticity in the range - 0.10 to - 0.15. The time series data alone do not provide clear information on the effects of other factors. However various kinds of cross-section data indicate long-term income elasticities. Comparisons of high and low income households give values in the range + 0.18 to + 0.43; comparisons of countries with different income levels gives + 0.55 to + 0.60 (Fig. 1). It is concluded that the effect of income growth on the time series data correspondsto an elasticity of the order of + 0.20. Such growth as cannot be explained by petrol price or income changes is likely to be due to changes in purchase cost and in speeds. An effect of speed is suggested by the higher kilometres per car for cars in rural households than for those in urban households and by the high elasticity in the international comparison.

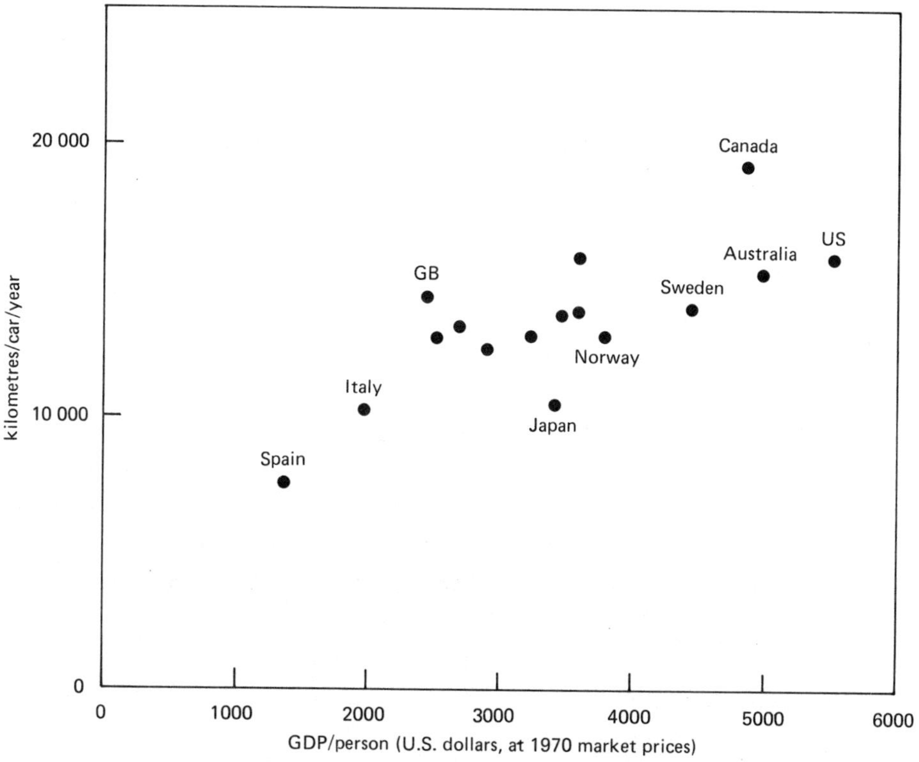

Fig. 1 KILOMETRES PER CAR IN 1976 – 77 IN VARIOUS COUNTRIES, RELATED TO INCOME PER PERSON

3. A SIMPLE FORECASTING MODEL

The model proposed here is a modification of that used in previous forecasts[1,2,3,4]. It explains past trends in terms of income growth, car purchase costs, petrol prices and car speeds. It is assumed that the elasticity of kilometres per car with respect to each of these quantities is constant; this assumption is unlikely to be true, but it is probably not critical over the range of the variables considered.

The most natural way to fit such a model to time series data is by linear regression on the logarithms of the variables. This however gave a negligible income elasticity, and it was therefore decided to use the value + 0.20 suggested in Section 2. With this constraint, regression analysis on data for 1953-1978 (excluding 1956, 1957, 1963) gave

$$\text{km/car} = K(\text{income})^{0.20}(\text{purchase cost index})^{0.02}(\text{petrol price})^{-0.12}(\text{speed})^{0.38}.$$

The purchase cost elasticity is low and not significant, but it has the expected sign and will be left in. The petrol price and speed elasticities are similar to those suggested in Section 2. Fig. 2 shows how the model compares with the data since 1953. The model satisfactorily reproduces the general trend and to some extent the ups and downs since 1970. However it does not reproduce properly other year to year fluctuations in the actual figures, though this is hardly surprising.

To use this model in a forecasting mode, the constant term K is adjusted so that it gives the correct figure in the base year for the forecasts. Also the petrol price is taken to be a petrol cost per car-kilometre instead of per gallon, to allow for future changes in fuel efficiency of cars.

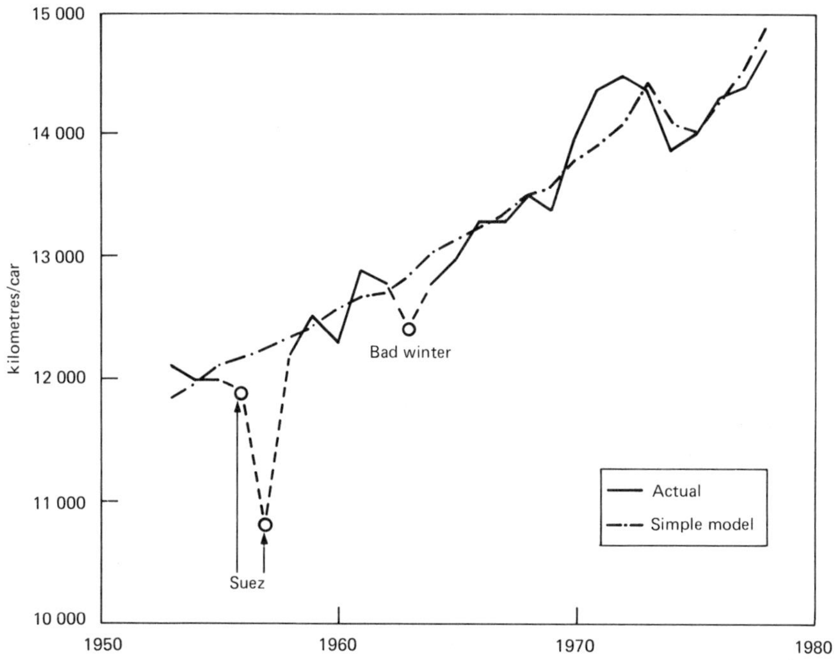

Fig. 2 COMPARISON OF TRENDS IN KILOMETRES PER CAR SINCE 1953 WITH ESTIMATES FROM A SIMPLE MODEL

4. A CAUSAL MODEL OF CAR OWNERSHIP AND USE

4.1 General

The data in section 2 have given some indication of the effects of various factors on annual kilometres per car and these have been used to set up the simple model of section 3. However they have not shown clearly the way in which these factors should be combined to give a formula that can be used for forecasting. In particular, they do not indicate the level that could be expected in the distant future or under extreme conditions. Nor do they indicate what effect if any may result from changes in such factors as car ownership and public transport provision and fares. It may therefore be useful to set up a model, necessarily over-simplified, which represents what are believed to be the most important factors in a way that may correspond to rational behaviour. Because car ownership and car use reflect different aspects of what in the long run is essentially a single decision-making process, the model will cover both aspects of it.

The approach has similarities to the General Motors approach[6,7,8], which has recently been adopted by Zahavi[9]. However the details of the present model are substantially different.

4.2 Theory

The object is to produce a model which explains why ownership and use should take the values that they do, in terms of rational self-interested response by individuals to the relevant external factors. The model described here is based on the view that car ownership and car use are inter-related: whether a household acquires a car depends on the use that it would make of it if it did and on the benefit that such use would provide. If owning and using a car is, overall, better for a household, then it will own and use one. The model is a very crude one, in that it ignores many relevant factors and grossly over-simplifies others; thus great care is needed in calibration and interpretation.

The initial specification of the model relates to one point in time, with such quantities as the price of petrol being regarded as single known quantities. The model is intended to represent the way in which car ownership and use would respond at that point in time to changes in quantities such as the price of petrol. Provided that all the external variables can be properly adjusted to allow for trends, the model should at least in principle be able to produce forecasts of changes in car ownership and use over time.

The model assumes that decisions about car ownership and use are taken by independent decision-making units which consist of two adults. No account is taken of other kinds of household, or of children. Further, it is assumed that all such households are identical except in respect of their income, and of a 'propensity' to travel, a concept not yet defined. The model will assert that at a given level of the remaining variables, a car will be owned by all those households with more than a certain propensity level and not by those below this level; a second car will be acquired at a higher propensity level. Changes in ownership will result from changes in numbers of households at different propensity levels, for example as a result of economic growth, and from changes in the propensity at which ownership becomes worthwhile; changes in use will depend on changes in use at a given propensity level, as well as on changes in propensity level at which ownership becomes worthwhile and on the numbers of households at each propensity level.

Propensity is not a directly observable quantity. It is the summation of all factors, other than the general level of travel costs, which determine whether an individual household will acquire a car. It represents both propensity to acquire a car and propensity to travel, whether by car or otherwise; no distinction is drawn between these. In fact it will be assumed that all aspects of travel behaviour depend on propensity rather than on income: propensity is to be thought of as effective income, in the travel context. For the purposes of this paper it is measured in income units. If

i = income (£ per year)
j = propensity (£ per year)

then the model asserts that

$$\log j = \log i + \varepsilon$$

or

$$j = i\, e^{\varepsilon}, \qquad (1)$$

where ε represents all non-income elements in propensity and is assumed to be normally distributed with mean zero and variance σ_o^2. ε is assumed to be independent of i.

It is further assumed that incomes follow a lognormal distribution, log i being normally distributed with mean μ_1

and variance $\sigma_1{}^2$. It follows that log i and log j are jointly normally distributed, with correlation coefficient

$$r = \frac{\sigma_1}{(\sigma_o^2 + \sigma_1{}^2)^{\frac{1}{2}}} \quad . \tag{2}$$

The mean and variance of log j are

$$\mu_2 = \mu_1$$
$$\sigma_2^2 = \sigma_o{}^2 + \sigma_1{}^2 \tag{3}$$

so that

$$\sigma_1 = r\sigma_2 \quad . \tag{4}$$

It is assumed that travel behaviour depends on the generalised costs of travel and on propensity j. Generalised cost means money cost plus a money equivalent of time spent. Money equivalents of time are calculated on the assumption that the value of time is a constant proportion of the propensity; the average value of time for an income group will then be proportional to income. The value of time is intended to represent travel choices near the margin: the amount of money that an individual would be willing to pay to save one unit of time.

Only two kinds of travel are distinguished: car travel and the remainder. The generalised cost of car travel is denoted by p_1 in units of £ per car-kilometre, where

$$p_1 = \mu + \nu p + \gamma jk/v \tag{5}$$

in which

μ = money cost (£) of car travel per kilometre, excluding fuel cost
ν = fuel consumption of cars (gallons/km)
p = price of fuel (£ per gallon)
k = occupancy, adults per car

γ = value of car travel time (£ per hour) as proportion of annual propensity
v = speed of car travel (km/h) .

The cost of car ownership is denoted by £q per car per year; this is a money cost, with no time element. As explained in Section 4.3, q will be expressed as g + hj.

Non-car travel is assumed to have a generalised cost p_2 per person-kilometre, where

$$p_2 = \alpha + \beta j/u \qquad (6)$$

in which

α = money cost (£) of non-car travel per kilometre,
β = value of non-car travel time (£ per hour) as a proportion of annual propensity,
u = speed of non-car travel (km/h).

The model is built round two fundamental concepts. The first is that of utility maximisation, and the second is that of a constant travel 'budget'. The model is defined in utility-maximising terms in such a way that the constant budget requirement emerges naturally.

There is empirical evidence[10] that generalised travel expenditure incurred by a household is proportional to income, but is independent of generalised travel costs per kilometre. This is broadly true when comparing households with different incomes or living in different areas at one point in time. Over a period of years, however, generalised expenditure per unit of income appears to have been slowly rising. A possible explanation is that as people use cars more and public transport less, and as most forms of transport, especially car, become more comfortable, easier and more the social norm, so the value of car time as a proportion of income may diminish; the analysis assumes however that the value of time remains strictly proportional to income.

Utility theory postulates that the utility a person or household achieves is a function of the amounts of the various kinds of goods and services that it consumes, and that the amounts of these goods and services consumed are such as to maximise this function, subject to their total cost not exceeding the income available. If the utility function is taken to be a weighted sum of the logarithms of the amounts of the various goods, then its maximisation leads to the result that expenditure on each kind of goods is independent of all prices and is a constant proportion of incomes if incomes change. This case corresponds precisely with the evidence for constant travel expenditures and therefore this form of utility function will be assumed here.

Four kinds of 'goods' will be distinguished:

travel by person 1 on his own = x km
travel by person 2 on his own = y km
travel by both persons together = w+w km
all other goods and services.

Thus it will be assumed that the constancy of travel expenditure applies to each kind of travel separately; there is no conclusive evidence for or against this. The utility function may be written

$$U = a \log x + a \log y + 2b \log w + (1-2a-2b) \log (j\text{-cost})$$

where j is the household's propensity, 'cost' is the generalised cost of making the x + y + 2w kilometres of travel, and a and b are parameters to be determined.

This utility function is intended to represent just a household's relative preferences for alternative patterns of consumption, not an absolute measure of benefit derived. Hence it is legitimate to define the weights of the logarithms so that they add up to unity. It also makes the use of propensity j rather than income i more plausible; the function just says that in a travel context the household behaves as if its total income was j.

In determining the travel pattern of a household that will maximise its utility, it is necessary to consider separately what would happen if it has 0, 1 or 2 cars. One determines the maximum utility in each case, and then assumes that the highest indicates the number of cars actually owned.

If one car is owned, it is assumed that only person 1 drives it. Thus it will be used for travel x and travel w+w, but travel y by person 2 will use the non-car mode. Thus

$$U = a \log x + a \log y + 2b \log w + (1-2a-2b)\log(j-q-p_{11}x - p_{12}w-p_2y)$$

where p_{11} is the cost per kilometre of running a car with one occupant and p_{12} the cost with two occupants. This is maximised when

$$x = a(j-q)/p_{11}\ , \quad y = a(j-q)/p_2\ , \quad w = 2b(j-q)/p_{12}\ .$$

Total travel expenditure is

$$B = q + (2a + 2b)(j - q)$$

and

$$U(\max) = \log(j-q) + 2a \log a + 2b \log 2b + (1-2a-2b)\log(1-2a-2b) - a \log p_{11} - a \log p_2 - 2b \log p_{12}\ .$$

Average car occupancy is

$$\frac{ap_{12} + 4bp_{11}}{ap_{12} + 2bp_{11}}$$

and kilometres per car are

$$(j-q)\left(\frac{a}{p_{11}} + \frac{2b}{p_{12}}\right) \quad .$$

Similar but slightly simpler expressions apply if 0 or 2 cars are owned.

The effect of these assumptions is to give a model in which the annual cost of car ownership is regarded as a reduction in income, and therefore in propensity, and in which other travel expenditure is proportional to this reduced propensity but is independent of unit costs.

It may readily be shown that for propensity j less than J_1 no car will be owned, that for propensities between J_1 and J_2 one car will be owned, and that above J_2 two cars will be owned, where

$$\log\left(\frac{J_1 - q}{J_1}\right) = a \log\left(\frac{p_{11}}{p_2}\right) + 2b \log\left(\frac{p_{12}}{2p_2}\right) \qquad (7)$$

$$\log\left(\frac{J_2 - 2q}{J_2 - q}\right) = a \log\left(\frac{p_{11}}{p_2}\right) . \qquad (8)$$

Because q, p_{11}, p_{12} and p_2 are functions of J_1 in the first equation, and J_2 in the second, it is not possible to solve these equations for J_1 and J_2 explicitly.

Using the bivariate lognormal distribution of i and j, it is now possible to write down expressions for car ownership probabilities P_o, P_1, P_2, average kilometres per car, and average car occupancy, both for households at a given income level i and averaged over all households; examples of the expressions are:

$$P_1(i) = \Phi\left(\frac{\log J_2 - \log i}{\sigma_o}\right) - \Phi\left(\frac{\log J_1 - \log i}{\sigma_o}\right)$$

$$P_1 = \Phi\left(\frac{\log J_2 - \mu_2}{\sigma_2}\right) - \Phi\left(\frac{\log J_1 - \mu_2}{\sigma_2}\right)$$

km/car (i,1) =

$$\int_{\frac{\log J_1 - \log i}{\sigma_o}}^{\frac{\log J_2 - \log i}{\sigma_o}} \left(ie^{x\sigma_o} - q\right)\left(\frac{a}{p_{11}} + \frac{2b}{p_{12}}\right) \phi(x)\; dx/P_1(i)$$

km/car (1) =

$$\int_{\frac{\log J_1 - \mu_2}{\sigma_2}}^{\frac{\log J_2 - \mu_2}{\sigma_2}} \left(e^{\mu_2 + x\sigma_2} - q\right)\left(\frac{a}{p_{11}} + \frac{2b}{p_{12}}\right) \phi(x)\; dx/P_1.$$

In the above expressions

$$\phi(x) = e^{-\frac{1}{2}x^2} / \sqrt{2\pi}\;,$$

$$\Phi(X) = \int_{-\infty}^{X} \phi(x)\; dx,$$

p_{11}, p_{12} and q are those that apply when

$$j = ie^{x\sigma_o}$$

for expressions relating to a given i, and

$$j = e^{\mu_2 + x\sigma_2}$$

for expressions averaged over the distribution of i.

4.3 Calibration

The main objectives are to calculate the car ownership probabilities and average kilometres per car averaged over all households. These depend, directly or indirectly, on the quantities listed below, except for σ_o, which is needed to relate travel budgets as a proportion of income to those as a proportion of propensity. These various quantities must be estimated for the base year, which it is convenient to take as 1978.

Most of these quantities can be estimated outside the model and without making use of data on car ownership or use. The way in which the various quantities will be estimated for the base year is as follows:

$\sigma_o, \mu_2, \sigma_2$	from observed income distribution and variation in car ownership probabilities with income. These also give estimates of J_1 and J_2.
g,h	from actual expenditures of car-owning households. Makes use of σ_o. Later adjusted so that equations (7) and (8) defining J_1 and J_2 are satisfied.
a,b	from information on travel expenditure budgets and average occupancy levels. Makes use of σ_o, g and h.
μ, ν, p, γ, v, α, β, u	from external sources. γ is later adjusted.

Table 1 shows numbers of households at different gross household income levels in the 1978 FES[11], and the percentages owning different numbers of cars. The incomes follow a lognormal distribution fairly closely giving, for annual income,

$\mu_1 = 8.435 \qquad \sigma_1 = 0.688$

The increases in the ownership percentages with income follow lognormal relations fairly closely; this yields estimates of σ_o, J_1 and J_2 which after slight adjustment to J_2 to allow for households with 3 or more cars are

$\sigma_o = 0.800 \qquad J_1 = 3672 \qquad J_2 = 14\ 499$

It follows from (2) and (3) that

$\mu_2 = 8.435 \qquad \sigma_2 = 1.054 \qquad r = 0.653$

Table 1

Car ownership data from 1978 Family Expenditure Survey

Income range (£ per household per week)	Number of households	Percentage of households with	
		1 or more cars	2 or more cars
0 - 50	1690	16.7	0.5
50 - 100	1965	53.0	5.9
100 - 150	1814	75.7	11.6
150 - 200	933	84.1	25.8
200 -	599	92.3	49.1
All	7001	57.6	12.4

Expenditure on car purchase appears to rise almost linearly with income from £3 per car per week at low incomes to £7 per car per week at high incomes, excluding expenditure incurred by employers on company cars. In addition, certain running costs are incurred on an annual basis. These include licences and some aspects of insurance, garaging and maintenance; £2 per week will be assumed for these. If it is supposed that the annual cost of keeping a car on the road is a linear function of propensity, then in order to reproduce the observed variation with income the relation must be approximately

$$q = 188 + 0.0196\ j.$$

This reproduces the correct £260 for low incomes, for which j is close to J_1; it also reproduces the right growth with income at high incomes: 0.0196 is 0.027, the slope of cost with income, times $e^{-\frac{1}{2}\sigma_o^2} = 0.7261$, since from (1) the average value of j for a given i is

$$E(ie^{\varepsilon}) = ie^{\frac{1}{2}\sigma_o^2}.$$

In fact the relation with i is not strictly linear if the relation with j is, since expenditure q must be averaged only over those households at a given income that actually own cars. Nevertheless the approximation is perhaps not a crucial one and it will therefore be assumed provisionally that

$$g = 188 \qquad h = 0.0196$$

These figures will be reconsidered below.

The model asserts that total generalised travel expenditure per 2-adult family per year depends on propensity and on how many cars are owned, as follows:

No car $(j<J_1)$	$B = (2a+2b)j$
1 car $(J_1<j<J_2)$	$B = (2a+2b)j + (1-2a-2b)q$
2 cars $(J_2<j)$	$B = (2a+2b)j + 2(1-2a-2b)q$

where $q = g + hj$. With the actual values above for J_1, J_2, g and h, a reasonable approximation to this step function is

$$B = [.04 + .96\ (2a+2b)]\ j \quad .$$

Hence for an income group i, for which the mean propensity is $ie^{\frac{1}{2}\sigma_0^2}$ the average generalised travel expenditure is about

$$[.04 + .96\ (2a+2b)]\ ie^{\frac{1}{2}\sigma_0^2} \quad .$$

Now the evidence[10] implies that this is about 0.14 i, excluding expenditure reimbursed by employers but including time on employer's business. If employers' expenditure puts this up to 0.16 i, it follows that 2a + 2b = 0.079 .

The average occupancy of cars in the 1975-76 NTS was 1.76 persons, or about 1.5 adults. The model gives about this figure when a = b, and it will therefore be assumed that

$$a = b = 0.020.$$

Fuel consumption is obtainable from fuel sales, and price per gallon from various surveys. Non-fuel operating costs are derived from expenditures given in the FES report and in the National Income and Expenditure Blue Book. These sources yield

μ = 0.006 £ per car-km
ν = 0.021 gallons per km
p = 0.75 £ per gallon .

The value of γ is estimated by assuming that car travel time is valued at 25 per cent of the wage rate. If the household income is £i per year, then the average propensity per year is £$ie^{\frac{1}{2}\sigma_0^2}$, the average income per person per year is £½i, or £i/4000 per hour (assuming 2000 working hours per year), and the value of travel time therefore £i/16 000 per hour. Hence

$$\gamma = \frac{e^{-\frac{1}{2}\sigma_0^2}}{16\ 000} = 0.000045 \ .$$

This value will be reconsidered shortly. The car speed v is 45.6 km/h.

The values for α,β and μ have to be a compromise between values for the various non-car modes, including bus, rail, walking and cycling. For bus and rail, the average fare was 2 to 2½ pence per km; the average door-to-door speed was perhaps about 10km/h. The value of time as a proportion of income is higher, at least as regards walking and waiting. The following values will be used:

α = 0.02 £/km
β = 0.000090
μ = 8 km/h .

This completes the specification of the model for 1978, except that the parameter values do not necessarily satisfy the equations (7) and (8) defining J_1 and J_2, nor does the model necessarily yield the correct kilometres per car for 1978. After a certain amount of trial and error it was concluded that the most plausible and simple way of dealing with this was to modify the values for g, h and γ, which determine the annual cost of owning a car and the value of car travel time. The corrected values are

$$g = 207$$
$$h = 0.0175$$
$$\gamma = 0.000\,040 \quad .$$

These values are well within the uncertainties inherent in the data on which the original values were based and in the interpretation placed on these values by the model.

Several kinds of check and correction may be necessary before the model specified above may be used for forecasting. Firstly it must adequately reproduce 1978 levels of ownership and use. Secondly it must give acceptable elasticities in respect of 1978 conditions. Thirdly it must adequately reproduce trends observed in recent years. To the extent to which these requirements are not met, the model must be further adjusted.

4.4 The base year model

The first run of the model was for 1978, with all the parameters at the values described above. This yielded the results in Table 2. These and subsequent calculations were done on a programmable desktop calculator, using a straightforward numerical integration procedure.

The car ownership probabilities agree well enough with those in Table 1:

	Table 2 (model)	Table 1 (data)
0	41	42.4
1	45	45.2
2+	14	12.4
	100	100.0

Table 2

Results of 1978 base run of causal model

Income (£/year)	Car ownership			Kilometres per car (thousands)			Average occupancy			Cars per adult
	P_0	P_1	P_2	1 car	2 cars	All	1 car	2 cars	all	
0	1.00	0.00	0.00	7.5	13.6	7.5	1.64	1.42	1.64	0.00
2 000	0.78	0.22	0.01	10.9	15.7	11.2	1.62	1.41	1.61	0.12
4 000	0.46	0.49	0.05	12.1	16.4	12.8	1.62	1.41	1.57	0.30
6 000	0.27	0.60	0.14	12.8	16.9	14.1	1.62	1.40	1.54	0.43
8 000	0.16	0.61	0.23	13.4	17.3	15.1	1.61	1.40	1.51	0.53
10 000	0.11	0.57	0.32	13.9	17.8	15.9	1.61	1.40	1.49	0.61
12 000	0.07	0.52	0.41	14.2	18.2	16.6	1.61	1.40	1.47	0.67
14 000	0.05	0.47	0.48	14.6	18.6	17.2	1.61	1.39	1.45	0.72
∞	0.00	0.00	1.00	20.0	34.2	34.2	1.42	1.33	1.33	1.00
All	0.41	0.45	0.14	12.7	17.9	14.7	1.62	1.40	1.52	0.36

Also the overall cars per adult figure of 0.3618 is close to the actual figure of 0.3640. Based on the new December licence data the actual figure is 0.3460, so a correction factor of 0.3460/0.3618 = 0.9563 will need to be applied.

The overall average kilometres per car is 14 676, which has been made to agree with the actual figure of 14 700. The recent revision of the vehicle-kilometre estimates and the change to December licences mean that the latest official estimate[12] of this figure is 15 300, and a correction factor of 15 300/14 676 = 1.0425 will be required.

The kilometres per car for 1-car and 2-car households are not in good agreement with the data, which show only a small difference (10 per cent higher for 2-car households, compared with 41 per cent from the model). The reason is that the model equates propensity to own a car to the propensity to travel; thus multi-car households are just those which will spend more than a certain amount on travel. In practice there is a difference between propensity to own cars and propensity to travel (though these are correlated) and this can be expected to give rise to the effects found. There is thus an element of mis-specification in the model; however this has been done to avoid the technical difficulties and/or internal inconsistencies that would arise in trying to improve the specification. For example, if ownership were assumed to depend on a propensity to own, but expenditure were assumed to be proportional to income, it would be difficult to explain in terms of rational economic behaviour why some poor people purchased cars they could not afford and then did not use them much. This is a matter that needs further consideration.

Despite the exaggerated difference in kilometres per car between 1-car and 2-car households, for each kind of household the variation in kilometres per car with income is about right: a faster increase with income for 1-car than for 2-car households.

The mean occupancy of 1.52 adults is about right. No comparisons have been made with actual data broken down by income and number of cars but the differences shown look plausible.

4.5 Elasticities in the base year

The calibrated model just described has been applied to show the effect in 1978 of changing selected parameters. For alternative 1978 predictions, the quantities varied were mean income, three aspects of car cost and two of cost by other modes. Normally a 10 per cent change has been assumed in one variable, with the remaining variables unchanged. The results are given in Table 3 as elasticities, i.e., the percentage change in kilometres per car or in cars per adult for each one per cent change in the parameter concerned, averaged over the 10 per cent range considered.

Generally the elasticities in Table 3 are reasonably plausible, but some are perhaps somewhat larger than might have been anticipated. Looking first at the car ownership elasticities, that with respect to income is 0.8 compared with 0.6 for the middle LR 799[13] model at about the same ownership level. The purchase and running cost elasticities are both large: if both increase by the same percentage, the total elasticity is -1.3, compared with about -0.25 in LR 799. Car journey times have little effect, and so do public transport fares. Non-car journey times have a surprisingly and perhaps unreasonably large effect.

Elasticities are generally smaller for kilometres per car. With respect to income the elasticity is 0.18, which compares well with the estimate 0.20 in section 2. The purchase cost elasticity of 0.37 seems high; its effect on car traffic is counterbalanced by the ownership elasticity of -0.87. The elasticities with respect to petrol price and journey time are somewhat greater than the estimates in section 2 : -0.26 compared with -0.15 and -0.38 compared with -0.30.

Before reaching any conclusions on how it may be desirable to modify the model so as to bring the elasticities more in line with expectation, the ability of the model to reproduce past growth must be examined. What will be required is a set of modifications which will simultaneously adjust the elasticities and give a good fit to past trends.

Table 3
Sensitivity of causal model to parameter values, 1978 levels

	Elasticities	
	kilo-metres per car	cars per adult
Mean income increased by 10 per cent	+ 0.18	+ 0.78
Car purchase cost increased by 10 per cent	+ 0.37	- 0.87
Money cost of running car increased by 10 per cent	- 0.35	- 0.43
Non-fuel cost/km	- 0.10	- 0.12
Fuel price/gallon	- 0.26	- 0.31
Fuel consumption/km	- 0.26	- 0.31
Car journey times increased by 10 per cent	- 0.38	- 0.15
Non-car costs increased by 10 per cent	- 0.07	+ 0.14
Non-car journey times increased by 10 per cent	- 0.21	+ 0.42

4.6 Ability of model to reproduce past trends

The objective here is to see whether the model can satisfactorily reproduce trends over the period 1953-1978, and to the extent that it cannot, to derive methods of improving the model. Parameters of the model that have varied in a measurable way over the period are income, car purchase costs, running costs and speeds, and non-car costs and speeds. Initially the remaining parameters (σ_0, σ_2, a, b, ν, γ, β) will be assumed to have remained constant.

Fig. 3 shows how the values of car ownership and kilometres per car given by the model compare with the actual figures since 1953. The agreement is not good in either case. Growth of car ownership is under-estimated by the model; growth over the 25-year period was by a factor of 4.9, while the model accounts for only a factor of 3.1. Also the predicted fluctuations since 1970 did not actually happen. For kilometres per car the comparison is equally unsatisfactory: although the shapes of the curves since 1970 have similarities, the growth up to 1970 is not reproduced by the model. Also the model predicts a much more rapid rise since 1974-75 than has actually occurred.

In order to make the model fit the trends better, adjustments are required which will not disturb the performance of the model in the base year 1978. This is achieved by adjusting not the model itself but the data inputs to it. First in order to improve the trends over the period as a whole it has been assumed that the value of γ has not been constant but has decreased linearly from 0.000060 in 1953 to 0.000040 in 1978. This quantity is the value of car time as a proportion of the annual wage rate, and it is perhaps not unreasonable to suppose that such a change in attitudes to motoring has occurred.

Secondly, in order to reduce the possibly undue sensitivity of the model to short-term fluctuations in incomes, purchase costs and running costs, it has been assumed that the effects of these variables are lagged. Thus the model, which is based on cross-section data, is assumed to give the correct long-term response to these variables, but in the short term the response is less. The amount of lag assumed reflects what seems reasonable in relation to the decision processes involved, what is needed to make the model fit, and what is needed to bring the short term elasticities down to what seems reasonable, or in the case of petrol prices, to the observed short term elasticity of around -0.15.

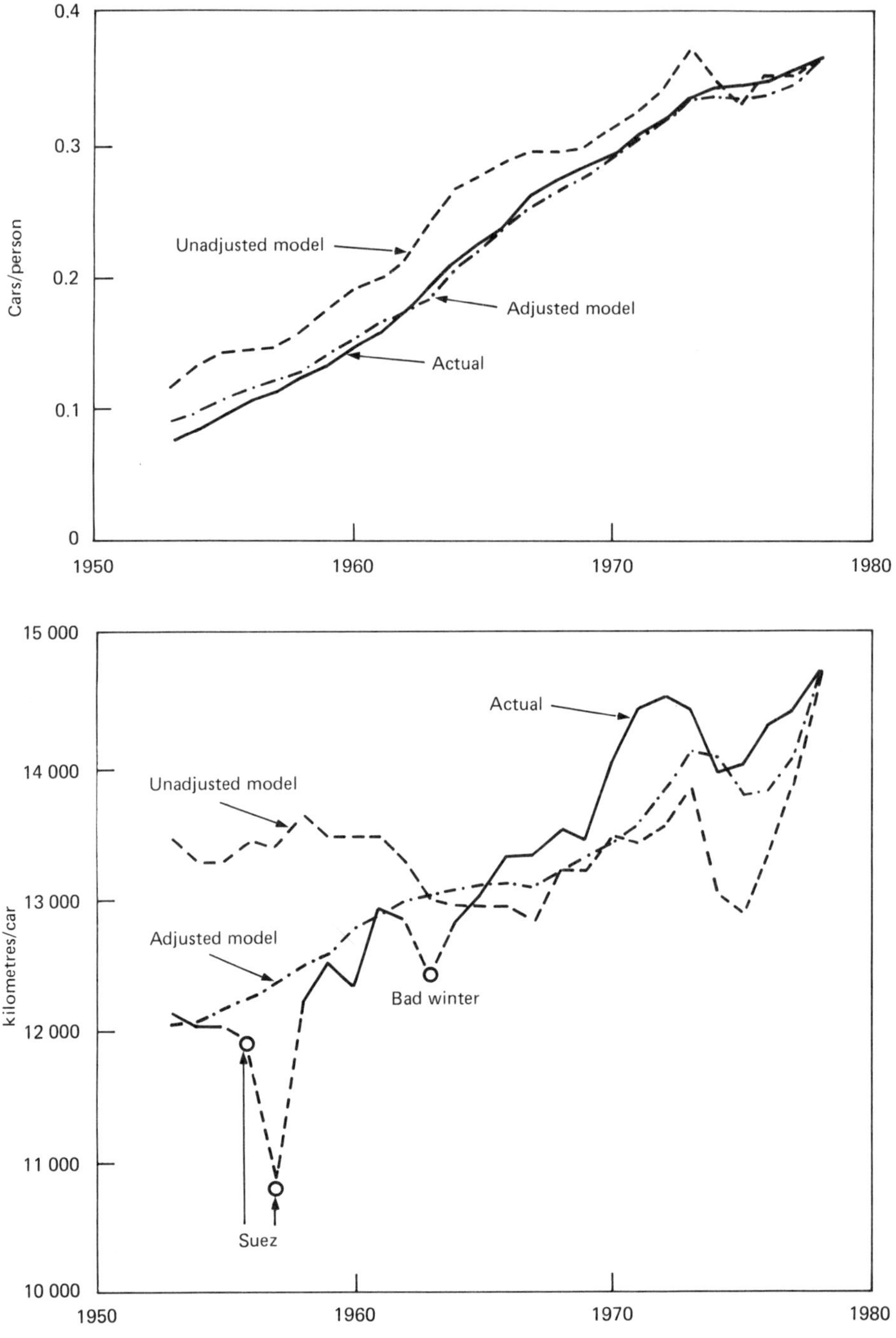

Fig. 3 COMPARISON OF CAUSAL MODEL WITH DATA, 1953 – 1978

For incomes, people are assumed to respond to a weighted average of (log) incomes in the current and two previous years, with weights 3,2,1; these weights are just guesses. For petrol prices and other operating costs, weights 4,3,2,1 are used for the current year and 3 previous years. This reduces the short-term response to about 4/10 of the long term effect,i.e.,-.10 for kilometres per car and -.12 for cars per adult according to the elasticities in Table 3. For car purchase costs, weights 10,9,8,7,6,5,4,3,2,1 are used. This heavy damping is based firstly on the counter-intuitive high elasticities in Table 3 in relation to short-term effects and on the view that income influences car ownership at least in part through such long term decisions as learning to drive and choices of location of homes and of jobs. After weighting these variables in this way, the time series were scaled so as to leave the 1978 values unchanged.

The effects of the weighting and of the trend in value of time are shown in the adjusted trends in Fig. 3. The general trends are much improved, although in detail the fit of the model to the data is not as good as might be wished. In particular, while the adjusted model fits better than the original model since 1970, it does not satisfactorily reproduce the actual fluctuations in kilometres per car.

5. USE OF MODELS FOR FORECASTING

To provide forecasts of future levels of kilometres per car is the main objective of the models, but not of this paper. Such forecasts have been calculated with both models, for a wide range of values of the independent variables, a wider range than might be considered reasonably possible. The resulting levels of kilometres per car vary widely : for the constant elasticity model they vary between 13 300 and 18 900 in the year 2 000, compared with the (revised basis) 1978 figure of 15 300. For the causal model they vary from 10 400 to 24 000. For middle assumptions, neither model predicts much change. Generally, as remarked earlier, the causal model is more sensitive to the independent variables than is the simple one.

The causal model produces forecasts of car ownership that are somewhat lower than those produced by other methods. This is due at least in part to the high elasticity with respect to purchase and running costs coupled with assumed departures from past trends in these costs. Little change is predicted in average car occupancy.

6. DISCUSSION

The analysis of data in Section 2 and the simple model of Section 3 can be regarded as a natural updating and extension of previous work. As such it raises few problems except of a detailed nature. The same cannot be said of the more complex model of Section 4. In some respects this model has what are usually regarded as desirable attributes: it is a disaggregate model, it is a causal or behavioural model in the sense that it models not just what people do but why they do it, it is policy-sensitive, and within limits it reproduces the current and past situations. However the overall outcome is not altogether satisfactory and it is perhaps worth enquiring why this should be so.

The symptoms indicating weaknesses in the model include the difficulty in reproducing the year-to-year pattern of variation observed, the high values implied for some of the elasticities and the arbitrary nature of a number of the assumptions made, especially in forcing the model to fit past trends. It is suggested that the causes of the problems are twofold. Firstly the model is built round a set of hypotheses about individuals' decision-making processes that are not directly verifiable, are inevitably over-simplifications of actual decision processes, and may in some respects not capture the essence of these processes. Secondly there are at present insufficient data about the real world to allow satisfactory estimation of the various parameters, even if the model were structurally satisfactory. Another way of expressing this is to say that the model extrapolates in too many dimensions from actual experience. All extrapolation is dangerous, and extrapolation using uncertain relationships is particularly dangerous.

What then could have been done, or could be done in future? Perhaps one should start by clarifying the questions which the model is designed to answer. One may well need different models to meet different requirements.

For example, if there were a policy issue concerned with the effect of a change in petrol tax, one would necessarily have to use a model that was sensitive to this factor but which might have poor predictive ability in a time series sense. On the other hand for highway investment appraisal, at its simplest, all that matters is what will happen in the future, not what might have happened under different assumptions. It could be argued that understanding of kilometres per car is not yet sufficiently advanced to allow a single model to be used both for time-series forecasting and for a wide range of policy testing.

Perhaps it may be possible to develop a model for kilometres per car that is a compromise between the two presented in this report. This would have to be more causal/disaggregate than the simple model, but not so much so that it could not be effectively calibrated. One move in this direction might be to alter the structure of the simpler model so that it combined petrol price, speed and other elements of car travel times and costs into a conventional generalised cost.

7. CONCLUSIONS

The report develops two models for forecasting kilometres per car. The broad conclusion reached is that the simpler one provides a better basis for the calculation of forecasts; the more complex causal approach requires more understanding of travel behaviour than is at present available. In more detail, it is concluded that:

(i) Kilometres per car have risen by about 20 per cent over the past 25 years. Recent fluctuations correspond to some extent with petrol price fluctuations.

(ii) The effect of income levels on kilometres per car is shown most clearly by household cross-section data, which give elasticities in the range 0.18 to 0.43.

(iii) Comparisons of richer and poorer countries give a higher elasticity, suggesting that the quality of the road system available has an effect.

(iv) A simple constant-elasticity forecasting model has been fitted to time-series data, using an imposed income elasticity of 0.20. This gives a petrol price elasticity of -0.12, a speed elasticity of 0.38 and a non-significant purchase cost elasticity.

(v) Applying this simple model to a wide range of alternative futures gives kilometres per car in 2000 varying between 13 300 and 18 900, compared with 15 300 in 1978.

(vi) A disaggregate causal model of car ownership and use is then developed. This is based on the idea that a household acquires a car or cars if by doing so its utility function would be increased. This utility function is specified so as to give rise to a travel expenditure budget which when expressed in terms of generalised cost is approximately proportional to a 'travel propensity' which is correlated with but not identical with income.

(vii) After calibration and adjustment, this model adequately reproduces the current situation and past trends, except that its sensitivity to certain parameters seems rather high. Lagged effects are introduced to reduce the impact of these.

(viii) Forecasts produced with this model cover a wider range than those from the simple model, between 10 400 and 24 000 kilometres per car in 2000. At the middle of the ranges of the independent variables little change from present levels is foreseen.

(ix) This model produces forecasts of car ownership that also cover a wide range, and tend to be somewhat lower than other recent car ownership forecasts, though this depends on the assumptions made. Little change in average car occupancy is forecast on any of the assumptions used.

(x) On the whole the disaggregate model in its present form is less satisfactory as a forecasting tool than the simple model of kilometres per car and the previous models of car ownership. However models of this kind provide useful insights and may have considerable scope for development.

ACKNOWLEDGEMENTS

The work described in this paper forms part of the programme of the Transport and Road Research Laboratory and the paper is published by permission of the Director.

REFERENCES

1. TANNER J C. Forecasts of vehicles and traffic in Great Britain: 1974 revision. Department of the Environment, TRRL Report LR650. Crowthorne 1975 (Transport and Road Research Laboratory).

2. DEPARTMENT OF THE ENVIRONMENT. "Standard forecasts of vehicles and traffic." DOE Directorate-General Highways Technical Memorandum H3/75. London 1975 (Department of the Environment).

3. DEPARTMENT OF TRANSPORT. "Interim Memorandum: National Traffic Forecasts." London 1978 (Department of Transport).

4. DEPARTMENT OF TRANSPORT. National road traffic forecasts. London 1980 (Department of Transport).

5. TANNER J C. Methods of forecasting kilometres per car. Department of the Environment Department of Transport TRRL Report LR968, Crowthorne 1981 (Transport and Road Research Laboratory).

6. BECKMANN M J, R L GUSTAFSON and T F GOLOB. "Locational factors in automobile ownership decisions." Annals of Regional Science, 1973, 7, 1-12.

7. BURNS L D, T F GOLOB and G C NICOLAIDIS. A theory of urban households automobile ownership decisions. Research Publication GMR-1785. Warren, Michigan 1975 (General Motors Corporation).

8. BURNS L D and T F GOLOB. "The role of accessibility in basic transportation choice behaviour." Transportation, 1976, 5, 175-198.

9. ZAHAVI Y. The UMOT project. US Department of Transportation Report DOT-RSPA-DPB-20-79-3. Washington 1979 (Department of Transportation).

10. TANNER J C. Expenditure of time and money on travel. Department of the Environment, Department of Transport TRRL Report SR466. Crowthorne 1979 (Transport and Road Research Laboratory).

11. DEPARTMENT OF EMPLOYMENT. Family Expenditure Survey 1978. London 1979 (Her Majesty's Stationery Office).

12. DEPARTMENT OF TRANSPORT. Transport statistics Great Britain 1968-1978. London 1979 (Her Majesty's Stationery Office).

13. TANNER J C. Car ownership trends and forecasts. Department of the Environment Department of Transport TRRL Report LR 799. Crowthorne 1977 (Transport and Road Research Laboratory).

Saturation Flow at Signalized Intersections
Through a Magnifying Glass

by

Stan Teply
University of Alberta
Edmonton, Alberta

The paper presents the results of an extensive study of saturation flows of signalized intersections in Alberta in the period 1978-1980. The objective was to determine the dependence of saturation flow rate on the following factors:

(a) Length of green
(b) Degree of oversaturation
(c) Weather conditions
(d) Community type
(e) Light conditions

An optimization method was applied to determine passenger car unit equivalents and the magnitudes of saturation flow for portions of green periods.

The results of the study significantly change the traditional concept of saturation flow at signalized intersections, expecially in respect to the duration of green and community type. The impact of weather conditions is defined.

The graphs presented in this paper have already been instrumental in the analysis and design of signal controls in Edmonton, especially in conditions close to saturation or in oversaturated situations.

The original interest in saturation flow was generated by practical problems associated with winter driving conditions in Alberta [1][2] and by the need to design signal controls for the Edmonton Light Rail Transit at-grade crossings [3][4]. It was found that the accuracy of computer programs, such as TRANSYT [5], developed by the British Transport and Road Research Laboratory or SINTRAL [6], developed by the University of Alberta in cooperation with the City of Edmonton, was very sensitive to saturation flow estimates.

The results of two independent surveys from 1975-1977 [2][7] were summarized in 1978 [8]. Strong indications that the rates of saturation flow depended not only on ambient conditions but on the duration of green and on the type of community (rather than community size) were revealed.

Subsequently, it was determined to study these issues in detail.

1. Saturation Flow as a Design Parameter

The analysis and design of signalized intersections concentrates on those approaches or lanes which are most important for a given signal group or a phase. This method has recently been termed "critical lane analysis" (9). The criticality of an approach or a lane is determined from its degree of saturation, expressed as

$$y = \frac{V}{S}$$

where

y = degree of saturation

V = approach or lane traffic volume (passenger car units/hour) (p.c.u./h)

S = saturation flow for that approach or lane (p.c.u./h green)

Traffic volume is determined by the sum of the products of individual vehicle categories and their passenger car unit equivalents, i.e,

$$V = \sum^{k} a_k X_k$$

where

V = traffic volume (p.c.u./h)

k = vehicle category

a_k = number of vehicles of category k

X_k = passenger car unit equivalent for vehicle category k

Saturation flow for an approach or a lane is the number of passenger car units which can be discharged at the stop-line of that approach or that lane during an hour of green. In the "Saturation Flow Manual" of the City of Edmonton and The University of Alberta [10] it is defined as the product of the basic saturation flow and an appropriate function of adjustment factors, i.e.

$$S = S_{BASIC} \{F_1, F_2 \ldots F_n\}$$

where

S = adjusted saturation flow (p.c.u./h green)

S_{BASIC} = The number of passenger car units which can discharge at a stop-line of an ideal intersection approach lane (width 3.0 m to 3.5 m) moving straight ahead (i.e. no turning movements are involved), with no additional traffic friction (e.g. no bus stops) under the given set of weather conditions during the given length of the green interval. In Alberta, this value has been termed "Basic Saturation Flow".

$F_1 \ldots F_n$ = adjustment factors, accounting for the effect of

A. Geometric conditions
   a. lane width
   b. turning radius
   c. grades
   d. queueing space
   e. discharge space

B. Traffic conditions
   f. traffic composition
   g. bus stops
   h. on-street parking
   i. left turns
   j. right turns
   k. length of green

$\{F_1 \ldots F_n\}$ = Composite effect of the combination of more than one adjustment condition. In some instances, a product is appropriate, such as a combination of lane width and grades. In other cases, such as a combination of lane width and left turns with opposing flow, a selection of the most critical factor yields a more sensible result.

It should be noted that expressing the values of saturation flow in passenger car units per hour of green represents a bias. It does not mean that there is a continuous hour of green but assumes a "stop-and-go" operation for the usual range of green periods.

In the determination of the degree of saturation, some researchers [9] concentrated on adjusting volume rather than the saturation flow. Although both approaches should yield identical results, adjustment of the basic saturation flow is preferred in Alberta, because it is possible to verify both parts of the fraction by direct surveys. This cannot be accomplished if an artificial volume and a generalized saturation flow are used.

The conventional basic saturation flow concept [11] is illustrated in Figure 1 by the solid line. It assumes that after the initial hesitation following the beginning of the green period, traffic discharges at a constant rate until shortly after the beginning of amber. After that point in time a sharp drop occurs.

The dashed line represents the same concept in a "Cumulative Average Rate of Flow" format. Since every green interval must include its beginning, points on the curve show the average value of saturation flow from the beginning of green interval to that point, as indicated by the dotted lines (e.g. 0 to 10, 0 to 20 seconds). The values of the saturation flow on the "cumulative average" curve are of direct importance to a designer, since they represent the capacity of any given green interval (provided that amber utilization is considered). The two conceptual portraits are easy to compare.

This traditional concept of saturation flow assumes that its rate for all portions of the green interval (except the beginning and the end) is constant, regardless of the duration of green. Consequently, the cumulative average curve would asymptotically climb to that rate, since for longer green intervals the effect of the initial loss is less pronounced. Alberta surveys, however, suggest that the

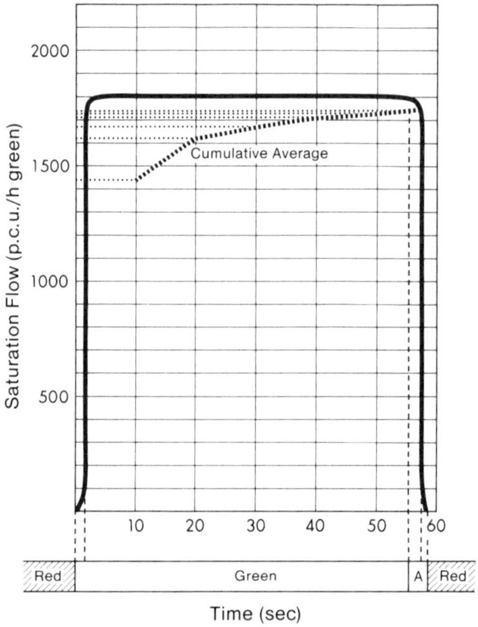

Figure 1: The conventional basic saturation flow concept(solid line) and its "Cumulative Average Rate of Flow" representation (dashed line). While the conventional illustration presents the saturation flow rate for individual portions of the green interval (e.g. for the period 20 to 30 sec), the cumulative curve shows the average rate of saturation flow from the beginning of the green interval to a given point (e.g. from 0 to 30 sec).

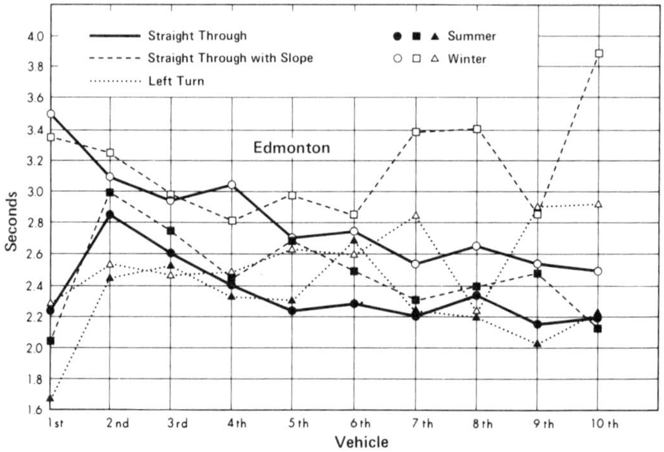

Figure 2: Average headways for passenger cars in relation to their position at the stopline as surveyed at five Edmonton locations during summer of 1976 and winter of 1977.

conventional concept is valid only as a rough approximation. The difference may account for as much as 25%.

2. Previous Surveys (1975-1977)

The objective of both the Edmonton and Calgary studies was to determine saturation flow magnitudes for summer and winter conditions. The definitions of typical "summer" and "winter" driving conditions represented a problem of their own.

"Summer" driving conditions were defined by:

- dry weather (i.e. no rain or snow)
- temperatures from about -10°C to well above 0°C
- dry pavement (no snow residues)
- good visibility

"Winter" driving conditions were characterized as:

- dry weather (no snowfall)
- clear or cloudy sky
- temperatures from about -10°C to -30°C
- dry pavement with some packed snow residues but well sanded
- no snowbanks at intersections
- good visibility

"Severe winter" conditions were not used during the 1975-1977 surveys. They applied to some situations during the 1978-1980 study. They were defined by:

- clear or cloudy sky
- very low temperatures (lower than -30°C)
- visibility obstructed by exhaust fumes

or by:

- heavy snowfall or shortly after, before snow has been cleared
- temperatures below 0°C.

As can be seen, the definitions do not provide for a clear seasonal distinction. "Summer" driving conditions may occur during the winter season while "winter" or "severe winter" conditions may be experienced in the spring or fall. Generally, however, the definitions characterized typical long lasting "summer" and "winter" conditions. Measurements of pavement friction indicated "winter" values lower than those for wet asphalt or concrete but well above those for icy roads or for packed snow polished by traffic. Freezing rain situations are rather rare in Alberta cities. Since they may cause infinite headways (in terms of one or several signal cycles), they were not considered in the studies.

About 97 to 98% of the traffic surveyed in 1975-1977 consisted of vehicles of the usual "Passenger Car" category, i.e. passenger cars ranging from "mini's" to large North American cars, vans (up to 9 passenger capacity or their commercial or recreational equivalents) and pick-up trucks (1/2 to 3/4 ton capacity). For that reason, no adjustments regarding passenger car equivalents were made. Headways for consecutive vehicles were computed for intervals which included "Passenger Cars" only.

Five locations (intersection approach lanes) in Edmonton and eight locations in Calgary were evaluated [8]. All locations were comparable regarding intersection type, type of signal control, locations within the urban area and lane width (3.35 m to 3.65 m or 11 to 12 feet). While the Calgary study [2] used the full length of saturated green periods per lane, the Edmonton study [7] employed saturated green intervals in increments of 10 seconds. Both surveys defined headways as measured at the point in time when the front bumper of a vehicle passes the stopline. Uphill approach lanes featured an average slope of about +4% to +5% within a 60 m distance from the stoplines. All left turn movements had a radius of more than 12 m.

More than 2,000 vehicles in Edmonton and over 14,000 vehicles in Calgary were observed during these surveys. Most individual surveys were secured at 95% or higher confidence levels.

Figure 2 illustrates the average headways for consecutive "passenger cars" crossing the stopline at Edmonton survey locations. The following observations were made:

(a) The shape of the curve for straight through

movement in "summer" follows closely those known from elsewhere.

(b) The shape of the curve for straight through movement in "winter" differs significantly from that for "summer' in the time lag required by the first vehicle at the stopline. As expected, the headways are generally higher than those for "summer" conditions. An asymptotic trend, however, is similar.

(c) The effect of uphill slopes is more pronounced for winter conditions, especially for those vehicles which have been waiting farther from the stopline.

(d) Left turns indicate a similar trend and magnitude for "summer" and "winter" conditions. Value fluctuations in "winter" are higher for vehicles which have been waiting farther from the stopline.

(e) The effect of long green intervals cannot be observed since the evaluation covered only about 22 to 30 seconds of green.

Figure 3 interprets the data for intersection approach lanes featuring saturation flow at four Edmonton and five Calgary locations. All locations are comparable. The points on the curves are average values from a number of surveys and follow the "Cumulative Average Rate of Saturation Flow" format.

The following observations can be made:

(a) There is a significant difference between "summer" and "winter" conditons. The average for all Edmonton locations (regardless of the length of green) was 1545 p.c.u. per hour green per lane in "summer", 1300 in "winter". For Calgary locations, the average values were 1780 in "summer" and 1350 in "winter". "Winter" saturation flows were, on the average, about 20% lower than those in "summer".

(b) Maximum saturation flow occurred with green intervals of 40 seconds for both "summer" and "winter" conditions. Uphill approaches in "winter" conditions featured the maximum saturation flow for

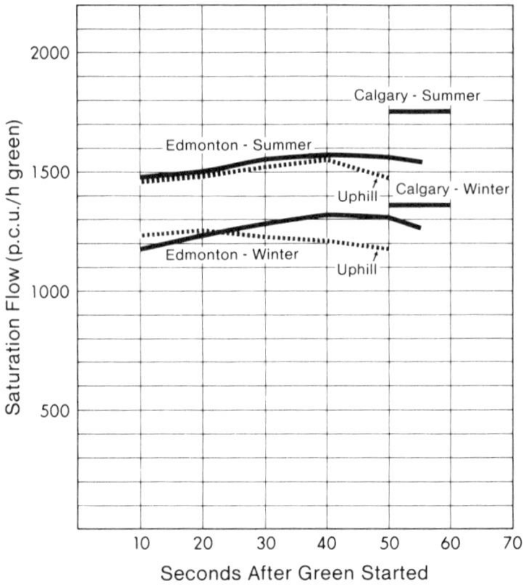

Figure 3: One-lane cumulative average rates of saturation flow for four Edmonton locations (two for level, two for uphill approach lanes) and for 50 to 70 seconds long green intervals at five Calgary locations during 1975-1978 surveys.

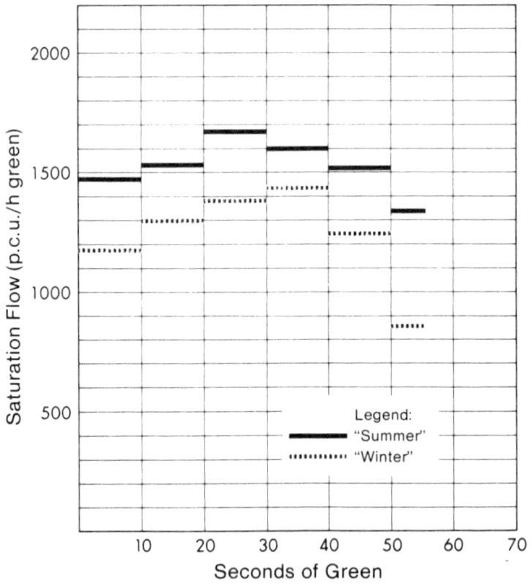

Figure 4: One lane 10-second average saturation flow rates for portions of the green interval, derived from Figure 3 for two Edmonton locations (no slopes, no turning movements) during 1975 surveys.

green intervals of about 20 seconds' duration. This observation has important implications regarding intersection capacity.

(c) Calgary locations indicated generally higher saturation flows than similar locations in Edmonton. These cities have about the same population size and are geographically separated by only about 300 km.

(d) After the conversion of data into headways, it was found that the average headway for all survey locations (i.e. including both Edmonton and Calgary) was 2.14 seconds for "summer" and 2.70 for "winter" conditions.

Figure 4 illustrates the same data as Figure 3 for Edmonton intersection approach lanes (i.e. no significant slopes) in a traditional interpretation format. In comparison to Figure 1, the survey results indicate a significantly different saturation flow behaviour in respect to time. The averages for 10 second long portions of the green interval are not constant but show a maximum region for both "summer" and "winter" conditions.

Although the results of the 1975-1977 surveys were successfully applied to a number of practical problems in Edmonton, it was determined that a closer look at the findings was appropriate. As a result, the 1978-1980 study attempted to validate the previous indications and secure them on a larger statistical basis.

3. 1978-1980 Study Method

The objectives of the methodology were specified as follows:

(a) simple data collection
(b) easy summarization of data
(c) computerized data manipulation
(d) accounting for the composite effects of traffic
(e) no "a-priori" determination of passenger car unit equivalents.
(f) understandable interpretation of the results.

The objectives related to computational procedures have, to a degree, determined the survey design and result presentation.

In the light of objectives (d) and (e), the usual method of determining passenger car equivalents from headways required by different vehicle categories was not acceptable. A uniform application of separately determined passenger car equivalents would neither account for the composite effects of traffic nor reflect the local and ambient conditions.

The optimization approach offered a suitable solution. Its application to a saturation flow problem was described by Fischer in 1958 [12]. A computer program titled PCU2 was written in Edmonton in 1978 and a number of associated data manipulation programs were developed.

3.1 The Principle

The program algorithm is based on the assumption that, under equal conditions and for uninterrupted homogeneous traffic flow, the number of passenger car units crossing the stopline of a single lane of an intersection approach must be constant during green periods of equal duration. This assumption can be more narrowly applied to portions of green periods which, commencing at the beginning of the green signal display, have equal duration. These portions of green have been termed "intervals".

If this assumption is valid, then the variations in the number of vehicles that cross a stopline under specified conditions can be attributed to the differences in behaviour of individual vehicle categories. Consequently, the problem can be defined as finding the values of passenger car unit equivalents which minimize the variations in the number of passenger car units for green intervals of equal duration.

Mathematically, the task represents an application of the least squares method.

Let vehicle categories be k.

For passenger cars $k = 0$

small trucks $k = 1$

medium trucks $k = 2$

etc. $\vdots$

The task is to determine equivalents X for each vehicle category, i.e. X_k.

During a saturated interval i, a given stopline is crossed by a_{ik} vehicles of category k. The flow during that interval can be called v_i. The arithmetic mean of saturated traffic flows during intervals $v_1, v_2, v_3 \ldots v_n$ is marked as $\bar{v}$.

Figure 5 illustrates that even during equally long and fully saturated intervals variations of v_i occur. The problem is defined as finding the equivalents for each vehicle category which minimize the sum of squared differences of individual flows v_i and the mean flow $\bar{v}$.

Since passenger cars are the basis for comparison,

$$X_o = 1.0 \; .$$

The number of passenger cars in the ith interval (a_{io}) will further be denoted as b_i.

The least squares condition is:

$$\text{Minimize} \quad F = \sum_{i=1}^{n} (v_i - \bar{v})^2$$

where

$$v_i = \sum_{k=1}^{k} a_{ik} X_k + b_i \qquad (i=1,2,3\ldots n)$$

and

$$\bar{v} = \frac{\sum_{j=1}^{n} v_j}{n} \qquad (j=1,2,3\ldots n) .$$

The condition of minimization is then

$$\frac{\partial F}{\partial X_\ell} = 0 \qquad (\ell=1,2\ldots k) .$$

The solution results in the following equation:

$$\frac{\partial F}{\partial X_\ell} = 2 \left(\sum_{k=1}^{k} \sum_{i=1}^{n} X_k \bar{a}_{ik} \bar{a}_{i\ell} + \sum_{i=1}^{n} \bar{b}_i a_{i\ell} \right) = 0$$

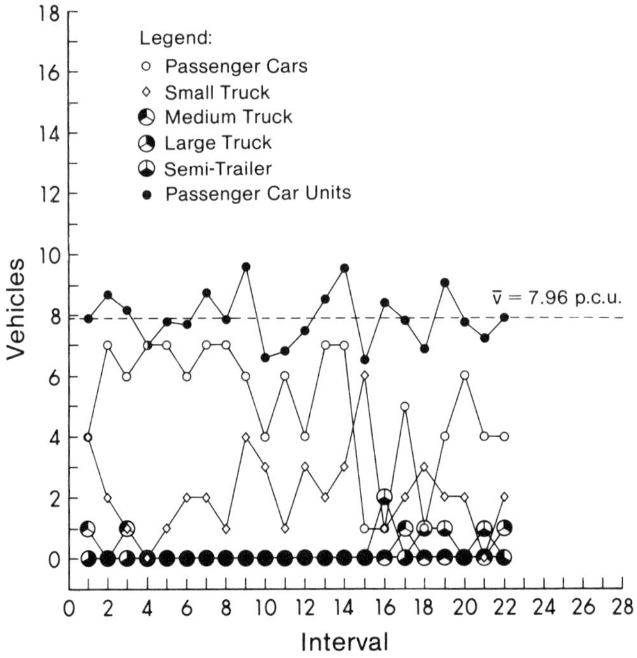

Figure 5: One-lane saturation flows during 20-second portions of the green period at 137th Avenue and 97th Street in Edmonton, straight through eastbound, centre lane, p.m. peak, "winter" conditions, January 23, 1980. Saturation flow ($\bar{v}$) was determined by the PCU2 computer program.

where, for reasons of simplicity

$$\bar{a}_{i\ell} = na_{i\ell} - \sum_{j=1}^{n} a_{j\ell} \quad \text{and}$$

$$\bar{b}_i = nb_i - \sum_{j=1}^{n} b_j .$$

Fisher (12) achieved further simplification by substituting

$$\sum_{j=1}^{n} \bar{a}_{ik} \, \bar{a}_{i\ell} = A_{k\ell} = A_{\ell k} \quad \text{and}$$

$$\sum_{k=1}^{n} \bar{b}_i \, a_{i\ell} = B_\ell$$

The resulting minimum condition is then

$$\sum_{k=1}^{k} X_k \, A_{k\ell} + B_\ell = 0$$

which yields a symmetrical system of k equations with k unknowns.

3.2 Surveys

The survey procedures consisted of two steps.

The first step included field data collection. In total, over 2000 green periods involving more than 25,000 vehicles at 18 locations (11 in Edmonton, 5 in Calgary and 2 in Red Deer) were included.

The surveyors used tape recorders and noted events in the observed lane as they happened. Simple mnemonic codes were used for the beginning of green, individual vehicles of designated categories, end of saturation and the end of green. Unusual events were described. The vehicles were recorded at the moment when their front bumpers passed over the stopline. A time check (usually a 15 second interval) was included at the beginning and the end of each survey.

The second step involved validation of the time check and transcription of the data. First, all codes were recorded on a time sheet with 10 second marks. Thus, written evidence for every green period was established. After that, the number of vehicles for all intervals (0 to 10 sec., 0 to 20, 0 to 30 sec. etc.) was transferred into data input sheets and entered into the computer data system.

Vehicle categories defined for the surveys were as follows:

(a)	Passenger Car	All types and sizes of passenger cars up to six persons' seating capacity, including taxicabs, limousines, station-wagons and ambulances (not in emergency mode).
(b)	Small Truck	Two axle pick-up trucks with four tires (1/2 to 3/4 ton capacity), vans (both for passenger or cargo applications), mini-buses (up to nine passengers).
(c)	Medium Truck	Two axle commercial vehicles with dual tires at the rear, for cargo transport. This category included larger motorhomes and minibuses (9 to 15 passengers).
(d)	Large Truck	Commercial cargo vehicles with three axles.
(e)	Semi-Trailer	Commercial vehicles with three or more axles, articulated.
(f)	Bus	Two or more axle vehicles for fifteen or more passengers.
(g)	Motorcycle	Two-wheeled vehicles for no more than two passengers.
(h)	Special Vehicle	Tractor-trailer combinations, special transports (construction machinery, car carriers, etc.).

The computation was applied not only to these categories but to their combinations as well. Since the differences in the results were small, all of the following graphs show the saturation flow functions for larger categories. These are:

(a) Passenger car and small truck
(b) Medium truck and bus
(c) Large truck and semi-trailer
(d) Special vehicles (if present)
(e) Motorcycles (if present)

It should be kept in mind that under Alberta conditions and at those locations selected as comparable and representative for the basic saturation flow, passenger cars and small trucks are the prevailing categories. Typically, the composition of traffic flow was as follows:

	Edmonton and Red Deer		Calgary	
Passenger cars	59.9% to	62.4%	70.5% to	85.2%
Small trucks	28.9%	30.6%	24.5%	10.1%
Medium trucks	4.5%	3.1%	2.8%	1.6%
Large trucks	2.4%	1.2%	1.8%	0.8%
Semi-trailers	2.8%	2.2%	0.4%	0.0%
Other	1.5%	0.5%	0.0%	2.3%

Calgary locations generally featured a higher percentage of passenger cars and a smaller percentage of small trucks than Edmonton or Red Deer. The spread of the percentage for individual categories was much higher for Calgary.

3.3 Evaluation of the Method

The surveys required some practice and a great deal of concentration, but no special problems were experienced.

Over 7,400 individual intervals in more than 800 PCU2 computer runs were evaluated.

Performance of the PCU2 computer program was excellent and the resulting saturation flow values were well within the range of those determined by previous surveys (which used a different method). Despite the fact that the program relies solely on a mathematical routine, few illogical, i.e. negative, very small or very large, passenger car equivalents were produced. Generally, this happened only if the samples were small (less than 15 to 20 intervals) and/or with insignificant representation of certain vehicle categories. A unique solution was obtained in all but a few cases.

Being a statistical procedure, the stability and, consequently, the reliability of the results depended on sample sizes. It has been found that the surveys with more than 30 intervals, and a reasonable representation of those vehicle categories included in the analysis, produced results well within the range of pragmatic values. The stability of saturation flow values was generally higher than that of passenger car unit equivalents.

The program was easy to use and, since it also produced basic parameters for significance testing (Student t-test was used), it assisted in the evaluation of different conditions.

An example of the program output format is shown in Figure 6. Most of the results subsequently were shown in graphs produced by a computerized plotter. The diagrams in this paper were redrafted in order to eliminate the distorted scale, designed to assist the evaluation.

4. Findings

In the course of the research, the original objectives have been expanded slightly. The findings will be discussed in the following sequence:

- The impact of the length of green period on saturation flow.
- Differences between saturation flows under saturated and oversaturated conditions (traffic "pressure" during peak and off-peak periods)
- Variations of saturation flow caused by weather conditions.
- The effect of community type
- The impact of light conditions (day versus night)
- Passenger car unit equivalents and their stability
- Amber utilization

Representative results are presented in graphical form.

ENTER TITLE:

112 AVE-97 ST,N BOUND-ST THRU,SEPT 30,3:50PM,PERIOD END=20SEC.

ENTER VEHICLE TYPES:

PASSENGER CAR+SMALL TRUCK,MEDIUM TRUCK,LARGE TRUCK,MOTORCYCLE,

ENTER INTERVAL SIZE, NO. OF INTERVALS AND NO. OF VEHICLE TYPES:

ENTER VEHICLE COUNTS PHASES 1 TO 26

112 AVE-97 ST,N BOUND-ST THRU,SEPT 30,3:50PM,PERIOD END=20SEC.

INPUT DATA					OUTPUT DATA		
1	4	1	1	0	7.450	I	*
2	7	0	0	0	7.000	I	*
3	8	0	0	0	8.000	I	*
4	7	0	0	0	7.000	I	*
5	6	1	0	0	7.450	I	*
6	8	0	0	0	8.000	I	*
7	7	0	0	0	7.000	I	*
8	7	0	0	0	7.000	I	*
9	7	0	0	1	7.783	I	*
10	8	0	0	1	8.783	I	*
11	5	0	0	1	5.783	I	*
12	7	0	0	0	7.000	I	*
13	8	0	0	0	8.000	I	*
14	7	0	0	0	7.000	I	*
15	9	0	0	0	9.000	I	*
16	7	0	0	0	7.000	I	*
17	7	0	0	0	7.000	I	*
18	6	1	0	0	7.450	I	*
19	8	0	0	0	8.000	I	*
20	7	0	0	0	7.000	I	*
21	7	0	0	0	7.000	I	*
22	7	0	0	0	7.000	I	*
23	8	0	0	0	8.000	I	*
24	7	0	0	0	7.000	I	*
25	8	0	0	0	8.000	I	*
26	8	0	0	0	8.000	I	*
%	96.4	1.6	0.5	1.6			

COEFFICIENTS

TYPE 1	TYPE 2	TYPE 3	TYPE 4
1.0	1.45	2.01	0.79

SATURATION FLOW= 7.45
1341.90 UNITS/HOUR

AVERAGE HEADWAY= 2.68 SEC/UNIT

SAMPLE VARIANCE= 0.46

T-VALUES:
T90= 1.708 T95= 2.060 T99= 2.787

Figure 6: An example of the PCU2 computer program output sheet.

4.1 Length of Green

The hypothesis that the rate of saturation flow depends also on the length of green period has been validated. The trends indicated in Figures 3 and 4 were obvious at a number of locations during the 1978-1980 surveys. A full confirmation was obtained at the intersection of 122nd Avenue and 97th Street in Edmonton, which had a long green period. The findings of the four surveys at that intersection are illustrated in Figures 7 and 8.

Figure 7, which features the "Cumulative Average" format, indicates that the maximum effectiveness can be achieved with green intervals of about 40 seconds. It shows also a relative stability of the saturation flow phenomenon since 3 of the four surveys are almost identical. The fourth survey, although carried out under identical "summer" conditions produced somewhat higher results (6% to 8%) but the same trend. The average of the four surveys is closely identical to a combined result of the surveys, i.e. when all the surveyed intervals were evaluated as one large sample.

Figure 8 illustrates the average saturation flows for individual portions of the green period for the average survey values. The maximum region is clearly pronounced and occurs in the 20 to 30 second interval.

4.2 Traffic "Pressure"

Figures 9 and 10 show the results of a comparison between peak and off-peak conditions in Edmonton, Calgary and Red Deer. Although some locations and some intervals indicate slightly lower values of saturation flow for off-peak conditions, this conclusion cannot be applied generally since the magnitudes lack the necessary stability. A similar lack of common magnitude was found when comparing work peak traffic with shopping peak traffic at one location. In Edmonton, morning peak traffic consistently featured slightly higher (but not significantly) saturation flows than the afternoon peak or off-peak traffic. The end of green paradox is discussed in Section 4.7.

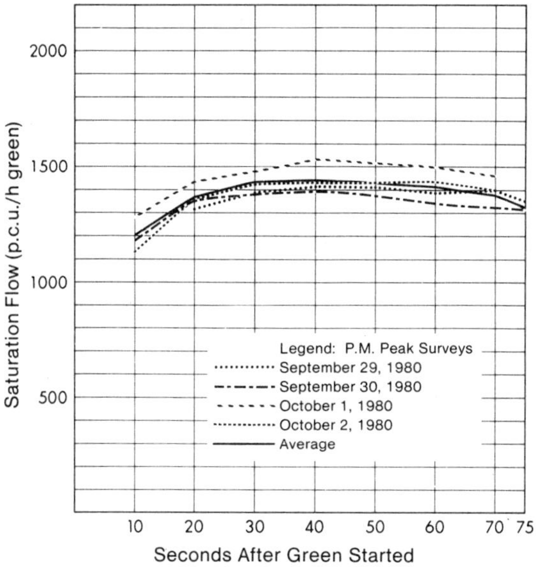

Figure 7: One-lane cumulative average rates of saturation flow for four surveys at 122nd Avenue and 97th Street in Edmonton, "summer" conditions.

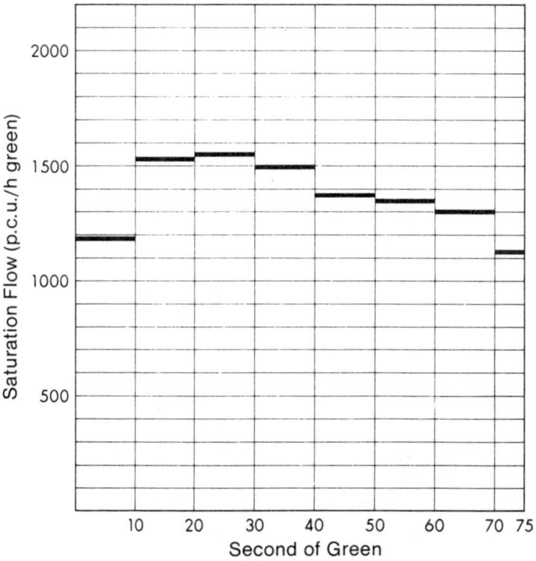

Figure 8: One-lane, 10-second average saturation flow rates for four surveys at 122nd Avenue and 97th Street in Edmonton, "summer" conditions.

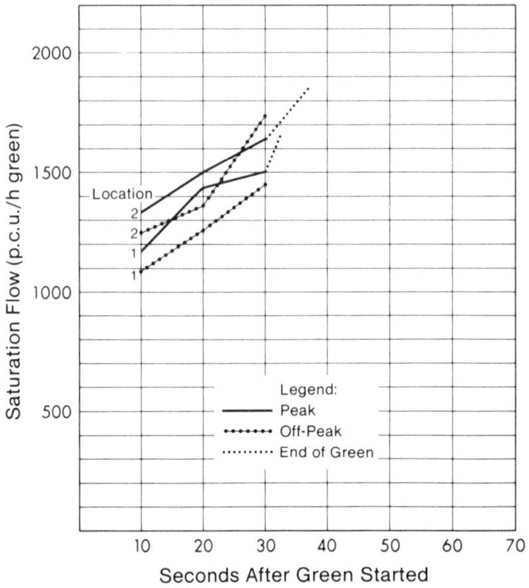

Figure 9: A comparison of one-lane cumulative average saturation flow rates for two typical Edmonton locations.

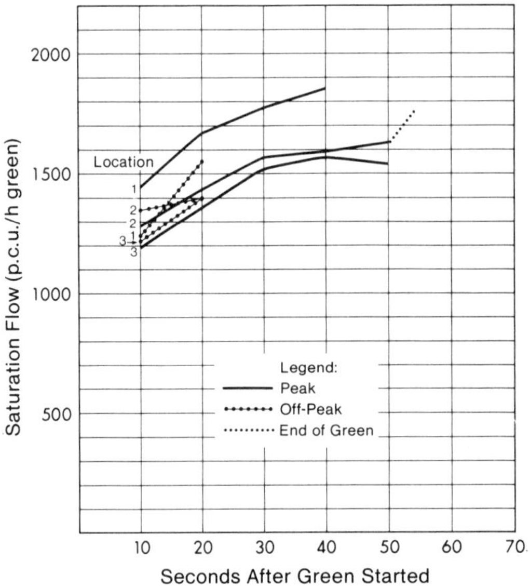

Figure 10: A comparison of one-lane cumulative average saturation flow rates for two typical Calgary [1,2] and one Red Deer [3] locations.

4.3 Weather

As indicated during the 1975-1977 surveys (see Figures 3 and 4) the differences between "summer" and "winter" conditions are significant. The current project confirmed this finding and added a set of conditions for "severe winter". Figures 11 and 12 illustrate the differences for several Edmonton locations. While Figure 11 shows the results of individual surveys during all three weather types at three typical locations, Figure 12 represents the blended values of saturation flow rates for 10 Edmonton locations. The reduction in the saturation flow rate cannot be observed here since most of the surveyed green periods were short. Nevertheless, in the average saturation flow rate interpretation for portions of the green period, a decrease after 20 seconds of green would be observed for "severe winter" conditions.

4.4 Community Type

The conclusion of the 1975-1977 studies that the difference between the basic saturation flow rates in Edmonton and Calgary was significant was confirmed during this project. In addition, attempts were made to include comparable locations from several smaller communities, such as Fort McMurray, Camrose and Red Deer. Only Red Deer locations, however, featured a statistically sufficient basis for evaluation. The other two communities did not provide enough saturated portions of green intervals.

Figures 13 and 14 illustrate some characteristic survey results from comparable locations in Edmonton, Red Deer and Calgary for "summer" and "winter" conditions. They illustrate the range of values which may be expected, even if the location characteristics and the conditions appear to be the same.

Figure 15 shows the average values of saturation flow for the three cities for "summer", and Figure 16 for "winter" conditions. While the rates for "summer" conditions seem to be different in each case, it must be kept in mind that the average values for Red Deer are based on

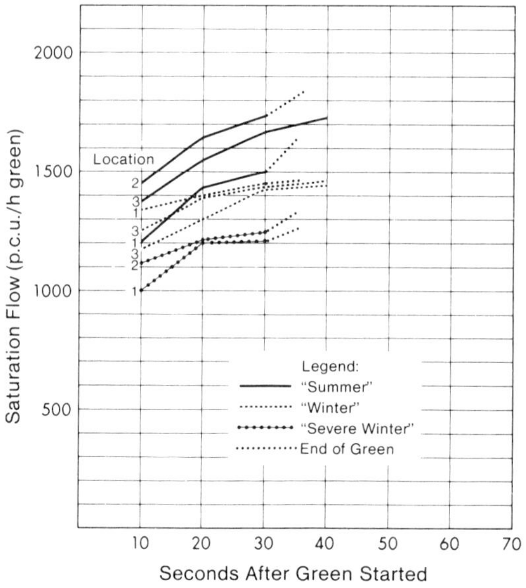

Figure 11: A comparison of one-lane cumulative average saturation flow rates for three typical weather conditions at three Edmonton survey locations.

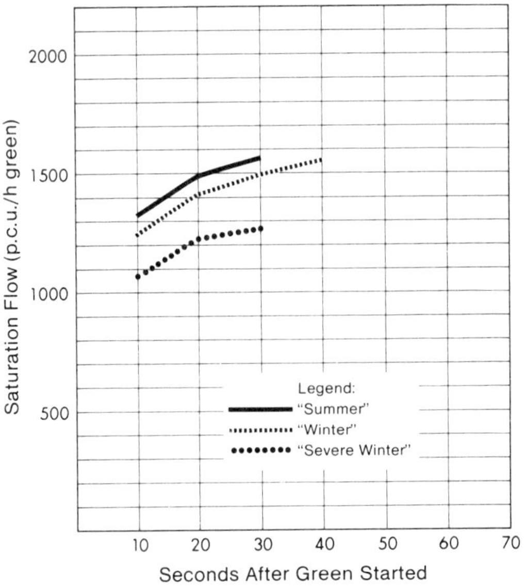

Figure 12: A comparison of one-lane cumulative average saturation flow rates for three typical weather conditions for combined survey data from 10 Edmonton locations.

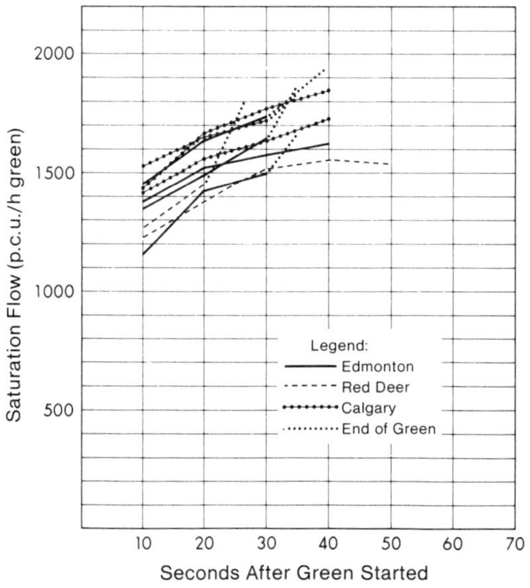

Figure 13: Characteristic one-lane cumulative average saturation flow rates for eight survey locations in Edmonton, Calgary and Red Deer during "summer" conditions.

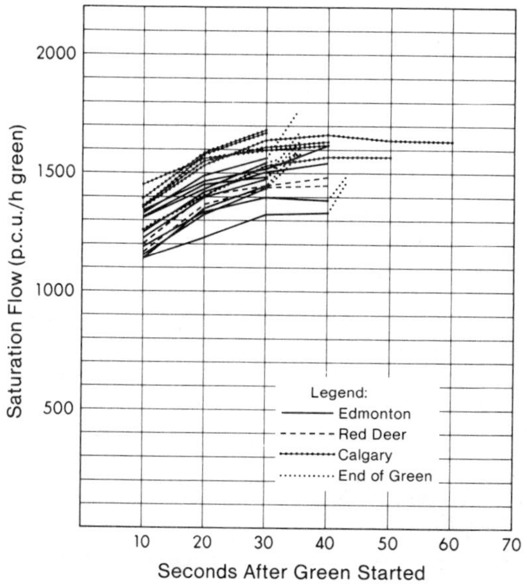

Figure 14: Characteristic one-lane cumulative average saturation flow rates for nine survey locations in Edmonton, Calgary and Red Deer during "winter" conditions.

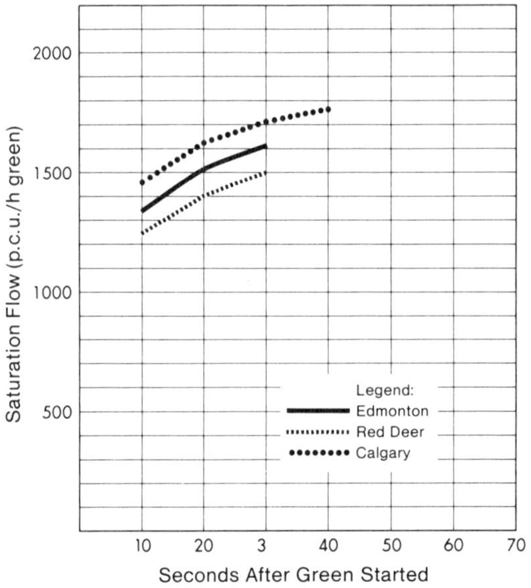

Figure 15: Average one-lane cumulative average saturation flow rates for Edmonton, Calgary and Red Deer during "summer" conditions.

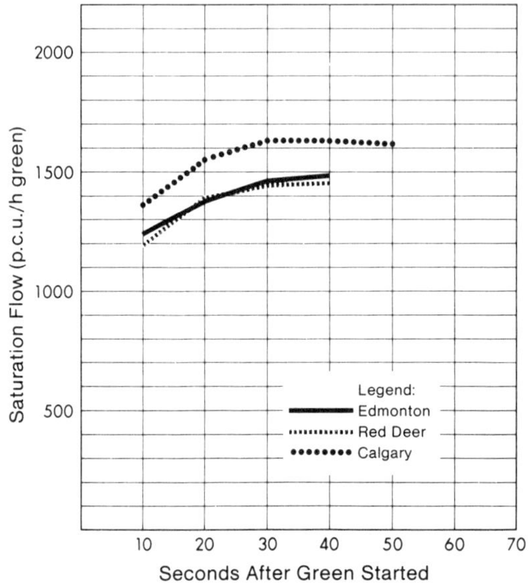

Figure 16: Average one-lane cumulative average saturation flow rates for Edmonton, Calgary and Red Deer during "winter" conditions.

smaller samples. Consequently, the difference between Edmonton and Red Deer is not statistically significant while the differences between Calgary and Red Deer, as well as those between Edmonton and Calgary, are significant. The same is also true (and evident from the graph) for "winter" conditions.

The trend to decreasing saturation flow rates for longer green intervals is especially visible under "winter" conditions for all three communities.

Figure 17 is a revised representation of design recommendations regarding saturation flows in Alberta from the "Saturation Flow Manual" of the City of Edmonton and the University of Alberta [10]. It combines the impact of weather conditions for three major cities in Alberta. The curves are based on typical magnitudes and trends. Although some overlaps occur, the ranges of saturation flow are distinctly different.

An attempt was made to correlate the differences in saturation flows (which reflect local traffic behaviour) with major socio-economic characteristics of the three communities. Nevertheless, examination of Table 1 does not reveal any significant differences. On the contrary, an amazing similarity can be found. The only relevant parameters appear to be the rate of population increase and the percentage of migrants from outside of Alberta and Canada. Other parameters, such as age distribution, retail trade, construction and other activities, also did not indicate significant differences. Some other behaviour and traffic flow related causal relationships are obviously at work here. In any case, the hypothesis that saturation flow depends only on the size of a community can be rejected.

4.5 Day and Night

Saturation flows during daylight and usual street illumination conditions were examined in detail at one Edmonton location (109th Street and 82nd Avenue) for both "summer" and "winter" conditions. The averaged results are

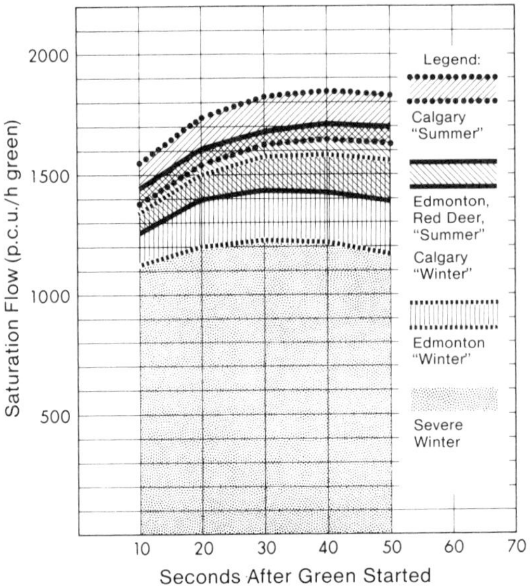

Figure 17: Typical ranges of one-lane cumulative average saturation flow rates for Edmonton, Calgary and Red Deer during "summer", "winter" and "severe winter" conditions.

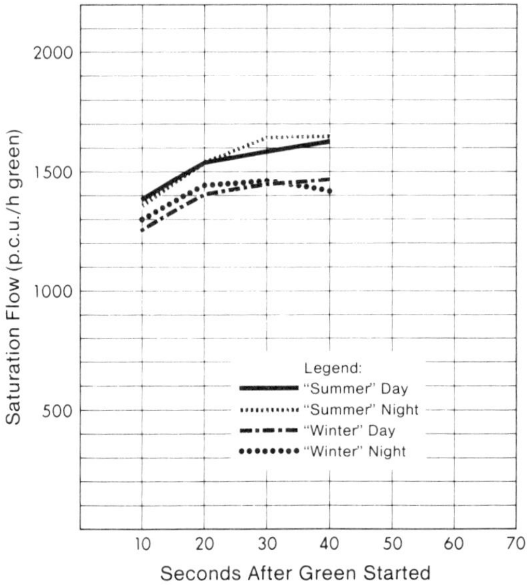

Figure 18: Average one-lane cumulative average saturation flow rates during daytime and street illumination conditions at the intersection of 109th street and 82nd Avenue in Edmonton in "summer" and "winter".

Table 1: Demographic characteristics of Edmonton, Calgary and Red Deer (Statistics Canada 1971 and 1976)

	Edmonton	Calgary	Red Deer
Population (1976)	461,361	469,917	32,184
Sex: Male	49.9%	49.8%	49.3%
Female	50.1%	50.2%	50.7%
Mean Age (years)	30.4	29.7	30.2
Median Age (years)	26.2	26.5	25.7
Marital status: single	47.3%	46.2%	49.8%
married	44.4%	45.9%	42.8%
Schooling: less than grade 9	15.6%	11.9%	15.9%
grade 9-13	45.0%	45.9%	49.8%
post-secondary	39.4%	42.2%	34.3%
Migrants over 5 years of age from other province or country	14.6%	19.3%	14.6%
Employment:			
Wage earners: male/female	94.9%/94.7%	94.7%/94.4%	not available
Self-employed: male/female	4.8%/1.5%	5.3%/1.9%	not available
Agriculture, Forestry, Mining, Oil	2.4%	6.9%	"
Manufacturing	11.2%	11.2%	"
Trade	17.8%	17.0%	"
Finance, Real Estate	4.5%	5.5%	"
Community, Business and Personal Services	28.3%	28.0%	"
Public Administration	10.4%	7.0%	"

presented in Figure 18. As can be seen, the trends and magnitudes do not indicate any differences. Some significant differences were produced by surveys carried out during "severe winter" conditions. A closer examination showed that these can be attributed to the changed ambient conditions (such as snow ploughing or sanding between the surveys) rather than light conditions.

4.6 Passenger Car Unit Equivalents

The PCU2 computer program employed passenger car unit equivalents as the tool for optimization. For that reason, their magnitudes differed somewhat from survey to survey and location to location. In order to produce more uniform results and to test the stability of their magnitudes, two "artificial" samples were analyzed for both the detailed and larger categories of vehicles (see Section 3.2). The results are graphically presented in Figures 19 to 22.

The first sample involved the data from all but one Edmonton location during both "summer" and "winter" conditions. Sample sizes (number of intervals) varied from about 450 to 150. The results were tested against a sample involving one location in Edmonton (122nd Avenue and 97th Street) under "summer" conditions (150 to 130 intervals).

Figures 19 and 20 show passenger car unit equivalents for traditional Alberta vehicle categories in samples covering green intervals of 0 to 10, 0 to 20, 0 to 30 seconds, etc. Expectedly, variations occur. The similarity of magnitudes in both figures indicate the significance of the results and the stability of passenger car unit equivalents. This stability is rather surprising considering the differences in the percentages of individual vehicle categories. Greater fluctuations in the magnitudes of passenger car unit equivalents for motorcycles and buses may be explained by small representation of these vehicle categories.

Similar conclusions may be reached for the passenger car unit equivalents for detailed vehicle categories as presented in Figures 21 and 22. The magnitudes of passenger car unit equivalents for the "small truck" category smaller than 1.0 are rather surprising although they were also consistently produced by individual surveys. This may reflect not only better performance of small trucks because of the driver's higher vantage point, but a deterioration of performance of passenger cars in a traffic stream with a large proportion of small trucks, i.e. the effect of the composition of traffic flow. In other words, taken absolutely, the value of the base unit varies.

The results of Calgary surveys indicated slightly higher magnitudes for both the "combined" and "detailed"

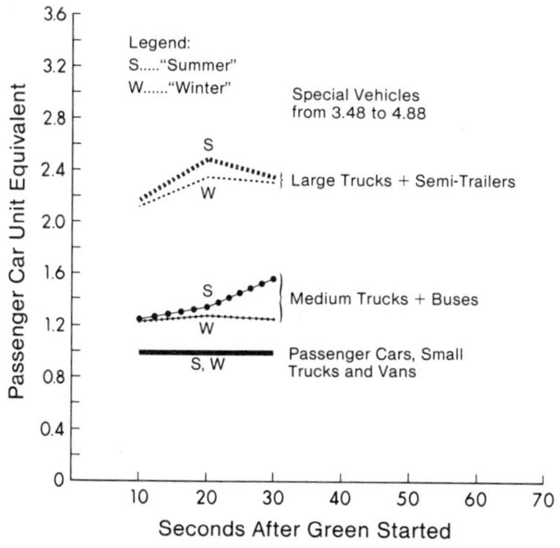

Figure 19: Passenger car unit equivalents for "passenger cars", "medium trucks and buses" and "large trucks and semi-trailers" in the cumulative average format for ten Edmonton locations under "summer" and "winter" conditions.

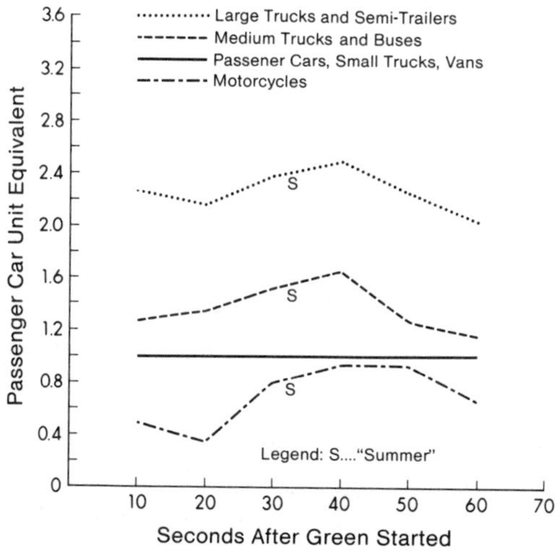

Figure 20: Passenger car unit equivalents for "passenger cars", "medium trucks and buses", "large trucks", "semi-trailers" and "motorcycles" in the cumulative average format for the intersection of 122nd Avenue and 97th Street in Edmonton for the combined surveys under "summer" conditions.

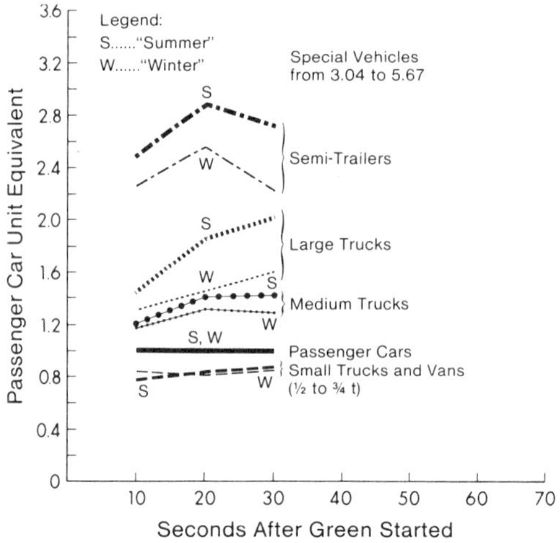

Figure 21: Passenger car unit equivalents for "passenger cars", "small trucks and vans", "medium trucks", "large trucks" and "semi-trailers" in the cumulative average format for ten Edmonton locations under "summer" and "winter" conditions.

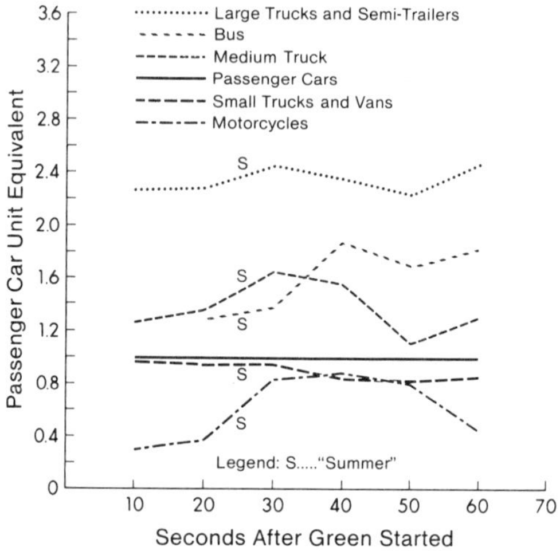

Figure 22: Passenger car unit equivalents for "passenger cars", "small trucks and vans", "medium trucks", "buses", "large trucks and semi-trailers" and "motorcycles" in the cumulative average format for the intersection of 122nd Avenue and 97th Street in Edmonton for the combined surveys under "summer" conditions.

vehicle categories. "Small trucks" also featured P.C.U. equivalents smaller than 1.00 but mostly for "winter" conditions. In "summer", their magnitudes fluctuated around about 1.10. Because of the smaller representation of the "bus", "medium truck", "large truck" and "semi-trailer" categories, the magnitudes of passenger car unit equivalents for Calgary fluctuate more than those for Edmonton. Consequently, their credibility is lower.

The values of passenger car unit equivalents recommended for Alberta as a result of this study are shown in Table 2. They coincide well with the findings of a recent study by the Transport and Road Research Laboratory [13].

4.7 Amber Utilization

Most surveys did not indicate a higher rate of saturation flow at the end of green and during the initial period of amber. Nevertheless, several locations in Edmonton and Calgary consistently featured much higher rates during this portion of the signal cycle. The increase was higher than the full utilization of amber for a corresponding rate of saturation flow (for that duration of the green period). This happened at locations with a high degree of oversaturation and with a higher percentage of large vehicles. Figures 9, 10, 11, 13 and 14 show this phenomenon. During the 1978-1980 study, it was not possible to explain the abnormality. So far, it has been speculated that it may be caused by frequent violations of all-red period (which have been experienced at these locations) or/and by the cases when a large vehicle with a high passenger car unit equivalent enters the intersection as the last vehicle during amber.

5. Conclusions

The surveys which supplied data for this study provided a satisfactory statistical basis. The method of data evaluation and interpretation has proved to be reliable and produced consistent results which do not contradict the logic of traffic engineering experience. The following

Table 2: Recommended magnitudes of passenger car unit equivalents for Alberta

Vehicle Category	P.C.U. Equivalents	
	Individual Values	Combined Values
Passenger Car	1.0	1.0
Small Truck, Van	0.9	
Medium Truck	1.3	1.4
Bus	1.4	
Large Truck	1.8	2.4
Semi-Trailer	2.6	
Motorcycle	0.6	0.6
Special Vehicle	3.5	3.5

conclusions may therefore be drawn:

(a) The traditional concept of saturation flow at signalized intersections which assumes a constant rate of vehicle discharge regardless of the duration of the green period must be modified.

(b) Saturation flow rate features a maximum region in the period between 20 to 40 seconds which results in most effective green periods of 30 to about 50 seconds' duration.

(c) The degree of oversaturation may influence the saturation flow at some locations.

(d) There are significant differences in saturation flow rates caused by weather fluctuations typical for long lasting seasonal conditions in a given geographic region.

(e) Saturation flows are dependent on the size of a community only if it reflects local traffic behaviour. Nevertheless, local traffic behaviour may be associated with other community characteristics (such as the community "dynamics") which are not necessarily reflected in the population size.

(f) The impact of illumination conditions (daylight and the usual street lighting) on Saturation Flows appears to be negligible in Alberta.

(g) The true passenger car unit equivalents for signalized intersections are generally lower than those assumed in traffic engineering practice.

(h) Amber utilization at some signalized intersections may be much higher than assumed in the traditional saturation flow concept (although this may be a reason for safety concerns).

The major lesson learned from this study confirms a major recommendation of the "Saturation Flow Manual" of The City of Edmonton and The University of Alberta [10]: for detailed analysis and design of critical intersections, local investigations of saturation flows are highly advisable.

Acknowledgements

The author wishes to gratefully thank the Natural Sciences and Engineering Research Council of Canada (NSERC) and The City of Edmonton for financial support of this research. The survey data and PCU2 computer programming were provided by TETRA Systems Ltd. of Edmonton in contract to The City of Edmonton. University of Alberta students Tim Choy, Allan Lo, Peter Barlow and George Teply assisted greatly in the surveys and data processing. Sandi Schnablegger analysed the socio-economic data and Doug Hunt wrote a series of data manipulation programs.

References

1. "A Transportation Management System for The City of Edmonton", The City of Edmonton, January 1977.

2. Bolger, D., "Planning for Intersection Capacity", M.Eng. Thesis, The University of Calgary, 1977.

3. Schnablegger, J., Teply, S., "Traffic Control Measures and Grade Railway Crossings with Provisions for Light Rail Transit", Transportation Research Record No. 628, Transportation Research Board, Washington, D.C., 1977, pp. 6-12.

4. Schnablegger, J., Teply, S., "System Design for Light Rail Transit Improves Intersection Performance", ITE Journal, Institute of Transportation Engineers, Washington, D.C., June 1978, pp. 36-39.

5. Robertson, D.I., TRANSYT, Transport and Road Research Laboratory Report No. LR253, Crowthorne, 1969.

6. Schnablegger, J., Stephenson, B., Teply, S., "Edmonton Version of Critical Lane Analysis: SINTRAL, An Interactive Computer Program System", accepted for publication in the ITE Journal, Spring 1981.

7. Teply, S., Veljkovic, D., "Ambient Effects on Intersection Capacity", The University of Alberta, Internal Research Report prepared for the City of Edmonton, 1977.

8. Teply, S., "A Canadian Problem--Winter Intersection Capacity", Proceedings of the Third Annual Conference of the Canadian District of the Institute of Transportation Engineers, Calgary, June 1978, pp. 39-50.

9. Interim Materials on Highway Capacity, Transportation Research Circular No.212, Transportation Research Board, Washington, D.C., January 1980.

10. "Saturation Flow Manual", The City of Edmonton and The University of Alberta, Edmonton, Alberta, April 1980.

11. Webster, F.V., Cobbe, B.M., Traffic Signals, Transport and Road Research Laboratory Report No. 56, Crowthorne, 1966.

12. Fischer, B., "Intersection Capacity in Mixed Traffic" (in German), Strasse und Autobahn, Cologne, No. 9, 1958.

13. "Passenger Car Equivalents of Goods Vehicles of Different Lengths at a Signal-Controlled Junction", Transport and Road Research Laboratory Leaflet No. LF686, Crowthorne, 1978.

COMPUTER SIMULATION AS AN AID TO THE DESIGN OF PRIORITY INTERSECTIONS

MARIAN TRACZ

CRACOW TECHNICAL UNIVERSITY

Cracow, Poland

This paper reports the author's recent simulation studies of capacity of priority intersections. In the first part the basic characteristics of traffic simulation models are described for three types of priority intersections, demonstrating new elements of these models. Attention is then focussed on the level of service concept for priority intersections to follow the US pattern for signalized intersections. Special attention has been given to the quantification of level of service and the service volumes determination measures. The basic assumptions of the capacity calculation method are briefly described.

Subsequently some results of several simulation runs are presented. The effect of turning movements and adjacent traffic signals on capacity of minor road approaches and other simulation results are discussed. The results presented have been used as a framework of the capacity estimation method for priority intersections.

1. INTRODUCTION

The major-minor priority intersection is by far the most common intersection layout in current use. There is a considerable variety of layouts including three-way "T" intersections and crossroads or staggered intersections. The joining roads can cover virtually all combinations of one- or two-way traffic, unaffected or affected by traffic signals, and the size of these intersections can range from unmarked minor intersections in housing estates to complex channelized intersections on high-speed dual-carriageway roads. Such intersections can also form an important part of some simple grade-separated intersections.

Consideration of methods of estimating capacity of priority intersections[1,2,3,4] shows that despite the considerable amount of research which has already been completed, the possibility of effectively designing such intersections is still limited. There are a number of factors affecting capacity and measures for increasing the capacity, which require further quantitative analysis, namely:

- turning flows on minor roads - particularly if separate left- and right-turning lanes are not available,
- platooning on the major road resulting from traffic signals upstream,
- storage space available in the central reserve of the dual-carriageway for crossing in two stages,
- other factors (composition and speed of traffic, signing, lighting, number of major road lanes, turning radius, gradients, visibility etc.).

Therefore it is not possible except in the very crudest manner to estimate - given the layout details and pattern of flows - the capacity of the minor road and delay to the minor road vehicles.

Simulation models and the computer program package developed at Cracow Technical University on the basis of extensive measurements of headways, lane choice, gap-acceptance, move-up times, delays, stops and queues, make it possible to calculate the effects of the most important geometric and traffic factors which affect capacity and cause delay to minor road traffic at the priority intersection. The results of simulation runs, examples of which are given in this paper, have been the basis for development of a capacity calculation method for such intersections[5].

2. SIMULATION MODELS

Computer simulation models have been widely used for studying traffic behaviour at intersections [1,5,6,7]. Although the development, calibration and validation of these models is complex, they do offer the possibility of convenient study of the effects of changes of traffic parameters and driver characteristics. The author decided to develop a computer simulation program package including interchangeable subprograms. Such an approach has made it possible to include the operation of the following priority type intersections, crossings and "T" intersections:

(a) intersection with undisturbed, random arrivals of vehicles (Fig.3 a),

(b) intersection located near or between adjacent traffic signals (Fig.3 b,c),

(c) intersection with storage space within the intersection in the central reserve (Fig.4 a).

The simulation model and computer program (a) are included as basic elements in the computer programs for intersections of types (b) and (c), which themselves overlap. In this model, both the event-by-event method and decision points are used. The program is written in FORTRAN IV and has been implemented on a CYBER-72 machine on which it takes 5 to 15 s to represent one hour of real time.

a. The isolated priority intersection

In the simulation model the generation process is assumed to be represented by the following distributions:

arrival headways Δt_i - shifted double exponential distribution or shifted exponential distribution, depending on traffic volume,

composition (vehicle type and manoeuvre) - discrete distribution.

The computer package provides the lane configurations for major road approaches shown in Figure 1; the first three can also be used for minor road approaches.

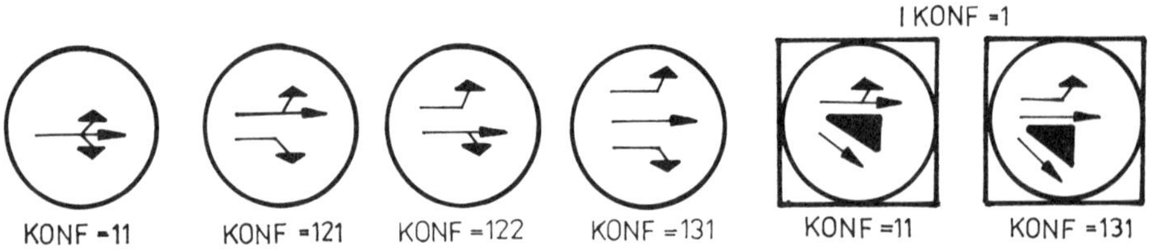

Figure 1. Configurations of lanes on major and minor (11,121,122) road approaches

The most important parameter in the departure process is the gap in major road traffic that the driver of a minor road vehicle will accept in order to perform his manoeuvre. It is assumed that for each driver there is a critical gap such that he will accept any gap larger than this and reject any gap smaller than it. In accordance with my own measurements and Ashworth's results[6], a normal distribution of critical gaps over the population of drivers has been assumed in the model rather than a step function. The determination of critical gaps t_{gr} depends primarily on vehicle manoeuvre, type of vehicle, approach speed and approach geometry. Validation based on minor road delay has indicated that these model assumptions lead to acceptable delay estimates. The model of gap acceptance by queues of vehicles assumes different move-up times for the first three minor road vehicles in each gap.

Although it is a fact that vehicles turning right off the major road which share the lane with straight moving vehicles (↱) do not conflict with vehicles from the adjacent minor road, they do have an inhibiting effect on the minor road approach capacity. Observations conducted on approaches having a small radius for right turn manoeuvres were the basis for introducing the model parameter Θ. This parameter lengthens the major road headway Δt, between a preceding straight-moving vehicle and a following right-turning major road vehicle - when analysing the possibility of start-up for a minor road vehicle (Fig. 2). For instance, for vehicle 4L turning left in Fig. 2:

$$\Delta t(4) = \Delta t_{23} + \Theta \qquad \text{if} \quad \Delta t_{12} > \Theta$$
$$\Delta t(4) = \Delta t_{23} + \Delta t_{12} \qquad \text{if} \quad \Delta t_{12} \leqslant \Theta$$

Minor road drivers do not decide to start until they are certain, Θ seconds before the end of the right turn

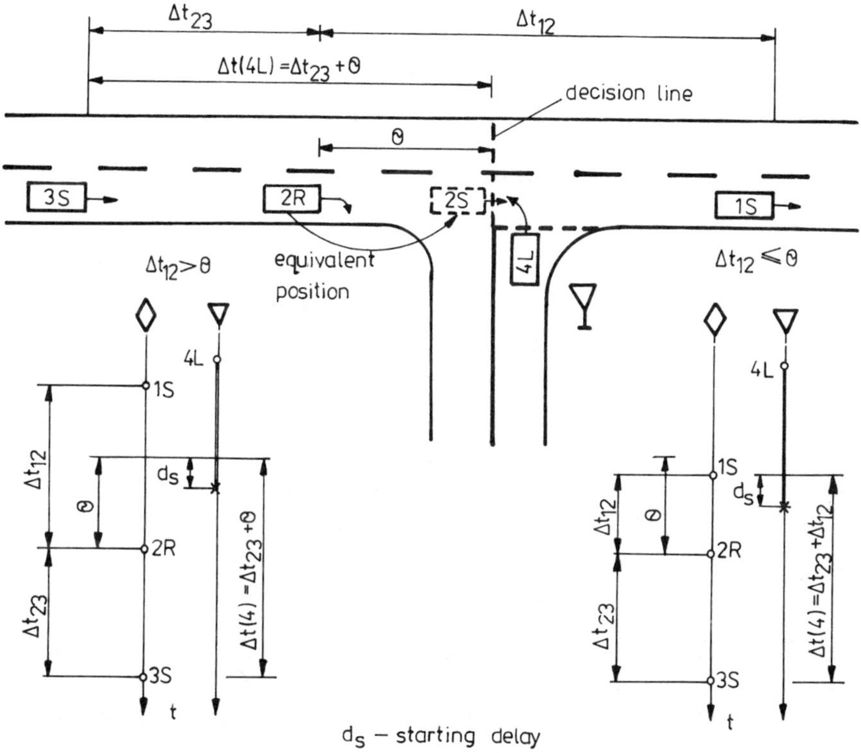

Figure 2. The inhibiting effect of right-turning vehicle (2R) on a minor road vehicle's (4L) gap acceptance.

manoeuvre, that the major road vehicle is turning right. The median value of the parameter θ derived from observations was 2.7 s.

The results of simulation runs based on the described models have confirmed Jessen's assumption[1,2] that the inhibiting effect of vehicles turning right off the major road (↑↗) may be represented by an equivalent volume of straight stream traffic equal to half of the right-turning stream.

b. Intersection located near or between traffic signals

Three types of inflows, as shown in Figure 3, have been considered for studying the effect of traffic signals on the capacity of an adjacent priority intersection. They include a priority intersection C with major road inflows which may be:

(a) undisturbed (random arrivals),

(b) disturbed from one direction (traffic signals at I),

(c) disturbed from both directions (traffic signals at I and II with possible offset $\Psi_{I,II}$ changes).

The major road flows N_A and N_B are a combination of the flows N_S and N_W - approaching the adjacent traffic signals. To examine the very important effect of crossroad inflows on capacity, two variants of traffic control on the approach S have been provided: allowing right turns off the approach S during the green period (Fig.3 d) or during the green and red periods (Fig.3 e). It has been assumed that there is no dispersion of platoons leaving signals I and II.

c. Priority intersection with storage space in the central reserve

At an intersection on a dual-carriageway the shortage space in the central reserve allows crossing and left-turning vehicles from minor road approaches C and D (Fig.4 a) to

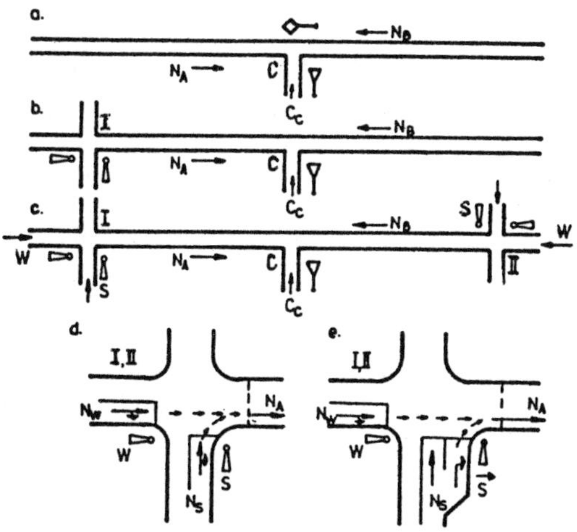

Figure 3. Schema of construction of major road flows N_A and N_B approaching the priority intersection C

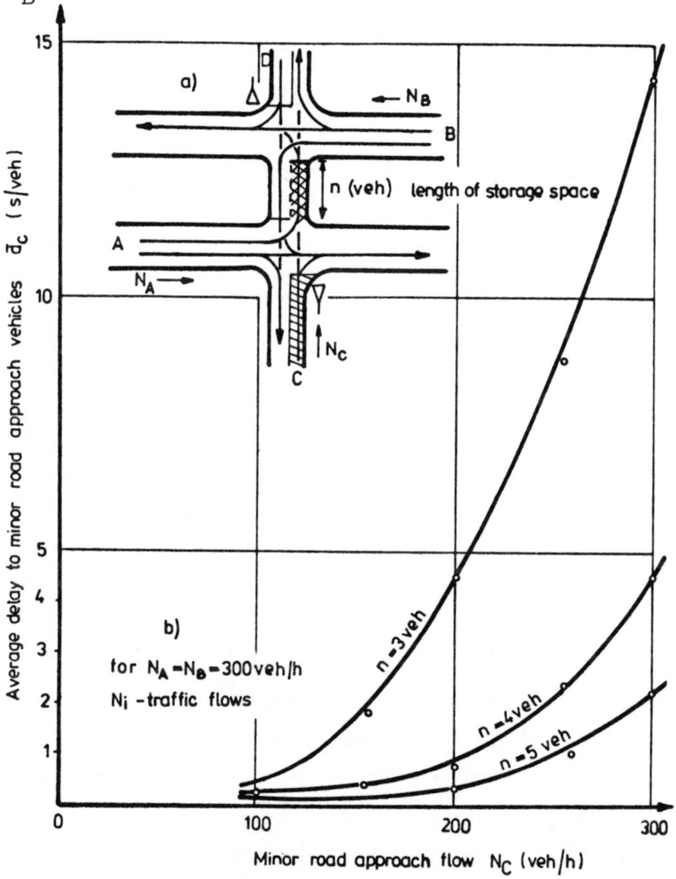

Figure 4. Relationship between the average delay d_C to minor road vehicles and the storage space n in the central reserve

be made in two stages; so the capacity of the minor road approach also depends on the length of storage space and degree of congestion on this "secondary" approach (Fig.4).

To determine the above mentioned relations, a simulation model has been developed whose most interesting element is perhaps the modelling of the interdependence between primary and secondary traffic flows. That is why, in the model, the effect of the opposite minor road approach D on operation of the "secondary" C approach should be taken into consideration when considering the capacity of minor road approach C. The following model solution has been chosen from a few under consideration: the model randomly determines whether a vehicle turning left off the "secondary" approach into a suitable gap in the major road flow N_B from the approach B will also be able to accept the available gap in the minor road flow N_D. The discrete distribution is assumed to be the best for this scanning with regard to the degree of congestion on approach D.

3. LEVEL OF SERVICE CONCEPT FOR PRIORITY INTERSECTION

The simulation program has made it possible - by conducting a specially programmed simulation run - to obtain graphs of basic relationships between the capacity of a minor road approach and major road flows for given critical gaps and also to obtain the effect of major road congestion, separate turning lanes and other factors on the minor road capacity.

The first step in the capacity calculation is the determination of the basic capacity of the minor road stream, which should be corrected according to real traffic conditions at a particular intersection. Calculation of the basic capacity is done by considering each minor stream individually in relation to the traffic flow that it has to cross and to the critical gap, which depends upon the various parameters of junction design.

Two sets of capacity charts for different values of the mean critical gap t_{gr} (corrected as described by Ashworth [6]) have been prepared on the basis of simulation runs. Figure 5 (left side) shows charts for one-way major road flow. Capacities are higher for two-way major roads owing to overlapping of major road vehicles from both directions. This effect is reduced, however, because the gap is larger when the main road is wider.

For design purposes, most European methods introduce a practical or a design capacity. Design capacity is defined as some fraction usually 70 to 90% of possible (absolute) capacity and this value is related to a specified delay (for instance 50 s/veh).

It can be seen from Table 1 and Figure 5 that such an approach does not provide a suitable capacity reserve ΔC, particularly at high major road volumes, so is not useful practically. For this or other reasons it seems advisable to extend the level of service concept of the <u>US Highway Capacity Manual</u> to priority intersections, although the number of levels is controversial [1,2].

Traffic conditions at an intersection may be determined by the average delay, percentage of stopped vehicles, flow to capacity ratio and reserve capacity. Taking into consideration our former design experience, four levels of service based on limiting values of average delay have been specified. These and others are summarized in Table 1.

Figure 5 (right side) shows the basic service volume N^i_{ko} and capacity C_o charts based on average delays to minor road vehicles $\bar{d}$ and percentage of stops $\bar{z}$. A good agreement

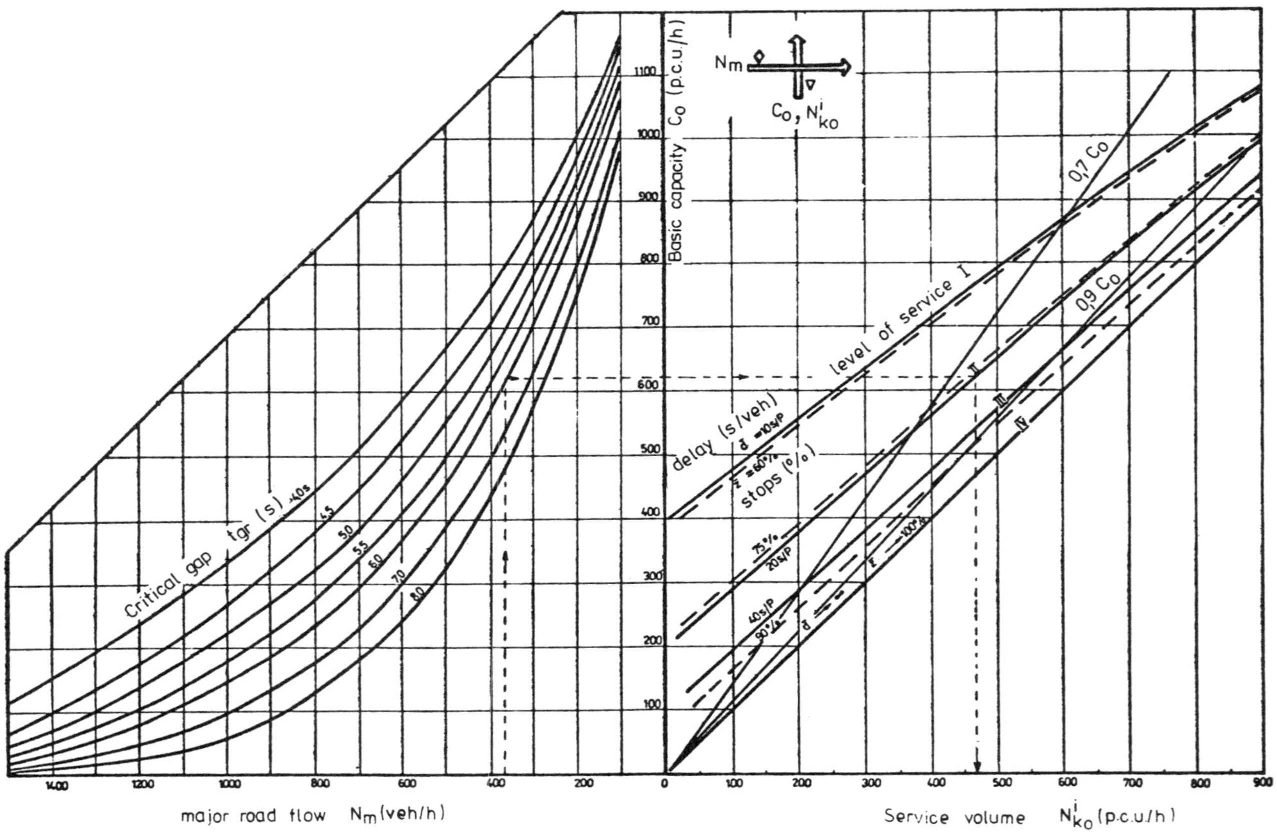

Figure 5. Basic capacity C_o and service volumes N_{ko}^{i} of minor road streams

Table 1
Definition of levels of service for priority intersections

Level of service	Average delay to minor road vehicles $\bar{d}$ (s/veh)	Percentage of stops of minor road vehicles $\bar{z}$ (%)	Reserve capacity C (veh/h) $\Delta C = C_o - N_{ko}^i$
I	< 10	< 60	180 - 400
II	10 - 20	60 - 75	95 - 200
III	20 - 40	75 - 87	40 - 100
IV	> 40	87 - 100	0 - 50

in the whole range of volumes can be seen. Reserve capacity $C = (C_o - N_{ko}^i)$ varies in a wide range for the particular level of service (I-IV) depending on major road volume N_m. Basic capacity C_o or service volume N_{ko}^i can be read from these graphs for a given major road flow N_m and a critical gap t_{gr}. These basic values of capacity have to be corrected for different geometric and traffic factors [5].

4. THE EFFECT OF TURNING MOVEMENTS ON THE CAPACITY OF MINOR ROADS

a. Left turning flows from the major road

Traffic streams turning left off the major road, having priority over minor road streams, may reduce the basic capacity of the minor road entry (as read from Figure 5) when major road traffic becomes congested. To enter the intersection, vehicles of the minor road stream have to wait till queued major road left-turning vehicles move into gaps in the opposite major road streams. The range of the basic capacity reduction depends on the ratio φ between the actual left turning flow on the major road N_B and the capacity of this stream C_B. On the basis of simulation runs, two graphs of the correction factor f_d have been determined (Fig. 6) for two cases: when separate left-turning lanes on the major road approach B are available or are not available. These and two other similar graphs have to be applied to capacity calculation for any minor road traffic stream likely to be impeded by such congestion.

b. The provision of special lanes for turning movements

Reconstruction of intersections for improvement of traffic safety and capacity usually consists of channelization of the intersection and construction of turning lanes, first for left-turning streams and secondly for right turning streams. To study the efficiency of turning lanes, a set of simulation runs has been conducted to determine the minor road approach capacity graphs.

An example of the simulation results is given in Figure 7. Curves of the capacity of minor road entries and separate minor road turning lanes were determined for three typical lane configurations; the layout without separate lanes is included for comparison. The results show that provision of a separate left-turning lane from major roads (curve b) gives a rather limited increase in the capacity of the minor road approach (average increase of about 60 veh/h). Thus the design in this case should also be based on safety benefits and progression of main road traffic movement. On the other hand, the greater the number of minor road entry lanes (curves c), the greater is the increase in entry capacity.

5. OTHER SIMULATION RESULTS

Separation of the two directions of traffic by a median storage space in the central reserve, allowing crossing in two stages, increases the capacity of a minor road. This

graphs of the correction factor f_d have been determined (Fig. 6) for two cases: when separate left-turning lanes on the major road approach B are available or are not available. These and two other similar graphs have to be applied to capacity calculation for any minor road traffic stream likely to be impeded by such congestion.

b. The provision of special lanes for turning movements

Reconstruction of intersections for improvement of traffic safety and capacity usually consists of channelization of the intersection and construction of turning lanes, first for left-turning streams and secondly for right turning streams. To study the efficiency of turning lanes, a set of simulation runs has been conducted to determine the minor road approach capacity graphs.

An example of the simulation results is given in Figure 7. Curves of the capacity of minor road entries and separate minor road turning lanes were determined for three typical lane configurations; the layout without separate lanes is included for comparison. The results show that provision of a separate left-turning lane from major roads (curve b) gives a rather limited increase in the capacity of the minor road approach (average increase of about 60 veh/h). Thus the design in this case should also be based on safety benefits and progression of main road traffic movement. On the other hand, the greater the number of minor road entry lanes (curves c), the greater is the increase in entry capacity.

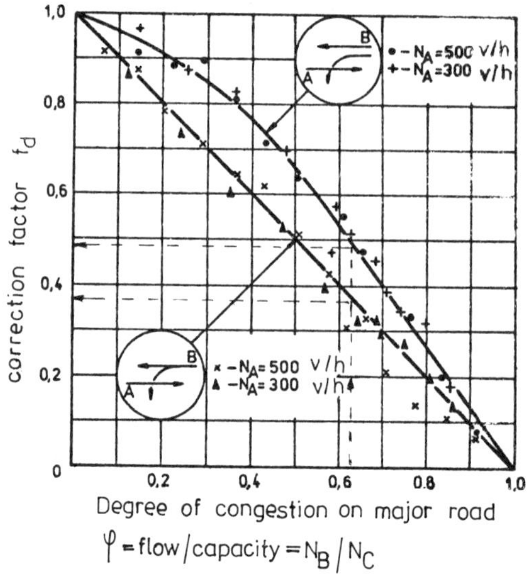

Figure 6. The effect of major road congestion on capacity of minor road

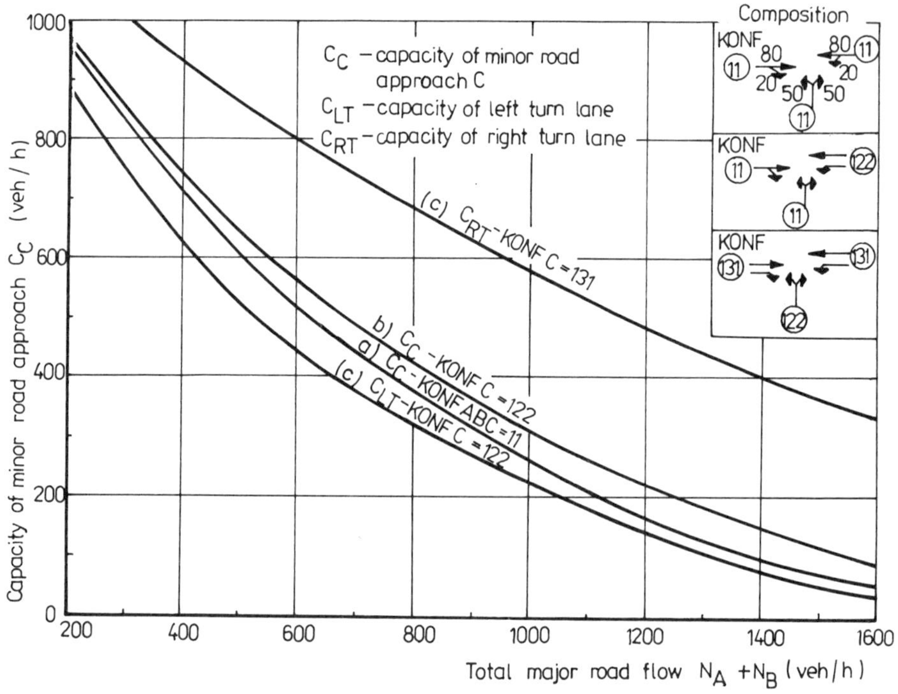

Figure 7. The effect of separate major and minor road turning lanes on minor road capacity C_c

increase of capacity and decrease of delay to minor road vehicles depends on the length of storage space. Simulation results (example in Figure 4) made it possible to include this type of intersection (using special corrections) in one method of capacity calculation for different priority intersections [5].

One example of the results derived from simulation studies of the effect of adjacent traffic signals on the capacity and delay to vehicles at priority intersections is shown in Figure 8. The effect of some traffic control factors has been included: offset Ψ, cycle time T, split, proportion of traffic inflows in major road flow. Charts of minor road capacity show the range of capacity increase of minor road entries for a few cases of major road flow constructions in comparison with similar, but isolated priority intersections. There is a substantial effect of offset on capacity.

The results of simulation studies show the necessity for the separate treatment of priority intersections located between traffic signals and indicate the need for detailed analysis of the benefits of installing traffic signals at such intersections[7]. Simulation is a powerful tool of research which can be effectively used to estimate the capacity of this type of intersection.

6. SUMMARY AND CONCLUSIONS

In the development of design procedures, special attention should be given to developing a more accurate capacity calculation method for use by practising traffic engineers. The level of service concept has been accepted as useful useful for the design of priority intersections. It should be emphasized that criteria based on delay, which have been accepted by the author in the calculation procedure,

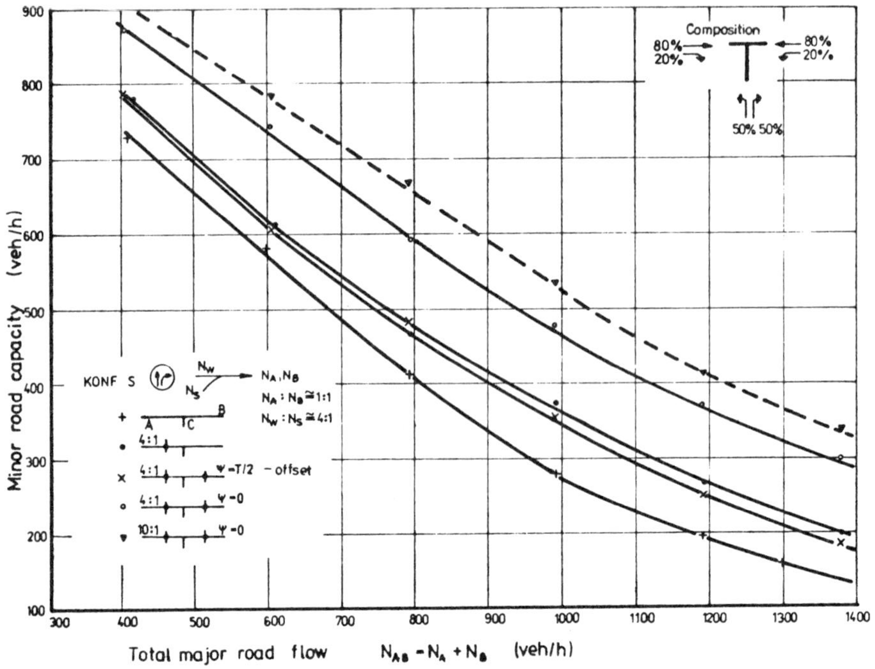

Figure 8. The effect of traffic signals on capacity of adjacent priority intersections.

conflict with criteria described by the load factor and by some fraction of possible capacity (design capacity).

This paper presents a complex simulation model approach to the problem of estimating capacity. As can be seen from the above description, use of the simulation package involves a considerable amount of field measurement, careful analysis and detailed preparation. It is reasonable to point out the most important problems relevant in such a simulation approach, namely:

- difficulties in generalization of the measurement results in the model calibration,
- satisfactory model validation,
- estimation of sample size or length of simulation run and the number of repetitions of the run.

Generally speaking, there is still a need for further studies and for the exchange of ideas and techniques in this field.

REFERENCES

1. O.E.C.D. Research Group, "Capacity of at-grade junctions". Paris, O.E.C.D. (1974).
2. Forschungsgesellschaft für das Strassenwesen, "Merkblatt für Lichtsignalanlagen an Landstrassen". Ausgabe 1972, Köln.
3. Hansson A., "Swedish capacity manual - 2; Unsignalized intersections." TRB Annual Meeting (1978).
4. Kimber R.M. and Coombe R.D., "The traffic capacity of major/minor priority junctions." TRRL Supplementary Rep. SR 582, Transport and Road Research Laboratory, England (1980).
5. Roads, Railways and Bridges Institute CTU, "Capacity of unsignalized intersections." Central Office of Public Roads, Warszawa (1979).
6. Ashworth R. and Bottom C., "Driver gap-acceptance behaviour at priority-type intersections." Res.Report R 63, Sheffield (1975).
7. Hauer E., "Location and capacity of unsignalized exits and crossings in between coordinated signalized intersections", in Transportation and Traffic Theory, Proceedings of the Sixth International Symposium on Transportation and Traffic Theory, D.J. Buckley (ed.). A.H. and A.W. Reed, Sydney (1974).

VALIDATION OF THE ME2 MODEL FOR ESTIMATING TRIP MATRICES FROM TRAFFIC COUNTS

D. VAN VLIET
Institute for Transport Studies, Leeds University

and

L.G. WILLUMSEN
Transport Studies Group, University College London

This paper reports on empirical investigations into the accuracy of the ME2 model for the estimation of origin-destination (O-D) matrices from traffic counts. The model, derived by entropy maximisation, produces the most likely trip matrix consistent with the information contained in the link flows and, optionally, a prior estimation of the trip table. The main properties of the model are discussed as well as the conditions required for its application. The validation of the model is based on a comprehensive dataset collected by the Transport and Road Research Laboratory in the town of Reading using license plate matching; this included O-D and route choice information as well as traffic counts. Considerable temporal variability was found in the sampled O-D matrices. However, the ME2 model was found to be reasonably robust (e.g. it was not very sensitive to the route choice model used) and to yield trip matrices of comparable accuracy to those observed.

1. INTRODUCTION

The ME2 model is a method of either estimating a 'best fit' trip matrix purely from the data contained in a series of traffic counts or, alternatively, updating a previous trip matrix using the same traffic counts. It uses the concept of entropy maximisation, already well established in traffic and transportation theory as one way of deriving constrained trip distribution models, to establish the 'most likely' trip matrix consistent with the observed traffic counts. The basic advantage of the technique is that traffic counts may be obtained accurately, easily and with minimal disruption, whereas conventional techniques tend to be expensive, disruptive and statistically inaccurate.

The basic underlying theory is contained in Section 2; more detailed derivations and examples of various applications have been given in a number of previous publications by Willumsen[1][2], Hall, Van Vliet and Willumsen[3] and Van Zuylen and Willumsen[4]. The purpose of this paper is to describe a series of validation exercises using a very detailed data base gathered by the Transport and Road Research Laboratory (TRRL) in the town of Reading, England in October 1976[5].

This data base (see Section 3) was obtained from a number plate matching survey in the centre of Reading and is particularly well suited to our purposes. Not only does it contain information on the number of trips between any two zones in the network, it also contains information on the precise routes used by these trips. In addition these data have been obtained for four different days of the same week. Thus in this paper we examine both the daily variation in the sampled trip matrices (Section 4) and how well the matrices estimated by ME2 compare with the 'true' matrices (Sections 5 and 6). We show that the daily variations are in fact considerable and that even making the very simple assumption of all-or-nothing route choice still allows ME2 to estimate a trip matrix whose 'error' is roughly comparable to the day-to-day variation. Thus in certain respects the estimated matrix is 'as good as' an observed matrix.

Finally in Section 7 we draw a number of preliminary conclusions from the assignment models which have been tested to date. Later publications will consider more refined assignment procedures.

2. THEORY: THE ME2 MODEL

Consider an area of interest represented by a network with Z centroids (zones), N nodes and L one-way links joining some of these nodes. One may consider the flow $V_{\ell m}$ on link ℓm to be the result of a series of trips T_{ij} between all the origin-destination (O-D) pairs ij which use the link. This can be expressed as

$$V_{\ell m} = \sum_{ij} T_{ij} p_{ij}^{\ell m} \qquad (1)$$

where $p_{ij}^{\ell m}$ represents the proportion of trips between zone i and zone j using (travelling on) link ℓm , $0 \le p_{ij}^{\ell m} \le 1$.

This proportion $p_{ij}^{\ell m}$ is obviously related to the choice of route between i and j and may be estimated for each link and O-D pair using an appropriate assignment model. The simplest case is provided by situations where route choice may be considered to be independent of the link volumes, thus allowing the set of $p_{ij}^{\ell m}$ to be calculated beforehand using a suitable proportional assignment model (see Robillard[6]) such as all-or-nothing or a stochastic assignment technique. We shall consider proportional assignment conditions first before relaxing (somewhat) this assumption in a later publication.

If the $p_{ij}^{\ell m}$ matrix can be estimated beforehand the Equations (1) will form a system of <u>linear equations</u>, one for each traffic count available. We shall assume here that these equations are consistent (i.e. the system of equations has a solution in non-negative T_{ij}) and that the counts are (provisionally) error free. In most cases, however, the number of independent Equations (1) will be much less than the number of unknowns T_{ij} even if all the links in a network have been counted. The problem can then be said to be underspecified in the sense that there will be more than one O-D matrix which will reproduce the observed flows $V_{\ell m}$. Different authors have suggested alternative ways of tackling this underspecification problem; for a review see Willumsen[1] and more recently[7].

The model developed by the authors[1][2][3][4] selects from these feasible matrices the one that maximises an entropy measure of the trip matrix. This results in the following mathematical programme:

$$\text{Maximise } S = - \sum_{ij} T_{ij}(\log T_{ij}/t_{ij} - 1) \qquad (2)$$

$$\text{subject to } \sum_{ij} T_{ij}p_{ij}^{\ell m} - V_{\ell m} = 0 \quad \text{for observed links } \ell m \qquad (1)$$

$$\text{and} \qquad T_{ij} \geq 0 \qquad (3)$$

where t_{ij} is a prior estimate of the number of trips between i and j.

The prior estimates t_{ij} may be obtained from an outdated trip matrix or a study of a larger area. If no prior information is available one may consider all O-D pairs to be equally likely by setting $t_{ij} = 1$ for all i,j.

The objective function (2) is obviously concave and provided there is a feasible solution in the constraints (1) and (3) the programme can be solved using Lagrangian methods resulting in

$$T_{ij} = t_{ij} \prod_{\ell m} X_{\ell m}^{p_{ij}^{\ell m}} \quad \text{for each } i,j \qquad (4)$$

where $X_{\ell m}$ is a variable associated with the Lagrange multiplier corresponding to the Equations (1) (for a more detailed description see [4]).

An alternative interpretation of this solution can be obtained by changing the objective (2) into the equivalent objective

$$\text{Minimise } R = \sum_{ij} T_{ij} \log T_{ij}/t_{ij} - T_{ij} + t_{ij} \qquad (5)$$

subject to the same constraints. As Murchland[8] has shown R may be considered as a measure of the difference between T_{ij} and t_{ij}, zero only if equal and otherwise positive. Minimising R subject to the constraints (1) and (3) results in the same formal solution (4) and may be interpreted as finding the matrix which reproduces the observed link flows and is as 'similar' as possible to the prior estimates $[t_{ij}]$.

Murchland[9] has also studied models of the form in (4) which he calls 'multi-proportional problems' and has put forward an algorithm for their solution. In our case this algorithm requires substituting Equation (4) into Equation (1) for each (counted) link in turn, and modifying the corresponding $X_{\ell m}$ factor so that modelled flows become equal to the observed ones. A single iteration processes the whole set of observed links; after a suitable number of iterations the difference between estimated and observed link volumes can be reduced to a reasonable level.

It can be seen that the method is only an extension of the Kruithof/Furness balancing factor algorithm for bi-proportional problems. Indeed, the $X_{\ell m}$ factors play a role analogous to the balancing factors in a doubly constrained gravity model and the exponents $p_{ij}^{\ell m}$ are simply weights attached to each link/O-D pair set. For all-or-nothing route choice proportions $p_{ij}^{\ell m}$ take only the values 0 or 1; an O-D pair is only modified by an $X_{\ell m}$ factor if it uses that particular link ℓm.

A series of computer programs have been written in FORTRAN IV to implement this approach and have been termed by the authors the ME2 (Maximum Entropy Matrix Estimation) model. These programs are based on the TRADVV suite of assignment models developed at Leeds but they can readily be adapted for use with other packages.

The conditions for the existence of a feasible solution for constraints (1) and (3) can be expressed in the following practical terms.

(a) The observed link volumes should satisfy, where relevant, normal flow continuity conditions:

$$\sum_{\ell} V_{\ell m} - \sum_{n} V_{mn} = 0, \quad \text{for all} \quad m. \qquad (6)$$

A computer program has been written to test this condition and to modify the observed link flows so that they do satisfy it using a maximum likelihood method[4].

(b) The prior trip matrix $[t_{ij}]$ should be consistent with the observed flows and route choice proportions. Note that, if $t_{ij} = 0$, so must T_{ij} and a very sparse prior matrix may make it impossible to generate a trip table which loads enough trips on certain links. Empty cells in a prior matrix may be seeded[3] to reduce this problem.

(c) The route choice model used to estimate the proportions $[p_{ij}^{\ell m}]$ should be consistent with the observed flows. This condition is violated, for example, when a poor route choice model assigns no trips onto a link where a positive flow has been observed. The best approach to this problem is to use the route choice model which is most appropriate to the study area conditions.

Given these conditions ME2 can estimate a trip matrix which when loaded onto the network reproduces the observed link flows to any reasonable degree of accuracy. While this is a desirable property for a model of this kind, it does not constitute a validation of the model. In order to approach this problem, it is necessary to have access to a data set which includes both traffic counts and an independently estimated trip matrix.

3. THE DATA BASE

The TRRL through its Urban Networks Division carried out a comprehensive license plate survey in the central area of Reading[5] and this has served as a basis for testing the ME2 model. The survey area covered approximately 4 km^2, see Figure 1, and 82 observation points were allocated to it so that a unique route for each recorded vehicle could be identified.

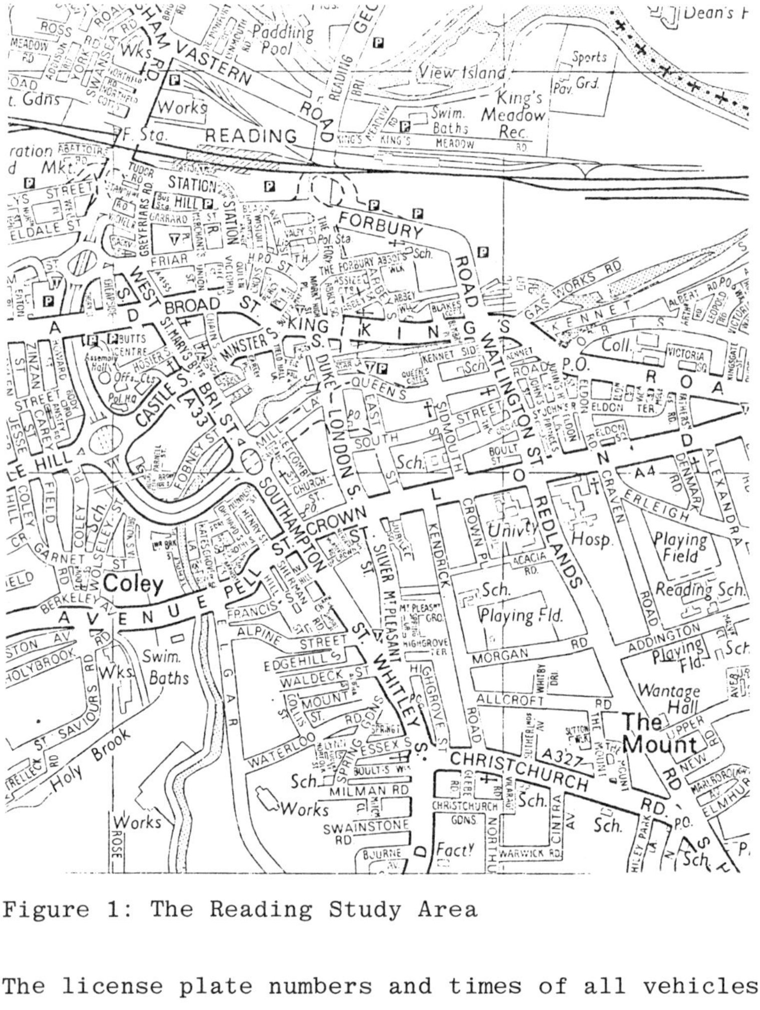

Figure 1: The Reading Study Area

The license plate numbers and times of all vehicles (buses and lorries were recorded but not identified separately) whose number ended in a particular digit were recorded at each observation point. The survey period covered four successive afternoon peak periods (16:10 - 18:10 hrs) starting on Monday 18 October 1976. Software written at TRRL was used to reconstruct the journey (route) of each observed vehicle through the survey area. This software also allowed minor interpolation and adjustment of records due to missing or misrecorded observations. The data thus validated has been used internally by TRRL as well as being made available to the authors to validate ME2.

The road system in the study area was coded into a network with 39 centroids, 80 nodes and 159 one-way links. Further validation runs were made to:

- remove journeys using bus only links (buses or cars on illegal routes)
- split into two or more trips those journeys which visited nodes more than once (looping journeys).

Around 5 per cent of the original journeys were modified in this way. The observations were then processed through the network to produce:

- an 'observed' or sampled trip matrix
- a set of observed route choice proportions
- set of 159 observed link volumes

for each of the 4 periods surveyed. These samples were not grossed up and all the tests refer to this unexpanded level which represented a 7 per cent sample of all the movements in the area.

4. DAILY VARIATIONS OF TRIP MATRICES

It is well known that link volumes are subject to variations over time, for example hourly, daily and seasonal variations, but no equivalent studies on the daily variations of trip matrices seem to have been carried out due to the high cost of obtaining trip matrices for different days. The Reading survey provides a reasonable and possibly unique data base to this end. Although only 7 per cent of all vehicles were observed the sample represents about 70 per cent of a particular sub-group of them, namely, those with a license plate number ending in the chosen digit. Checks on the reliability of the data were made by TRRL using observations with floating vehicles and automatic traffic counters.

The following indicators of the differences between matrices were found to be useful in this comparison.

(a) Mean Absolute Difference defined as

$$MAE = \sum_{ij} |T^a_{ij} - T^b_{ij}| / n \qquad (7)$$

where T^a_{ij} and T^b_{ij} are the two matrices and n is the number of cells.

(b) Relative MAE

$$\% \ MAE = \sum_{ij} |T^a_{ij} - T^b_{ij}| / \sum_{ij} T^a_{ij} \qquad (8)$$

(c) Coefficient of determination R^2

$$R^2 = 1 - \frac{\sum_{ij} (T^a_{ij} - T^b_{ij})^2}{\sum_{ij} (T^a_{ij} - \bar{T}^a)^2} \qquad (9)$$

when $\bar{T}^a = (\sum_{ij} T^a_{ij})/n$.

Note that when used in this context R^2 may take negative values for large differences.

(d) Root Mean Square Error

$$RMSE = \sqrt{\frac{\sum_{ij} (T^a_{ij} - T^b_{ij})^2}{n}} \qquad (10)$$

(e) Relative RMSE

$$\% \ RMSE = RMSE * 100 / \bar{T}^a \qquad (11)$$

Analogous indicators were computed for daily variations of the link flows by replacing T_{ij} by $V_{\ell m}$ above. A summary of the comparisons of Monday-Tuesday and Tuesday-Wednesday appear in Table 1. Comparisons of other days gave similar results.

Table 1: Comparison of trip matrices and sampled flows on different days

Indicator	Monday 18-Tuesday 19		Tuesday 19-Wednesday 20	
	Matrix	Flow	Matrix	Flow
MAE	0.9	10.7	0.8	12.0
% MAE	76	11	68	13
R^2	0.56	0.91	0.68	0.94
RMSE	1.9	19.9	1.7	19.1
% RMSE	160	21	135	20

It is not surprising to observe that the variations are much greater at the trip matrix level than at the flow level as some of the variations average out at the more aggregate level. However, the daily variations at the level of the sampled trip matrices were greater than the authors had expected. It can be seen that the % RMSE is around 20 per cent for the sampled flows but around 150 per cent for the sampled trip matrix.

In order to have a better feel for the variations of observed or sampled trip matrices artificial random errors of different magnitudes were introduced into one of them. It was found that the observed daily variations were approximately equivalent to the introduction of a normally distributed error with $\sigma = 1.5 + 0.3\ \bar{x}$ (the cell value). This is a very high figure which can only reinforce doubts about using samples from single days for the purpose of estimating trip matrices. If daily variations are so large as these samples suggest one must question the use of expensive techniques to estimate them.

5. TESTS WITH OBSERVED ROUTE CHOICE

During the data validation stage a record was kept of all the observed routes for each origin-destination pair, and therefore a set of observed $p_{ij}^{\ell m}$ values could be obtained for each of the four days surveyed. These values represent the exact proportions which result in the observed flows, that is perfect knowledge of the real traffic assignment model. The observed $p_{ij}^{\ell m}$ were then used as input to ME2 to estimate trip matrices from traffic counts. In all the tests all the cells in the prior trip matrix were made t_{ij} = 1.25 approximately the mean cell value.

Runs with two sets of observations are reported here; first tests using Tuesday 19 data and secondly tests using the aggregate data for the four surveyed days. This aggregate data can be deemed to represent some average of the first four days of the week. Two cases are tested for each set of data:

(a) only counts on the 159 real links are considered

(b) counts on the real links plus counts on the 39 centroid connectors are used.

This second case represents conditions where information on trip ends is also available. In all the tests iterations were stopped when all the modelled flows were within 5 per cent of the observed ones.

The results of these tests are summarised in Table 2, the inspection of which leads to the following comments.

(i) Trip end information appears to be valuable in both cases. The trip matrices estimated using both link counts and trip end counts are much closer to the observed trip matrices than those estimated with real link counts only.

(ii) The estimated trip matrices are not very close to the

sampled ones despite the fact that complete knowledge of route choice is assumed. It must be concluded that the actual destination choices are a result of a process whose complexity is not fully reflected in the resulting link flows.

(iii) However, the ME2 model performs better when the data for four days is used. This can be explained partially by the larger sample size and partially because some of the daily variations tend to cancel out so that the resulting average is better modelled by the entropy maximising framework.

(iv) Moreover, the trip matrices estimated without trip end information seem to be almost as close to the observed ones as the observed (sampled) trip matrix of one day is close to that of another day of the same week. When trip end information is taken into account the ME2 model seems to perform better than using the sampled O-D matrix of one day to model that of another day.

Table 2: Comparison of observed versus estimated trip matrices, known routes

Indicator	Tuesday 19 October		4 Days	
	No trip ends	With trip ends	No trip ends	With trip ends
MAE	1.2	0.8	3.5	3.0
% MAE	101	69	71	61
R^2	0.61	0.66	0.50	0.69
RMSE	1.8	1.7	7.2	5.6
% RMSE	149	134	144	112
Iterations	11	26	15	33
CPUTIME (sec)	2.18	4.91	5.97	10.39

6. TESTS WITH ALL-OR-NOTHING ASSIGNMENT

The second set of tests was performed with the simplest type of assignment, all-or-nothing. Minimum path routes were calculated with all links at free flow costs (travel time was the only cost element considered in all these tests). The level of congestion in Reading can be illustrated by the fact that during the survey period more than 1/3 of the links were operating at flow levels corresponding to more than half their nominal capacity. The same two data sets and two cases were again tested and the results are reported in Table 3.

Table 3: Comparison of observed versus estimated trip matrices using all-or-nothing assignment

Indicator	Tuesday 19 October		4 Days	
	No trip ends	With trip ends	No trip ends	With trip ends
MAE	1.1	1.1	4.0	3.7
% MAE	92	86	80	75
R^2	0.47	0.50	0.49	0.54
RMSE	2.1	2.1	7.5	7.3
% RMSE	172	169	150	146
Iterations	13	24	18	25
CPUTIME (sec)	1.82	3.56	2.27	3.53

(i) Again trip end information appears to be quite valuable and again the results from four days are closer than those from a single day.

(ii) Compared with the results in Table 2 (known route choice) the ME2 model performs surprisingly well. The rather crude assumptions behind a set of minimum cost trees (free flow conditions) do not reduce the performance of the model much. In other words, although a better route choice model is likely to lead to a better estimated matrix the use of a simpler one may be justified if only an approximation is sought.

7. CONCLUSIONS

The four major conclusions to come out of our validation tests on ME2 to date are the following.

1. The day-to-day variations in the samples of observed trip matrices obtained at a high level of detail are relatively high, implying that a 'true' trip matrix is extremely difficult to measure.

2. Trip matrices estimated by ME2 using observed routes (so that one important source of error is effectively eliminated) are not very close to the sampled matrices observed indicating that traffic counts alone are not sufficient to estimate matrices.

3. Nonetheless, the estimated matrices are roughly speaking within the day-to-day variations of the sampled trip tables indicating that the model does indeed give 'reasonable' answers.

4. Even with a very simple assignment model, all-or-nothing which we know to be fraught with errors, ME2 still yields matrices which are only marginally outside the above mentioned day-to-day variations.

Thus it appears that ME2 is a reasonably robust model, which , to a certain extent, can give as good an estimate of the trip matrix as a very expensive one-day survey of license plate numbers.

Improvements in the performance of the ME2 model may come from different sources.

(a) The use of better route choice models to estimate the proportions $[p_{ij}^{\ell m}]$.

(b) The use of a better prior estimate of the trip matrix $[t_{ij}]$ either from an outdated matrix, a larger study or even a travel demand model, perhaps of a simple gravity type.

(c) The inclusion of other information in the form of extra (linear) constraints, for example a trip length distribution.

Clearly a large number of important tests remain to be carried out on the data base. For example we wish to examine the potential improvement from more sophisticated assignment models, or the effect of using, say, Monday's observed trip matrix as an a priori guess for Tuesday's.

Indeed it may well emerge that the 'best' method of obtaining a trip matrix in practice would be to obtain a rough estimate of the matrix using some conventional technique, such as roadside interviews or number plate matching, and to update that matrix using traffic counts of a much higher statistical reliability obtained over a very much longer time period. The 'art' of the technique would then be to determine the most cost effective mix of the expensive conventional trip matrix observations and the much cheaper traffic counts.

ACKNOWLEDGEMENTS

We are pleased to acknowledge the generous help and co-operation of Dennis Robertson and his staff in the Urban Networks Division of Transport and Road Research Laboratory.

REFERENCES

1. L.G.WILLUMSEN, 'Estimation of an O-D matrix from traffic counts - a review.' Institute for Transport Studies Working Paper 99, University of Leeds, 1978.

2. L.G.WILLUMSEN, 'O-D matrices from network data; a comparison of alternative methods for their estimation.' Proceedings PTRC Summer Annual Meeting, Transport Models Stream. PTRC Education and Research Services Ltd., London, 1978.

3. M.D.HALL, D.VAN VLIET and L.G.WILLUMSEN, 'SATURN - a simulation-assignment model for the evaluation of traffic management schemes.' Traffic Engineering & Control, 21, 168-176, 1980.

4. H.J.VAN ZUYLEN and L.G.WILLUMSEN, 'The most likely trip matrix estimated from traffic counts.' Transportation Research, 14B, 281-293, 1980.

5. D.R.LEONARD and J.B.TOUGH, 'Validation work on CONTRAM - a model for use in the design of traffic management schemes.' Proceedings PTRC Summer Annual Meeting, Traffic and Environmental Management, pp135-153. PTRC Education and Research Services Ltd., London, 1979.

6. P.ROBILLARD, 'Estimating an O-D matrix from observed link volumes.' Transportation Research, 9, 123-128, 1975.

7. L.G.WILLUMSEN, 'Simplified models based on traffic counts.' Transport Studies Group, University College London, 1981 (to be published in Transportation).

8. J.D.MURCHLAND, 'What are the errors we should be seeking to minimise in the estimation process?' Institute for Transport Studies Working Paper 126, University of Leeds, (1980).

9. J.D.MURCHLAND, 'The multi-proportional problem.' Transport Studies Group, University College London JDM-263, 1977 (unpublished).

SOME IMPROVEMENTS IN THE ESTIMATION OF AN OD MATRIX FROM TRAFFIC COUNTS

Henk J. van Zuylen
Verkeersakademie, Tilburg, Netherland

An OD matrix may be estimated from traffic counts. The most likely OD matrix is calculated by a minimisation of the information contained in the matrix. The methods developed recently have some illogical features. Multiple counted trips are systematically overestimated in one of the methods, while the other method is sensitive to the inclusion of traffic counts that actually do not add new information.

These features are removed in a modified estimation method. Further it is indicated how additional information can be used in calculating the most likely OD matrix.

1. INTRODUCTION

Studies to obtain direct information on origin-destination (OD) matrices are mostly expensive and complicated. There is a need for simple methods to calculate OD matrices from traffic counts, and other data that are readily available.

Since the number of unknown quantities (elements of the OD matrix) for most networks is larger than the number of constraints (flows on the links), no unique OD matrix can be found from traffic counts. Willumsen (1978) describes several methods to find yet a single solution that satisfies the constraints and seems more or less likely. One class of methods reduces the number of parameters. It is assumed that the elements of the OD matrix can be written as a function of several parameters, such as socio-economic data and the distance between origin and destination. The various coefficients of this function are chosen in such a way that an OD matrix is calculated which produces flows that are as close as possible to the observed flows.

Van Zuylen and Willumsen [7] developed two new, rather similar models. These can be used to estimate the most likely OD matrix, using all available information. The basis of their model is the assumption that the most likely OD matrix contains a minimum of information. The information which is available in the traffic counts is not sufficient to determine a unique matrix, thus every OD matrix will contain more information than is actually known.

In the next section a review of the method is given. It is shown that the models have some illogical features. A revised estimation method results in a model which apparently does not have these illogical aspects. A further section shows how information next to traffic counts can be used in the estimation of the most likely OD matrix.

2. THE MOST LIKELY OD MATRIX

Volume counts impose a set of linear constraints on the OD matrix T_{ij}

$$\sum_{ij} T_{ij} \, p_{ij}^{a} = V_a \tag{1}$$

in which V_a is the volume on link a, T_{ij} is the number of trips from i to j and p_{ij}^{a} is the fraction of these trips that pass over link a. The information contained in the constraints (1) is not sufficient to find a unique matrix. Van Zuylen and Willumsen define the most likely OD matrix as the matrix that satisfies constraints (1) and contains minimum information. Depending on the measure for the information they find

$$T_{ij} = t_{ij} \prod_{a} X_a^{p_{ij}^{a}} \tag{2}$$

or $$T_{ij} = t_{ij} \prod_{a} X_a^{p_{ij}^{a}/g_{ij}} \qquad (g_{ij} = \sum_{a} p_{ij}^{a}), \tag{3}$$

where t_{ij} is an a priori guess for the OD matrix and X_a are factors that have to be calculated by the substitution of T_{ij} (2 or 3) in eq. (1). Formula (2) is found when the negative entropy is used as a measure for the information and the total number of trips is assumed to be known. Formula (3) is found when the information measure of Brillouin (1956) is used for each link separately and the information for all links is added up.

Both models have some illogical features which can be demonstrated in the following simple examples.

Example 1.

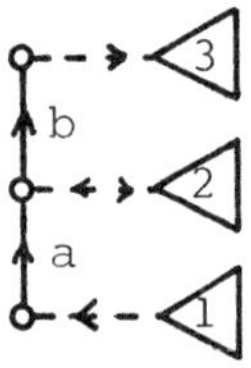

a priori OD matrix	volumes
$t_{12} = 3$ $t_{13} = 5$ $t_{23} = 4$	$V_a = 16$ $V_b = 18$

fig. 1

The actual OD matrix is twice the a priori matrix. Formula (3) gives the correct answer. Formula (2) however gives

$$T_{12} = 4.4,\ T_{13} = 11.6 \text{ and } T_{23} = 6.4.$$

The trips that are counted only once are underestimated. This is obvious from formula (2). When the factors X_a are all larger than 1, the effect on trips that are counted twice is larger than on trips that are counted only once. This gives an overestimation of long distance trips. Due to g_{ij} in the exponent of eq. (3) this effect does not occur for this model.

This effect has also been found in an OD study for the following simplified network of the village Twello (Smit 1980). When for t_{ij} the unit matrix is assumed, the following deviations are found between estimated and observed trips (fig.3). The overestimation of the twice counted trips disappears when the initial OD matrix is

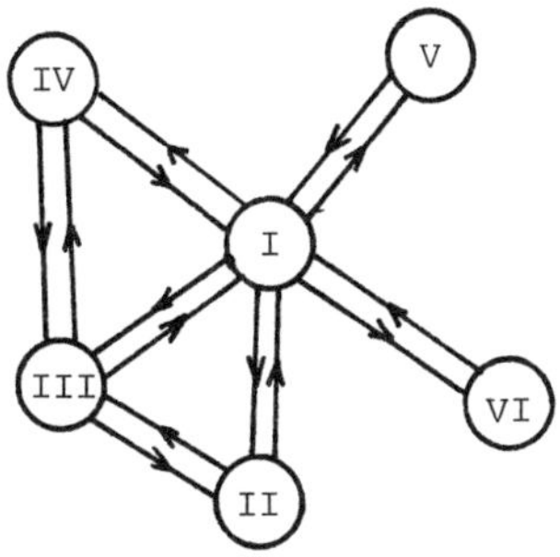

fig. 2
Simplified road network of the village of Twello (Smit)

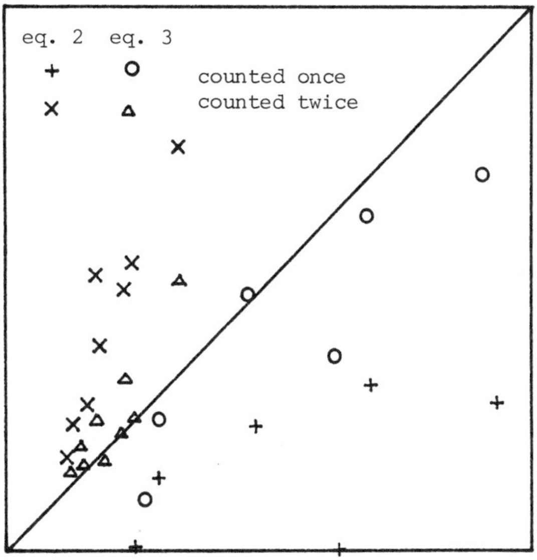

fig. 3. The difference between observed trips and trips estimated with eqs. (2) and (3).

multiplied by a factor such that $\Sigma t_{ij} = \Sigma T_{ij}$. This is of course only possible when the total number of trips is known a priori.

Willis and Chan (1980) came to the same conclusion. They calculated the most likely OD matrix for a larger network and found that eq. (2) gave rather large deviations when the total number of trips in the a priori matrix was not equal to the actual total number of trips.

Example 2.

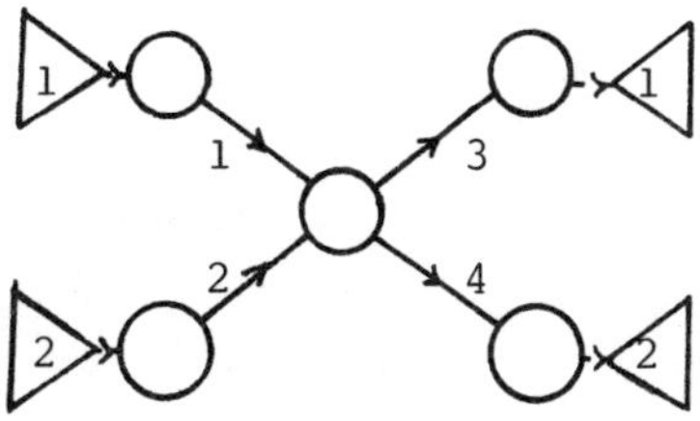

fig. 4
Network with dependent flows.

In this network the information contained in the four links is not independent. When one of the flows has not been counted, the total information contained in the traffic counts should not be less. The following a priori and actual OD matrix are assumed.

TABLE 1. A priori and actual matrix and flows for fig. 4

t_{11} = 10	T_{11} = 15	V_1 = 25
t_{12} = 15	T_{12} = 10	V_2 = 28
t_{21} = 8	T_{21} = 8	V_3 = 23
t_{22} = 20	T_{22} = 20	V_4 = 30

Formula (2) gives estimates for the OD matrix that do not change when redundant information is included. Formula (3) is sensitive to the inclusion of information from traffic counts on a link that actually is completely dependent on the counts on the other links.

TABLE 2. Estimated OD matrix for example 2

	Formula 2	Formula 3				
		counted links:				
		all	1,2,3	1,2,4	1,3,4	2,3,4
T_{11}	12.5	12.5	12.60	12.39	11.64	13.36
T_{12}	12.5	12.5	12.40	12.61	13.36	11.64
T_{21}	10.5	10.5	10.40	10.61	11.36	9.64
T_{22}	17.5	17.5	17.60	17.39	16.64	18.36

So both formulas (2) and (3) show an illogical feature. Formula (2) is sensitive to the total number of trips in the a priori matrix. It gives systematic errors when this number is not equal to the actual total number of trips. In formula (3) all counted traffic volumes are considered to give new information, even on links the volumes of which are completely dependent on the volumes on other links.

3. THEORETICAL CONSIDERATIONS

Equation (2) was derived by Willumsen (Van Zuylen and Willumsen) by the maximisation of the entropy of the OD matrix under the assumption that the total number of trips is fixed. It is this assumption that gives the illogical feature shown in the previous section. A derivation of a new formula should be made without this assumption.

Entropy maximisation is equivalent to information minimisation. In the derivation of eq. (2) the negative value of the entropy can be considered as the information contained in the distribution of T trips over all possible origins and destinations. Formula (3) was derived in a similar way, but here the counted vehicles are distributed over possible origins and destinations. The fact that a trip from i to j occurs on the average g_{ij} times in this distribution procedure gave rise to the factor g_{ij} in the exponent of eq. (3).

In the derivation of eq. (3) the distribution of counted vehicles over origins and destinations was done for each link separately. In fact this is only possible when the flows on all links are mutually independent. This is generally not true, however.

The definition of information according to Brillouin is

$$I = - \ln M, \qquad (4)$$

where M is the number of possible ways to realise the OD matrix, under the assumption that all realisations are equally likely. For a trip matrix T_{ij} from which the trips are counted on some links with volume V_a, the number of possible realisations is calculated in the following way. When the total number of trips is equal to T, the number of possible ways to select T_{ij} trips from i to j out of these T trips is equal to

$$\frac{T.(T - 1)\ldots(T - T_{ij})}{T_{ij}\,!} \,. \qquad (5)$$

When the T_{ij} trips have their route over link a, the connection between these trips and the counted vehicles V_a can be made in the following number of ways:

$$\frac{V_a.(V_a - 1)\ldots(V_a - T_{ij}\, p_{ij}^a)}{(T_{ij}\, p_{ij}^a)!} . \qquad (6)$$

A route between origin i and destination j is defined as a set of roads, r. The number of realisations to select trips from T and from several routes is

$$\frac{T.(T - 1)\ldots(T - T_{ij}) \prod_r \prod_{a \varepsilon r} V_a.(V_a - 1)\ldots(V_a - T_{ij}p_{ij}^a)}{T_{ij}!} . \qquad (7)$$

For the whole OD matrix the number of realisations is

$$M = \frac{T! \prod_a V_a!}{\prod_{ij} T_{ij}!} \qquad (8)$$

In eq. (8) all links are included, also the dependent ones. When the assumption that all realisations are equally likely is replaced by the assumption that the probability of the realisation of a trip from i to j is given by q_{ij}, the information of the OD matrix becomes

$$I = - \ln T! \;_{a} V_a! \prod_{ij} q_{ij}{}^{T_{ij}} \;/\; T_{ij}!. \qquad (9)$$

The probability q_{ij} can be defined as

$$q_{ij} = t_{ij} \;/\; \sum_{ij} t_{ij}. \qquad (10)$$

Apart from the minus sign and the factor $\prod_a V_a$, the information (9) is equal to the entropy as defined by van Zuylen and Willumsen.

The most likely OD matrix is found by the minimisation of the information under the constraints of the link flows (1). The following Lagrangian is defined, with Lagrange multipliers λ_a.

$$\begin{aligned} L &= I + \sum_a \lambda_a \left(V_a - \sum_{ij} p^a_{ij} T_{ij}\right) \\ &\simeq \sum_a (V_a \ln V_a - V_a) + \sum_{ij} T_{ij} \ln \sum_{ij} T_{ij} \\ &\quad + \sum_{ij} T_{ij} \ln q_{ij}/T_{ij} + \sum_a \lambda_a \left(V_a - \sum_{ij} p^a_{ij} T_{ij}\right). \end{aligned} \tag{11}$$

Differentiation of L gives the solution

$$\begin{aligned} \frac{T_{ij}}{\sum_{ij} T_{ij}} &= \frac{t_{ij}}{\sum_{ij} t_{ij}} \prod_a \exp\left(-\lambda_a p^a_{ij}\right) \\ &= \frac{t_{ij}}{\sum_{ij} t_{ij}} \prod_a X_a^{p^a_{ij}}, \text{ with } e^{-\lambda_a} = X_a. \end{aligned} \tag{12}$$

Solution (12) can be written as

$$T_{ij} = t_{ij} X_o \prod_a X_a^{p^a_{ij}}, \tag{13}$$

$$\text{with } X_o = \sum_{ij} T_{ij} \Big/ \sum_{ij} t_{ij}. \tag{13a}$$

X_o and X_a are to be solved from

$$V_a = \sum_{ij} p^a_{ij} t_{ij} X_o \prod_a X_a^{p^a_{ij}} \tag{14a}$$

$$\text{and } \sum_{ij} t_{ij} = \sum_{ij} t_{ij} \prod_a X_a^{p^a_{ij}}. \tag{14b}$$

An algorithm to find a solution for eqs. (14 a,b) is a slightly modified algorithm to find solution (2), as presented by van Zuylen and Willumsen.

(i) Set number of iterations $n = 0$,

(ii) set $X^n_a = 1$ for all links,

(iii) set $X^n_o = \sum_a V_a \Big/ \sum_a \sum_{ij} p^a_{ij} t_{ij}$,

(iv) set $a = 0$,

(v) increase the link counter a by one. Calculate X^{n+1}_a by solving

$$V_a = \sum_{ij} p^a_{ij} X^n_o t_{ij} \left(\prod_{a'} X^n_{a'}{}^{p^{a'}_{ij}}\right) Y_a^{p^a_{ij}} \tag{15}$$

and making $X_a^{n+1} = X_a^n . Y_a$,

(vi) if the link list has not been exhausted, proceed to step (v),

(vii) calculate $X_o^{n+1} = X_o^n \sum_{ij} t_{ij} \prod_a (X_a^{n+1})^{p_{ij}^a} / \sum_{ij} t_{ij}$,

(vii) if all the flows calculated from the estimated OD matrix are suitably near the observed ones, stop; otherwise increment n by 1 and proceed to step (iv).

The convergence of this algorithm appears to be very slow. The reason is that X_o has an oscillating behaviour. A smoothening of X_o gives a reasonably good convergence,

$$X_o^{n+1} = \tfrac{1}{2} (X_o^n + \hat{X}_o^{n+1}) , \qquad (16)$$

where $\hat{X}_a^{n+1}$ is calculated according to step (vii).

The modified algorithm gives a solution that satisfies the constraints on the volume counts (1) and does not have the illogical features of the previous methods. The estimate T_{ij} (13) is insensitive to a multiplication of the a priori matrix with a factor α. Replacing t_{ij} by α t_{ij} gives

$$T_{ij} = \alpha \, t_{ij} \, X_o' \prod_a X_a'^{p_{ij}^a} , \qquad (17)$$

with

$$V_a = \sum_{ij} p_{ij}^a \, X_o' \, \alpha \, t_{ij} \prod_a X_a'^{p_{ij}^a} , \qquad (18)$$

which has the same solution for T_{ij} as eq. (14) with

$$X_o' = X_o / \alpha$$

and

$$X_a' = X_a .$$

This shows that the total number of trips in the a priori matrix is irrelevant for the estimate. The illogical solution for the example of fig. 1 does not exist in the new model.

The information about the actual total number of trips T may still be important. This information can be written as

$$T = \sum_{ij} T_{ij}. \tag{19}$$

When this is added as a constraint, solution (13) changes to

$$T_{ij} = t_{ij} X_o X_t \prod_a X_a^{p_{ij}^a} \tag{20}$$

and eqs. (14 a,b) to

$$V_a = \sum_{ij} p_{ij}^a t_{ij} X_o X_t \prod_a X_a^{p_{ij}^a} \tag{21a}$$

$$T = \sum_{ij} t_{ij} X_o X_t \prod_a X_a^{p_{ij}^a} \tag{21b}$$

$$\sum_{ij} t_{ij} = \sum_{ij} t_{ij} X_t \prod_a X_a^{p_{ij}^a} \tag{21c}$$

or, from eqs (21 b,c)

$$X_o = T / \sum_{ij} t_{ij}. \tag{22}$$

Inclusion of information about the total number of trips will generally increase the reliability of the estimate. The additional information can be treated as contained in an imaginary link used by all trips. When more information from counted links is included, the set of feasible solutions will become smaller, except when the new information is redundant.

A general proof that the systematic over- and under-estimation of multiple counted links does not exist in this model cannot be given. The reason for this phenomenon in the original model (2) was that an a priori matrix that is systematically too small is multiplied by factors X_a that are larger than one. The more a trip has been counted, the more factors X_a are multiplying the a priori trips and the larger the estimate becomes. This effect disappears by the factor X_o multiplying all trips. The resulting factors X_a are expected to be larger as well as smaller than one, so that systematic errors are not expected in the revised model.

The introduction of redundant information from dependent links does not change solution (13). The factors X_a for such a link can be made equal to unity. The resulting solution is identical to the original one and satisfies the constraints on the flows. Murchland (1977) shows, for the multiproportional problem, that when it has a solution, the solution is a unique one. The solution in terms of X_a is not unique, of course.

Although the new algorithm is no longer sensitive to one aspect of the a priori matrix, it should be recognised that the a priori matrix is still essential for the estimate. The information minimisation used in this method is very similar to the entropy maximisation used in the calculation of the gravity model. Snickars and Weibull (1977) showed that, when the gravity model is used with a uniform a priori distribution, 60 to 70 percent of the flows are estimated incorrectly. When historical data are used for an a priori distribution, the errors are about 10 percent. Similar results have been found for the estimation model (13). For example, in a certain urban network the error in the estimates decreased by 30 percent when an historical OD matrix was used as an a priori matrix instead of an equal a priori probability for all origin-destination pairs.

It is safe to use the estimation model as a means to update an old OD matrix. Further research is needed to determine how useful this technique is for a situation without a priori information.

4. ADDITIONAL INFORMATION

Often more information is available about an OD matrix than just an a priori matrix and traffic counts. Long distance trips are less probable than shorter trips and sometimes information is available on the average trip length. In some parts of the network the trip production may be known, possibly only approximately.

Additional information may be introduced in two ways. The first method is the inclusion in the a priori OD matrix. Some mathematical model may be assumed for t_{ij}, for instance

$$t_{ij} = A_i\ B_j\ f(d_{ij}), \tag{23}$$

where A_i and B_j are some characteristics of the origin and destination zone resp., for instance A_i is the number of households and B_j the employment and $f(d_{ij})$ is some distance function. There is no need to normalise the a priori matrix, since this is done by the algorithm described in the previous section.

A possibility suggested by Willis and Chan is to start with some model for t_{ij} including socio-economic data and some model parameters. The parameters are estimated by a minimisation of the difference between the observed and the calculated volumes, as done by Hogberg (1976) for instance. The OD matrix obtained in this way can be further "refined" when it is used as an a priori matrix for an estimation according to eq. (13).

Especially in areas where socio-economic data and distances are expected to give information about an OD matrix, this method is likely to give much better estimates than a method starting with no a priori information at all.

Another possibility to obtain an appropriate a priori matrix is to estimate it from a partial OD matrix, assuming some model.

In the second method, additional information is added as linear equations:

$$\sum_{ij} a^m_{ij}\ T_{ij} = b_m \tag{24}$$

and these are considered as additional constraints. The solution then is

$$T_{ij} = t_{ij}\ X_o \prod_a X_a^{p^a_{ij}} \prod_m Y_m^{a^m_{ij}}. \tag{25}$$

The factors X_o, X_a and Y_m have to be solved by the substitution of eq.(25) in eqs. (1), (13a) and (24).

Suppose for example that the average trip length is known. This can be written as

$$\sum_{ij} T_{ij} d_{ij} = \ell \sum_{ij} T_{ij}. \tag{26}$$

ℓ is the average trip length and d_{ij} the distance between i and j. The solution can be written as

$$T_{ij} = t_{ij} X_o e^{-(\ell - d_{ij})\beta} \prod_a X_a^{p^a_{ij}}. \tag{27}$$

In this equation $e^{-\beta}$ has been used instead of a Y-factor.

Sometimes the information contained in the linear constraints is inconsistent. The inconsistencies can be interpreted as a loss of flow. The sum of volumes going into and out of a region is not equal to the sum of attraction and production of the region. For intersections Van Zuylen and Willumsen solved the removal of inconsistencies. The inconsistencies in additional information can be removed in a similar way. When it is assumed that the counts on a link have a Poisson distribution, the flows going into and out of an intersection can be corrected by means of the following formula:

$$V_a = \hat{V}_a (1 + \sum_i \delta_{ai} \eta_i)^{-1}, \tag{28}$$

where V_a is the corrected flow, $\hat{V}_a$ the observed flow and $\delta_{ai} = 1$ for flows going into node i and - 1 for flows out of node i. η_i has to be solved by the substitution of eq. (28) in

$$\sum_a V_a \delta_{ai} = 0. \tag{29}$$

Suppose that some information about a set of flows is available, for instance the number of trips from zone i, O_i, and the number of trips to zone i, D_i. The flows going out of and coming into zone i should satisfy the constraint

$$\sum_a \delta_{ai} V_a = O_i - D_i, \tag{30}$$

where δ_{ai} have the same meaning as in eq. (28). In most cases O_i and D_i are not known exactly. They are results of an estimate. Inconsistencies in eq. (30) are very likely to occur. They are the results of both estimation errors and errors in the traffic counts.

The errors in the estimates of $O_i - D_i$ are assumed to have a normal distribution. When $O_i - D_i$ is written as B_i, the probability density $P(\hat{B}_i)$ is

$$P(\hat{B}_i) = \frac{1}{(2\pi)^{\frac{1}{2}}\sigma_i} \exp\left(-(\hat{B}_i - B_i)^2/2\sigma_i^2\right), \tag{31}$$

where B_i is the expectation value of $\hat{B}_i$ and σ_i^2 is the variance . The likelihood of finding a set of counts $\hat{V}_a$ and a set $\hat{B}_i$ is now given by

$$L = \prod_a \left(\frac{e^{-V_a} V_a^{\hat{V}_a}}{\hat{V}_a!}\right) \prod_i \left(\frac{1}{(2\pi)^{\frac{1}{2}}\sigma_i} \exp\left(-(\hat{B}_i - B_i)^2/2\sigma_i^2\right)\right). \tag{32}$$

Maximisation of L (32) with the constraints (30) gives a consistent set of volumes V_a, to be solved from the following equations:

$$V_a = \hat{V}_a (1 + \sum_i \eta_i \delta_{ai})^{-1} \tag{33}$$

and $$\sum_a \hat{V}_a \delta_{ai} (1 + \sum_j \eta_j \delta_{aj})^{-1} = \hat{B}_i + \eta_i \sigma_i^2. \tag{34}$$

Constraints for flows going into and out of an intersection are a special case with $\hat{B}_i = 0$ and $\sigma_i = 0$.

With this method all kinds of information, even inconsistent information, can be included in the estimation of an OD matrix.

5. CONCLUSION

Two methods to estimate an OD matrix from traffic counts have been considered critically. It appears that the definition of available knowledge and information is crucial for the estimation of the OD matrix. When the a priori assumptions are not correct, systematic errors in the OD matrix result.

An example is the systematic overestimation of trips that are counted more than once and the underestimation of trips that are counted only once. There are indications that similar errors occur in other estimation methods in which an OD matrix is estimated from traffic counts or other information obtained on roads (Roos, 1980).

The revised algorithm has no longer the illogical properties, however. Together with the possibility to use additional information next to the traffic counts, it seems to provide a good basis for a cheap estimation method for OD matrices.

6. LITERATURE

1. Brillouin, L. (1956) Science and Information Theory, Academic Press, New York.
2. Hogberg, P. (1976) "Estimation of parameters in models for traffic prediction: a non-linear regression approach," Transpn Res 10, 263 - 265.
3. Murchland, J. (1977) "The multiproportional problem," University College London, Research Note JDM 263 (unpublished).
4. Roos, J.P. (1980) "Het opstellen van HB-tabellen op basis van wegenquetes," in P.H.L. Bovy et al. (ed.), Colloquium Vervoersplanologisch Speurwerk-1980, Delft.
5. Smit, J.G. (1980) "Het schatten van een herkomst-bestemmings-matrix," in P.H.L. Bovy et al (red.), Colloquium Vervoersplanologisch Speurwerk-1980, Delft.
6. Snickars, F. and J.W. Weibull (1977) "A minimum information principle," Regional science and urban economics, vol. 7 (1/2) 137 - 168.
7. van Zuylen, H.J. and L.G. Willumsen (1980) "The most likely trip matrix from traffic counts," Transpn Res B 14B, 281 - 293.

8. Willis, A.E. and V.K. Chan (1980) "Deriving Origin-Destination Information from Routinely Collected Traffic Counts, Phase II, Comparison of techniques for single path and multipath networks," University of California, Berkeley Institute of Transportation Studies Working Paper 80-10.
9. Willumsen, L. (1978) "Estimation of an OD matrix from traffic counts: a review," Institute for Transport Studies, Working Paper 99, Leeds University.

HYSTERESIS AND COLLAPSE: A THEORY OF PARKING ENFORCEMENT

Christopher Wright
Middlesex Polytechnic, London, England

In order to limit traffic demand and to keep the road network operating efficiently, most large towns rely on parking controls which are enforced by the police or by traffic wardens. The enforcement is not always effective. In London for example, it has broken down: many drivers park indiscriminately on yellow lines, and those who park at meters often do so without paying.

This paper puts forward a simple model of the interaction between enforcement activity and driver behaviour, and this interaction turns out to be an unstable process which may 'collapse' rather suddenly under certain conditions. Essentially, the system is unstable because a traffic warden needs more time to inspect and deal with an illegally parked vehicle than with a legally parked one. If for some reason the deterrent 'pressure' on drivers is reduced below a certain threshold, e.g., fewer wardens are employed, or the impact of the penalty is eroded through inflation, more drivers will park illegally and wardens will no longer be able to cover their 'beats'.

In theory, a collapse is not easily reversed: either one has to employ more wardens, or increase parking penalties. (By analogy, to reverse the polarity of a bar magnet one has to apply a large current to overcome 'hysteresis' effects.) However, the theory shows that a system can be restored to normal simply by concentrating manpower within selected areas and gradually expanding the patrolled areas over a period of time.

1. Introduction

In large towns and cities the potential demand for travel usually exceeds the capacity of the road network, and some degree of restraint must be applied to keep the road network running efficiently. Although they rarely publicise the fact, many local authorities achieve this by deliberately limiting the supply of parking spaces. This sort of control is only feasible if drivers can be discouraged from parking illegally on the streets. Enforcement is therefore an essential practical ingredient in urban transport policy.

In many countries, traffic wardens are given special responsibility for patrolling the streets and issuing fixed penalty notices to offending drivers. The patrols are organised according to a schedule to ensure full and regular coverage of the road network. The main offences are:

(i) Parking illegally on a yellow line (certain vehicles may be exempt, for instance, lorries which are delivering goods to a nearby store)

(ii) Overstaying, or failing to pay, at a parking meter.

(iii) Parking illegally in a resident's parking bay or some other on-street space reserved for a particular category of road user.

In London, and in many other cities throughout the world (for example, Paris), street parking facilities are widely abused. A notable feature of the situation in London is the rate at which the number of offences has risen during the last few years. Between 1966 and 1978 the proportion of drivers committing offences of one kind or another at parking meters increased from 10% to 32% (Greater London Council [3]). More remarkably, in just one year (1978-79) the number of illegal parking acts on yellow lines in a particular district increased by 30% [1]. It is now widely recognised that the system in London has broken down.

The problem is by no means a trivial one. A large city may employ over 1000 traffic wardens, to patrol several hundred kilometers of streets. Illegal parking on yellow lines alone may account for a large proportion of all parking acts in a central area (about $\frac{2}{3}$ in Central London; see Elliott and Bursey [2]), and the costs of enforcement, including administrative tasks such as the collection of fines, tracing drivers who fail to pay fines, and prosecuting drivers in court, are considerable.

The abuse of parking regulations has certain undesirable consequences for the transport system as a whole. First, if a commuter occupies a parking meter all day, he deprives several short-stay parkers of a valuable facility. The people affected may be businessmen, doctors, or other road users with urgent and legitimate reasons for using a car during the day. High occupancy levels at any parking facility cause frustration and delay, and result in extra travel costs while drivers search for empty spaces. Second, widespread non-compliance is both inequitable and discouraging for those who are willing to comply, and it may damage the public's respect for traffic regulations generally. Third, parking on yellow lines often impedes traffic flow and creates additional accident risks. Fourth, and perhaps most important, illegal parking generates extra journeys which may cause additional congestion. It has been estimated that if the regulations had been properly enforced, traffic flows throughout Central London would have been reduced by about 20% in 1978. This in turn would have yielded roughly a 20% increase in journey speeds [2]. Consequently, by streamlining the enforcement process one could achieve traffic reductions comparable with those claimed for restraint measures such as Area Licensing.

There appear to be four main factors which influence parking compliance generally: the number of traffic wardens or policemen employed on patrol, the size of the fixed penalty, the pressure of demand in relation to the supply of legitimate parking spaces, and the general attitude of the motoring public to parking offences. These factors, and others, were studied by a UK government review body set up in 1977 to consider the problem of deteriorating enforcement in London; the results have been reported by Whetnall [5].

The objectives of this paper are to provide a theoretical basis for understanding the parking enforcement/compliance process, to investigate possible sources of instability within the system, and to put forward some suggestions for optimal operation, in particular, suggestions for restoring a system to normal operation after a collapse.

2. Basic assumptions of compliance theory

We shall adopt the following basic hypothesis:

> Each driver weighs the cost (to himself) of a legitimate parking act against the expected cost of an illegal parking act, and opts for the cheaper of the two.

The user cost $C(h)$ of a legitimate parking act of h hours is taken to include not only the money tariff $C_t(h)$, but also an allowance C_w for the walking time to the driver's ultimate destination together with the time spent searching for the parking space, manoeuvering into it, and paying the fee. Thus

$$C(h) = C_t(h) + C_w . \qquad (1)$$

We shall refer to C(h) as the cost of compliance for a parking act of duration h. The expected cost C'(h) of an illegal parking act of duration h hours will include a walking time and parking search component C'_w corresponding to C_w above, together with the expected cost of being caught and fined. This expected cost is equal to the probability p of being caught and fined, multiplied by the 'effective penalty' F. We shall define the effective penalty as the money cost of the fine (assumed to have a fixed value), plus an allowance for the embarrassment and inconvenience involved. In small provincial towns and villages, where parking demand rarely exceeds supply and people are not in too much of a hurry, drivers tend to avoid breaking the law on principle and F is effectively very large. But in larger cities where attitudes are more cynical and there is competition for parking space, F is probably about equal to the money value of the fixed penalty. No doubt, different drivers have different valuations, but for present purposes it is sufficient to incorporate a single, average value for F in our 'model', thus

$$C'(h) = pF + C'_w . \qquad (2)$$

We shall refer to C'(h) as the cost of non-compliance for a parking act of duration h.

Note that C'(h) does in fact vary with the duration of stay: the longer the stay, the greater the probability p of detection and the greater the average cost. This fact is well known to drivers, who behave accordingly (see for example, Moore and Callahan [4]). Even drivers who are normally scrupulous in obeying the regulations will sometimes park illegally for two or three minutes, apparently in the belief that a short stay is not morally reprehensible in the same way that a long stay would be.

A traffic warden is assumed to patrol each potential parking space just I times during the parking 'day' of H hours, at regular intervals. Consequently, each space is inspected at regular intervals of H/I hours. Any vehicle parked illegally for more than H/I hours is certain to be detected and 'booked', but if the duration of stay h is less than H/I hours, detection is not certain: the probability of detection will in fact be Ih/H. Consequently,

$$C'(h)\begin{cases} = \dfrac{IFh}{H} + C'_w, \text{ if } h \leq H/I & (3) \\ = F + C'_w, \text{ if } h \geq H/I & (4) \end{cases}$$

We can now specify two main quantities which we shall be interested in and define them in mathematical terms. The first important variable is the 'non-compliance ratio', defined as the proportion of the total parking time which is accounted for by non-complying vehicles, and denoted by P. Note that this is identical to the proportion of vehicles on the street at any one moment which are parked illegally. We shall naturally want to assess the effect on P of changing the traffic wardens' operating schedule and methods. The second important variable is the 'pressure' exerted on drivers to comply. For our purposes, this can be summarised entirely in terms of the variables I and F. As it happens, for drivers whose parking durations are less than H/I, their effects are multiplicative (see eq 3). Consequently, for these drivers we can characterise the enforcement pressure solely in terms of the product IF. In the following section, we examine the non-compliance ratio P plotted against the 'enforcement pressure' IF in order to illustrate the effects of changes in the operating system.

3. Parking meters - a simple case

Consider a parking area containing B parking meter bays. We are interested in drivers who fail to pay the tariff (other forms of non-compliance are ignored). A driver is required to pay a tariff if he uses a meter at any time during the 'parking day', which is usually, but not always, 24 hours. We assume that the demand exceeds the supply, so that each meter is occupied all day, except for very short periods while one driver leaves and another enters. Suppose that all drivers wish to park for the same duration h_0. We assume that h_0 is less than H/I so that eq (3) applies. The cost of compliance $C(h_0)$ will be equal to $C_t(h_0) + C_w$. Now, according to our assumptions, if the cost $C'(h_0)$ of non-compliance is greater than $C(h_0)$, all drivers will comply with the regulations and pay at the meters. Otherwise, none will. Hence

$$P = \begin{cases} 0, \text{ if } \frac{IFh_0}{H} + C'_w > C_t(h_0) + C_w \\ 1, \text{ if } \frac{IFh_0}{H} + C'_w < C_t(h_0) + C_w \end{cases} \qquad (5)$$

Now, for meter parking under saturated conditions it is reasonable to assume that the walking distance and parking time will be the same both for complying drivers and for non-complying drivers. If in addition we ignore the extra time taken by a complying driver to insert money in the meter we can then put $C_w = C'_w$. The above relationship then simplifies to

$$P = \begin{cases} 0, \text{ if } IF > HC_t(h_0)/h_0 \\ 1, \text{ if } IF < HC_t(h_0)/h_0 \end{cases} \qquad (6)$$

This relationship is plotted in Fig 1. The non-compliance ratio switches suddenly between zero and one as the enforcement pressure passes through the critical value.

Now, this relationship does not give an entirely clear picture of cause and effect, because the enforcement pressure itself depends on the non-compliance ratio. If compliance is good, traffic wardens will be able to inspect more vehicles per day, the average interval between inspections will be small, and the probability of detection of an offender high. On the other hand if compliance is bad, the wardens will not be able to cover so many vehicles because of the extra time required to book offenders. The probability of detection will therefore be lower. We really want to measure the intensity of enforcement in terms of the raw effort (i.e., warden man-hours and penalty fines) put into the system. This can be done by a simple transformation of the enforcement variable. Let the enforcement effort E_0 be the product of the effective penalty F and the 'nominal inspection frequency' I_0. The latter is simply the value taken by I in the hypothetical case where all drivers comply with the regulations, and P is zero. If there are W wardens on patrol at all times of the day, and it takes a warden t hours to inspect a legally parked vehicle and rt hours to inspect and book an illegally parked vehicle, then

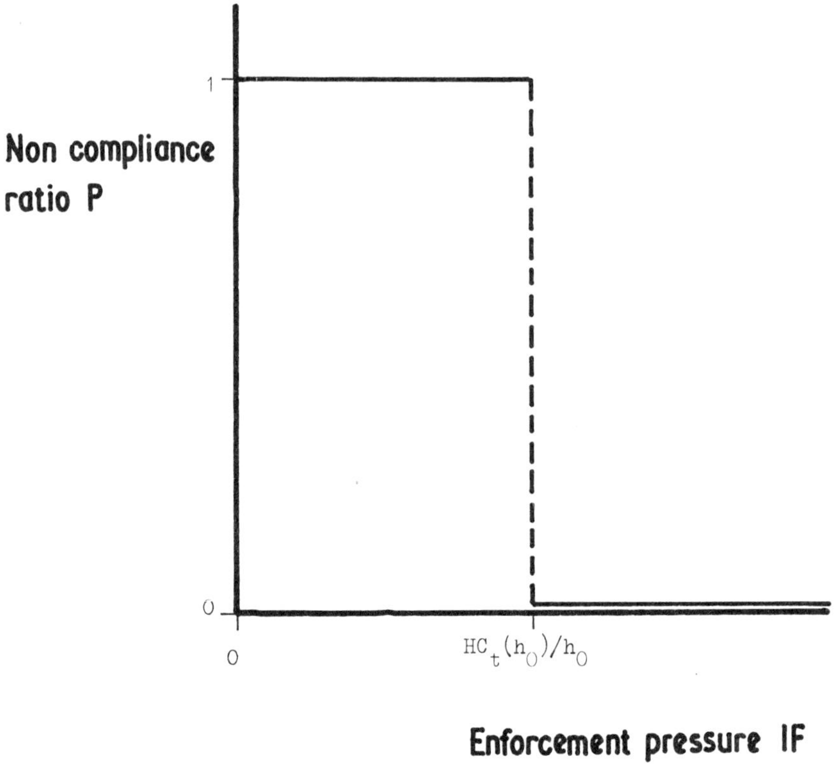

Fig. 1. Variation of non-compliance ratio with enforcement pressure in a simple case

Average time required to inspect one bay

$$= (1 - P)t + Prt$$

$$= t\left[1 + P(r-1)\right].$$

Total number of bay inspections per parking day

$$= \frac{WH}{t\left[1 + P(r-1)\right]}.$$

Hence the mean number I of inspections per bay per parking day is given by

$$I = \frac{WH}{Bt\left[1 + P(r-1)\right]}. \qquad (7)$$

When P = 0, this reduces to

$$I_0 = \frac{WH}{Bt}.$$

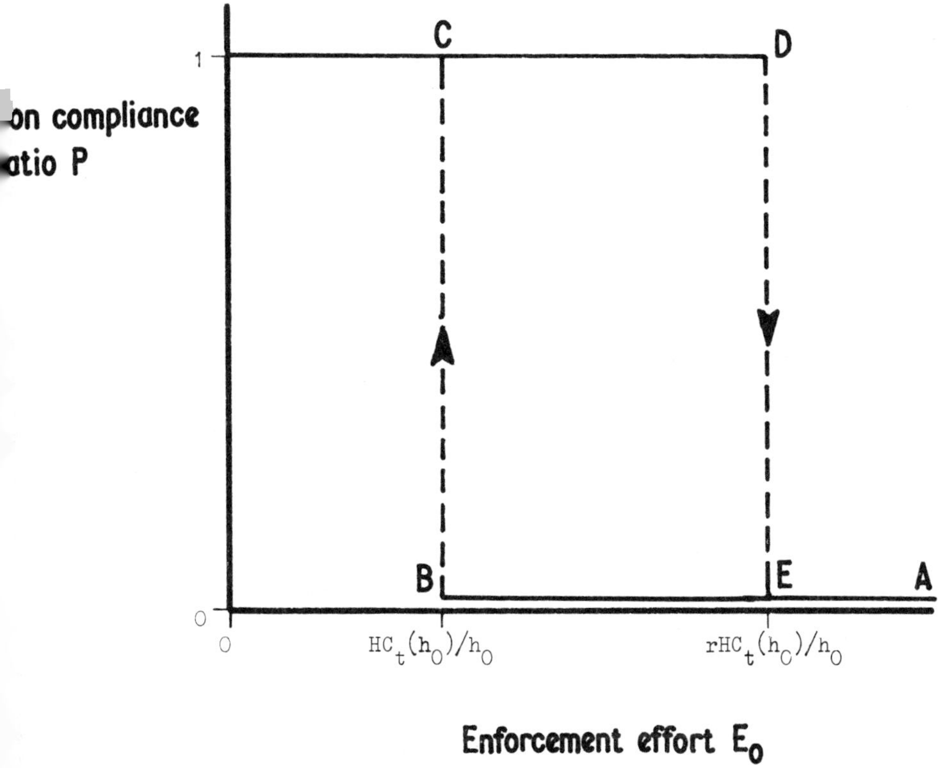

Fig. 2. Variation of non-compliance ratio with enforcement effort at parking meters - a simple case

Hence

$$E_0 = I_0F = IF\left[1 + P(r-1)\right]. \tag{8}$$

We can now transform Fig 1, replacing IF with E_0 on the horizontal axis. The result is shown in Fig 2. The new relationship is not single-valued, and a little thought will show that the transition between the upper and lower horizontal lines can only occur along the dotted lines, in the directions shown. Suppose that initially the system is operating in a 'normal' state represented by the point A in the diagram. We can allow the enforcement effort to drop almost as far as the point B in the diagram without affecting the non-compliance ratio. This might happen, for example, if the number of wardens were reduced, or it might occur unintentionally if inflation were permitted to erode the real cost of the penalty F (this pre-supposes that by contrast the legitimate parking tariff is indexed to keep pace with inflation; if not, inflation as such would not have very much effect since it is the cost difference which determines drivers' behaviour. The supposition is reasonable, since in real life, parking tariffs are up-dated much more frequently than penalties and fines).

In some ways, the point B represents an optimum operating scheme, but it is rather risky, because if E_0 falls slightly below this value the system will collapse (point C in the diagram). The mean inspection rate I will immediately decrease by a factor of r from WH/Bt to WH/Brt, and consequently the cost of non-compliance will fall by a factor of r. The system will not revert to normal operation until the non-compliance cost is restored to its former level (point D in the diagram). This implies that the number of wardens must be increased by a factor of r, or the penalty must be increased by a factor of r, or the two must be increased in conjunction. Once normal operation has been achieved (Point E), the enforcement can be relaxed again. The process of collapse and recovery broadly resembles the phenomenon of 'hysteresis' in a bar magnet whose polarity is reversed in a field coil. The reversal requires the expenditure of a certain amount of energy.

It is interesting to see how the number of penalty tickets issued by wardens changes during a collapse and recovery cycle. When P = 0, the number issued is zero, but when P = 1, the number issued is equal to the total number of vehicles inspected, assuming that $h_0 < H/I$ and hence no vehicle is inspected twice. The number issued per bay per day is then equal to the inspection rate I. Assuming that F remains fixed, the relationship between tickets issued per bay per day, and enforcement effort E_0, is as shown in Fig 3. Clearly, the maximum number of tickets will be issued after the system has collapsed, and has been pushed almost to the point of recovery. This has important implications for the administrative and legal authorities which support the work of the traffic wardens through the magistrates courts. It is also worth noting that the number of tickets issued by traffic wardens is not a good indicator of how successfully the system is operating. An increase in the number of tickets from one year to the next may herald a collapse in the system arising from reduced enforcement effort.

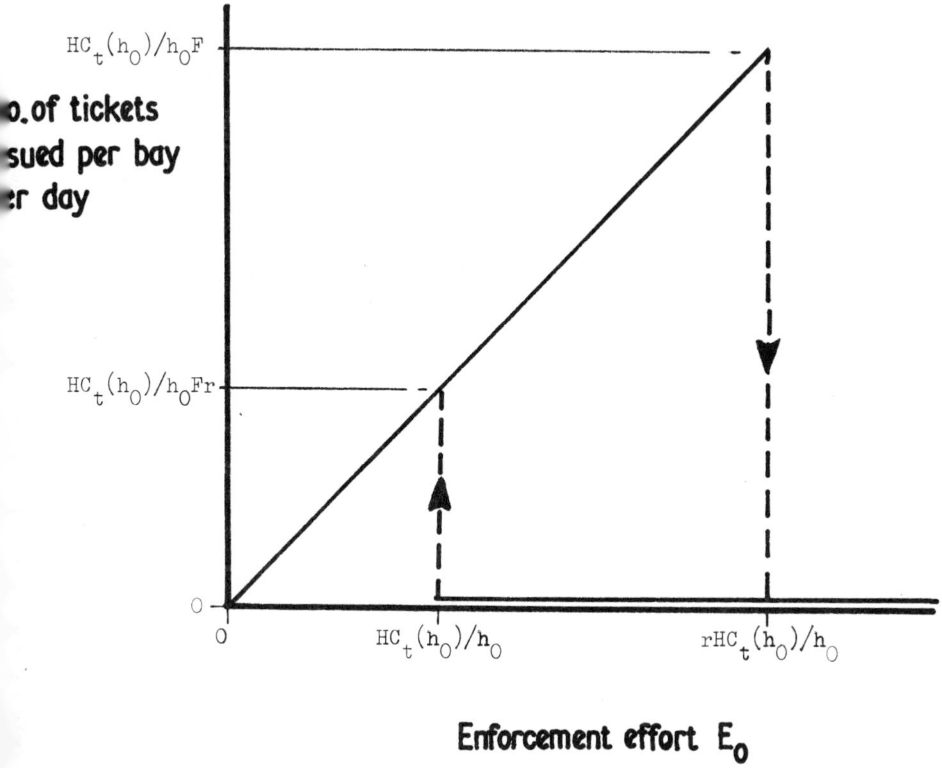

Fig. 3. The effect of enforcement effort on the number of penalty tickets issued to offending drivers at parking meters - a simple case

4. Parking meters - more realistic cases

In real life, different drivers park for different lengths of time. If the tariff varies with the duration of stay, some drivers may be better off parking legally while others are not. It may also occur that the demand is less than the supply, and only a certain proportion (u) of meters are used. Empty bays allow wardens to cover their beats more quickly; the inspection frequency and hence the enforcement pressure will rise. For the moment, we shall continue to assume that the meters are saturated, and $u = 1$.

Up to now, we have modelled the response of the driver population as a function of the relative costs of compliance and non-compliance. In the cases we are about to deal with, it is more convenient to work in terms of costs per unit time, or cost rates. The <u>cost rate of compliance</u> $K(h)$ for a stay of duration h is just $C(h)/h$. Thus from eq (1):

$$K(h) = \frac{C_t(h)}{h} + \frac{C_w}{h},$$

$$= T(h) + \frac{C_w}{h}, \qquad (9)$$

where $T(h)$ is the hourly tariff rate for a stay of duration h. The <u>cost rate of non-compliance</u> $K'(h)$ for a stay of duration h is obtained from eq (3) and eq (4) as:

$$K'(h) = C'(h)/h = \frac{IF}{H} + \frac{C'_w}{h}, \quad (h \leq H/I) \qquad (10)$$

$$= \frac{F}{h} + \frac{C'_w}{h} \quad (h \geq H/I) \qquad (11)$$

Let the p.d.f. of parking durations be f(.), and let the p.d.f. of the compliance cost rates which would be incurred by the driver population if all complied with the regulations be g(.). Then the proportion of drivers for whom the cost per hour of paying at a meter lies in the range, say, x to $x + \delta x$ will be $g(x)\delta x$. Let the average duration of stay corresponding to a compliance cost x be $h(x)$. Then the non-compliance ratio P can be calculated as the proportion of the total parking time accounted for by drivers for whom $K'(h)$ is less than $K(h)$, thus:

$$P = \int_{k_{min}}^{\infty} h(x)g(x)\Delta dx \Big/ \int_{k_{min}}^{\infty} h(x)g(x)dx \qquad (12)$$

where k_{min} is the lowest available cost rate for any duration of stay, and Δ is a variable which takes the value 0 if $K'[h(x)] \geq x$, and the value 1 if $K'[h(x)] < x$. As before we shall assume that C_w and C'_w are equal. Now, drivers fall naturally into two distinct groups: those whose duration of stay is less than H/I, who stand a positive chance of avoiding detection if they fail to pay, and drivers whose duration of stay is greater than H/I, who are certain to be detected. All drivers in the second group can be ignored for present purposes, because it is very unlikely that for these drivers the compliance cost would ever exceed the non-compliance cost: such an event would imply that the fixed penalty were less than the meter tariff. We will therefore put $\Delta = 0$ for all drivers in this category, and concern ourselves solely with drivers in the first category, for whom eq (10) applies. We can then write

$$P = \int_{IF/H}^{\infty} h(x) g_t(x)\,dx \Big/ \int_{k_{min}}^{\infty} h(x) g_t(x)\,dx$$

$$= \int_{IF/H}^{\infty} h(x) g_t(x)\,dx \Big/ \bar{h}, \qquad (13)$$

where $g_t(.)$ is the p.d.f. of parking tariff hourly rates which would be incurred if all drivers complied, and $\bar{h}$ is the overall average duration of stay.

This is a more-or-less general formula for non-compliance at a group of parking meters. In passing, it is worth noting another formula which can be derived along similar lines. The proportion P_d of drivers who fail to comply is given by

$$P_d = \int_{IF/H}^{\infty} g_t(x)\,dx. \qquad (14)$$

Assume now that the drivers' durations of stay are uniformly distributed between 0 and H, and that the meter tariff varies linearly from a minimum of $C_t(0)$ to a maximum of $C_t(H)$. We shall put $C_t(0) = (1-A)\overline{C}_t$, and $C_t(H) = (1+A)\overline{C}_t$, where $\overline{C}_t$ is the mean of $C_t(0)$ and $C_t(H)$, and A is a constant $(-1 \leq A \leq 1)$. We then have

$$\bar{h} = H/2,$$

$$T(h) = \bar{C}_t(1 - A + 2A\frac{h}{H}) \Big/ h,$$

$$h(x) = (1 - A) \Big/ \left[\frac{x}{\bar{C}_t} - \frac{2A}{H}\right],$$

$$g_t(x) = f\left[h(x)\right] \left|\frac{dh}{dx}\right|$$

$$= (1 - A) \Big/ \bar{C}_t\, H \left[\frac{x}{\bar{C}_t} - \frac{2A}{H}\right]^2 \qquad (15)$$

Substituting in eq (13) and integrating, we obtain

$$P = (1 - A)^2 \Big/ \left[\frac{IF}{\bar{C}_t} - 2A\right]^2 \qquad (16)$$

Finally, substituting for IF via eq (8) and re-arranging, we obtain

$$\frac{E_0}{\bar{C}_t} = = \left[\frac{1 - A}{P^{\frac{1}{2}}} + 2A\right] \left[1 + P(r-1)\right]. \qquad (17)$$

This formula represents only one segment of the non-compliance curve; there is an additional straight-line segment P = 1, which represents the 'collapsed regime'. The composite curve is shown in Fig 4 for the particular case where A = 0, which corresponds to a 'flat-rate' tariff system, i.e., a fixed price regardless of duration. Note that when a system reaches point W on the curve, a sudden collapse will occur directly to point X. Recovery can take place only via the route YV in the diagram.

There is no fundamental qualitative difference between this case and the simpler one discussed in the previous section. With the parameter values used for the example plotted in Fig 4, the collapse mechanism is essentially the same, and the hysteresis effect still clearly apparent. However, there are quantitative differences. The collapse itself is less sudden and less catastrophic, because not all drivers respond at the same time to a

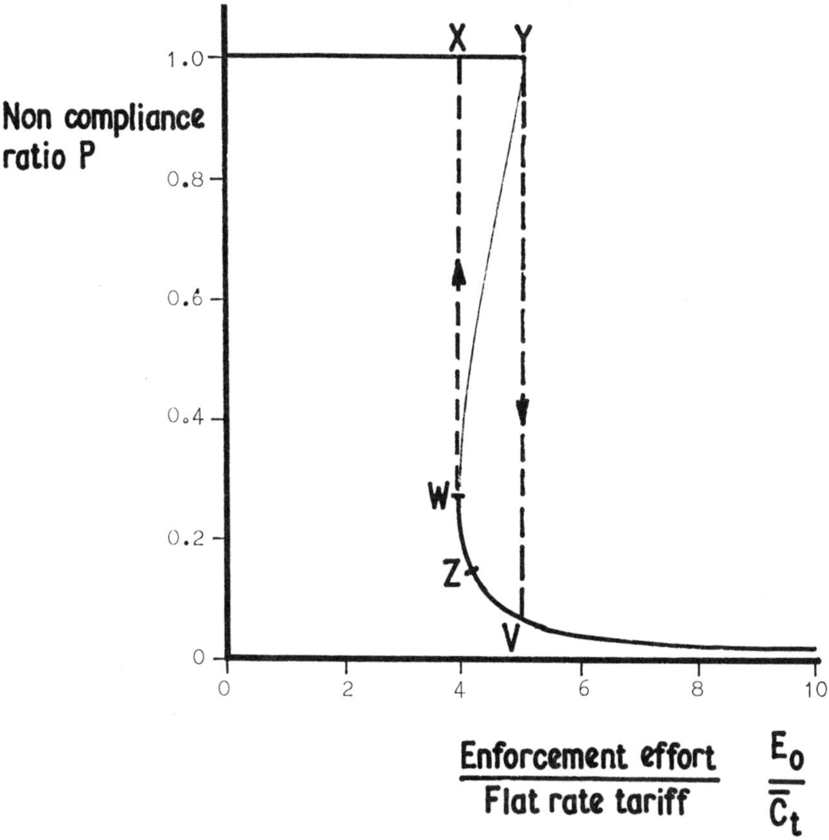

Fig. 4. Relationship between non-compliance ratio and enforcement effort for the case where parking durations are uniformly distributed, the parameter r is equal to 5, and a flat rate tariff system is used

change in enforcement pressure. Furthermore, the extra enforcement effort required to restore normal operation is not as large (in relation to the enforcement level at collapse) as it was for the simpler case.

An examination of a variety of non-compliance curves, plotted for a wide range of values of the parameters A and r, reveals in fact that the severity of collapse, and the extra effort required for recovery, vary widely according to the values of those parameters. In some cases, there is no collapse or hysteresis effect at all, and the system can be described as stable. Some examples are shown in Fig 5. By equating the differential of E_0 with respect

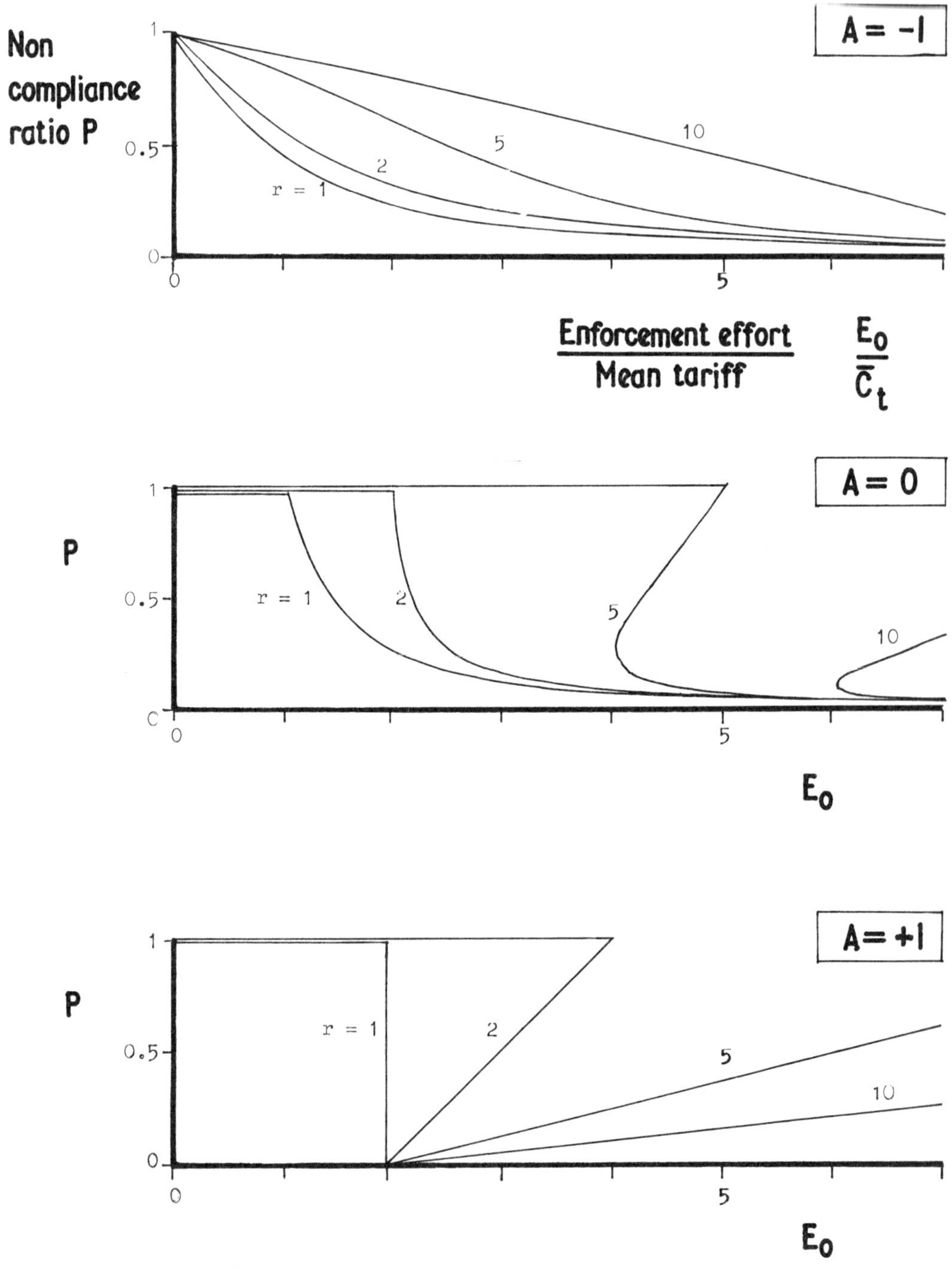

Fig. 5. Non-compliance curves for different values of the parameters A and r (parking durations uniformly distributed)

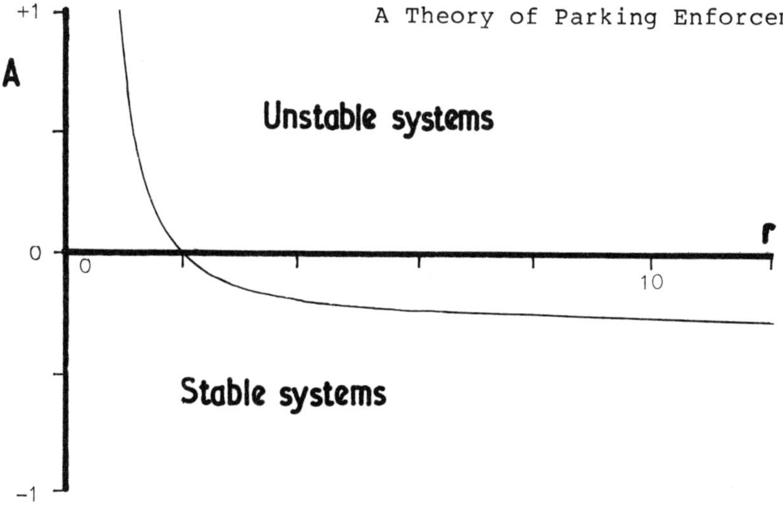

Fig. 6. Stability of the enforcement/compliance relationship at parking meters (uniformly distributed parking durations)

to P to zero, and putting P = 1, it can be shown quite easily that the condition for a stable system under the particular assumptions used in deriving this model can be expressed in the form

$$A > \frac{2 - r}{3r - 2} . \tag{18}$$

This condition is illustrated diagramatically in Fig 6.

In practice, it takes a warden in Central London two or three minutes to inspect and ticket an illegally parked vehicle, and perhaps 30 seconds to inspect a legally parked one; hence an appropriate value of r might be about 5. The distribution of parking durations is of course rather different from the uniform distribution assumed here, with a pronounced peak at about 2.2 hours and a relatively inconspicuous upper tail (most meters in Central London have a 2-hour time limit). The distribution of parking tariffs takes the form of a step function rather than the simple linear form assumed here, because meters accept only certain coins. Bearing in mind that the model bears only a crude resemblance to real life, the most appropriate value for the parameter A would appear to be somewhere in the middle of the range 0 to 1. With the values of the parameters A and r as specified, it would appear from Fig 6 that the enforcement system in Central London - and in other comparable cities - is likely to be unstable rather than stable.

As a matter of interest, it is possible to extend this model to obtain numerical results for the case where the distribution of parking durations is triangular in shape, with a peak at $h = 0$. This corresponds more closely to the known characteristics of on-street parking demand in large towns. The results are very similar to those derived with the uniform distribution model.

Before going on to discuss parking on yellow lines, it is worth mentioning a small refinement which can be made to the model to cover the case where the demand for parking meters is less than the supply, so that only a proportion, u say, of meters are in use at any one time. On average, a proportion 1-u of spaces which a warden inspects will be empty. We assume that the time taken to 'inspect' these spaces is the same as the time to inspect a complying vehicle, i.e., t hours. Then the average time to inspect a bay can easily be shown to be $t\left[1 + Pu(r-1)\right]$, and following the same procedure as was used to derive eq (7) and (8) earlier, we obtain the result

$$E_0 = IF\left[1 + Pu(r-1)\right]. \tag{19}$$

The formulation of the model is otherwise unchanged. We can therefore deduce the following result immediately from eq (16):

$$\frac{E_0}{\overline{C}_t} = \left[\frac{1 - A}{P^{\frac{1}{2}}} + 2A\right]\left[1 + Pu(r-1)\right]. \tag{20}$$

The empty bays effectively cause the (r-1) term in the original model eq (17) to be reduced by a factor of u. This is likely to reduce the severity of the collapse mechanism. However, for areas like Central London, where u is typically of the order of 0.9 during the working day, the effect is unlikely to be very noticeable.

5. Parking on yellow lines

At first sight, the enforcement of yellow line parking regulations appears to be quite a different process from the enforcement of meter parking regulations. At any one time, the majority of potential 'spaces' may be unoccupied: legitimate parkers will have found spaces elsewhere and will not be inspected by the wardens. However, in mathematical terms the analysis is virtually the same. Each driver has the choice of parking either in a legitimate parking space and paying the tariff (we assume there are sufficient spaces to go round), or of leaving his vehicle on a yellow line. The cost of compliance can be defined as before, except that the tariff values will now be averaged across the various types of legitimate facility which are available, including off-street car parks and parking meters. Let the total number of potential 'spaces' on yellow lines be Y, and let the total number of drivers parking in the area as a whole at any one time be D. Then the non-compliance ratio P can be interpreted as the proportion of parking hours accounted for by drivers who park on yellow lines, and formulated in exactly the same way as before (eq (12)).

Now, the total number of drivers parked on yellow lines at any one moment will be PD, and therefore the proportion of yellow line spaces which will be occupied at any one time by illegally parked vehicles will be PD/Y. Assuming that it takes a warden t hours to patrol an empty kerbside space, and rt hours to inspect and book a vehicle parked illegally, it can easily be deduced that the average time required to inspect one yellow line space is $t\left[1 + P(r-1)D/Y\right]$. Thus we obtain a modified form of eq (8):

$$E_0 = IF\left[1 + P(r-1)\frac{D}{Y}\right]. \tag{21}$$

Otherwise, the formulation of the model is unchanged. If we assume as before that C_w and C'_w are equal, we can derive a modified form of eq (17) in which the (r-1) term is multiplied by the factor D/Y. (This is not a very realistic assumption for yellow line parking, but it is the best we can do). Now in Central London, where the total potential parking capacity on yellow lines accounts for just over half the total parking stock, the ratio D/Y is about 2. This effectively doubles the value of (r-1) which suggests that if anything, yellow line parking enforcement is likely to be less stable as a system than meter parking.

6. Restoring a collapsed system to normal operation

The results we have obtained indicate that once a parking enforcement/compliance system has collapsed, a considerable investment in additional traffic wardens, or a considerable increase in parking fines, may be necessary to restore the system to normal operation. Neither of these alternatives may be practicable or desirable. Many of the additional traffic wardens will become superfluous once the system is working properly again, and parking penalties must be kept down to a level which is consistent with the penalties imposed for other types of offence.

However, one may be able to sidestep the problem by concentrating the enforcement effort in selected small areas, and then expanding these areas gradually as the drivers respond to pressure. Consider the non-compliance curve AB in Fig 7, and suppose that a collapse has occurred from point B to point C in the diagram. We now redeploy the traffic wardens so that their efforts are focussed on a limited proportion b of all the parking spaces in the area. The overall effort remains unchanged, but within the patrolled area the inspection frequency, and hence the enforcement effort, is increased by a factor of 1/b. Let the non-compliance ratio for this area be P*. The non-compliance ratio for the area outside, which is no longer patrolled, is assumed to be unity. The non-compliance ratio for the area as a whole can then be expressed by the formula

$$P_{overall} = bP^* + (1-b). \qquad (22)$$

The effect of the concentrated deployment is to reduce both the horizontal and vertical scales of the non-compliance/enforcement effort curve by a factor of b, and to shift the curve bodily upwards. Curve CDH in the diagram illustrates this transformation. In this particular case the value of b has been chosen so that the recovery point C on the new curve lies vertically above the collapse point B on the original one. If the existing traffic wardens were in fact deployed in such a way as to satisfy this condition, the system would in theory recover as far as point D in the diagram without any additional manpower being necessary. The recovery process could then be completed by expanding the patrolled area gradually; the compliance curve would then move downwards and to the right until it coincided with the original curve AEBG. The point D would move vertically down the line DB and finish up at B.

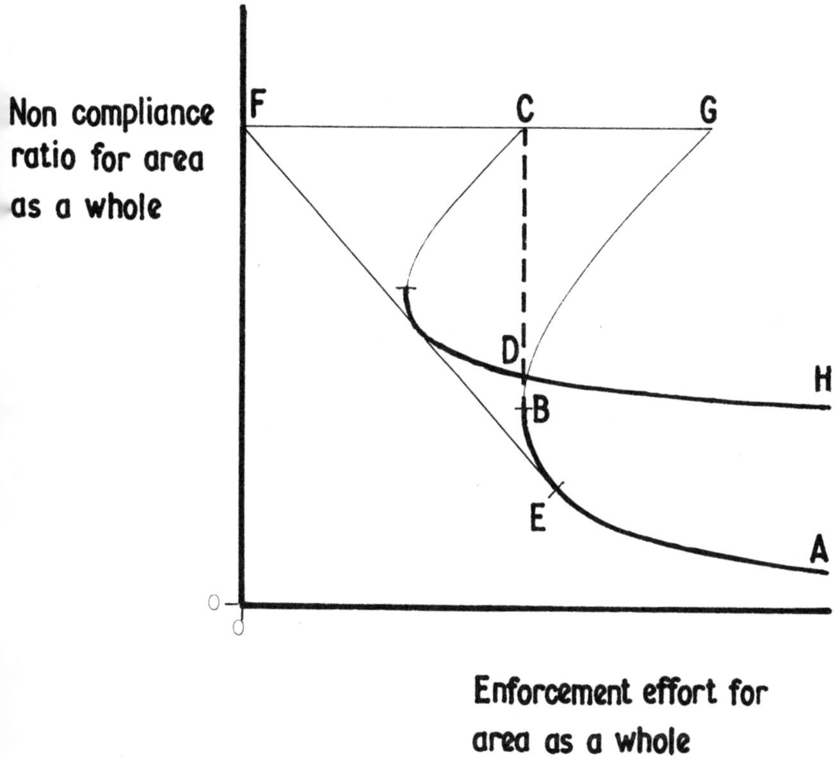

Fig. 7. Recovery from collapse by concentrating enforcement effort within small areas

As a matter of interest it is theoretically possible to operate the system anywhere within the area EFG, and this may be a useful option when enforcement resources are limited so severely that the minimum level for normal operation (corresponding to point B in the diagram) cannot be maintained. Clearly, for any given level of enforcement effort, it is preferable to achieve the lowest possible value of non-compliance; thus, we would wish to operate on the line EF which forms the envelope of the non-compliance curves.

7. Conclusions

It is useful to ask what are the criteria for 'optimal' operation of a parking enforcement/compliance system, and how optimum conditions can be achieved. Three possible criteria are:

(i) Robustness: the system should be stable, or at least resistant to collapse;

(ii) Maximisation of enforcement revenue in relation to costs: the revenue from parking penalties could be maximised in relation to the costs of running the system;

(iii) Cost-effectiveness: compliance should be as high as possible in relation to the enforcement effort.

As far as the first criterion is concerned, the system can be made robust by keeping the value of the parameter r small, and this means that the booking procedure for illegally parked vehicles should be quick and efficient. It also helps to set the fixed penalty at a high level so that the system can accommodate changes in the number of wardens, and can tolerate a certain amount of inflation before the enforcement pressure deteriorates to collapse point.

Criterion (ii) does not seem to be a practicable criterion because the maximum income from penalty fines occurs when the system is in a state of collapse. However, given the necessary data, it is theoretically possible to operate the system so as to maximise the combined revenue from penalties and legitimate parking tariffs in relation to the overall costs of parking provision and enforcement.

The third criterion can be specified in precise mathematical terms: the object would be to maximise the ratio $(1-P)/E_0$. For the particular case illustrated in Fig 4, the operating state which maximises this ratio is the point labelled 'Z'. In practice, one would probably want to operate with a rather higher enforcement effort, so as to minimise the risk of a sudden and unexpected collapse.

The relationship between enforcement effort and parking compliance as put forward in this paper can be interpreted as an example of catastrophe theory (Zeeman, 1977). It has the same general form as a cross-section of one of the seven basic types of catastrophe surface - the well-known 'cusp' surface. However, catastrophe theory is more concerned with qualitative explanations than quantitative ones, and it does not seem to help very much in making predictions about the particular problem under discussion.

To summarise, the theory outlined seems to account for the sharp deterioration in compliance with street parking regulations in London, and the results may be applicable to other large towns. In order to make the theory tractable, many aspects of parking behaviour have been grossly simplified, and the results can only be expected to give a broad qualitative indication of the nature of the enforcement/compliance process. Four major simplifications should be emphasised. First, all drivers parking for the same duration are assumed to have equal non-compliance costs. In practice, the perceived cost of non-compliance will vary from driver to driver, and this will reduce the severity of the collapse mechanism. Second, drivers who are caught breaking parking regulations do not necessarily pay the fixed penalty: they may elect to have their case heard in court (which may result in a heavier fine) or they may ignore the ticket and risk a summons on a more serious charge if they are successfully traced. This does not affect the basic structure of the model, but it does make it difficult to pin down the costs accurately. Third, we have not taken account of the wide range of options available to a meter parker, who may pay for a short parking duration and overstay the paid time, or who may 'feed' the meter to provide time above the permitted maximum duration of stay (normally two hours in London). Fourth, we have ignored market effects which might result from an overall change in the costs of parking in the centre of a town, in other words, we have assumed inelastic demand.

For quantitative applications, the model will probably need to be re-structured within a proper micro-economic framework, to take into account the elastic nature of parking demand and the many options which are available to drivers. Note that in theory, the more elastic the demand, the more severe the collapse mechanism is likely to be: when the enforcement system becomes overloaded, not only will existing parkers switch from legal to illegal parking, but additional non-complying parkers will be attracted to the area because of the general reduction in non-compliance costs.

Acknowledgements

The author is grateful to J.R. Elliott (Westminster City Council), for providing data and for making several useful suggestions.

References

1. ELLIOTT J R (1980) "Note on deterioration of parking enforcement in the last year," City of Westminster, City Engineer's Department (unpublished).

2. ELLIOTT J R and BURSEY N C (1979) "Traffic restraint and the effective use of existing parking controls in London," PTRC Summer Annual Meeting, University of Warwick, July 1979.

3. GREATER LONDON COUNCIL (1978) "Studies and surveys to determine reasons for car journeys in Central London," Central Area Planning Committee Report No CAP 252, 26 June 1978.

4. MOORE U and CALLAHAN C C (1943) "Law and learning theory: a study in legal control," Yale Law Journal, 53 (1), 1-136.

5. WHETNALL A D (1980) "A review of enforcement," PTRC Summer Annual Meeting, University of Warwick, July 1980, Paper No. K23.

6. WIGAN M R and BAMFORD T J (1974) Parking policy when enforcement is limited. Transport and Road Research Laboratory, Supplementary Report No. SR 75 UC.

7. ZEEMAN E C (1977) Catastrophe theory: selected papers 1972-77. Addison-Wesley.

Directions in Transport Research

Robert Herman, Moderator

Center for Studies in Statistical Mechanics
and
Civil Engineering Department
The University of Texas at Austin

I am extremely pleased to attend this symposium at the University of Toronto with my colleagues and many old friends. It is the eighth in a distinguished and long series that was born in the middle of a snowstorm in Warren, Michigan during December 1959. I did not offer a paper to this meeting for the first time in the series and hope to enjoy playing the role of an old senator. However, I admit that I was delighted to have been invited to act as moderator of this panel since it provides me with an opportunity to express my outlook and concerns about the field of vehicular traffic and transportation in whose development I have participated for a quarter of a century. The future of this subject area is of the utmost significance to the continued well-being of all societies around the world and therefore to us.

I regret that our friend and colleague, Reuben Smeed, is no longer with us, since it is precisely more of his general outlook, balanced in theory and practice, that, in my opinion, is urgently required. This is one of the main points I wish to discuss here.

With my background as a physicist, I am keenly aware that great progress has been and can be made through creative new ideas in rather narrow problem areas. However, as time goes on we become more and more aware that in the physical, chemical and biological worlds, as well as in the social world, there are profound problems involving systems of many variables that interact in a nonlinear fashion. We hope that for such highly complex systems there are collective effects that might result in generating some simple relations among a small number of the pertinent variables. Problems of this type can be called symplectic problems. The word symplectic derives from the Greek word symplektikon meaning intertwined or complex in the interrelational sense such as the Oedipus complex. Thus, many of our modern problems are highly complex from this point of view; a problem of this type might then be termed a Symplegma (Gr.).

The reason for mentioning the class of symplectic problems is that a purely theoretical approach is, in my opinion, highly problematical. I have always espoused the point of view that the science of traffic and transportation is fundamentally an empirical science. Over the years there has been, in general, far too little emphasis on the observational and experimental aspects of problems in our subject area. My personal hope is that through the examination of actual processes, it will prove possible to discover some of the simple "Ohm-like Laws" that may obtain in the domain of transportation.

Is it not remarkable that so much of our knowledge of the physical world has been capable of being codified in what are extremely simple-appearing relations such as: $E = IR$, $F = ma$, $E = h\nu$, $p = h/\lambda$, $E = mc^2$, etc.? These extremely 'simple-appearing' equations very likely express collective effects. For example, Mach's Principle, which states that the inertial mass of a body is the result of all the forces exerted on it by all the other bodies in the universe, may well be at the core of a simple law of motion. Who can tell? John Muir in his book _In My First Summer in the Sierra_, 1916, wrote - "When we try to pick out anything by itself, we find it hitched to everything else in the universe." On rare occasions brilliant creative theory leads observation but, much more often than not, theoretical modelling lags. It is the sensitive interplay among theory, observation and experimentation that generally promotes a healthy advancing discipline.

Science and technology, as well as all other human efforts, are inextricably interrelated with one another. It is for this reason that I feel that what is perhaps most important is the style of our research rather than the specific results which derive from it. It is in this spirit that I have examined the papers in our series of symposium proceedings in order to see who has been writing the papers, to what extent various types of organizations are sponsoring the work, and, most important, what is the character of the research reported at our conferences.

In Table 1 we list the numbers and percentages of the authors who are affiliated with universities, organizations in the private sector, and with various levels of government. Over all the eight symposia the fraction of the authors in universities has risen about 40 percent and is now in the 90 percent range. The fraction of the authors from the private sector has fallen off considerably, while the fraction in government establishments seems to have remained approximately constant. Likewise, in Table 2, we see the distribution for numbers and percentages of papers among the same three categories prorated according to the affiliations of the authors. As to be expected, the fraction of the papers from the university sector has risen, the fraction from the private sector has fallen off considerably, while that from the government has remained essentially constant. Overall for symposia 1 through 8 the percentage of authors for university, private sector, and

TABLE 1. AUTHORS

Symposium		1	2	3	4	5	6	7	8	1-8
University	No.	10	17	43	38	30	38	45	42	263
	%	52.6	34.7	59.7	61.3	69.8	74.5	72.5	91.3	65.1
Private	No.	5	18	15	13	6	5	4	2	68
	%	26.3	36.7	20.8	21.0	14.0	9.8	6.5	4.3	16.8
Government	No.	4	14	14	11	7	8	13	2	73
	%	21.1	28.6	19.5	17.7	16.2	15.7	21.0	4.3	18.1
Total	No.	19	49	72	62	43	51	62	46	404
Authors per paper		1.27	1.44	1.60	1.51	1.43	1.46	1.55	1.59	1.50

TABLE 2. PAPERS

Symposium		1	2	3	4	5	6	7	8	1-8
University	No.	7 1/2	11 1/6	28 1/2	27	21 1/2	27 1/2	29 1/6	26	177 11/12
	%	50.0	32.8	62.4	65.9	71.7	78.6	72.9	89.7	66.2
Private	No.	4 1/2	10 1/3	6 2/3	7 1/2	3	2 1/2	2	1	37 6/12
	%	30.0	30.4	14.8	18.2	10.0	7.1	5.0	3.4	13.9
Government	No.	3	12 1/2	10 1/4	6 1/2	5 1/2	5	8 5/6	2	53 7/12
	%	20.0	36.8	22.8	15.9	18.3	14.3	22.1	6.9	19.9
Total	No.	15	34	45	41	30	35	40	29	269

government is 65.1, 16.8 and 18.1, respectively, and for papers is 66.2, 13.9 and 19.9, respectively. It is interesting to note that the number of authors per paper over the first eight symposia has remained about constant at approximately 1.5.

There are many factors which might account for the distributions in Tables 1 and 2. For example, university faculty publish, practitioners very often do not publish; the pressure toward relevance diminishes the effort in the more fundamental research areas, mainly in the private and government sectors; a poorer economic climate lessens support of the more basic research, etc. A clearer understanding of these various trends will require considerably more examination. However, all this brings forth the question of the character of the papers with respect to the relative emphasis on theory, experiment and observation.

While it is rather difficult to categorize papers, we have attempted to do so in Table 3. As indicated in the table we use five categories as follows: theory means essentially all theory; theory/(data) is mainly theory with some analysis of data obtained from various sources; Exper/Obs//(theory) is mainly experimental and/or observational with some theoretical discussion; Exper/Obs is essentially all experimental and/or observational; general refers to two discursive papers.

During the first eight symposia there were 161 out of a total of 269 papers or 59.9 percent that were essentially all theory, while the category theory including some discussion of data involved 52 papers or 19.3 percent. Taken together we have 213 papers or some 79.2 percent which can be said to be primarily theoretical with some attention to data from experiment or observation. In the category of experiment and observation with some theoretical discussion, there have been 29 papers or 10.8 percent, while for those papers which were essentially experimental/observational there were 25 papers or 9.3 percent. The two discursive general papers are rather unusual and require no special elaboration.

It would be most interesting to examine this question in considerably greater detail and depth. In addition, we should study in a similar manner papers appearing in the various journals and publications of different groups. This would provide an interesting and balanced view of what has been taking place in our research field all over the world. It is not possible here to go into all these matters properly. However, I should like to venture my personal impression which has not changed over a long time. I am convinced that there is far too little emphasis on conducting well-thought-out experiments and observations

TABLE 3. CATEGORIES OF PAPERS *

Symposium		1	2	3	4	5	6	7	8	1-8
Theory	No.	7	21	31	26	19	19	21	17	161
	%	46.6	61.7	68.8	63.4	63.3	54.3	52.5	58.6	59.9
Theory (Data)	No.	4	5	3	4	5	13	13	5	52
	%	26.7	14.7	6.7	9.8	16.7	37.1	32.5	17.3	19.3
Exper/Obs (Theory)	No.	1	4	3	5	4	2	6	4	29
	%	6.7	11.8	6.7	12.2	13.3	5.7	15.0	13.8	10.8
Exper/Obs	No.	3	2	8	6	2	1	0	3	25
	%	20.0	5.9	17.8	14.6	6.7	2.9	0	10.3	9.3
General	No.	0	2	0	0	0	0	0	0	2
	%	0	5.9	0	0	0	0	0	0	0.7
Total	No.	15	34	45	41	30	35	40	29	269

Theory	–	Exclusively
Theory (Data)	–	Mainly theory; some analysis of data from various sources
Exper/Obs (Theory)	–	Mainly experimental and/or observational; some theory
Exper/Obs	–	Essentially all experimental and/or observational
General	–	Discursive

that are integrated with a theoretical foundation. This is another way of saying what I indicated earlier, namely, we must study the actual processes that are taking place and attempt to find the simple relations that may exist. If we can accomplish this goal we might then have a basis on which to study problems in a much more general fashion rather than be in the situation which now often exists where most results are highly problem specific.

Please note that I have not taken upon myself the difficult task of passing judgement on the quality of the various research efforts. That is, as you all know, a rather subjective exercise. However, whether we like it or not, the significance of a given work is determined by the contributions it makes over time and that, in the final analysis, ends up as a collective judgement of the scientific community.

We cannot enter into all of the ramifications of the sundry questions touched upon in this brief excursion into where our research has been and is going. My own outlook is strongly imbedded in the idea that we must go forward with much greater attention to developing an observational and experimental basis for theory, and on the other hand generating a healthy theoretical framework out of which can come the directions for fruitful examination of real processes.

In conclusion, I should like to quote Leonardo DaVinci who touches so aptly on one aspect of the above questions: "Those who fall in love with practice without science are like a sailor who enters a ship without a helm or compass and who can never be certain whither he is going." May I presume to give another facet to this though in the following words - on the other hand, science without practice is like a culture medium in which no creative ideas can flourish.

Richard Allsop
Transport Studies Group, University College London

This contribution to the opening of our discussion raises several issues which are germane both to the content of transport research now and in the next few years, and to its relationship with practice. The issues are not new ones, but they seem to the writer to be particularly important at the present stage in the development of the subject. They are

1. variation over time
2. degree of aggregation
3. realism about data
4. responsibility to users of results.

1. Variation over Time

All who study the transport system and its use in relation to the activities it serves are aware of the importance of variation over time - from second to second, minute to minute, hour to hour, day to day, year to year, decade to decade. For purposes of analysis we often take cross-sections at particular instants in time and treat these as if they were steady states. In this way we have obtained many useful results, and some of these can be extended to represent variation over time by considering a succession of steady states in successive periods of time, often called time-slices. There are, however, cases in which this procedure cannot be used, for example, because some components of the system take a time comparable with or longer than any appropriate time-slice to reach any new equilibrium, or simply because the system is in an inherently non-steady state. In such cases, fully time-dependent analysis is required. The need for this in contexts such as traffic control has been mentioned in the Symposium, and it is also required on much longer timescales in the modelling of transport systems and travel patterns in relation to land-use.

The latter requirement has important practical implications because the tendency to analyse cross-sectional steady states has been associated in urban and regional transport planning with undue emphasis on the ultimate state that would be reached after implementing extensive changes in the transport system - a state which may never in fact be reached - and too little emphasis on conditions occurring during the implementation of the changes - conditions which will certainly arise in one form or another and which may last for 10 or 20 years. This imbalance of emphasis needs to be corrected by giving more attention to the modelling of transient conditions in land-use transport systems which are changing slowly over long periods.

2. Degree of Aggregation

Until about 10 years ago, most modelling of patterns of travel was of the aggregate type - concerned with total numbers of people travelling from one area to another by various routes or means of transport, with relatively little distinction between people of different kinds. Over the last decade, a new generation of disaggregate models has been developed, in which the travel choices of people of different kinds are modelled explicitly in terms of their personal characteristics as well as those of the transport system and the available destinations. For a mixture of reasons - some scientific, in that some of the impetus for development of disaggregate models stemmed from certain limitations of aggregate ones, others more to do with human nature - there has been an element of polarisation between aggregate and disaggregate modelling. This has been typified by generalised and sometimes dubious claims that one is more behavioural, more policy relevant or more data-hungry than the other.

What is now needed is a balanced synthesis of the two kinds of modelling, so that the validity of each is tested in terms of the other, and an appropriate degree of aggregation can be selected for any given purpose, taking into account the resources available for data-collection and analysis, and the kind of information that the modelling is intended to yield.

3. Realism about Data

Just as it was reasonable to approach the analysis of a time-varying situation by taking a cross-section in time, so it has been reasonable to set other boundaries to analysis of the complex systems found in transport, by considering certain variables as given and fixed whilst analysing interactions between other variables. For this and other reasons, many analyses assume perfect knowledge of certain quantities in the system under study. When the results of such analyses are used in practice, it is tempting to assume that the best available estimate or the best observations the user is prepared to make may be taken as a proxy for true values of the quantities that the analysis takes as given. Fortunately, there are many situations in which this assumption is probably more or less true, but there are others where it may well not be so, for example, situations in which there are substantial correlations between errors in different inputs. Such situations call for analysis to determine how much the lack of certainty about the inputs matters in practical terms. Where it is found to matter, there is a case for carrying out analyses in which the error-structure in the data is an explicit input and is carried through the analysis so that its implications for the reliability of the conclusions are also made explicit.

4. Responsibility to Users of Results

Practitioners in transport look to researchers for techniques that will help them to plan, design and operate the transport system. Decision-makers look to researchers for information that will help them to decide. These two groups are the main users of results of transport research. Ideally, the researchers' relationship with the users has four characteristics: the users know what they want, it is something that can be provided, the researchers provide it, and the users understand it well enough to be able to use it. In reality, things can go wrong in any of these four respects, and when they do go wrong it is up to the researchers to do their part in putting them right. This cannot just be left to others.

Unless the researchers are sure that the users know what they want, it is a research task in itself to get alongside them and find out. It is another research task to assess whether what they want can be provided at all or within any given budget. If not, it is the researcher's responsibility to say so clearly - hard though this message may be to communicate. It goes without saying that if the techniques or information can be provided it is the researcher's job to provide them. But more than this, it is his job to provide them in forms which the users can understand and use properly - and it may be another research task in itself to find out what these forms are.

This is not to say that every researcher is equally well-suited to every one of these research tasks. But it is to say that they are indeed part of research; they cannot be conveniently given a label like 'development' and left to somebody else. Again, it is not suggested that the researcher has the power to protect the world from all the follies of practitioners and decision-makers - only that he has no right to say 'I told you so' unless he has done so in the practitioner's or decision-maker's own language.

Denos C. Gazis
IBM Research Center
Yorktown Heights, New York

Before attempting to define directions in transport research, it is useful to look at its recent history. A wise man observed once that a measure of progress is not whether or not we have problems, but whether or not they are the same ones we had last year. Judged by this measure, transport research over the last 20 years has had a mixed record. Some of the problems that existed 20 years ago have been addressed successfully and a range of solutions to them is well in hand. However, some others are still the subject of investigation, for a good or a bad reason. The good reason is that the problem has been intractable so far and still merits study. The bad reason is that the transport research field sometimes suffers from a common ailment in scientific endeavour, the urge to revisit old themes in order to provide a new variation or a last hurrah. In the remainder of my remarks, I will try to describe what we have accomplished in a few areas of transport research with which I am familiar, and then give my opinion regarding desirable future directions.

1. Traffic Control

Twenty years ago, computerized design and management was in its infancy. Computers were relatively expensive and rather unfamiliar to traffic engineers. In design, research was groping with traffic light synchronization schemes and ad hoc adaptive control. In implementation, there was a raging debate over the relative merits of digital versus analog systems. The former were judged to be expensive, lacking modularity, "fail-hard," unfamiliar, but flexible.

The latter were modular, "fail-soft," familiar, but relatively difficult to change once installed. Very few cities, mostly large ones, were thought of as capable of putting up the investment required for a digital traffic control system.

The traffic light synchronization problem is one that has been successfully addressed. There is an impressive array of computer programs (e.g., SIGOP, TRANSYT, TRANSIGN) which produce satisfactory synchronization schemes. There is also adequate understanding of the limitations of synchronization in dealing with traffic congestion. Adequate attention has also been paid to the transition between synchronization schemes as traffic patterns change during the day.

In hardware, the progress has been nothing short of spectacular. The cost of computers has decreased dramatically, allowing the notion of Compact UTC (Urban Traffic Control), accessible to the smallest of urban areas involving a few tens of traffic lights. The advent of microprocessors has increased the modularity of digital traffic control. Measuring traffic is still a relatively costly process. But we have succeeded in making the most out of traffic measurements by developing methodologies for on-line measurement of delay, and using filtering techniques in order to translate speed and count measurements into density estimates, or detect stoppages.

Looking into the future, I see the following challenges. In control strategies we have so far concentrated largely on matching the network (e.g., traffic lights) to traffic. We must now pay increased attention to influencing traffic movement so that it can make optimum use of a network. Route control is in an embryonic state today, and I believe it is the key to substantial improvements in traffic management, particularly during periods of congestion. The modularity of microprocessor-based traffic control systems presents us with a new challenge of decentralized traffic control. This is an area that is not yet well understood. We have today, and we will have even more in the future, great flexibility in distributing processing capability over an urban area. We have a lot to learn about the proper organization of such a distributed traffic control system.

The use of microprocessors, and the concomitant sensor-based system in individual cars, is bringing closer to us a new dimension in traffic control. It is more possible now than before for an individual car to broadcast substantial information about its dynamical state as well as its destination. The problem of installing short range transmitters that will provide this information to the traffic system is more institutional than economic. The availability of such information, even on a sampling basis, could greatly facilitate both the proper setting of the network parameters to accommodate traffic, and the proper dissemination of information to the drivers to make best use of the traffic facilities. I present this as one of the most interesting challenges of the 1980s.

2. Transportation Network Planning

The procedures for planning transportation networks generally involve three steps: traffic generation, traffic distribution, and traffic assignment. The first step involves the use of ad hoc methods for obtaining estimates of traffic supply and demand on the basis of patterns of residence, employment and other social activities. The second step distributes these traffic quantities on the basis of somewhat arbitrary, but plausible, mathematical models such as the gravity model. A great deal of activity has also been devoted to the "modal split" problem which attempts to account for individual preferences for various alternative modes of transportation between an origin and a destination. Finally, the third step involves the use of network flow techniques for estimating the routing of traffic along the various links of a network.

Traffic assignment has received most of the attention during the recent past. Traffic assignment combines shortest route computations with ad hoc assumptions concerning the selection of routes by individuals in order to obtain assignments of routes for all traffic units. It is essentially a multi-commodity flow problem, since the various origin-destination requirements are not interchangeable. If, in addition, we have different modes of transportation serving the same origin-destination pair, we add a multi-modal dimension to the problem. The travel time along a given link is generally assumed to increase as traffic along the link increases.

John Wardrop started it all with his seminal papers in the early 1950s, defining "user equilibrium" as an assignment, such that the total origin-destination travel cost per unit of flow is the same on each path used by a given commodity and is not greater than that on each path feasible for, but unused by, the given commodity. An extensive literature has followed, treating extensions to the original traffic assignment concept. One particularly significant variant of traffic assignment has been referred to as "stochastic assignment." It treats link travel times as random variables distributed across the population of traffic units, trying to account for errors in drivers' perception of the travel times along a link. I believe that the number of contributions in this area have outstripped the ability of the traffic engineering profession to benefit from them. Rather than looking for more generalizations to the basic traffic assignment problem from now on, I would look for attention to some relatively neglected areas such as the following:

i) Sensitivity Analysis: This refers to the need for a convenient method for finding the effect of relatively small perturbations to the network (e.g., a small change in an arc capacity), in order to account for effects such as a partial breakdown in the system. This might also point out critical links in the system and lead to cost-effective methods for restoring system availability after breakdowns.

ii) Time Dependence (with or without regulation of input): There is still no satisfactory theory of dynamic traffic assignment, most important, namely during the rush period. Most likely, driver behavior is greatly influenced by this time dependence. My own work on store-and-forward networks gives a possible approach to this problem, but it remains largely untested in a sizeable traffic system.

iii) Elastic Demand: Ideally one would like to correlate traffic distribution and traffic assignment (as well as modal split), since they are not independent of each other.

iv) Economic Investment: Some work has been done in investigating the value of a unit of investment in improving (or destroying) a network by selective improvement (or destruction) of one or several links. This is different from sensitivity analysis, since it may involve more than a small perturbation in the system. It would be desirable to develop a methodology for treating investment questions of this type, particularly in conjunction with topic number (i) above.

As a final remark, I do hope that transfer of the results of transportation planning studies into practice will keep this area of endeavor from becoming a mere academic exercise.

3. New Frontiers of Transportation Research

It is unfortunate that this series of symposia has not drawn some of the contributions to new, growing areas of transport research. I will mention one such area here, that of Air Transportation. In view of the growing contribution of air transportation to the Gross National Product, and the constraints imposed by growing energy costs, I believe that the study of air transportation is one of the most important frontiers in transport research.

A recent MIT study by Odoni and Simpson reviews 230 "National Airspace System Models," most of them contributed during the 1970s. (An earlier MIT report had reviewed the pre-1970 models.) The MIT study presents the following taxonomy of models.

i) Capacity/Delay Models

a) Capacity-oriented models, applicable to runways, representing operations in the final approach/runway sequence (for landings), and on runways (for takeoffs), for various runway combinations and configurations, arrival/departure mixes, etc.

b) Delay-oriented models, applicable to runways, representing operations in the final approach/runway sequence (for landings), and on runways (for takeoffs), for various runway combinations and configurations, arrival/departure mixes, etc.

c) Models applicable to the complete airport, representing airport operations from final approach to apron gate, and back through completion of takeoff.

d) Models applicable to terminal airspace, representing operations such as holding, vectoring, sequencing, metering and spacing in the terminal area.

e) Models applicable to air route traffic, representing airway flows, airway intersections, en route flow control, communications workload of sectors.

f) Models applicable to en route sectors, representing controller actions, used to estimate controller workload and performance under various operating conditions.

g) Macroscopic models of all, or of major segments, of National Airspace System (NAS) covering departure, en route, and arrival phases of flight.

ii) Safety Models

Models applicable to various phases of flight, used to compute collision probabilities deviations from prescribed flight paths, and other safety-related measures.

iii) Noise Models

Models applicable to the vicinity of airports, used to compute noise-related measures.

In addition to the above models reviewed by Odoni and Simpson, several studies have been carried out on terminal building operations, addressing such problems as baggage handling, or the effect of airline schedules (and departures thereof) on terminal congestion.

One of the most striking aspects of air transportation research is that it has involved many of the people responsible for the management or regulation of air transportation, notably the U.S. Federal Aviation Administration. This bodes well for the possibility of transferring the usable results of such studies to practice.

This brings me to the end of my opening remarks, which I want to conclude by echoing a message that our distinguished moderator has driven home many times. The message is that transport research is an experimental science that should be carried out always with an eye to its eventual application to a real transport system. Otherwise it may degenerate into a fruitless, albeit erudite, pursuit of refinements of solutions to yesterday's problems.

Frank Haight
The Pennsylvania Transport Institute
Pennsylvania State University

I would just like to make a few observations on the subject matter of the Symposium, and to comment on the relationship between theory and application in the transportation context.

The persistence of these symposia argues strongly for the existence of a unified body of knowledge significantly distinct from that characterizing any other phenomenon. "Significantly distinct" would probably be accepted at least in traffic flow theory* and I would guess that the more general area of transportation could be similarly characterized - at least as sharply as, for example, operations research.

Moreover, the proliferation of journals, which so far shows no signs of tapering off, argues equally a need (at least in the eyes of acquisitions librarians) for the continued expansion of this body of doctrine.

Does the uniqueness and commercial viability of transportation studies imply that the subject is in some sense practical, that is, of technological benefit commensurate with its intellectual content? I wish to argue that this is not necessarily the case. In doing so, I do not intend to deplore but only to describe. If I am not mistaken, there has already been a good deal of ritually deploring the lack of applications at these symposia and I certainly do not wish to add to the chorus.

---

* For example by the flow-density diagrams provided in yesterday's excellent paper "On a Simulation Model for the Traffic Stream on Freeway Merging Area" by Makigami, Nakanishi, Toyama and Mizote.

It seems to me that the expression "applied mathematics" (or applied anything else) encompasses two intellectually separate activities linked by a common phenomenon; "two sides of the same coin" is in this case a beautiful metaphor, because the sides face in opposite directions.

On one face we have studies designed to abstract from the phenomenon those properties which are scientifically intrinsic, and to formalize the resulting structure. Thus we have many papers on vehicular speed but few if any on vehicular color, since color has not (as yet) been found to relate significantly to cars as transport units.

This process of abstracting a theory from a phenomenon is essentially equivalent to the scientific method and the results are "science" - in our case, as someone has (perhaps prematurely) remarked "transportation science."

The other side of the coin is obviously application: taking the theorems and other products of science (whether established specifically for the purpose, or pre-existing) and formulating programs based on a knowledge of these conclusions.

There is one caveat needed at this point. Bob Herman used the word "observational," and it is an excellent word to contrast with "practical." The best theory - some say the only worthwhile theory - should be glued to the phenomenon through "observational" studies. But observational material, although important for testing theory, should not be confused with practical work, even when they may be embedded in the same experimentation. The difference is at least in motive, and this is a not insignificant difference.

This brings me back to the symposia. In my view, the cries for more applied papers are misdirected. As Bob Herman's interesting tabulations suggest, we as a body are almost exclusively concerned with the other side of the coin.

What separates the two sides in fact is not the subject matter, or the importance of the work or the intellectual strength required, but rather temperament, especially as reflected in, or conditioned by, occupation. The fact that participants are by now over ninety percent academic tells the story. Few of us here are professionally involved with the practical side of transportation decision-making and we would surely be deceiving ourselves if we confused our natural interest in observational data with practical work.

It has been asked before: where is the professional cadre of applied transporation specialists? Do they take notice of our theoretical results? Do these reults need to be interpreted in concrete terms?

In the narrower field of road traffic, the answers must involve users of books like the Highway Capacity Manual. On the wider subject of transportation in general, there are now several textbooks which (although of varying quality) seem at least to agree on the main themes. Should there now be a transportation equivalent of the Highway Capacity Manual, in which the textbook topics are reduced to turnkey recipes providing access to modern theoretical results?

I leave these questions, not for you and me to consider, but for us to ignore as we continue our work on our own side of the coin.

W. Leutzbach
Institut fuer Verkehrswesen
Karlsruhe Technical University

Transport research requirements depend largely on one's personal point of view. I am a civil engineer; therefore in my opinion research is very much a means to an end. It not only serves to increase knowledge as such, but it must also assist in solving practical problems in a more efficient manner. This obviously brings one's regional environment into play. My thoughts are naturally influenced by the fact that I originate from the Federal Republic of Germany, and this is part of central Europe. Against this background, the definition of future research requirements is largely determined by the future development of transportation, and in this context, I would like to make three remarks.

First, extension of existing transport networks is coming to an end. From this there are also consequences for research, which, in the past, and at least within the engineering sciences, placed heavy emphasis on issues relating to the optimum development of transport networks. Such issues will, of course, not completely disappear in the future, but they will have to stand back in the face of all those questions which are connected, for example, with the problems of maintaining large networks. Another point is that any expansion of networks, essentially in order to eliminate still existing bottlenecks, will progress at a much slower pace owing to budget constraints. And, thus, the optimum utilization of existing infrastructures will gain considerable importance. This will most likely revive research into traffic flow, which although originally the main subject of our Symposium, has to some extent faded into the background over the past years.

Secondly, increases in energy prices put a higher price on mobility, not only by private automobile but also by public transport modes. Therefore, political priority will be given to those research activities which concern themselves with resulting effects. First, how will travel behaviour change? Secondly, will there be changes in commodity flows? And finally, how will land-use and economic patterns be affected? It is most likely that this will lead to a much closer interaction between the transport and the regional sciences.

Finally, the importance of urban public transport will grow. Where very great infrastructures exist, it will be economic and operational optimization within given financial constraints which will receive greatest attention. Where there is no, or only poorly developed, urban transit, perhaps the systems, for instance, PRT, may have to be changed. My personal feeling is that, in the past, we have slightly overlooked freight transportation, and I see a greater need to examine this in the near future. Lastly, deciding which fields of science will subsequently play the dominant role in research and development is certainly an interesting question. However, I shall refrain from answering it for the time being.

Synopsis of Discussion

Following Professor Leutzbach's remarks, the chairman opened the floor for an unstructured exchange of views. The transcript of the discussion is a complex weave of many threads which the editors have attempted to untangle to reveal the main pattern.

The contrast between what researchers do and what administrators consider urgent was highlighted by quoting from a recent "needs priorities" list. How should this disparity affect what researchers do, write about and present at symposia?

Some concern about the gap between research and its effect on practice is widely shared. Many argued that researchers should consciously attempt to work on the important problems of tomorrow. This, of course, opens further avenues of discussion. Who is to decide what the "important problems" are? How can a better alignment of science and practice be attained?

In this respect, two thoughts emerged in subsequent remarks. First, that not all issues and problems which are genuinely important can, in fact, benefit from research. Second, that it would be better for the research community to develop an agenda for research than to have one imposed on it by politicians and administrators. Is such exhortation sufficient to induce the researchers to work on the problems of today and tomorrow? This question was left dangling. However, concern was expressed that if we do not "get alongside the users" the freedom of doing research of one's choice may diminish in the future.

One device for the creation of a closer alignment between research and practice is through budgeted research programmes. This may tend to make research a commodity, its practice into business and symposia into advertising. To some, extreme commercialization of research would destroy its essence.

There were several specific suggestions as to what constitutes the "important problems", some of which are listed in the contributions by the panelists. During the discussion the sentiment was voiced that often in research castles are being built in thin air and some of the premises on which these edifices rest are of doubtful validity. Thus, "back to basics" empirical research is advocated. In a similar vein, the validity of the artificial division between transport "planning" and "operations" was questioned. Both are thought to be based on the same foundation.

Since this freewheeling discussion on research in transportation was the editors' brainchild, the task of describing the discussion is self-inflicted punishment. Whether it was all worthwhile will have to be judged by the proceedings of the next symposium.

AUTHORS

Professor Alan Abouchar
Dept. of Political Economy
University of Toronto
100 St. George St.
Toronto M5S 1A1, Canada

Professor Samir A. Ahmed
School of Civil Engineering
Oklahoma State University
Stillwater, Okla. 74078, U.S.A.

Professor Andre Babin
Centre de Recherche sur les Transports
Universite de Montreal
C.P. 6128, Succursale A
Montreal H3C 3J7, Canada

Dr. James H. Banks
Dept. of Civil Engineering
San Diego State University
San Diego, Cal. 92182, U.S.A.

Dr. Michael G.H. Bell
Transport Studies Group
University College London
Gower St.
London WCIE 6BT, U.K.

Mr. Jan L. Botha
c/o Professor Adolf D. May
Dept. of Civil Engineering
109 McLaughlin Hall
University of California
Berkeley, Cal. 94720, U.S.A.

Professor Avishai Ceder
Transportation Research Institute
Technion-Israel Institute of Technology
Technion City, 32000 Haifa, Israel

Professor Allen R. Cook
School of Civil Engineering &
Environmental Science
University of Oklahoma
Norman, Okla. 73019, U.S.A.

Professor Dr. M. Cremer
Universitaet Hamburg
Lohbruegger Kirchstrasse 65
2050 Hamburg 80, Germany (BRD)

Professor Stella Dafermos
Division of Engineering
Brown University
Providence, R.I. 02912, U.S.A.

Professor Carlos F. Daganzo
Dept. of Civil Engineering
109 McLaughlin Hall
University of California
Berkeley, Cal. 94720, U.S.A.

Mr. J. Darzentas
Mathematics Dept.
Royal Holloway College
University of London
Egham Hill, Surrey, U.K.

Dr. Leonard Evans
Transportation & Traffic Science
General Motors Technical Center
General Motors Research Laboratories
Warren, Mich. 48090, U.S.A.

Mr. Paul-Henri Fargier
Institut de Recherche des Transports
2, Avenue du General Malleret
Joinville, B.P. 34
94114 Arcueil, Cedex, France

Professor Michael Florian
Centre de Recherche sur les Transports
Universite de Montreal
C.P. 6128, Succursale A
Montreal H3C 3J7, Canada

Mr. Masaharu Fukuyama
Dept. of Transportation Engineering
Kyoto University
Sakyo-ku, Kyoto, Japan 606

Mr. John C. Golias
Transport Studies Group
Dept. of Civil & Municipal Engineering
University College London
Gower St.
London WCIE 6BT, U.K.

Professor Chris Hendrickson
Dept. of Civil Engineering
Carnegie-Mellon University
Schenley Park
Pittsburgh, Pa. 15213, U.S.A.

Professor M. Iwasaki
c/o Professor M. Koshi
Institute of Industrial Science
University of Tokyo
22-1, Roppougi 7 Chome
Minato-ku
Toyko 106, Japan

Professor Dr. Hartmut Keller
Technische Universitaet Muenchen
Arcisstrasse 21
D-8000 Muenchen 2, FDR

Mr. Howard R. Kirby
Institute for Transport Studies
University of Leeds
Leeds LS2 9JT, U.K.

Professor Ryuichi Kitamura
Dept. of Civil Engineering
University of California
Davis, Cal. 95616, U.S.A.

Professor M. Koshi
Institute of Industrial Science
University of Tokyo
22-1, Roppougi 7 Chome
Minato-ku, Tokyo 106, Japan

Professor Tenny Lam
Dept. of Civil Engineering
University of California
Davis, Cal. 95616, U.S.A.

Dr. Y. Makigami
Dept. of Engineering & Science
Ritsumeikan University
Toji-inkitamachi, Kita-ku
Kyoto, Japan 603

Mr. Patrice Marcotte
GERAD
Ecole des Hautes Etudes Commerciales
5255 avenue Decelles
Montreal H3T 1V6, Canada

Professor Adolf D. May
Dept. of Civil Engineering
109 McLaughlin Hall
University of California
Berkeley, Cal. 94720, U.S.A.

Professor R. Mizote
Dept. of Engineering & Science
Ritsumeikan University
Toji-inkitamachi, Kita-ku
Kyoto, Japan 603

Dr. J.D. Murchland
35 Rectory Park
Sanderstead
Surrey CR2 9JR, U.K.

Professor Daniel Nagin
Dept. of Civil Engineering
Carnegie-Mellon University
Schenley Park
Pittsburgh, Pa. 15213, U.S.A.

Professor Tsunehiko Nakanishi
Dept. of Science & Engineering
Ritsumeikan University
Toji-inkitamachi, Kita-ku
Kyoto, Japan 603

Professor Gordon Newell
University of California
416 McLaughlin Hall
Berkeley, Ca. 94720, U.S.A.

Professor I. Ohkura
Dept. of Civil Engineering
Yokohama National University
Tokiwadai 156, Hodogaya-ku
Yokohama, Japan 240

Mr. Edward Plank
Dept. of Civil Engineering
Carnegie-Mellon University
Schenley Park
Pittsburgh, Pa 15213, U.S.A.

Mr. Richard W. Rothery
Transportation & Traffic Science
General Motors Research Laboratories
General Motors Technical Center
Warren, Mich. 48090, U.S.A.

Mr. M.J. Smith
Dept. of Mathematics
University of York
Heslington
York YO1 5DD, U.K.

Mr. Heinz Spiess
Centre de Recherche sur les Transports
Universite de Montreal
C.P. 6128, Succursale A
Montreal H3C 3J7, Canada

Professor Helman I. Stern
Dept. of Industrial Engineering and Management
Ben Gurion University of the Negev
Beersheva, Israel

Mr. J.C. Tanner
Transport & Road Research Laboratory
Dept. of the Environment
Crowthorne, Berkshire RG11 6AU, U.K.

Professor S. Teply
Dept. of Civil Engineering
University of Alberta
Edmonton T6G 2G7, Canada

Professor M. Toyama
Dept. of Engineering & Science
Ritsumeikan University
Toji-inkitamachi, Kita-ku
Kyoto, Japan 603

Dr. M. Tracz
Roads, Railways & Bridges Inst.
Cracow Technical University
31-155 Krakow
Warszawska 24, Poland

Dr. D. Van Vliet
Institute for Transport Studies
University of Leeds
Leeds LS2 9JT, U.K.

Dr. H.J. van Zuylen
Glennstraat 24
5081 TK Hilvarenbeek, The Netherlands

Mr. L.G. Willumsen
Transport Studies Group
Dept. of Civil and Municipal Engineering
University College London
Gower St.
London WCIE 6BT, U.K.

Dr. C.C. Wright
Middlesex Polytechnic
Queensway
Enfield, Middlesex EN3 4SF, U.K.

PREVIOUS SYMPOSIA AND PROCEEDINGS

I General Motors Research Laboratory, Detroit, Michigan, 1959: *Theory of Traffic Flow*, Robert Herman, Editor, Elsevier Publishing Company, Amsterdam, The Netherlands, 1961

II Road Research Laboratory, London, England, 1963: *Proceedings of the Second International Symposium on the Theory of Road Traffic Flow, London 1963*, Joyce Almond, Editor. Organisation for Economic Co-Operation and Development (OECD), Paris, 1965

III Transportation Science Section ORSA, New York, 1965: *Vehicular Traffic Science*, L.C. Edie, R. Herman and R. Rothery, Editors. American Elsevier Publishing Company, Inc., New York, 1967

IV Technical University, Karlsruhe, Germany, 1968: *Beitrage zur Theorie des Verkehrsflusses*, Wilhelm Leutzbach and Paul Baron, Editors. Strassenbau und Strassenverkehrstechnik Heft 86, 1969

V University of California, Berkeley, 1971: *Traffic Flow and Transportation*, Gordon F. Newell, Editor. American Elsevier Publishing Company, Inc., New York, 1972

VI University of New South Wales, Sydney, Australia, 1974: *Transportation and Traffic Theory*, D.J. Buckley, Editor. A.H. & A.W. Reed Pty Ltd., Artarmon, New South Wales, 1974

VII The Institute of Systems Science Research, Kyoto, Japan, 1977: *Proceedings of the Seventh International Symposium on Transportation and Traffic Theory*, Tsuna Sasaki and Takeo Yamaoka, Editors. The Institute of Systems Science Research, Kyoto, 1977